AF358643

Rotaviruses and Rotavirus Vaccines

Rotaviruses and Rotavirus Vaccines

Guest Editors

Ulrich Desselberger
John T. Patton

Basel • Beijing • Wuhan • Barcelona • Belgrade • Novi Sad • Cluj • Manchester

Guest Editors

Ulrich Desselberger
University of Cambridge
Cambridge
UK

John T. Patton
Department of Biology,
Indiana University
Bloomington
USA

Editorial Office
MDPI AG
Grosspeteranlage 5
4052 Basel, Switzerland

This is a reprint of the Special Issue, published open access by the journal *Viruses* (ISSN 1999-4915), freely accessible at: https://www.mdpi.com/journal/viruses/special_issues/BR15SD6300.

For citation purposes, cite each article independently as indicated on the article page online and using the guide below:

Lastname, A.A.; Lastname, B.B. Article Title. *Journal Name* **Year**, *Volume Number*, Page Range.

ISBN 978-3-7258-2789-3 (Hbk)
ISBN 978-3-7258-2790-9 (PDF)
https://doi.org/10.3390/books978-3-7258-2790-9

Contents

About the Editors . . . . **vii**

John T. Patton and Ulrich Desselberger
Rotaviruses and Rotavirus Vaccines: Special Issue Editorial
Reprinted from: *Viruses* **2024**, *16*, 1665, https://doi.org/10.3390/v16111665 . . . . **1**

Jiajie Wei, Scott Radcliffe, Amanda Pirrone, Meiqing Lu, Yuan Li, Jason Cassaday, et al.
A Novel Rotavirus Reverse Genetics Platform Supports Flexible Insertion of Exogenous Genes
and Enables Rapid Development of a High-Throughput Neutralization Assay
Reprinted from: *Viruses* **2023**, *15*, 2034, https://doi.org/10.3390/v15102034 . . . . **8**

Anthony J. Snyder, Chantal A. Agbemabiese and John T. Patton
Production of OSU G5P[7] Porcine Rotavirus Expressing a Fluorescent Reporter via Reverse
Genetics
Reprinted from: *Viruses* **2024**, *16*, 411, https://doi.org/10.3390/v16030411 . . . . **21**

**Roman Valusenko-Mehrkens, Katja Schilling-Loeffler, Reimar Johne and Alexander
Falkenhagen**
VP4 Mutation Boosts Replication of Recombinant Human/Simian Rotavirus in Cell Culture
Reprinted from: *Viruses* **2024**, *16*, 565, https://doi.org/10.3390/v16040565 . . . . **34**

**Takahiro Kawagishi, Liliana Sánchez-Tacuba, Ningguo Feng, Harry B. Greenberg and
Siyuan Ding**
Reverse Genetics of Murine Rotavirus: A Comparative Analysis of the Wild-Type and
Cell-Culture-Adapted Murine Rotavirus VP4 in Replication and Virulence in Neonatal Mice
Reprinted from: *Viruses* **2024**, *16*, 767, https://doi.org/10.3390/v16050767 . . . . **51**

**Saori Fukuda, Masanori Kugita, Kanako Kumamoto, Yuki Akari, Yuki Higashimoto,
Shizuko Nagao, et al.**
Generation of Recombinant Authentic Live Attenuated Human Rotavirus Vaccine Strain
RIX4414 (Rotarix®) from Cloned cDNAs Using Reverse Genetics
Reprinted from: *Viruses* **2024**, *16*, 1198, https://doi.org/10.3390/v16081198 . . . . **63**

**Sarah Woodyear, Tawny L. Chandler, Takahiro Kawagishi, Tom M. Lonergan, Vanshika A.
Patel, Caitlin A. Williams, et al.**
Chimeric Viruses Enable Study of Antibody Responses to Human Rotaviruses in Mice
Reprinted from: *Viruses* **2024**, *16*, 1145, https://doi.org/10.3390/v16071145 . . . . **77**

**Ola Diebold, Shu Zhou, Colin Peter Sharp, Blanka Tesla, Hou Wei Chook, Paul Digard and
Eleanor R. Gaunt**
Towards the Development of a Minigenome Assay for Species A Rotaviruses
Reprinted from: *Viruses* **2024**, *16*, 1396, https://doi.org/10.3390/v16091396 . . . . **91**

Janine Vetter, Melissa Lee and Catherine Eichwald
The Role of the Host Cytoskeleton in the Formation and Dynamics of Rotavirus Viroplasms
Reprinted from: *Viruses* **2024**, *16*, 668, https://doi.org/10.3390/v16050668 . . . . **103**

Sarah L. Nichols, Cyril Haller, Alexander Borodavka and Sarah M. Esstman
Rotavirus NSP2: A Master Orchestrator of Early Viral Particle Assembly
Reprinted from: *Viruses* **2024**, *16*, 814, https://doi.org/10.3390/v16060814 . . . . **118**

Jinlan Wang, Xiaoqing Hu, Jinyuan Wu, Xiaochen Lin, Rong Chen, Chenxing Lu, et al.
ML241 Antagonizes ERK 1/2 Activation and Inhibits Rotavirus Proliferation
Reprinted from: *Viruses* **2024**, *16*, 623, https://doi.org/10.3390/v16040623 . . . . **137**

Arash Hellysaz and Marie Hagbom
Rotavirus Sickness Symptoms: Manifestations of Defensive Responses from the Brain
Reprinted from: *Viruses* 2024, *16*, 1086, https://doi.org/10.3390/v16071086 152

**Liguo Gao, Hanqin Shen, Sucan Zhao, Sheng Chen, Puduo Zhu, Wencheng Lin and Feng
Chen**
Isolation and Pathogenicity Analysis of a G5P[23] Porcine Rotavirus Strain
Reprinted from: *Viruses* 2024, *16*, 21, https://doi.org/10.3390/v16010021 166

Mengli Qiao, Meizhen Li, Yang Li, Zewei Wang, Zhiqiang Hu, Jie Qing, et al.
Recent Molecular Characterization of Porcine Rotaviruses Detected in China and Their
Phylogenetic Relationships with Human Rotaviruses
Reprinted from: *Viruses* 2024, *16*, 453, https://doi.org/10.3390/v16030453 179

**Sergei A. Raev, Maryssa K. Kick, Maria Chellis, Joshua O. Amimo, Linda J. Saif and
Anastasia N. Vlasova**
Histo-Blood Group Antigen-Producing Bacterial Cocktail Reduces Rotavirus A, B, and C
Infection and Disease in Gnotobiotic Piglets
Reprinted from: *Viruses* 2024, *16*, 660, https://doi.org/10.3390/v16050660 192

**Amy Strydom, Neo Segone, Roelof Coertze, Nikita Barron, Muller Strydom and Hester G.
O'Neill**
Phylogenetic Analyses of Rotavirus A, B and C Detected on a Porcine Farm in South Africa
Reprinted from: *Viruses* 2024, *16*, 934, https://doi.org/10.3390/v16060934 205

**Benilde Munlela, Eva D. João, Amy Strydom, Adilson Fernando Loforte Bauhofer, Assucênio
Chissaque, Jorfélia J. Chilaúle, et al.**
Whole-Genome Characterization of Rotavirus G9P[6] and G9P[4] Strains That Emerged after
Rotavirus Vaccine Introduction in Mozambique
Reprinted from: *Viruses* 2024, *16*, 1140, https://doi.org/10.3390/v16071140 218

**Maximilian H. Carter, Jennifer Gribble, Julia R. Diller, Mark R. Denison, Sara A. Mirza,
James D. Chappell, et al.**
Human Rotaviruses of Multiple Genotypes Acquire Conserved VP4 Mutations during Serial
Passage
Reprinted from: *Viruses* 2024, *16*, 978, https://doi.org/10.3390/v16060978 235

Chenxing Lu, Yan Li, Rong Chen, Xiaoqing Hu, Qingmei Leng, Xiaopeng Song, et al.
Safety, Immunogenicity, and Mechanism of a Rotavirus mRNA-LNP Vaccine in Mice
Reprinted from: *Viruses* 2024, *16*, 211, https://doi.org/10.3390/v16020211 253

**Dmitriy L. Granovskiy, Nelli S. Khudainazarova, Ekaterina A. Evtushenko, Ekaterina M.
Ryabchevskaya, Olga A. Kondakova, Marina V. Arkhipenko, et al.**
Novel Universal Recombinant Rotavirus A Vaccine Candidate: Evaluation of Immunological
Properties
Reprinted from: *Viruses* 2024, *16*, 438, https://doi.org/10.3390/v16030438 267

Catherine Montenegro, Federico Perdomo-Celis and Manuel A. Franco
Update on Early-Life T Cells: Impact on Oral Rotavirus Vaccines
Reprinted from: *Viruses* 2024, *16*, 818, https://doi.org/10.3390/v16060818 286

**Benjamin Morgan, Eleanor A. Lyons, Amanda Handley, Nada Bogdanovic-Sakran, Daniel
Pavlic, Desiree Witte, et al.**
Rotavirus-Specific Maternal Serum Antibodies and Vaccine Responses to RV3-BB Rotavirus
Vaccine Administered in a Neonatal or Infant Schedule in Malawi
Reprinted from: *Viruses* 2024, *16*, 1488, https://doi.org/10.3390/v16091488 299

About the Editors

Ulrich Desselberger

Ulrich Desselberger has been an Honorary Senior Visiting Fellow at the Department of Medicine, University of Cambridge, since 2014. He earned his MD degree at the Free University Berlin in 1967 and habilitation in medical microbiology at Hannover Medical School in 1976. From 1977 to 1979, he carried out postgraduate studies on influenza viruses at Mount Sinai School of Medicine. In 1980, he became a Senior Lecturer of Virology at the University of Glasgow and an Honorary Consultant of the Greater Glasgow Health Board. In 1987, he became Consultant Virologist and Director at the Regional Virus Laboratory, East Birmingham Hospital. From 1990, he served as a Consultant Virologist and Director of the Clinical Microbiology and Public Health Laboratory, Addenbrooke's Hospital, retiring in 2002. He spent the following 5 years as a Visiting Research Fellow at UMR Virologie Moléculaire et Structurale, CNRS, and the International Center for Genetic Engineering and Biotechnology, Italy. From 2007 to 2013, he worked as the Director of Research in the Department of Medicine, University of Cambridge, Addenbrooke's Hospital. His research interests comprise the molecular biology of rotaviruses and other major enteropathogenic viruses, the molecular biology of influenza viruses, factors determining viral pathogenicity, and viral immune responses. He is a member of the American Society of Virology, American Society of Microbiologists, Microbiology Society, and European Society of Virology.

John T. Patton

John T. Patton is the Lawrence M. Blatt Chair of Virology and a Professor in the Department of Biology at Indiana University. Dr. Patton received his Ph.D. at Virginia Tech in 1980 for research on parvovirus biology and then moved to University of North Carolina-Chapel Hill where he studied the replication of vesicular stomatitis virus as a postdoctoral fellow. He joined the faculty at the University of South Florida in 1983, where he began studies on the molecular biology of rotaviruses, infectious agents that cause acute gastroenteritis in young children. Dr. Patton moved to the University of Miami School of Medicine in 1987 and then to the Laboratory of Infectious Diseases at the National Institute of Allergy and Infectious Diseases, NIH, in 1996, where he was promoted to Senior Investigator and Chief of the Rotavirus Molecular Biology Section. In 2017, Dr. Patton joined the Department of Biology at Indiana University. Dr. Patton is an elected fellow of the American Academy of Microbiology and the American Association for the Advancement of Science. His research group studies rotavirus replication and assembly, rotavirus interaction with the innate immune system, and the development of next generation vaccines.

Editorial

Rotaviruses and Rotavirus Vaccines: Special Issue Editorial

John T. Patton [1,*] **and Ulrich Desselberger** [2]

[1] Department of Biology, Indiana University, 212 S Hawthorne Drive, Simon Hall 011, Bloomington, IN 47405, USA

[2] Department of Medicine, University of Cambridge, Addenbrooke's Hospital, Hills Road, Cambridge CB2 0QQ, UK; ud207@medschl.cam.ac.uk

* Correspondence: jtpatton@iu.edu

Citation: Patton, J.T.; Desselberger, U. Rotaviruses and Rotavirus Vaccines: Special Issue Editorial. *Viruses* **2024**, *16*, 1665. https://doi.org/10.3390/v16111665

Received: 15 October 2024
Accepted: 16 October 2024
Published: 24 October 2024

Species A rotaviruses (RVA) are a major cause of acute gastroenteritis in infants and young children and in the young of various mammalian and avian species [1–3]. Since 2006, several live attenuated RV vaccines have been licensed and are now components of childhood immunization programs in >100 countries worldwide [4,5]. However, vaccine efficacy/effectiveness (VE) varies in different countries [6,7]. As summarized elsewhere, much has been learned about many aspects of rotavirus biology, including its entry and assembly pathways, mechanism of genome replication, the structure and function of its proteins, and its basis of pathogenesis [2,3]. The study of rotavirus biology was significantly advanced about 6 years ago when reliable, plasmid-only-based reverse genetics (RG) systems for species A RVs (RVA) were established [8–10]. In this Special Issue of *Viruses* on *Rotaviruses and Rotavirus Vaccines*, novel data on the use of RVA RG, RVA replication, RVA molecular epidemiology, and RVA next-generation (NG) RV candidate vaccine developments are presented in 21 contributions and assessed in the larger context.

RG procedures have been used to rescue recombinant (r)RVAs of simian, human, murine, porcine, bovine and avian origin, as well as reassortants thereof, as summarized in Table 1 [8,9,11–27]. Murine RVs (MuRVs) replicate well in their natural host and are therefore an attractive animal model system. Using RG procedures, Kawagishi et al. [18] succeeded in rescuing recombinant RVs containing all 11 RNA segments of murine origin (rMuRV). Oligo-reassortants of rMuRV containing human VP7 and VP4 genes were shown to replicate efficiently in mice and to elicit a robust antibody response to human rotavirus antigens, whilst the control chimeric murine rotavirus did not [18,19]. Fukuda et al. [16] generated an rRV from the RV vaccine strain RIX4414 (Rotarix®) which was live attenuated in a mouse model. The porcine RV OSU strain (G5P[7]) was completely rescued as an rRVA and also engineered to express the fluorescent reporter protein UnaG [21]. Recently, an rRV of the Chinese Lanzhou lamb RV LLR was established [23]. These constructs can be used in the gnotobiotic (gn) piglet model ([28,29], see below) and will permit the development of urgently needed, next-generation porcine RV vaccines. Fluorescing rRVs can be used in large-scale neutralization assays to assess antibody responses in vaccinees ([8,30], see below). These developments in RVA RG research should be seen in the larger context of highly original recent work. Using a virus-like, codon-modified transgene, it was possible to generate a stable dsRNA virus as a potential vector [31]. An rRVA with a glycosylation-defective NSP4 gene showed attenuated replication in cultured cells and also less pathogenicity than its corresponding wildtype virus in a mouse model [32]. Furthermore, the combination of the structure–function-based molecular data of RVA proteins and RG techniques permitted the creation of an rRV which was infectious for only a single round of replication and thus an interesting candidate for a next-generation RV vaccine ([33], see below).

Table 1. Rescue of species A rotaviruses by reverse genetics.

Virus	References
Simian SA11 G3P[2]	[8,9]
Human rotavirus Ku G1P[8]	[11]
Human rotavirus Odelia G4P[8]	[12]
Human rotavirus CDC-9 G1P[8]	[13]
Human rotavirus G2P[4]	[14]
Human rotavirus RIX4414 G1P[8]	[15,16]
Simian rotavirus RRV G3P[3]	[17]
Murine rotavirus EW/ETD G3P[17]	[17–19]
Avian rotavirus PO-13 G18P[17]	[20]
Porcine rotavirus OSU G5P[7]	[21]
Bovine RV RF G6P[1]	[22]
Lanzhou lamb rotavirus LLR G10P[15]	[23]
Species A rotavirus reassortants (selection)	[17,19,23–27]

Regarding RV replication, the introduction of a point mutation into the VP4 gene of a human/simian triple reassortant rRVA strain led to increased replication in cell cultures; however, the effect was highly VP4 genotype-specific [34]. Human RVA clinical isolates initially replicate poorly in continuous cell lines, e.g., MA104 cells, but can be adapted to higher replication rates upon serial cell passage. This was observed with human RV isolates of five different genotype constellations upon passage: Illumina NG sequencing of the passaged viruses demonstrated the emergence of multiple point mutations, which, however, were partially conserved for the VP4 genes throughout genotypes [35].

In search of antivirals specifically blocking cellular compounds, a small molecule, ML241 (an ATPase inhibitor) was identified which blocked RVA replication at an IC_{50} of 22.5 uM in vitro and in vivo (suckling mouse model) via inhibition of MAPK signaling, leading to activation of the NF-kB pathway [36].

Germfree gnotobiotic (gn) piglets were colonized with apathogenic-bacteria-expressing histo-blood group antigens ($HBGA^+$) that are well recognized in regard to binding to RVAs. Upon RV infection, the $HBGA^+$ colonized animals showed reduced disease severity and virus shedding compared to the $HBGA^-$ animals. This finding can lead to further identification of bacteria with substantial probiotic activity [37].

The formation and kinetics of viroplasms during RVA infection has been shown to depend on interactions with cytoskeleton proteins (microtubules, kinesin E5, intermediate filaments, myosin a.o.) [38]. In this context, the interaction of RV NSP5, a viroplasm building block, with the tailless complex polypeptide I ring complex (TRiC), a cellular chaperonin involved in the folding of cellular proteins, was discovered [39]. Inhibition of TRiC expression was shown to reduce RV replication. There is also the interaction of various components of cellular lipid metabolism with viral proteins not only during the replication of RVAs but also with the proteins of many other RNA viruses, a discovery with the potential to help in developing broadly active antivirals [40].

RV-encoded NSP2 interacts with NSP5 by liquid–liquid phase separation to form the biomolecular condensates of viroplasms [41]. NSP2 and NSP5 have enzymatic activities and are involved in RNA-RNA interactions during assortment in early RV morphogenesis; in this, NSP2 acts as a viral chaperone for the viral RNAs [42]. There are still questions about how the enzymatic activities of the viroplasm proteins are correlated with their functions during RVA replication and how assortment of the 11 pregenomic ss(+)RNA segments is controlled [42]. The groups of E Gaunt and P Digard joined forces to analyze the molecular details of the transcription/replication machinery of species A rotaviruses [43]. They constructed a mini-replicon assay which can serve as an important tool in determining the functional correlates of the viral RNA-dependent RNA polymerase (RdRp; [43]).

Hellysaz and Hagbom [44] reviewed what is known about the correlation of rotavirus disease symptoms with pathophysiological data, emphasizing CNS responses to the enteric infection.

The co-expression of RVA NSP5 with either NSP2 or VP2 in uninfected cells was found to lead to the formation of viroplasm-like structures (VLSs) several decades ago [45]. The knowledge gap regarding this structure–function relationship in non-RVA viroplasms is due to the lack of specific antibodies and suitable cell culture systems. In a recent study, the ability of the NSP5 and NSP2 of non-RVA species to form VLSs was explored [46]. While co-expression of these two proteins led to globular VLSs in RV species A, B, D, F, G, and I, in RVC, filamentous VLSs were formed. Remarkably, the co-expression of the NSP5 and NSP2 of the RV species H and J did not result in VLS formation. Interestingly, interspecies VLSs were formed between the relevant components of closely related RV species B with G and D with F [46]. This innovative data set is considered to form the basis of numerous follow-up experiments.

Rapid progress in nucleic acid sequencing techniques allows for more detailed analyses of the molecular epidemiology of RVAs. Species A, B, and C RV isolates were obtained from a pig farm in South Africa. Among thr 12 RVA isolates of G5 genotype, reassortment with three different P genotype genes (encoding P[6], P[13], and P[23]) were detected on an otherwise unchanged genetic background [47]. A G5P[23] porcine RV isolated from a pig farm in China carried several genes closely related to the cogent (analogous) genes of human RV isolates, strongly suggesting that it had emerged from a human–porcine RV reassortment event [48]. In China, >20,000 RVA strains were isolated from pigs during 2022, with G9P[23]I5 virus strains being the most prevalent [49]. Some of the porcine isolates were closely related genetically to human RVA strains, strongly suggesting that zoonotic transmissions occurred [49]. Similar results were recorded in other areas of the world [50]. G9P[6] and G9P[4] strains were isolated from children in Mozambique after the introduction of RVA vaccines [51].

RG procedures were used to insert a reporter gene (encoding green fluorescent protein, GFP, or others) downstream of NSP1 (or NSP5), separated by the 2A sequence from the viral gene. These rRV were used in micro-neutralization assays, permitting screening of large numbers of sera [30]. The procedure also demonstrated the presence of pre-existing immunity to rotaviruses in the sera of humans and various animal species (rhesus monkey, rabbit, mouse, guinea pig, and cotton rat).

Existing RV vaccines are live attenuated, can reassort with co-circulating RV wildtype strains, and may revert to virulence. Therefore, present efforts for next-generation RV vaccines are focused on antigens of non-replicating RVs [33,52,53]. In the search for a broad-spectrum, non-replicating RV vaccine candidate with high immunogenicity and cross-protection, a peptide containing a VP8* neutralizing epitope of a P[8] RV strain and a 150 amino-acid long consensus sequence derived from a wide range of human P[8] strains was constructed [54], expressed, and purified from recombinant *E. coli* [55]. This fusion peptide was recognized by a large number of RV-infected patients' sera, and its immunogenicity was proven in pilot mouse experiments. The authors are aware that testing for protective efficacy will be required and will be the subject of future work [55]. In this context, it should be remembered that an rRV with VP6 mutations has been created which does not produce infectious viral progeny while expressing viral proteins in immunogenic concentrations [33].

High titers of maternal, RV-specific antibodies may be a factor contributing to a decrease in vaccine efficacy in low- and middle-income countries. This possibility was investigated in >500 mother–infant pairs as part of an RV3-BB vaccine trial in Malawi [56]. Maternal anti-RV IgG, but not IgA levels, were correlated with reduced takes after several doses of the neonatal vaccine, but not at the end of the study, leading to the conclusion that the RV3-BB 3-dose neonatal vaccine schedule may have the potential to protect against severe RV disease [56].

Chimeric human–mouse recombinant rotaviruses [18,19] provide a new strategy for studying human rotavirus-specific immunity and can be used to investigate factors causing variability in rotavirus vaccine efficacy. The genome of the human RV vaccine strain RIX4414 (Rotarix®) was used with the aim of generating an authentic live attenuated rRV [15] with biological characteristics very similar to those of the parental virus 89-12 (G1P[8]) [16]; this will enable the identification of attenuation mutations and the rational design of NG RVA vaccines.

The use of mRNA-based therapeutics and vaccines [57,58] has led to the recent successes of the mRNA-1273 SARS-CoV-2 [59] and the BNT162b2 mRNA COVID-19 [60] vaccines. Lu et al. [61] constructed a recombinant plasmid containing a wildtype VP7 (G1) gene as insert from which RV RNA was transcribed and capped in vitro, followed by purification and enclosure into lipid nanoparticles (LNPs). In mice, the VP7-LNP elicited RV- specific antibodies and activated T cells. Testing for protective efficacy is underway [61]. Recently, mRNA-based RV-vaccine candidates were shown to exert partial protection in the gn piglet model of RV Wa (G1P[8]) infection [62].

In order to understand the dynamics of systemic and mucosal immune responses early in life, studies on T cells in mice have been initiated. It was shown that early life T cells have a low capacity to generate long-term memory compared to those of adult mice. Similarly, T cells in human neonates are immature and evolve only with age. At present, it is not clear how T cell diversification in early life affects the development of clinical symptoms after infection [63].

As co-guest editors, we are acutely aware that the >20 contributions to RV research collated in this SI of RVs and RV vaccines represent only a relatively small sector of a multitude of ongoing original work, implying that RV research still faces large gaps in regard to knowledge and understanding. However, with the discovery and application of new biological tools, including refined reverse genetics systems, many aspects of rotavirus biology will become amenable to study. These include:

- Developing rotavirus vaccine candidates that are more immunogenic and grow to high titers;
- Probing the usefulness of rotaviruses as vaccine vector systems through their capacity to incorporate and express heterologous sequences;
- Understanding the molecular basis for partial gene duplications, and their selection, in RVA isolates in vivo and in vitro;
- Revealing the contributions of cellular components to the RV–host relationship and pathogenesis;
- Expanding rudimentary knowledge of the biology, epidemiology, and pathogenesis of the non-species A RVs;
- Establishing the basis for the difference in efficiency of RV vaccines in different parts of the world, an effort that is vital to improving protection in low socio-economic regions where children are most at risk due to rotavirus disease.

Conflicts of Interest: The authors declare no conflict of interest.

References

1. Crawford, S.E.; Ramani, S.; Tate, J.E.; Parashar, U.D.; Svensson, L.; Hagbom, M.; Franco, M.A.; Greenberg, H.B.; O'Ryan, M.; Kang, G.; et al. Rotavirus infection. *Nat. Rev. Dis. Primers.* **2017**, *3*, 17083. [CrossRef]
2. Caddy, S.; Papa, G.; Borodavka, A.; Desselberger, U. Rotavirus research: 2014–2020. *Virus Res.* **2021**, *304*, 198499. [CrossRef]
3. Crawford, S.E.; Ding, S.; Greenberg, H.B.; Estes, M.K. Rotaviruses. In *Fields Virology*, 7th ed.; Volume 3: RNA Viruses; Howley, P.M., Knipe, D.M., Damania, B.A., Cohen, J.I., Whelan, S.P.J., Freed, E.O., Eds.; Wolters Kluwer: Philadelphia, PA, USA, 2023; pp. 362–413.
4. Burnett, E.; Parashar, U.D.; Tate, J.E. Real-world effectiveness of rotavirus vaccines, 2006–2019: A literature review and meta-analysis. *Lancet Glob. Health* **2020**, *8*, e1195–e1202. [CrossRef] [PubMed]
5. Bergman, H.; Henschke, N.; Hungerford, D.; Pitan, F.; Ndwandwe, D.; Cunliffe, N.; Soares-Weiser, K. Vaccines for preventing rota-virus diarrhoea: Vaccines in use. *Cochrane Database Syst. Rev.* **2021**, *11*, CD008521.
6. Desselberger, U. Differences of Rotavirus Vaccine Effectiveness by Country: Likely Causes and Contributing Factors. *Pathogens* **2017**, *6*, 65. [CrossRef]
7. Parker, E.P.K.; Ramani, S.; A Lopman, B.; A Church, J.; Iturriza-Gómara, M.; Prendergast, A.J.; Grassly, N.C. Causes of impaired oral vaccine efficacy in developing countries. *Future Microbiol.* **2017**, *13*, 97–118. [CrossRef]

8. Kanai, Y.; Komoto, S.; Kawagishi, T.; Nouda, R.; Nagasawa, N.; Onishi, M.; Matsuura, Y.; Taniguchi, K.; Kobayashi, T. Entirely plas-mid-based reverse genetics system for rotaviruses. *Proc. Natl. Acad. Sci. USA* **2017**, *114*, 2349–2354. [CrossRef]

9. Komoto, S.; Fukuda, S.; Ide, T.; Ito, N.; Sugiyama, M.; Yoshikawa, T.; Murata, T.; Taniguchi, K. Generation of Recombinant Rotaviruses Expressing Fluorescent Proteins by Using an Optimized Reverse Genetics System. *J. Virol.* **2018**, *92*, e00588-18. [CrossRef]

10. Kanai, Y.; Kobayashi, T. Rotavirus reverse genetics systems: Development and application. *Virus Res.* **2021**, *295*, 198296. [CrossRef] [PubMed]

11. Komoto, S.; Fukuda, S.; Kugita, M.; Hatazawa, R.; Koyama, C.; Katayama, K.; Murata, T.; Taniguchi, K. Generation of Infectious Re-combinant Human Rotaviruses from Just 11 Cloned cDNAs Encoding the Rotavirus Genome. *J. Virol.* **2019**, *93*, e02207-18. [CrossRef] [PubMed]

12. Kawagishi, T.; Nurdin, J.A.; Onishi, M.; Nouda, R.; Kanai, Y.; Tajima, T.; Ushijima, H.; Kobayashi, T. Reverse Genetics System for a Human Group A Rotavirus. *J. Virol.* **2020**, *94*, e00963-19. [CrossRef] [PubMed]

13. Sánchez-Tacuba, L.; Feng, N.; Meade, N.J.; Mellits, K.H.; Jaïs, P.H.; Yasukawa, L.L.; Resch, T.K.; Jiang, B.; López, S.; Ding, S.; et al. An Optimized Reverse Genetics System Suitable for Efficient Recovery of Simian, Human, and Murine-Like Rotaviruses. *J. Virol.* **2020**, *94*, e01294-20. [CrossRef] [PubMed]

14. Hamajima, R.; Lusiany, T.; Minami, S.; Nouda, R.; Nurdin, J.A.; Yamasaki, M.; Kobayashi, N.; Kanai, Y.; Kobayashi, T. A reverse genetics system for human rotavirus G2P[4]. *J. Gen. Virol.* **2022**, *103*, 001816. [CrossRef]

15. Philip, A.A.; Agbemabiese, C.A.; Yi, G.; Patton, J.T. T7 expression plasmids for producing a recombinant human G1P[8] rotavirus comprising RIX4414 sequences of the RV1 (Rotarix, GSK) vaccine strain. *Genome Announc.* **2023**, *12*, e0060323. [CrossRef] [PubMed]

16. Fukuda, S.; Kugita, M.; Kumamoto, K.; Akari, Y.; Higashimoto, Y.; Nagao, S.; Murata, T.; Yoshikawa, T.; Taniguchi, K.; Komoto, S. Generation of recombinant authentic live attenuated human rotavirus vaccine strain RIX4414 (Rotarix®) from cloned cDNAs using reverse genetics. *Viruses* **2024**, *16*, 1198. [CrossRef]

17. Sánchez-Tacuba, L.; Kawagishi, T.; Feng, N.; Jiang, B.; Ding, S.; Greenberg, H.B. The Role of the VP4 Attachment Protein in Rotavirus Host Range Restriction in an In Vivo Suckling Mouse Model. *J. Virol.* **2022**, *96*, e0055022. [CrossRef]

18. Kawagishi, T.; Sánchez-Tacuba, L.; Feng, N.; Greenberg, H.B.; Ding, S. Reverse Genetics of Murine Rotavirus: A Comparative Analysis of the Wild-Type and Cell-Culture-Adapted Murine Rotavirus VP4 in Replication and Virulence in Neonatal Mice. *Viruses* **2024**, *16*, 767. [CrossRef]

19. Woodyear, S.; Chandler, T.L.; Kawagishi, T.; Lonergan, T.M.; Patel, V.A.; Williams, C.A.; Permar, S.R.; Ding, S.; Caddy, S.L. Chimeric Viruses Enable Study of Antibody Responses to Human Rotaviruses in Mice. *Viruses* **2024**, *16*, 1145. [CrossRef]

20. Kanda, M.; Fukuda, S.; Hamada, N.; Nishiyama, S.; Masatani, T.; Fujii, Y.; Izumi, F.; Okajima, M.; Taniguchi, K.; Sugiyama, M.; et al. Establishment of a reverse genetics system for avian rotavirus A strain PO-13. *J. Gen. Virol.* **2022**, *103*, 001760. [CrossRef]

21. Snyder, A.J.; Agbemabiese, C.A.; Patton, J.T. Production of OSU G5P[7] Porcine Rotavirus Expressing a Fluorescent Reporter via Reverse Genetics. *Viruses* **2024**, *16*, 411. [CrossRef]

22. Diebold, O.; Gonzalez, V.; Venditti, L.; Sharp, C.; Blake, R.A.; Tan, W.S.; Stevens, J.; Caddy, S.; Digard, P.; Borodavka, A.; et al. Using Species a Rotavirus Reverse Genetics to Engineer Chimeric Viruses Expressing SARS-CoV-2 Spike Epitopes. *J. Virol.* **2022**, *96*, e0048822. [CrossRef] [PubMed]

23. Liu, X.F.; Li, S.; Yu, J.J.; Chai, P.D.; Pang, L.L.; Li, J.S.; Zhu, W.Y.; Ren, W.H.; Duan, Z.J. Estab-lishment of a reverse genetics system for rotavirus vaccine strain LLR and developing vaccine candidates carrying VP7 gene cloned from human strains circulating in China. *J. Med. Virol.* **2024**. *under review.*

24. Fukuda, S.; Hatazawa, R.; Kawamura, Y.; Yoshikawa, T.; Murata, T.; Taniguchi, K.; Komoto, S. Rapid generation of rotavirus sin-gle-gene reassortants by means of eleven plasmid-only based reverse genetics. *J. Gen. Virol.* **2020**, *101*, 806–815. [CrossRef] [PubMed]

25. Kanai, Y.; Onishi, M.; Kawagishi, T.; Pannacha, P.; Nurdin, J.A.; Nouda, R.; Yamasaki, M.; Lusiany, T.; Khamrin, P.; Okitsu, S.; et al. Reverse Genetics Approach for Developing Rotavirus Vaccine Candidates Carrying VP4 and VP7 Genes Cloned from Clinical Isolates of Human Rotavirus. *J. Virol.* **2020**, *95*, e01374-20. [CrossRef]

26. Patzina-Mehling, C.; Falkenhagen, A.; Trojnar, E.; Gadicherla, A.K.; Johne, R. Potential of avian and mammalian species A rota-viruses to reassort as explored by plasmid only-based reverse genetics. *Virus Res.* **2020**, *286*, 198027. [CrossRef]

27. Falkenhagen, A.; Patzina-Mehling, C.; Gadicherla, A.K.; Strydom, A.; O'Neill, H.G.; Johne, R. Generation of Simian Rotavirus Re-assortants with VP4- and VP7-Encoding Genome Segments from Human Strains Circulating in Africa Using Reverse Genetics. *Viruses* **2020**, *12*, 201. [CrossRef]

28. Saif, L.J.; Ward, L.A.; Yuan, L.; Rosen, B.I.; To, T.L. The gnotobiotic piglet as a model for studies of disease pathogenesis and immunity to human rotaviruses. *Arch. Virol. Suppl.* **1996**, *12*, 153–161.

29. Yuan, L. *Vaccine Efficacy Evaluation: The Gnotobiotic Pig Model*; CRC Press Taylor & Francis Group: Boca Raton, FL, USA, 2022; p. 189.

30. Wei, J.; Radcliffe, S.; Pirrone, A.; Lu, M.; Li, Y.; Cassaday, J.; Newhard, W.; Heidecker, G.; Rose, W., II; He, X.; et al. A Novel Rotavirus Reverse Genetics Platform Supports Flexible Insertion of Exogenous Genes and Enables Rapid Development of a High-Throughput Neutralization Assay. *Viruses* **2023**, *15*, 2034. [CrossRef]

31. Kanai, Y.; Onishi, M.; Yoshida, Y.; Kotaki, T.; Minami, S.; Nouda, R.; Yamasaki, M.; Enoki, Y.; Kobayashi, T. Genetic engineering strategy for generating a stable dsRNA virus vector using a virus-like codon-modified transgene. *J. Virol.* **2023**, *97*, e0049223. [CrossRef]

32. Nurdin, J.A.; Kotaki, T.; Kawagishi, T.; Sato, S.; Yamasaki, M.; Nouda, R.; Minami, S.; Kanai, Y.; Kobayashi, T. N-Glycosylation of Ro-tavirus NSP4 Protein Affects Viral Replication and Pathogenesis. *J. Virol.* **2023**, *97*, e0186122. [CrossRef]

33. Kotaki, T.; Kanai, Y.; Onishi, M.; Minami, S.; Chen, Z.; Nouda, R.; Nurdin, J.A.; Yamasaki, M.; Kobayashi, T. Generation of single-round infectious rotavirus with a mutation in the intermediate capsid protein VP6. *J. Virol.* **2024**, *98*, e0076224. [CrossRef] [PubMed]

34. Valusenko-Mehrkens, R.; Schilling-Loeffler, K.; Johne, R.; Falkenhagen, A. VP4 Mutation Boosts Replication of Recombinant Human/Simian Rotavirus in Cell Culture. *Viruses* **2024**, *16*, 565. [CrossRef] [PubMed]

35. Carter, M.H.; Gribble, J.; Diller, J.R.; Denison, M.R.; Mirza, S.A.; Chappell, J.D.; Halasa, N.B.; Ogden, K.M. Human Rotaviruses of Multiple Genotypes Acquire Conserved VP4 Mutations during Serial Passage. *Viruses* **2024**, *16*, 978. [CrossRef] [PubMed]

36. Wang, J.; Hu, X.; Wu, J.; Lin, X.; Chen, R.; Lu, C.; Song, X.; Leng, Q.; Li, Y.; Kuang, X.; et al. ML241 Antagonizes ERK 1/2 Activation and Inhibits Rotavirus Proliferation. *Viruses* **2024**, *16*, 623. [CrossRef]

37. Raev, S.; Kick, M.; Chellis, M.; Amimo, J.; Saif, L.; Vlasova, A. Histo-Blood Group Antigen-Producing Bacterial Cocktail Re-duces Rotavirus A, B, and C Infection and Disease in Gnotobiotic Piglets. *Viruses* **2024**, *16*, 660. [CrossRef] [PubMed]

38. Vetter, J.; Lee, M.; Eichwald, C. The Role of the Host Cytoskeleton in the Formation and Dynamics of Rotavirus Viroplasms. *Viruses* **2024**, *16*, 668. [CrossRef]

39. Vetter, J.; Papa, G.; Tobler, K.; Rodriguez, J.M.; Kley, M.; Myers, M.; Wiesendanger, M.; Schraner, E.M.; Luque, D.; Burrone, O.R.; et al. The recruitment of TRiC chaperonin in rotavirus viroplasms correlates with virus replication. *mBio* **2024**, *15*, e0049924. [CrossRef]

40. Desselberger, U. Significance of Cellular Lipid Metabolism for the Replication of Rotaviruses and Other RNA Viruses. *Viruses* **2024**, *16*, 908. [CrossRef]

41. Geiger, F.; Acker, J.; Papa, G.; Wang, X.; E Arter, W.; Saar, K.L.; A Erkamp, N.; Qi, R.; Bravo, J.P.; Strauss, S.; et al. Liquid–liquid phase separation underpins the formation of replication factories in rotaviruses. *EMBO J.* **2021**, *40*, e107711. [CrossRef]

42. Nichols, S.L.; Haller, C.; Borodavka, A.; Esstman, S.M. Rotavirus NSP2: A Master Orchestrator of Early Viral Particle Assembly. *Viruses* **2024**, *16*, 814. [CrossRef]

43. Diebold, O.; Zhou, S.; Sharp, C.P.; Tesla, B.; Chook, H.W.; Digard, P.; Gaunt, E.R. Towards the Development of a Minigenome Assay for Species A Rotaviruses. *Viruses* **2024**, *16*, 1396. [CrossRef] [PubMed]

44. Hellysaz, A.; Hagbom, M. Rotavirus Sickness Symptoms: Manifestations of Defensive Responses from the Brain. *Viruses* **2024**, *16*, 1086. [CrossRef] [PubMed]

45. Fabbretti, E.; Afrikanova, I.; Vascotto, F.; Burrone, O.R. Two non-structural rotavirus proteins, NSP2 and NSP5, form viroplasm-like structures in vivo. *J. Gen. Virol.* **1999**, *80*, 333–339. [CrossRef]

46. Lee, M.; Cosic, A.; Tobler, K.; Aguilar, C.; Fraefel, C.; Eichwald, C. Characterization of viroplasm-like structures by co-expression of NSP5 and NSP2 across rotavirus species A to J. *J. Virol.* **2024**, *98*, e0097524. [CrossRef] [PubMed]

47. Strydom, A.; Segone, N.; Coertze, R.; Barron, N.; Strydom, M.; O'neill, H.G. Phylogenetic Analyses of Rotavirus A, B and C Detected on a Porcine Farm in South Africa. *Viruses* **2024**, *16*, 934. [CrossRef]

48. Gao, L.; Shen, H.; Zhao, S.; Chen, S.; Zhu, P.; Lin, W.; Chen, F. Isolation and Pathogenicity Analysis of a G5P[23] Porcine Ro-tavirus Strain. *Viruses* **2024**, *16*, 21. [CrossRef]

49. Qiao, M.; Li, M.; Li, Y.; Wang, Z.; Hu, Z.; Qing, J.; Huang, J.; Jiang, J.; Jiang, Y.; Zhang, J.; et al. Recent Molecular Characterization of Porcine Rotaviruses Detected in China and Their Phylogenetic Relationships with Human Rotaviruses. *Viruses* **2024**, *16*, 453. [CrossRef]

50. Brnić, D.; Čolić, D.; Kunić, V.; Maltar-Strmečki, N.; Krešić, N.; Konjević, D.; Bujanić, M.; Bačani, I.; Hižman, D.; Jemeršić, L. Rotavirus A in Domestic Pigs and Wild Boars: High Genetic Diversity and Interspecies Transmission. *Viruses* **2022**, *14*, 2028. [CrossRef]

51. Munlela, B.; João, E.D.; Strydom, A.; Bauhofer, A.F.L.; Chissaque, A.; Chilaúle, J.J.; Maurício, I.L.; Donato, C.M.; O'neill, H.G.; de Deus, N. Whole-Genome Characterization of Rotavirus G9P[6] and G9P[4] Strains That Emerged after Rotavirus Vaccine Introduction in Mozambique. *Viruses* **2024**, *16*, 1140. [CrossRef]

52. Fix, A.; Kirkwood, C.D.; Steele, D.; Flores, J. Next-generation rotavirus vaccine developers meeting: Summary of a meeting sponsored by PATH and the bill & melinda gates foundation (19–20 June 2019, Geneva). *Vaccine* **2020**, *38*, 8247–8254. [CrossRef]

53. McAdams, D.; Estrada, M.; Holland, D.; Singh, J.; Sawant, N.; Hickey, J.M.; Kumar, P.; Plikaytis, B.; Joshi, S.B.; Volkin, D.B.; et al. Concordance of in vitro and in vivo measures of non-replicating rotavirus vaccine potency. *Vaccine* **2022**, *40*, 5069–5078. [CrossRef] [PubMed]

54. Kondakova, O.A.; Ivanov, P.A.; Baranov, O.A.; Ryabchevskaya, E.M.; Arkhipenko, M.V.; Skurat, E.V.; Evtushenko, E.A.; Nikitin, N.A.; Karpova, O.V. Novel antigen panel for modern broad-spectrum recombinant rotavirus A vaccine. *Clin. Exp. Vaccine Res.* **2021**, *10*, 123–131. [CrossRef]

55. Granovskiy, D.L.; Khudainazarova, N.S.; Evtushenko, E.A.; Ryabchevskaya, E.M.; Kondakova, O.A.; Arkhipenko, M.V.; Kovrizhko, M.V.; Kolpakova, E.P.; Tverdokhlebova, T.I.; Nikitin, N.A.; et al. Novel Universal Recombinant Rotavirus A Vaccine Candidate: Evaluation of Immunological Properties. *Viruses* **2024**, *16*, 438. [CrossRef] [PubMed]

56. Morgan, B.; Lyons, E.A.; Handley, A.; Bogdanovic-Sakran, N.; Pavlic, D.; Witte, D.; Mandolo, J.; Turner, A.; Jere, K.C.; Justice, F.; et al. Rotavirus-specific maternal serum antibodies and vaccine responses to RV3-BB rotavirus vaccine administered in a ne-onatal or infant schedule in Malawi. *Viruses* **2024**, *16*, 1488. [CrossRef]
57. Barbier, A.J.; Jiang, A.Y.; Zhang, P.; Wooster, R.; Anderson, D.G. The clinical progress of mRNA vaccines and immunotherapies. *Nat. Biotechnol.* **2022**, *40*, 840–854. [CrossRef]
58. Qin, S.; Tang, X.; Chen, Y.; Chen, K.; Fan, N.; Xiao, W.; Zheng, Q.; Li, G.; Teng, Y.; Wu, M.; et al. mRNA-based therapeutics: Powerful and versatile tools to combat diseases. *Signal Transduct. Target Ther.* **2022**, *7*, 166. [CrossRef]
59. Baden, L.R.; El Sahly, H.M.; Essink, B.; Kotloff, K.; Frey, S.; Novak, R.; Diemert, D.; Spector, S.A.; Rouphael, N.; Creech, C.B.; et al. Efficacy and Safety of the mRNA-1273 SARS-CoV-2 Vaccine. *N. Engl. J. Med.* **2021**, *384*, 403–416. [CrossRef] [PubMed]
60. Polack, F.P.; Thomas, S.J.; Kitchin, N.; Absalon, J.; Gurtman, A.; Lockhart, S.; Perez, J.L.; Pérez Marc, G.; Moreira, E.D.; Zerbini, C.; et al. Safety and Efficacy of the BNT162b2 mRNA COVID-19 Vaccine. *N. Engl. J. Med.* **2020**, *383*, 2603–2615. [CrossRef]
61. Lu, C.; Li, Y.; Chen, R.; Hu, X.; Leng, Q.; Song, X.; Lin, X.; Ye, J.; Wang, J.; Li, J.; et al. Safety, Immunogenicity, and Mechanism of a Rotavirus mRNA-LNP Vaccine in Mice. *Viruses* **2024**, *16*, 211. [CrossRef]
62. Hensley, C.; Roier, S.; Zhou, P.; Schnur, S.; Nyblade, C.; Parreno, V.; Frazier, A.; Frazier, M.; Kiley, K.; O'Brien, S.; et al. mRNA-Based Vaccines Are Highly Immunogenic and Confer Protection in the Gnotobiotic Pig Model of Human Rotavirus Diarrhea. *Vaccines* **2024**, *12*, 260. [CrossRef]
63. Montenegro, C.; Perdomo-Celis, F.; Franco, M.A. Update on Early-Life T Cells: Impact on Oral Rotavirus Vaccines. *Viruses* **2024**, *16*, 818. [CrossRef] [PubMed]

Article

A Novel Rotavirus Reverse Genetics Platform Supports Flexible Insertion of Exogenous Genes and Enables Rapid Development of a High-Throughput Neutralization Assay

Jiajie Wei [1,*], Scott Radcliffe [2], Amanda Pirrone [1], Meiqing Lu [1], Yuan Li [1], Jason Cassaday [1], William Newhard [1], Gwendolyn J. Heidecker [1], William A. Rose II [2], Xi He [1], Daniel Freed [1], Michael Citron [1], Amy Espeseth [1] and Dai Wang [1]

[1] Department of Infectious Diseases and Vaccines, Merck & Co., Inc., West Point, PA 19486, USA; amanda.pirrone@merck.com (A.P.); meiqing_lu@merck.com (M.L.); yuan_li@merck.com (Y.L.); jason_cassaday@merck.com (J.C.); william_newhard@merck.com (W.N.); gwendolyn_heidecker@merck.com (G.J.H.); xi_he@merck.com (X.H.); dan_freed@merck.com (D.F.); michael_citron@merck.com (M.C.); espeseth@comcast.net (A.E.); dai_wang@merck.com (D.W.)

[2] Department of Quantitative Biosciences, Merck & Co., Inc., West Point, PA 19486, USA; scott.radcliffe@merck.com (S.R.); william.rose@merck.com (W.A.R.II)

* Correspondence: jiajie.wei@merck.com

Abstract: Despite the success of rotavirus vaccines, rotaviruses remain one of the leading causes of diarrheal diseases, resulting in significant childhood morbidity and mortality, especially in low- and middle-income countries. The reverse genetics system enables the manipulation of the rotavirus genome and opens the possibility of using rotavirus as an expression vector for heterologous proteins, such as vaccine antigens and therapeutic payloads. Here, we demonstrate that three positions in rotavirus genome—the C terminus of NSP1, NSP3 and NSP5—can tolerate the insertion of reporter genes. By using rotavirus expressing GFP, we develop a high-throughput neutralization assay and reveal the pre-existing immunity against rotavirus in humans and other animal species. Our work shows the plasticity of the rotavirus genome and establishes a high-throughput assay for interrogating humoral immune responses, benefiting the design of next-generation rotavirus vaccines and the development of rotavirus-based expression platforms.

Keywords: rotavirus; high-throughput screening; microneutralization assay; neutralizing antibody; pre-existing immunity

Citation: Wei, J.; Radcliffe, S.; Pirrone, A.; Lu, M.; Li, Y.; Cassaday, J.; Newhard, W.; Heidecker, G.J.; Rose II, W.A.; He, X.; et al. A Novel Rotavirus Reverse Genetics Platform Supports Flexible Insertion of Exogenous Genes and Enables Rapid Development of a High-Throughput Neutralization Assay. *Viruses* **2023**, *15*, 2034. https://doi.org/10.3390/v15102034

Academic Editors: Ulrich Desselberger and John T. Patton

Received: 8 September 2023
Revised: 27 September 2023
Accepted: 28 September 2023
Published: 30 September 2023

1. Introduction

Although rotavirus vaccines have substantially reduced rotavirus-related childhood morbidity and mortality worldwide [1], rotaviruses remain one of the most common causes of diarrheal diseases in children, with a higher disease burden in developing countries. Rotaviruses are responsible for 128,500–215,000 deaths annually in children under 5 years old [2,3]. The mechanisms underlying rotavirus vaccine-induced protection are not fully understood, partially due to limitations in current animal models of rotavirus infection and disease. The lower vaccine efficacy observed in low-income countries has been attributed to multiple factors, including higher levels of maternally derived antibodies, different intestinal microbiome resulting from chronic enteropathy, and/or poor nutritional status [4]. To develop the next-generation vaccines with improved safety and efficacy, a better understanding of pre-existing immunity, including neutralizing antibodies in human and animal models and its impact on vaccine efficacy, is needed.

Rotaviruses are double-stranded, segmented RNA viruses. The 11 genome segments encode 12 viral proteins, including six non-structural proteins (NSP1-NSP6) and six structural proteins (VP1–VP4, VP6, and VP7). Each segment encodes one open reading frame (ORF), except segment 11, with NSP5, which contains an internal ORF for NSP6. Rotavirus particles consist of three concentric proteins shells (VP2, VP6, and VP7) and a spike protein, VP4, which spans the VP6 and VP7 layers and extends out from the particle. Serologically, rotaviruses are grouped into distinct serogroups based on VP6 reactivity. Group A rotaviruses (RVA), further classified into serotypes defined by VP7 (G) and VP4 [P], cause the majority of disease in human. RVA strains are also found in animals, and infection has been shown to be highly species-specific.

Rotavirus research has been hindered by the lack of a reverse genetics system, which would allow for the generation and engineering of defined viral particles. Since the first publication of a plasmid-only-based reverse genetics system for the simian rotavirus strain SA11 [5], multiple groups have utilized the system to rescue recombinant rotaviruses with different properties. These include the simian RRV strain; human CDC-9 strain; a murine-like RV strain [6]; human Odelia strain [7]; human KU strain [8]; human HN126 strain [9]; bovine RF strain [10]; avian PO-13 strain [11]; chimeric strains with SA11 backbone and VP4, VP7, and/or VP6 genes from human clinical samples [12]; and SA11 carrying NSP2 phosphorylation mutation [13]. Heterologous protein expression from rotavirus was explored by replacing part of the NSP1 ORF with foreign genes [5]. Later, it was shown that genome segment 7 could be re-engineered to encode NSP3 fused to a fluorescent protein [14]. Additionally, domains of SARS-CoV-2 spike protein were also expressed downstream of NSP3 [15], suggesting that rotaviruses may serve as a vector for gene delivery. Recently, the concept of using recombinant rotaviruses expressing norovirus capsid proteins as a dual vaccine was established in an infant mouse model [16]. A solid understanding of pre-existing immunity in human and animal models would further explore the feasibility of using rotavirus to deliver therapeutic proteins or vaccine antigens.

Here, in addition to the previously published NSP3 site, we have demonstrated two additional genomic locations in NSP1 and NSP5 of the simian rotavirus strain SA11 for expressing heterologous proteins by reverse genetics. We utilized a recombinant rotavirus expressing GFP from NSP1 (rSA11-GFP) to establish a high-throughput rotavirus microneutralization assay. This assay enabled us to determine the presence of pre-existing neutralizing antibodies in human and other animal serum samples. Among the 50 human donor samples [17] that were surveyed, 41 had detectable neutralization titers against SA11, a serotype G3 virus. Additionally, all African green monkeys and rhesus monkeys examined showed pre-existing immunity. In contrast, other animal models, such as rabbit, mouse, guinea pig and cotton rat, either had much lower levels or no detectable neutralizing antibodies. Through the identification of novel approaches for heterologous gene expression, we have demonstrated the plasticity of the rotavirus genome and developed a high-throughput neutralization assay based on a GFP expressing virus.

2. Materials and Methods

2.1. Cell Culture

CV1, MA104, and baby hamster kidney cells expressing T7 RNA polymerase (BHK-T7) were maintained in Dulbecco's Modified Eagle's Medium (DMEM) with 10% fetal bovine serum (FBS) and 1% penicillin-streptomycin. All cultures were grown at 37 °C in a 5% CO_2 incubator.

2.2. Plasmid Construction

Sequences of all 16 plasmids used for the generation of the wild-type SA11 strain were obtained from Addgene (https://www.addgene.org/Takeshi_Kobayashi/ (accessed on 29 September 2023)) [5]. pUC19 and pV1Jns served as the backbones of 11 plasmids, each encoding one rotavirus genome segment (pT7/VP1SA11, pT7/VP2SA11, pT7/VP3SA11, pT7/VP4SA11, pT7/VP6SA11, pT7/VP7SA11, pT7/NSP1SA11, pT7/NSP2SA11, pT7/NSP3SA11, pT7/NSP4SA11, and pT7/NSP5SA11) and 5 helper plasmids (pCMV/NSP2, pCMV/NSP5, pCMV/NBVFAST, pCMV/D12L, and pCMV/D1R), respectively. For the expression of GFP, NSP open reading frames were fused with 2A peptide GSGEGRGSLLTCGDVEENPGP and then GFP. After the stop codon, an NSP open reading frame sequence repeat was inserted when indicated. All plasmids were synthesized using Genewiz.

2.3. Recombinant Rotavirus Rescue

Recombinant SA11 (rSA11) strains were generated by reverse genetics, as described previously with modifications [5]. Monolayers of BHK-T7 cells in 6-well plates (1×10^6 cells/well) were used for transfection. 16 plasmids (0.75 µg/plasmid except 0.015 µg pCMV/NSVFAST) in 150 µL of Opti-MEM were added to 150 µL of Opti-MEM containing 12.5 µL of Lipofectamine 2000. Transfection complexes were incubated at room temperature for 20 min and then added to BHK-T7 cells drop-wise. 24 h post transfection, the culture medium was changed into serum-free DMEM. 48 h post transfection, 1.5×10^5 CV1 cells were added to transfected cells, and TPCK-treated trypsin (1 mg/mL stock) was added to culture medium to achieve a final concentration of 1 µg/mL. The transfection reaction was monitored daily and harvested when a complete cytopathic effect (CPE) was observed (typically three days after the addition of CV1 cells). To generate recombinant viruses with heterologous genes, pT7/NSP1SA11, pT7/NSP2SA11, pT7/NSP3SA11, pT7/NSP4SA11, and pT7/NSP5SA11 were replaced with the corresponding plasmid with heterologous gene insertion. rSA11/GFP was then plaque-purified for three rounds on MA104 cells.

2.4. Virus Infection

Recombinant viruses were treated with 10 µg/mL TPCK-treated trypsin at 37 °C for 1 h. Monolayers of CV1 or MA104 cells were washed with serum-free DMEM three times and then infected with trypsin-treated viruses in serum-free DMEM at 37 °C. After 1 h of incubation, inoculums were removed.

2.5. Virus Stock Preparation

MA104 cells were infected with recombinant viruses and cultured in serum-free DMEM containing 1 µg/mL trypsin. Once complete CPE was observed, viruses were harvested by subjecting the infected cells to three freeze–thaw cycles. The virus suspension was then filtered through 0.2 micro filters. Virus concentration and purification were performed using an Amicon Ultra-15 50,000 NMWL centrifugal filter unit.

2.6. Plaque Assay

MA104 cells were infected with recombinant viruses and overlaid with phenol-red-free MEM containing 0.8% agarose and 0.5 µg/mL trypsin. After 4 days of incubation, plaques were visualized by adding 5 mg/mL MTT or picked directly for plaque purification.

2.7. Flow Cytometry and Data Analysis

CV1 cells were counted, infected with recombinant viruses, and cultured in DMEM supplemented with 10% FBS. After overnight incubation, cells were harvested, fixed with 4% paraformaldehyde, and stained with a primary antibody anti-RotaVP6 (UK1, ThermoFisher) and then a secondary antibody Alexa Fluor 647 AffiniPure Goat Anti-Mouse IgG (H + L) (Jackson Immuno Research Labs, West Grove, PA, USA). Staining and washing steps were performed using Perm/Wash buffer (BD Biosciences, San Jose, CA, USA). Flow cytometric data were acquired using a BD LSRII flow cytometer (BD Biosciences, San Jose, CA, USA) and gated on single cells. Data analysis was conducted using FlowJo version 10 software (FlowJo LLC, San Jose, CA, USA). In the flow cytometry-based infectivity assay, the percentage of VP6-positive cell population was numerated by FlowJo. Based on the Poisson distribution, the infectious unit (IU)/mL was calculated as IU/mL = (# of cells at infection) × [MOI/(ml of viral stock used at infection)].

2.8. Growth Kinetics

MA104 cells were infected with recombinant viruses at a multiplicity of infection (MOI) of 0.01 IU/cell and cultured in serum-free DMEM containing 1 µg/mL trypsin. At 24, 48, and 72 h post-infection, the viruses were harvested by subjecting the infected cells to three freeze–thaw cycles. The virus titer was determined by a flow cytometry-based infectivity assay (Supplementary Figure S1).

2.9. Genetic Stability

Viruses were serially passaged on MA104 cells. Monolayers of MA104 cells were infected with viruses and cultured in serum-free DMEM containing 1 µg/mL trypsin. When complete CPE was observed, the cell culture supernatant was used directly for the next round of infection with 1:1000 final dilution. Viral RNA was extracted from 140 µL of the supernatant using QIAamp viral RNA kit, and 15 µL RNA was used in the SuperScript IV one-step RA-PCR system with forward primer 5′-CAACGGAGGAACTGATTGAAATGAAGAA-3′ and reverse primer 5′-TTGCCAGCTAGGCGCTACT-3′ following manufacturers' instructions. PCR reactions were analyzed by 1.2% E-gel (ThermoFisher) along with E-Gel 1 Kb Plus Express DNA Ladder. Sanger sequencing reactions were conducted by Genewiz using primers 5′-GCTACTGATCTCCAACTCAGAAGATG-3′ and 5′-TAGTCTGGACGGTCTT GTGA-3′.

2.10. ELISA

96-well assay plates were coated with wild-type SA11 (10^5 PFU/well) in DMEM at 4 °C overnight. The plates were then washed once with 300 µL/well Washing Buffer (PBS + 0.05% Tween 20) and blocked with 200 µL/well Blocking Buffer (Alfa Aesar, Haverhill, MA, USA) at 4 °C overnight. Blocked plates were incubated with a series of 3-fold diluted sera in Blocking Buffer (100 µL/well) at 4 °C overnight. Following sera incubation, the plates were washed three times with 300 µL/well Washing Buffer and incubated with 100 µL/well 1:4000 diluted alkaline phosphatase conjugated Goat anti-Rhesus IgG H&L (used for both African green monkey and human, Southern Tech, Ardmore, OK, USA) in Blocking buffer with 0.1% Tween 20 for 1.5 h at room temperature. After washing the plates three times with Washing Buffer, 100 µL/well Tropix CDP-Star Sapphire II substrate (Applied Biosystem, Waltham, MA, USA) were added. After incubation at room temperature for 10 min, the chemiluminescent signal from each well was read on PHERAstar. The threshold value was 25 times the mean plate background. Interpolated titers were calculated by drawing a line between the last point above the threshold and the first point below the threshold and solving for the fold dilution where that line crosses the threshold.

2.11. rSA11-GFP Based Micro-Neutralization Assay

CV1 cells were seeded into 96-well plate (4×10^4 cells/well) and cultured overnight. Serum samples were heat-inactivated at 56 °C for 30 min and serial diluted in PBS. rSA11-GFP (used at MOI = 1) was activated with 10 µg/mL TPCK-treated trypsin at 37 °C for 1 h and then mixed with serial diluted animal serum samples at 37 °C for 1 h on an orbital shaker. Pre-seeded CV1 cells in the 96-well plates were washed three times with serum-free DMEM, infected with virus/serum mixtures at 37 °C for 1 h, and then cultured in phenol-red-free DMEM supplemented with 10% FBS. For the 384-well neutralization assay, CV1 cells were harvested and washed in serum-free DMEM. 1×10^4 CV1 cells in suspension were added into virus/serum mixtures directly and incubated at 37 °C for 1 h. FBS was then added to the plate to achieve a final concentration of 10%. For both 96 well and 384 well plates, after overnight incubation at 37 °C, plates were read by an Acumen high content screening (HCS) reader (TTP LabTech, Melbourn, UK) at 488 nm to determine numbers of GFP positive cells in each well. The percentages of inhibition were calculated based on control wells to which no animal serum was added. A sample 384-plate map is provided in Supplemental Table S1. The plate had the following characteristics: (1). The first column of the plate was reserved for "no serum" controls. (2). The last column of the plate contained mock infected controls. (3). Serum samples were diluted threefold with an 11-point titration. (4). One serum sample was tested for each animal. (5). Technical duplication was performed for each sample. (6). A positive control, cynomolgus serum sample (enQuire Bioreagents AB155109, Littleton, CO, USA), was included on each plate. This sample had an NT50 value around 100. NT50 was calculated by nonlinear four-parameter curve fitting using Prism 8 (GraphPad, La Jolla, CA, USA) to determine the serum dilution that resulted in a 50% reduction in GFP-positive cells compared to the control.

2.12. Animal and Human Serum Samples

The serum samples used in the assay were obtained from various sources. Human serum samples were reported before [17]. Serum samples from rabbit, mouse, guinea pig and cotton rat were obtained from animals housed in the animal facility in the research laboratories at Merck & Co., Inc., West Point, PA, USA. Simian serum samples were collected from animals at the New Iberia Primate Research Center (NIRC, New Iberia, LA, USA). All studies were conducted in accordance with relevant guidelines using protocols approved by NIRC and the Institutional Animal Care and Use Committee of Merck & Co., Inc., Kenilworth, NJ, USA.

2.13. Serum Purification

African green monkey serum was diluted 1:25 in binding buffer and purified using Nab™ Protein G Spin Kit (Thermo Scientific Cat# 89949, Waltham, MA, USA). Diluted samples were incubated in immobilized protein G spin columns at room temperature with end-over-end mixing for 10 min. Spin columns were then centrifuged for 1 min at $5000 \times g$, washed three times with binding buffer, and eluted three times with Elution Buffer into Neutralization Buffer via centrifugation. Elution fractions were pooled and concentrated using Amicon Ultra 0.5 mL centrifugal filters.

3. Results

3.1. The Utilization of Three Positions in Rotavirus Genome for Heterologous Gene Expression

We aimed to construct rotaviruses expressing heterologous proteins in addition to the full set of rotavirus proteins. A plasmid-only system to rescue simian rotavirus strain SA11 enables the engineering of rotaviruses [5]. Because naturally occurring human rotaviruses that contain rearranged genomes can package up to 1800 additional base pairs in virus particles [18], and rearranged genome segments 7 and 11 with open reading frame (ORF) sequence repeat at the C terminus of NSP3 and NSP5, respectively, are preferentially packaged into rotaviruses [19], we reasoned that these two positions could tolerate foreign gene insertion.

To systematically explore genomic positions that support heterologous protein expression, we fused a green fluorescent protein (GFP) after a 2A self-cleaving peptide to the C terminus of every nonstructural protein (Figure 1A) except NSP6, given that NSP6 is encoded by an internal ORF inside of NSP5. To keep the potential genome packaging signal at the 3′ end of ORF, after the stop codon, a 450 bp fragment of the NSP ORF 3′ sequence was repeated upstream of the 3′UTR.

We tested each modified genome segment by replacing the corresponding wild-type (WT) SA11 genome segment in the reverse genetics system [5]. We transfected cells with plasmids, used supernatant to infect the CV-1 cells, and then determined the expression levels of GFP and rotavirus VP6 proteins from CV-1 cells by flow cytometry (Figure 1B–H). As expected, compared to mock infected cells, CV-1 cells infected with the WT virus showed a VP6 signal (Figure 1B,C). We failed to generate recombinant rotaviruses that contained modified NSP2 or NSP4, as VP6-positive cells were not observed after infecting CV-1 cells with the corresponding reverse genetics reactions (Figure 1E,G). In contrast, double-positive cell populations that expressed VP6 and GFP were observed when the rescue reactions contained modified NSP1, NSP3, or NSP5, indicating that we successfully generated a recombinant rotavirus with a GFP insertion at each of these locations (Figure 1D,F,H). We thus found three positions for heterologous gene insertion in rotavirus genome. Among NSP1, NSP3, and NSP5, modified NSP1 and NSP5 led to a higher percentage of VP6/GFP double positives (Figure 1D), suggesting a higher efficiency of virus rescue.

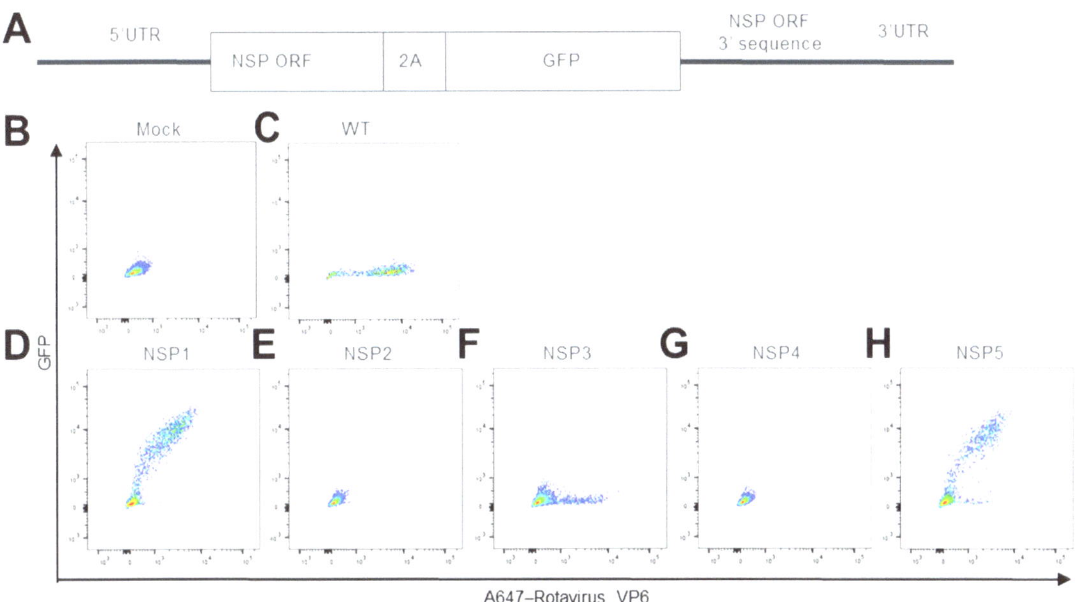

Figure 1. Rescue effort of rSA11 strains expressing GFP from five genome locations. (**A**) Schematic representation of plasmids used for the rescue of rSA11 viruses encoding GFP. (**B–H**) Representative flow cytometry analysis of CV-1 cells infected with rSA11 rescue products. GFP sequence was inserted at the indicated positions. Expressions levels of GFP and rotavirus protein VP6 were examined. Pseudocolor plots were shown using color to denote areas of high and low population density.

We then deleted the NSP1 and NSP5 ORF 3′ sequence repeat after the stop codon to avoid potential recombination events (Figure 2A). We still observed the double-positive cell population that expressed GFP and VP6 (Figure 2B), indicating virus packaging does not require the additional repeated sequence. Similar to recombinant rotaviruses rescued in Figure 1, after deleting the repeated sequence, the NSP1 position still resulted in a higher percentage of VP6/GFP double-positive population than NSP5 (Figure 2B). We thus focused on viruses containing GFP downstream of NSP1 without a 3′ sequence repeat for further characterization and referred to it as rSA11-GFP.

Using flow cytometry to examine VP6 expression, we also established a flow-cytometry-based infectivity assay for determining rotavirus titers (Supplementary Figure S1). After infection, CV-1 cells were cultured in media supplemented with FBS, but not trypsin, to prevent multiple rounds of infection. Based on Poisson distribution, the multiple of infection (MOI) can be calculated after determining the percentage of VP6-positive cells. The titer, as infectivity unit per ml (IU/mL), can then be calculated based on the number of cells and the amount of viral stock used at infection.

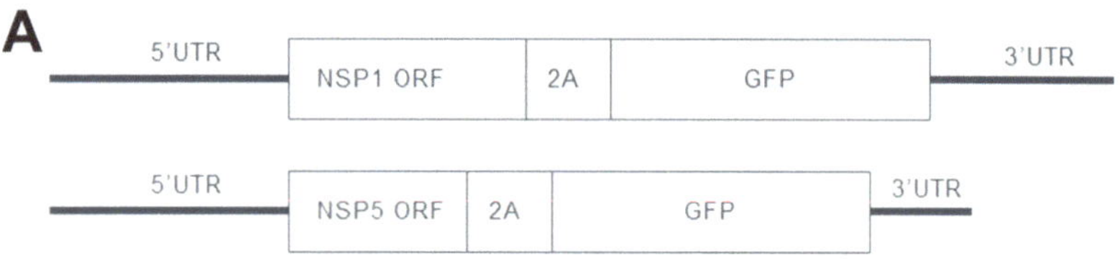

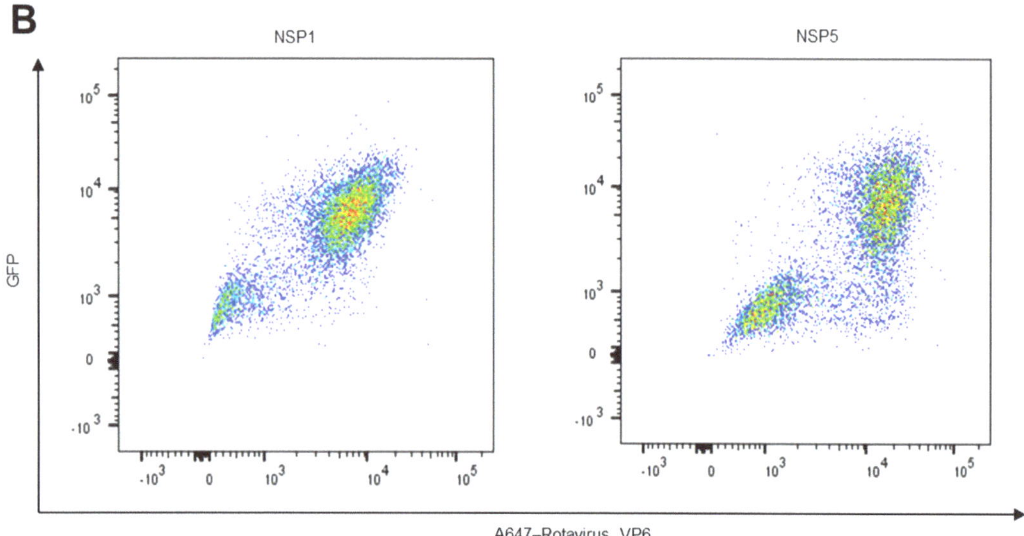

Figure 2. Generation of rSA11 strains expressing GFP at the C terminus of NSP1 or NSP5 without 3′ ORF repeats. (**A**) Schematic representation of plasmids used for the recovery of rSA11 viruses encoding GFP downstream of NSP1 or NSP5. (**B**) Representative flow cytometry analysis of CV-1 cells infected with rSA11 strains generated with plasmids shown in (**A**). Expression levels of GFP and rotavirus protein VP6 were examined. Pseudocolor plots were shown using color to denote areas of high and low population density.

3.2. Characteristics of rSA11-GFP

To examine the genetic stability of rSA11-GFP, we passaged rSA11-GFP and rSA11-WT on MA104 cells ten times, extracted viral RNA from passage one (reverse genetics product) and ten, and performed RT-PCR using primers flanking the insertion site. RT-PCR products were visualized by gel electrophoresis and sequenced by Sanger sequencing. Fragments migrated to expected sizes (Figure 3A), and sequencing reactions showed that no mutations were generated for ten passages. Our results thus indicated that rSA11-GFP was genetically stable.

We also compared the growth kinetics of rSA11-GFP with rSA11-WT. MA104 cells were infected with viruses at an MOI of 0.01 IU/cell and harvested at 24, 48, and 72 h post infection (Figure 3B). The growth curves of rSA11-GFP and rSA11-WT were indistinguishable, indicating that the insertion of GFP did not affect the fitness of the recombinant virus in vitro. In addition, plaques formed by rSA11-GFP and rSA11-WT were of similar sizes (Figure 3C), further supporting that the insertion of GFP downstream of NSP1 had no effects on rotavirus replication.

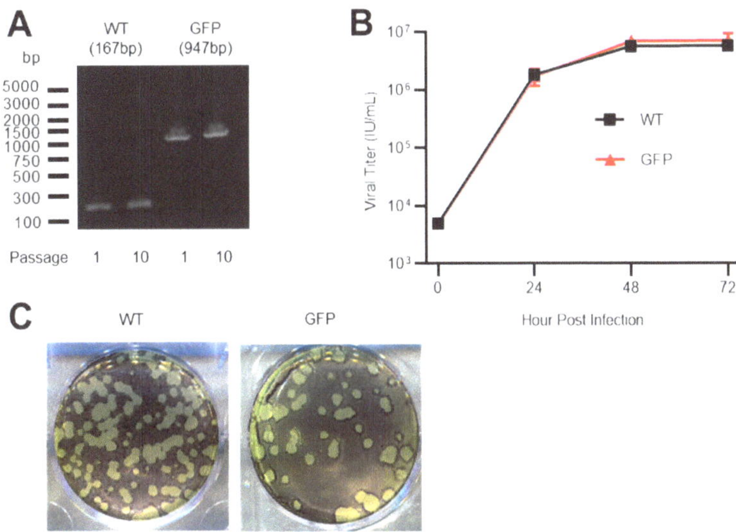

Figure 3. Properties of rSA11-GFP. (**A**) Genetic stability of rSA11-GFP. rSA11 and rSA11-GFP were serially passaged ten times on MA104 cells and analyzed by RT-PCR using primers flanking the insertion site. The expected band sizes are indicated in parentheses. (**B**) Growth kinetics of rSA11 and rSA11-GFP. MA104 cells were infected with viruses at an MOI of 0.01 IU/cell and harvested at indicated time points. Virus titer was determined in a flow cytometry-based infectivity assay using CV1 cells. Data are expressed as the mean and range of duplicates. (**C**) Plaque formation on MA104 cells by rSA11and rSA11-GFP. Data representative of three independent experiments.

3.3. rSA11-GFP Based Microneutralization Assay

Because traditional neutralization assays, such as plaque reduction neutralization test (PRNT) and fluorescent foci reduction neutralization test (FRNT), that rely on antibody staining are time-consuming and labor intensive, we sought to develop a microneutralization assay based on the GFP signal using rSA11-GFP. It is known that immunity against rotavirus exists naturally in some monkey colonies [20]. We examined four rhesus monkey serum samples in the 96-well format microneutralization assay and found that, as expected, all four samples neutralized rSA11-GFP, with two showing higher neutralizing capacity (Figure 4A).

We then converted the assay to a higher throughput format by adapting it to 384-well plates and eliminating the CV-1 pre-seeding step (Figure 4B and Supplementary Table S1). In the 384-well plate format, CV-1 cells in suspension were applied directly to the virus/serum mixtures for infection. We used this high-throughput assay to examine 12 African green monkeys in our animal facility (Figure 4C). All animals showed pre-existing antibodies against rotavirus based on neutralization assay and IgG ELISA assay that used SA11 for coating. We observed a wide range of antibody titers with NT_{50} titers ranging from 9 to 545 and ELISA titers ranging from 5444 to 323,096. Our results indicated that all African green monkeys we examined were seropositive for SA11. It is unlikely that we are detecting maternal antibodies as all monkeys are 2–3 years old. To verify that the neutralization ability was antibody-dependent, we performed polyclonal IgG purification on the serum samples from seven African green monkeys using protein G beads. We obtained similar titers before and after purification while the flow-through showed limited neutralization capacity (Supplementary Figure S2). The high level of correlation (r = 0.9247, $p < 0.0001$) between the neutralization titer and ELISA titer (Figure 4C) suggested that either almost all antibodies captured by ELISA were neutralizing antibodies or the proportions of rotavirus antibodies with neutralization capacity were consistent among African green monkeys.

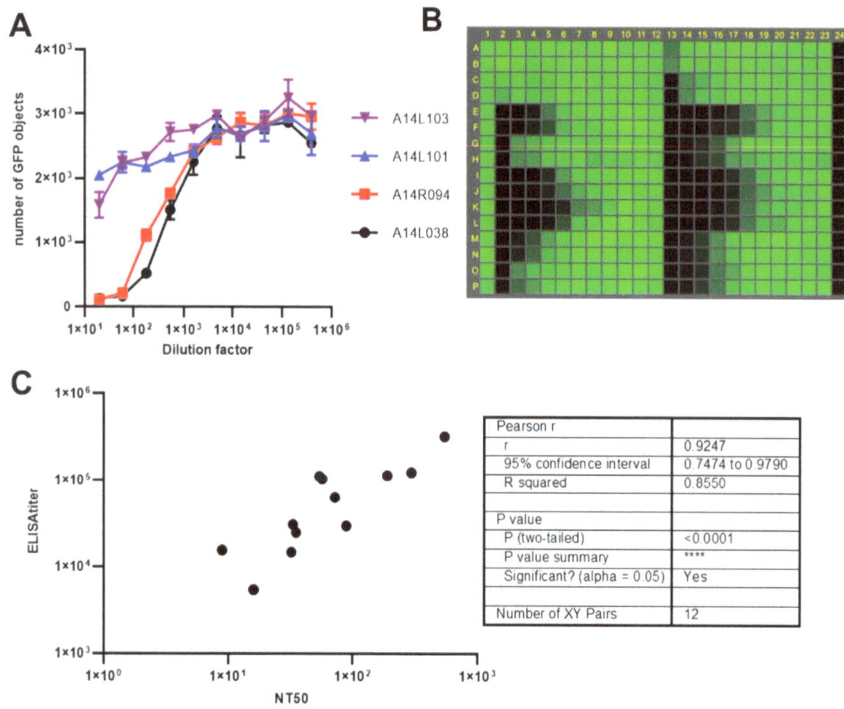

Figure 4. rSA11-GFP based microneutralization assay. (**A**) Representative serum neutralization curves of four rhesus monkeys. rSA11-GFP, preincubated with serial diluted serum samples, was used to infect CV-1 cells. After overnight incubation, numbers of GFP positive cells were numerated. Median is shown from triplicated wells. (**B**) The conversion of the neutralization assay to a high-throughput platform. rSA11-GFP, preincubated with serial diluted serum samples, was used to infect CV-1 cells. After overnight incubation GFP positive cells were numerated by Acumen. A representative review of a 384-well plate is shown. The plate map is provided in Supplemental Table S1. No serum samples were used in the first column and the last column contained mocked infected cells. Serum samples were diluted threefold with 11-point titration. Wells are colored based on the numbers of GFP positive cells. Average for no serum, infected controls was ~100 objects/well while average for cells only was ~3. (**C**) The correlation of neutralization titers and ELISA titers of serum samples from 12 African green monkeys and its statistical analysis. **** means *p* value less than 0.0001.

3.4. Pre-Existing Immunity in Human and Other Animal Species

We next determined neutralizing antibodies in human donors by the rSA11-GFP-based microneutralization assay (Figure 5A). Group A rotavirus contains more than 40 G (VP7) serotypes and more than 55 P (VP4) serotypes according to Rotavirus Classification Working Group (RCWG [21]). SA11 was originally isolated from a healthy African green monkey and belongs to G3P5B[2]. Serotypes G1, 2, 3, 4, 9, and 12 are epidemiologically important for human. Although both VP4 and VP7 can elicit neutralizing antibodies, this assay specifically reveals G3-specific antibodies as there is no known P5B[2] human strain. Out of the twenty samples we initially examined, only one did not show a neutralization titer above the limit of detection, suggesting the wide prevalence of G3 antibodies in human population. The titers were similar to those of African green monkeys in the animal facility and higher than those of the 11 rhesus monkeys we examined. We further tested 30 human serum samples by ELISA binding and neutralization assays (Supplementary Figure S3). As expected, we observed a weaker correlation between the neutralization titers and ELISA titers from human (r = 0.5322, *p* = 0.0025) compared to African green monkey. SA11 is a simian RV origin isolated from an asymptomatic African green monkey. African green monkeys are potentially infected with strains similar to SA11 naturally, while there are many human rotavirus strains with different serotypes. The weaker correlation revealed the complexity of human rotavirus immune status as repeated infections are common, with secondar infections often involving different serotypes [22]. The observation that all rhesus monkeys examined showed neutralization titers could be a result of natural infection as rhesus monkey rotavirus RRV also belongs to G3 [23]. It is worth noting that the possibility of heterotypic neutralization cannot be ruled out, as multiple studies of human and murine neutralizing antibodies showed heterotypic immunity against both VP7 and VP4 [24–27].

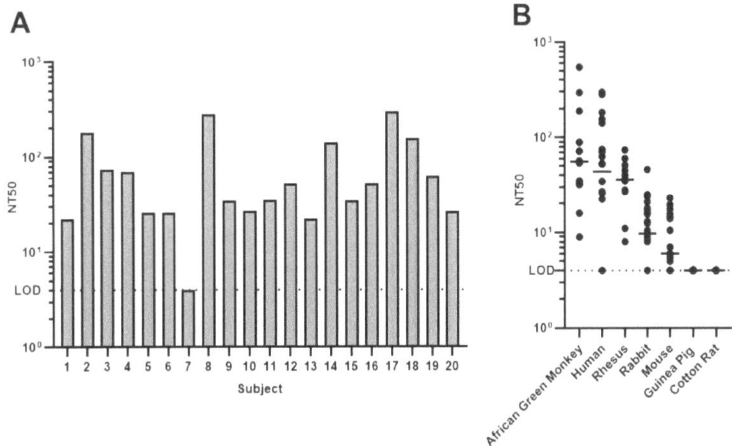

Figure 5. Pre-existing immunity in human and other animal species. (**A**) Serum neutralization titers of 20 human donors. (**B**) Serum neutralization titers of animal samples from indicated species. The bars indicate the median.

Rotaviruses are a group of viruses impacting a variety of animals, including common animal models used for vaccine or drug development. The microneutralization assay also allowed us to evaluate rabbit, mouse, guinea pig and cotton rat serum samples (Figure 5B). Several rabbit rotaviruses are G3 serotype viruses, and we indeed revealed neutralizing antibodies in many of the rabbits (15 out of 22), although the titers were much lower than those of human or simian origin. Although there is no known mouse rotavirus strain in the same serotype as SA11 based on VP4 and VP7, we observed neutralization titers in some of the mouse serum samples (16 out of 25) that could be caused by heterotypic immunity or undiscovered mouse strains. Guinea pig and cotton rat are widely used in infectious disease and vaccine research. No rotavirus has been reported in those two species. We did not discover any neutralizing antibodies against SA11 in any guinea pig or cotton rat we examined. In summary, the rSA11-GFP-based microneutralization assay enabled us to evaluate pre-existing immunity in several animal species, including humans, in a high-throughput manner.

4. Discussion

We extend prior studies [5,7] on rotavirus reverse genetics systems to demonstrate three positions in monkey rotavirus strain SA11 genome that are permissive for the insertion and expression of heterologous genes. Using SA11 expressing GFP, we develop a high-throughput microneutralization assay that enables a rapid evaluation of pre-existing antibodies in different animal species, including humans.

Our results show that rotavirus nonstructural proteins NSP1, NSP3, and NSP5 can tolerate heterologous gene insertion at their C-termini without altering any rotavirus protein-coding sequences. The original publication on rotavirus reverse genetics explored the modification of the C-terminus of NSP1 to express reporter proteins by partially deleting NSP1 protein to accommodate split GFP fragment GFP11 or NanoLuc [5]. In our strategy, we encoded the entire GFP after NSP1 along with a 2A cleavage sequence while keeping NSP1 intact as it plays a role in intestinal viral replication, pathogenesis, and transmission [28]. In another publication on reverse genetics for rotavirus, the C-terminus of NSP3 was fused to heterologous antigens, including partial SARS-CoV-2 spike protein [14,15]. However, our experiments showed that rescuing NSP3-GFP was less efficient compared to NSP1-GFP and NSP5-GFP, suggesting that the insertion of foreign sequences at this position renders rotavirus more genetically unstable. The largest insertion at NSP3 has been around 2 kb [15]. NSP1 and NSP5 positions may tolerate larger insertions, increasing the capability of rotaviruses as an expression platform. Additionally, for the first time, we demonstrate that NSP5 can also be manipulated to express a heterologous gene.

Neutralizing antibodies play a crucial role in the humoral immune response. Traditional rotavirus neutralization assays, such as the plaque reduction neutralization test (PRNT) or fluorescent foci reduction neutralization test (FRNT), are laborious and time-consuming, limiting the in-depth examination of neutralization antibodies in human and animal models. Microneutralization assays based on GFP or other reporter proteins have been established for a variety of viruses, including human metapneumovirus, human cytomegalovirus, and respiratory syncytial virus [29–31]. These assays support a high-throughput format compatible with robotic system, facilitating basic research and vaccine development.

Neutralizing antibodies against rotaviruses have been discovered against VP4 and VP7, two viral proteins that determine the serotype. Certain antibodies specific to VP6 have demonstrated the ability to neutralize rotaviruses intracellularly [32]. In our study, we developed a microneutralization assay based on rSA11-GFP, a serotype G3P5B[2] virus. The same concept and method can be applied to interrogate other serotypes. Strains, including several human strains (KU [8], Odelia [7], CDC-9 [6] and HN126 [9]), have been rescued by reverse genetics, and chimeric rotaviruses bearing VP4 and VP7 from clinical isolates have been generated using SA11 as the backbone [12]. Alongside our findings, these results imply the possibility of either constructing rotaviruses from different serotypes expressing GFP or generating a panel of rotaviruses containing SA11 NSP1-GFP but with VP4, VP7 and VP6 from other strains/serotypes to tease out neutralizing antibodies against each component in animal and human samples. Viruses with different serotypes expressing other reporter proteins, such as BFP and RFP, can potentially be used in parallel with a GFP virus, enabling multiplexed neutralization assays that detect NT50s against more than one serotype in one assay.

The high level of correlation between ELISA titers and neutralization titers in our African green monkey experiments suggest that the microneutralization assay can be used as a surrogate assay for ELISA. By using this assay, we can bypass the requirement of species-specific secondary antibodies in ELISA and compare neutralizing titers of sera from different animal species. Consistent with previous reports [20], all simians we tested showed neutralization titers above the limit of detection. Interestingly, the neutralizing titers of human samples are similar to those of simian with only nine negative samples out of 50, indicating the prevalence of G3-serotype-specific antibody in humans. In contrast, no neutralizing antibodies were detected in cotton rats and guinea pigs, while more than half of the mouse samples contained neutralizing antibodies above the limit of detection, although the titers are lower than those of rabbit, in which several rotaviruses are of G3 type. Our results suggest either heterotypic neutralizing immunity is prevalent in mice, or there might be mouse rotavirus strains with serotypes G3 or P5B[2] that have yet to be discovered.

A high-throughput method to measure neutralizing antibodies enables two types of studies. First, once a panel of recombinant rotaviruses with different serotypes containing NSP1-GFP is established, it provides a toolset for seroepidemiology studies to characterize the prevalence of different serotypes of rotavirus in humans. Second, the method also facilitates the selection of appropriate animal models for preclinical studies. These studies will provide insights into the mechanisms underlying the low efficacy of rotavirus vaccines in developing country, enabling the rational design of next-generation rotavirus vaccines. Importantly, although SA11 is unlikely to be a suitable platform, now that it is established that rotavirus can express heterologous proteins, these studies will explore the potential of using rotaviruses as a vector for delivering vaccine antigens or therapeutic payloads.

Supplementary Materials: The following supporting information can be downloaded at: https://www.mdpi.com/article/10.3390/v15102034/s1, Figure S1: Flow cytometry-based infectivity assay; Figure S2: Neutralization titers of African green monkey serum samples before and after antibody purification; Figure S3: The correlation of neutralization titers and ELISA titers of 30 human serum samples and its statistical analysis; Table S1: A sample 384-plate map.

Author Contributions: J.W. and D.W. designed the study and wrote the manuscript. J.W. designed and constructed the panel of rSA11-GFP. S.R. performed the 384-well microneutralization assay. Y.L. and X.H. contributed to the growth curve experiment. M.L., A.P., J.C. and W.N. developed and performed the ELISA assay. G.J.H. carried out virus passage. W.A.R.II, D.F. and M.C. collected animal serum samples. D.W. and A.E. supervised the study. All authors have read and agreed to the published version of the manuscript.

Funding: This work was supported by Merck Sharp & Dohme Corp., a subsidiary of Merck & Co., Inc., Kenilworth, NJ, USA.

Institutional Review Board Statement: The animal study protocols were approved by NIRC and the Institutional Animal Care and Use Committee of Merck & Co., Inc., Kenilworth, NJ, USA.

Informed Consent Statement: Not applicable.

Data Availability Statement: The data presented within this study are available within the manuscript.

Conflicts of Interest: All authors are employees of Merck Sharp & Dohme Corp., a subsidiary of Merck & Co., Inc., Kenilworth, NJ, USA. A provisional patent application on the discoveries of this work has been filed.

References

1. Burnett, E.; Parashar, U.D.; Tate, J.E. Global Impact of Rotavirus Vaccination on Diarrhea Hospitalizations and Deaths Among Children <5 Years Old: 2006–2019. *J. Infect. Dis.* **2020**, *222*, 1731–1739. [CrossRef] [PubMed]
2. Troeger, C.; Khalil, I.A.; Rao, P.C.; Cao, S.; Blacker, B.F.; Ahmed, T.; Armah, G.; Bines, J.E.; Brewer, T.G.; Colombara, D.V.; et al. Rotavirus Vaccination and the Global Burden of Rotavirus Diarrhea Among Children Younger Than 5 Years. *JAMA Pediatr.* **2018**, *172*, 958–965. [CrossRef] [PubMed]
3. Tate, J.E.; Burton, A.H.; Boschi-Pinto, C.; Parashar, U.D.; World Health Organization-Coordinated Global Rotavirus Surveillance Network. Global, Regional, and National Estimates of Rotavirus Mortality in Children <5 Years of Age, 2000–2013. *Clin. Infect. Dis.* **2016**, *62* (Suppl. S2), S96–S105. [CrossRef] [PubMed]
4. Varghese, T.; Kang, G.; Steele, A.D. Understanding Rotavirus Vaccine Efficacy and Effectiveness in Countries with High Child Mortality. *Vaccines* **2022**, *10*, 346. [CrossRef] [PubMed]
5. Kanai, Y.; Komoto, S.; Kawagishi, T.; Nouda, R.; Nagasawa, N.; Onishi, M.; Matsuura, Y.; Taniguchi, K.; Kobayashi, T. Entirely plasmid-based reverse genetics system for rotaviruses. *Proc. Natl. Acad. Sci. USA* **2017**, *114*, 2349–2354. [CrossRef]
6. Sanchez-Tacuba, L.; Feng, N.; Meade, N.J.; Mellits, K.H.; Jais, P.H.; Yasukawa, L.L.; Resch, T.K.; Jiang, B.; Lopez, S.; Ding, S.; et al. An Optimized Reverse Genetics System Suitable for Efficient Recovery of Simian, Human, and Murine-Like Rotaviruses. *J. Virol.* **2020**, *94*, e01294-20. [CrossRef]
7. Kawagishi, T.; Nurdin, J.A.; Onishi, M.; Nouda, R.; Kanai, Y.; Tajima, T.; Ushijima, H.; Kobayashi, T. Reverse Genetics System for a Human Group A Rotavirus. *J. Virol.* **2020**, *94*, e00963-19. [CrossRef]
8. Komoto, S.; Fukuda, S.; Kugita, M.; Hatazawa, R.; Koyama, C.; Katayama, K.; Murata, T.; Taniguchi, K. Generation of Infectious Recombinant Human Rotaviruses from Just 11 Cloned cDNAs Encoding the Rotavirus Genome. *J. Virol.* **2019**, *93*, e02207-18. [CrossRef] [PubMed]
9. Hamajima, R.; Lusiany, T.; Minami, S.; Nouda, R.; Nurdin, J.A.; Yamasaki, M.; Kobayashi, N.; Kanai, Y.; Kobayashi, T. A reverse genetics system for human rotavirus G2P[4]. *J. Gen. Virol.* **2022**, *103*, 001816. [CrossRef]
10. Diebold, O.; Gonzalez, V.; Venditti, L.; Sharp, C.; Blake, R.A.; Tan, W.S.; Stevens, J.; Caddy, S.; Digard, P.; Borodavka, A.; et al. Using Species a Rotavirus Reverse Genetics to Engineer Chimeric Viruses Expressing SARS-CoV-2 Spike Epitopes. *J. Virol.* **2022**, *96*, e0048822. [CrossRef]

11. Kanda, M.; Fukuda, S.; Hamada, N.; Nishiyama, S.; Masatani, T.; Fujii, Y.; Izumi, F.; Okajima, M.; Taniguchi, K.; Sugiyama, M.; et al. Establishment of a reverse genetics system for avian rotavirus A strain PO-13. *J. Gen. Virol.* **2022**, *103*, 001760. [CrossRef] [PubMed]

12. Kanai, Y.; Onishi, M.; Kawagishi, T.; Pannacha, P.; Nurdin, J.A.; Nouda, R.; Yamasaki, M.; Lusiany, T.; Khamrin, P.; Okitsu, S.; et al. Reverse Genetics Approach for Developing Rotavirus Vaccine Candidates Carrying VP4 and VP7 Genes Cloned from Clinical Isolates of Human Rotavirus. *J. Virol.* **2020**, *95*, e01374-20. [CrossRef] [PubMed]

13. Criglar, J.M.; Crawford, S.E.; Zhao, B.; Smith, H.G.; Stossi, F.; Estes, M.K. A Genetically Engineered Rotavirus NSP2 Phosphorylation Mutant Impaired in Viroplasm Formation and Replication Shows an Early Interaction between vNSP2 and Cellular Lipid Droplets. *J. Virol.* **2020**, *94*, e00972-20. [CrossRef] [PubMed]

14. Philip, A.A.; Perry, J.L.; Eaton, H.E.; Shmulevitz, M.; Hyser, J.M.; Patton, J.T. Generation of Recombinant Rotavirus Expressing NSP3-UnaG Fusion Protein by a Simplified Reverse Genetics System. *J. Virol.* **2019**, *93*, e01616-19. [CrossRef] [PubMed]

15. Philip, A.A.; Patton, J.T. Rotavirus as an Expression Platform of Domains of the SARS-CoV-2 Spike Protein. *Vaccines* **2021**, *9*, 449. [CrossRef]

16. Kawagishi, T.; Sanchez-Tacuba, L.; Feng, N.; Costantini, V.P.; Tan, M.; Jiang, X.; Green, K.Y.; Vinje, J.; Ding, S.; Greenberg, H.B. Mucosal and systemic neutralizing antibodies to norovirus induced in infant mice orally inoculated with recombinant rotaviruses. *Proc. Natl. Acad. Sci. USA* **2023**, *120*, e2214421120. [CrossRef]

17. Adler, S.P.; Lewis, N.; Conlon, A.; Christiansen, M.P.; Al-Ibrahim, M.; Rupp, R.; Fu, T.M.; Bautista, O.; Tang, H.; Wang, D.; et al. Phase 1 Clinical Trial of a Conditionally Replication-Defective Human Cytomegalovirus (CMV) Vaccine in CMV-Seronegative Subjects. *J. Infect. Dis.* **2019**, *220*, 411–419. [CrossRef]

18. McIntyre, M.; Rosenbaum, V.; Rappold, W.; Desselberger, M.; Wood, D.; Desselberger, U. Biophysical characterization of rotavirus particles containing rearranged genomes. *J. Gen. Virol.* **1987**, *68 Pt. 11*, 2961–2966. [CrossRef]

19. Troupin, C.; Schnuriger, A.; Duponchel, S.; Deback, C.; Schnepf, N.; Dehee, A.; Garbarg-Chenon, A. Rotavirus rearranged genomic RNA segments are preferentially packaged into viruses despite not conferring selective growth advantage to viruses. *PLoS ONE* **2011**, *6*, e20080. [CrossRef]

20. Jiang, B.; McClure, H.M.; Fankhauser, R.L.; Monroe, S.S.; Glass, R.I. Prevalence of rotavirus and norovirus antibodies in non-human primates. *J. Med. Primatol.* **2004**, *33*, 30–33. [CrossRef]

21. RCWC. Available online: https://rega.kuleuven.be/cev/viralmetagenomics/virus-classification/rcwg (accessed on 29 September 2023).

22. Velazquez, F.R.; Matson, D.O.; Calva, J.J.; Guerrero, L.; Morrow, A.L.; Carter-Campbell, S.; Glass, R.I.; Estes, M.K.; Pickering, L.K.; Ruiz-Palacios, G.M. Rotavirus infection in infants as protection against subsequent infections. *N. Engl. J. Med.* **1996**, *335*, 1022–1028. [CrossRef]

23. Fields, B.N.; Knipe, D.M. *Fields Virology*; Lippincott Williams & Wilkins: Philadelphia, PA, USA, 2022; Volume 3.

24. Nair, N.; Feng, N.; Blum, L.K.; Sanyal, M.; Ding, S.; Jiang, B.; Sen, A.; Morton, J.M.; He, X.S.; Robinson, W.H.; et al. VP4- and VP7-specific antibodies mediate heterotypic immunity to rotavirus in humans. *Sci. Transl. Med.* **2017**, *9*, eaam5434. [CrossRef]

25. Mackow, E.R.; Shaw, R.D.; Matsui, S.M.; Vo, P.T.; Benfield, D.A.; Greenberg, H.B. Characterization of homotypic and heterotypic VP7 neutralization sites of rhesus rotavirus. *Virology* **1988**, *165*, 511–517. [CrossRef] [PubMed]

26. Matsui, S.M.; Offit, P.A.; Vo, P.T.; Mackow, E.R.; Benfield, D.A.; Shaw, R.D.; Padilla-Noriega, L.; Greenberg, H.B. Passive protection against rotavirus-induced diarrhea by monoclonal antibodies to the heterotypic neutralization domain of VP7 and the VP8 fragment of VP4. *J. Clin. Microbiol.* **1989**, *27*, 780–782. [CrossRef] [PubMed]

27. Taniguchi, K.; Hoshino, Y.; Nishikawa, K.; Green, K.Y.; Maloy, W.L.; Morita, Y.; Urasawa, S.; Kapikian, A.Z.; Chanock, R.M.; Gorziglia, M. Cross-reactive and serotype-specific neutralization epitopes on VP7 of human rotavirus: Nucleotide sequence analysis of antigenic mutants selected with monoclonal antibodies. *J. Virol.* **1988**, *62*, 1870–1874. [CrossRef] [PubMed]

28. Hou, G.; Zeng, Q.; Matthijnssens, J.; Greenberg, H.B.; Ding, S. Rotavirus NSP1 Contributes to Intestinal Viral Replication, Pathogenesis, and Transmission. *mBio* **2021**, *12*, e0320821. [CrossRef] [PubMed]

29. Shambaugh, C.; Azshirvani, S.; Yu, L.; Pache, J.; Lambert, S.L.; Zuo, F.; Esser, M.T. Development of a High-Throughput Respiratory Syncytial Virus Fluorescent Focus-Based Microneutralization Assay. *Clin. Vaccine Immunol.* **2017**, *24*, e00225-17. [CrossRef]

30. Wang, Z.; Mo, C.; Kemble, G.; Duke, G. Development of an efficient fluorescence-based microneutralization assay using recombinant human cytomegalovirus strains expressing green fluorescent protein. *J. Virol. Methods* **2004**, *120*, 207–215. [CrossRef]

31. Biacchesi, S.; Skiadopoulos, M.H.; Yang, L.; Murphy, B.R.; Collins, P.L.; Buchholz, U.J. Rapid human metapneumovirus microneutralization assay based on green fluorescent protein expression. *J. Virol. Methods* **2005**, *128*, 192–197. [CrossRef]

32. Caddy, S.L.; Vaysburd, M.; Wing, M.; Foss, S.; Andersen, J.T.; O'Connell, K.; Mayes, K.; Higginson, K.; Iturriza-Gomara, M.; Desselberger, U.; et al. Intracellular neutralisation of rotavirus by VP6-specific IgG. *PLoS Pathog.* **2020**, *16*, e1008732. [CrossRef]

Article

Production of OSU G5P[7] Porcine Rotavirus Expressing a Fluorescent Reporter via Reverse Genetics

Anthony J. Snyder [1], Chantal A. Agbemabiese [1,2] and John T. Patton [1,*]

[1]   Department of Biology, Indiana University, 212 S. Hawthorne Drive, Simon Hall 011,
     Bloomington, IN 47405, USA; anthsnyd@indiana.edu (A.J.S.); cagbemabiese@noguchi.ug.edu.gh (C.A.A.)
[2]   Department of Electron Microscopy and Histopathology, Noguchi Memorial Institute for Medical Research,
     College of Health Sciences, University of Ghana, Accra 00233, Ghana
*   Correspondence: jtpatton@iu.edu

Abstract: Rotaviruses are a significant cause of severe, potentially life-threatening gastroenteritis in infants and the young of many economically important animals. Although vaccines against porcine rotavirus exist, both live oral and inactivated, their effectiveness in preventing gastroenteritis is less than ideal. Thus, there is a need for the development of new generations of porcine rotavirus vaccines. The Ohio State University (OSU) rotavirus strain represents a *Rotavirus A* species with a G5P[7] genotype, the genotype most frequently associated with rotavirus disease in piglets. Using complete genome sequences that were determined via Nanopore sequencing, we developed a robust reverse genetics system enabling the recovery of recombinant (r)OSU rotavirus. Although rOSU grew to high titers (~10^7 plaque-forming units/mL), its growth kinetics were modestly decreased in comparison to the laboratory-adapted OSU virus. The reverse genetics system was used to generate the rOSU rotavirus, which served as an expression vector for a foreign protein. Specifically, by engineering a fused NSP3-2A-UnaG open reading frame into the segment 7 RNA, we produced a genetically stable rOSU virus that expressed the fluorescent UnaG protein as a functional separate product. Together, these findings raise the possibility of producing improved live oral porcine rotavirus vaccines through reverse-genetics-based modification or combination porcine rotavirus vaccines that can express neutralizing antigens for other porcine enteric diseases.

Keywords: rotavirus; reverse genetics; expression vector; porcine rotavirus

Citation: Snyder, A.J.; Agbemabiese, C.A.; Patton, J.T. Production of OSU G5P[7] Porcine Rotavirus Expressing a Fluorescent Reporter via Reverse Genetics. *Viruses* **2024**, *16*, 411. https://doi.org/10.3390/v16030411

Academic Editor: Subir Sarker

Received: 30 January 2024
Revised: 4 March 2024
Accepted: 5 March 2024
Published: 7 March 2024

1. Introduction

Reverse genetics systems have been developed for several *Rotavirus A* strains, including those that infect non-human primates (SA11 and RRV), humans (KU, CDC-9, HN126, Odelia, and RIX4414-like), cattle (RF), mice (rD6/2-2g), and birds (PO-13) [1–10]. These systems can be used for producing next-generation live oral vaccines, dual vaccine platforms that express foreign proteins, and diagnostic tools. Rotavirus (RV) accounts for a significant disease burden within important livestock, highlighting the need for effective animal RV vaccines. Nonetheless, no reverse genetics systems currently exist for any porcine RV, including those with the G5P[7] genotype, which represents the most frequent cause of disease in porcine populations [11,12].

RV is a major cause of acute gastroenteritis in piglets [11–13]. The virus is transmitted via the fecal–oral route and damages small intestinal enterocytes; as such, milk consumed by nursing piglets is not digested or absorbed into the intestines [12,13]. Moreover, RV is resistant to environmental factors, such as temperature, pH, and common disinfectants, which creates a persistent risk of infection [14–16]. Although RV-induced diarrhea is associated with low mortality and high morbidity, productivity losses create a significant economic burden on the global pork industry. Current vaccines only control diarrhea among infected populations [12,13]. Thus, a need exists for vaccines that are more effective.

Through modification of the RV genome though reverse genetics, the virus can be used as an expression vector of foreign proteins [1–3,17–26]. The RV genome is composed of 11 segments of double-stranded (ds)RNA. Each genome segment contains the coding sequence for a single protein except for segment 11, which expresses two proteins [27]. As one approach for using RV as an expression vector, the NSP3 open reading frame (ORF) in the segment 7 RNA is replaced with a modified ORF that encodes NSP3 fused to a foreign protein. In such modified segment 7 RNAs, the NSP3 stop codon is removed and a ~19 amino acid porcine teschovirus 2A translational stop-restart element is introduced between the coding sequences for NSP3 and the foreign protein. The 2A element (ATNFSLLKQAGDVEENPG/P) contains a canonical PGP (underlined) motif [28,29]. During translation of the 2A element, a peptide bond typically fails to form between the G and P residues of this motif. As a result, modified segment 7 RNAs generate two separate proteins: (i) NSP3 with 18 residual resides of the 2A element (NSP3-2A); (ii) a foreign protein that initiates with a P residue [2,3,19]. Modification of segment 7 RNAs in this manner has allowed for the generation of recombinant RVs that express fluorescent reporters, such as UnaG (green), mRuby (red), and TagBFP (blue), and other viral proteins, such as the norovirus VP1 capsid protein and the severe acute respiratory coronavirus-2 (SARS-CoV-2) S1 spike domain [2,3,17–22]. Notably, the segment 7 homolog of *Rotavirus C* strains encodes two separate proteins (NSP3 and dsRNA-binding protein) through the presence of an intervening naturally-occurring 2A element [30]. The modifications of RV segment 5 (NSP1) and 11 (NSP5) RNAs, in some cases involving the use of 2A elements, have revealed that recombinant *Rotavirus A* strains can express separate foreign proteins from at least three genome segments [23–26]. The ability of recombinant RVs to express foreign proteins may be useful in the generation of combination vaccines capable of inducing protective immune responses against RV and a second pathogenic virus.

We report a robust reverse genetics system for The Ohio State University (OSU) G5P[7] porcine RV. The sequences of its 11 dsRNA genome segments were determined via Nanopore sequencing of a laboratory-adapted OSU strain [31–33]. Using T7 expression plasmids designed using these sequences, we generated recombinant OSU (rOSU) and 11 rSA11/OSU monoreassortants. We also generated a rOSU isolate with a modified segment 7 RNA (rOSU-2A-UnaG) that expressed NSP3 and the UnaG reporter protein as separate products. Together, this work reveals (i) the first reverse genetics system for a porcine RV; (ii) a platform for making targeted genetic modifications of OSU; and (iii) a system that allows for the expression of foreign proteins during OSU infection. This system will enable detailed studies on the molecular mechanisms of porcine RV replication pathogenesis.

2. Materials and Methods

2.1. Cells and Virus

Embryonic monkey kidney (MA104) cells were grown in Dulbecco's modified eagle medium (DMEM) containing 5% fetal bovine serum (FBS, Gibco) and 1% penicillin (10,000 U/mL)–streptomycin (10 mg/mL) (Quality Biological) at 37 °C in a 5% CO_2 incubator [34]. Baby hamster kidney cells that constitutively express the T7 RNA polymerase (BHK-T7) were provided by Dr. Ulla Buchholz (Laboratory of Infectious Diseases, NIAID, NIH) and were grown in Glasgow minimum essential medium (GMEM) containing 5% heat-inactivated FBS, 1% penicillin–streptomycin, 2% 100X MEM-nonessential amino acids (NEAA) (Gibco), and 1% glutamine (200 mM) at 37 °C in a 5% CO_2 incubator [35]. BHK-T7 cells were grown in medium supplemented with 2% Geneticin (50 mg/mL) (Gibco) in every other passage.

RVA/Pig-tc/USA/1975/OSU/G5P[7] was provided by Dr. Taka Hoshino (Laboratory of Infectious Diseases, NIAID, National Institutes if Health, Bethesda, MD, USA). This virus was activated by adjusting to a final concentration of 10 µg/mL porcine pancreatic trypsin, type IX (Millipore Sigma, Burlington, MA, USA), and incubating at 37 °C for 60 min. The activated virus was then propagated in MA104 cells maintained in serum-free DMEM with

0.5 µg/mL trypsin. The infected cells lysates were clarified via low-speed centrifugation at $1500\times g$ for 15 min at 4 °C. Virus was isolated from the clarified lysates via extraction with an equal volume of Vertrel-XF (TMC Industries, Waconia, MN, USA) followed by ultracentrifugation at $100,000\times g$ for 2 h at 4 °C. The pelleted virus [OSU-tc(MA104)] was resuspended in 500 µL of Tris-buffered saline and stored at -80 °C.

2.2. Nanopore Sequencing

OSU-tc_(MA104) dsRNA was extracted from 250 µL of clarified infected cell lysate using a Direct-zol RNA Miniprep Kit (Zymo Research, Irvine, CA, USA) following the manufacturer's instructions. Prior to library preparation, the dsRNA was denatured with dimethylsulfoxide and poly(A)-tailed using New England Biolabs *Escherichia coli* Poly(A) polymerase following the manufacturer's instructions. The poly(A)-tailed RNA was subjected to library preparation using an Oxford Nanopore Technologies (Oxford, UK) direct cDNA sequencing kit (SQK-DCS109) following the manufacturer's instructions. The library was sequenced using an Oxford Nanopore Technologies MinION sequencer. Sequence assemblies of the 11 genome segments of OSU-tc_(MA104) were prepared using Geneious Primer software version 2023.2.1 [https://www.geneious.com/, accessed on 29 January 2024].

2.3. OSU Sequences Used in the Generation of T7 Expression Plasmids

The sequences of the OSU-tc_(MA104) genome segments were deposited in GenBank [https://www.ncbi.nlm.nih.gov/genbank/, accessed on 29 January 2024] under the accession numbers OP978238-OP978248. The sequences of modified segment 7 RNAs of OSU NSP3-2A (PP112343) and OSU NSP3-2A-UnaG (PP112344) were also deposited in GenBank.

2.4. Plasmids Used in This Study

Recombinant SA11 (rSA11) viruses were prepared using the plasmids pT7/SA11VP1, pT7/SA11VP2, pT7/SA11VP3, pT7/SA11VP4, pT7/SA11VP6, pT7/SA11VP7, pT7/SA11NSP1, pT7/SA11NSP2, pT7/SA11NSP3, pT7/SA11NSP4, and pT7/SA11NSP5 and pCMV/NP868R [1,3]. Recombinant OSU (rOSU) viruses were prepared using the plasmids pT7/OSUVP1, pT7/OSUVP2, pT7/OSUVP3, pT7/OSUVP4, pT7/OSUVP6, pT7/OSUVP7, pT7/OSUNSP1, pT7/OSUNSP2, pT7/OSUNSP3, pT7/OSUNSP4, and pT7/OSUNSP5, and pCMV/NP868R [3,35,36]. The pT7/OSU plasmids were made by Genewiz, Azenta Life Sciences (Waltham, MA, USA), based on OSU sequences determined via Nanopore sequencing. The plasmid pT7/SA11 NSP3-2A-UnaG was previously described [19]. The plasmids pT7/OSU NSP3-2A and pT7/OSU NSP3-2A-UnaG were produced by fusing DNA fragments for 2A or 2A-3xFLAG-UnaG, respectively, to the 3′-end of the OSU NSP3 open reading frame of pT7/OSU NSP3 using the In-Fusion cloning system (TaKaRa Bio, San Jose, CA, USA). Primer synthesis and plasmid sequencing were performed by EuroFins Scientific (Indianapolis, IN, USA) (Table 1).

2.5. Isolation, Amplification, and Analysis of Recombinant Viruses

RV reverse genetics was performed as previously described [3,35,37]. Briefly, BHK-T7 cells in 12-well plates were transfected with the 11 SA11 or OSU T7 plasmids, or combinations thereof, and with the capping enzyme plasmid, pCMV-NP868R, using Mirus TransIT-LT1 transfection reagent (Madison, WI, USA. Transfection mixtures contained 0.8 µg each of the pT7 plasmids, except for pT7/NSP2 and pT7/NSP5, which were used at 3-fold higher concentrations [2]. Two days post transfection, the BHK-T7 cells were overseeded with MA104 cells, and trypsin was added to the medium to a final concentration of 0.5 µg/mL. Three days later, the BHK-T7/MA104 cell mixtures were freeze–thawed thrice, and the lysates were clarified via low-speed centrifugation at $800\times g$ for 5 min at 4 °C. To amplify the recovered viruses, the lysates were adjusted to 10 µg/mL trypsin and incubated for 1 h at 37 °C. MA104 cells in 6-well plates were then infected with 300 µL of the trypsin-treated lysates and incubated at 37 °C in a 5% CO_2 incubator until all cells were lysed (typically 3–5 days). Recombinant viruses were recovered from the lysates via plaque isolation on MA104 cells [34,35]. Plaque-isolated viruses were initially grown on MA104 cells in 6-well plates and then, to generate larger pools, grown on MA104 cells in T175 tissue culture flasks at low multiplicity of infection (<1 plaque-forming unit [PFU]/cell). Briefly, 100 µL of the plaque-amplified lysates were activated via incubation with 10 µg/mL trypsin (final concentration). The activated lysates were diluted into 10 mL of serum-free DMEM and then used as inoculum to infect MA104 cells in T175 flasks. The flasks were placed at 37 °C in a 5% CO_2 incubator for 1 h with rocking to ensure equal coverage of the inoculum over the monolayers. Following adsorption, the inoculum was removed and 25 mL of serum-free DMEM containing 0.5 µg/mL trypsin was added to each flask. The flasks were returned to the incubator until all cells were lysed (typically 3–5 days). The infected cell lysates were collected, clarified via low-speed centrifugation at $800\times g$ for 5 min at 4 °C, and stored at −80 °C. Viral dsRNAs were recovered from the infected cell lysates via extraction with TRIzol (ThermoFisher Scientific, Waltham, MA, USA), resolved via electrophoresis on Novex 8% polyacrylamide gels (ThermFisher Scientific) in Tris-glycine buffer, detected via staining with ethidium bromide, and visualized using a Bio-Rad ChemiDoc MP imaging system (Hercules, CA, USA) [34,35,37]. Peak titers were determined from the infected cell lysates via plaque assay.

Table 1. Primers used in constructing OSU segment 7 expression platforms.

Constructed Plasmid	Template Plasmid for PCR	Primer	Sequence (5′ -> 3′)
pT7/OSU NSP3-2A	pT7/OSU NSP3	Destination Vector—Forward	TAGTCACATAATTTAAATATATTAA
		Destination Vector—Reverse	TAAATTATGTGACTAAGGACCGGGGTTTTCTTCCAC GTCTCCTGCTTGCTTTAACAGAGAGAAGTTC GTTGCGCCGGCGCCTTCATATGTACATTCGTAGT
pT7/OSU NSP3-2A-UnaG	pT7/OSU NSP3-2A	Destination Vector—Forward	TAGTCACATAATTTAAATATATTAA
		Destination Vector—Reverse	TTCATATGTACATTCGTAGT
	pT7/SA11 NSP3-2A-UnaG	Insert—Forward	GAATGTACATATGAAGGCGCCGGCGCAACGAAC
		Insert—Reverse	TAAATTATGTGACTATTCTGTGGCCCTTCTGTAGCTC

2.6. Plaque Assay

RV plaque assays were performed as previously described [34,35]. At 5 days post infection, MA104 monolayers with agarose overlays were incubated overnight with phosphate-buffered saline (PBS) containing 3.7% formaldehyde. The agarose overlays were then removed, and the monolayers were stained with 1% crystal violet in 5% ethanol for 3 h. The fixed and stained monolayers were rinsed with water and air dried. The plaque diameters were measured using ImageJ software [38]. Statistically significant differences in titer and plaque size were determined using ANOVA (GraphPad Prism, Boston, MA, USA).

2.7. Immunoblot Analysis

RV infections were performed as previously described [21,35]. Briefly, MA104 cells in 6-well plates were infected with 5 PFU/cell of the indicated viruses. At 9 h post infection, the infected cells were scraped into cold PBS. The infected cells were then washed with cold PBS, pelleted via centrifugation at $5000\times g$ for 5 min at 4 °C, and lysed via incubation with nondenaturing lysis buffer (300 mM NaCl, 100 mM Tris-HCl [pH 7.4], 2% Triton X-100, and 1× EDTA-free Roche protease inhibitor cocktail [Sigma Aldrich, St. Louis, MO, USA]) for 30 min on ice. For immunoblot analysis, lysates were resolved via electrophoresis on 10% polyacrylamide gels in Tris-glycine buffer and transferred to nitrocellulose membranes [35,37]. After blocking with PBS containing 0.1% Tween-20 and 5% nonfat dry milk, the blots were probed with FLAG M2 antibody (F1804, Sigma Aldrich, 1:2000), 2A antibody (NBP2-59627, Novus, Centennial, CO, USA; 1:1000), RV VP6 antibody (lot 53963, 1:2000), or β-actin antibody (D6A8, Cell Signaling Technology [CST], Danvers, MA, USA; 1:2000). The bound primary antibodies were detected using 1:10,000 dilutions of horseradish peroxidase (HRP)-conjugated secondary antibodies (goat anti-mouse IgG [CST], goat anti-guinea pig IgG [KPL/SeraCare, Milford, MA, USA], or goat anti-rabbit IgG [CST]) in 5% nonfat dry milk. HRP signals were developed using the Bio-Rad Clarity Western ECL substrate and developed using a Bio-Rad ChemiDoc imaging system [35,37].

2.8. Genetic Stability Analysis

Genetic stability experiments were performed as previously described [3,18–21]. Briefly, the indicated viruses were serially passaged five times using 1:100 dilutions of infected cell lysates that were prepared in serum-free DMEM. When the cytopathic effect reached completion (typically 3–5 days), the cells were freeze–thawed thrice. Viral dsRNAs were recovered from the infected cell lysates via TRIzol extraction [35,37], resolved by electrophoresis on 8% polyacrylamide gels in Tris-glycine buffer, detected by staining with ethidium bromide, and visualized using a Bio-Rad ChemiDoc MP imaging system.

2.9. Assessment of Infectivity via Infectious Particle Production

RV infections were performed as previously described [21,35]. Briefly, MA104 cells in 6-well plates were infected with 5 PFU/cell of the indicated viruses. Following adsorption, the infected cells were washed thrice with PBS and incubated in serum-free DMEM containing 0.5 µg/mL trypsin at 37 °C in a 5% CO_2 incubator. At the indicated times post infection, the infected cells were freeze–thawed thrice, clarified via low-speed centrifugation at $800\times g$ for 5 min at 4 °C, and analyzed via plaque assay [34,35]. Statistically significant differences in titer were determined using ANOVA (Graph Pad Prism).

2.10. Assessment of Fluorescent Reporter Expression

RV infections were performed as previously described [21,35]. Briefly, MA104 cells in 6-well plates were infected with 0.05 PFU/cell of rOSU and rOSU-2A-UnaG. Following adsorption, the infected cells were washed thrice with PBS and incubated in serum-free DMEM containing 0.5 µg/mL trypsin at 37 °C in a IncuCyte S3 Live-Cell Analysis System (Sartorius, Bohemia, NY, USA). At 28 h post infection, images were acquired at $10\times$ magnification under phase and green (excitation [440–480 nm], emission [504–544 nm]) channels.

2.11. Statistical Analyses

The results from all experiments represent three biological replicates. Horizontal bars indicate the means. Error bars indicate the standard deviations. *p*-values were calculated using one-way analysis of variance with Bonferroni correction (GraphPad Prism).

3. Results and Discussion

3.1. Recovery of OSU G5P[7] Porcine Rotavirus via Reverse Genetics

Genotype G5P[7] is representative of most RVs that cause acute gastroenteritis in suckling and weaned pigs, which leads to economic losses that plague the global pork industry [11,12]. Current treatments are generally ineffective at preventing disease [39–43]; thus, a need exists for robust molecular tools to develop next-generation porcine RV vaccines. Reverse genetics systems exist for several *Rotavirus A* strains [1–10]; however, no such system is available for a porcine RV. To address this knowledge gap, we utilized the well-studied Ohio State University (OSU) G5P[7] prototype strain [31–33]. Laboratory-adapted OSU (OSU-tc_(MA104)) was grown in MA104 cells, and viral dsRNA was extracted from infected cell lysates. The isolated RNA was then processed for Nanopore sequencing to obtain the complete sequences of all 11 genome segments (deposited in GenBank). Notably, the Nanopore sequences for 7 OSU-tc_(MA104) genome segments (VP2, VP3,

VP6, VP7, NSP2, NSP3, and NSP5) were identical to those of the virulent (RVA/Pig-tc/USA/1975/OSU/G5P7/virulent) and attenuated (RVA/Pig-tc/USA/1975/OSU/G5P7/attenuated) strains. In contrast, VP1, VP4, and NSP1 were >99% identical, whereas NSP4 was >95% identical (nucleotide and amino acid sequence comparisons) [33]. The OSU-tc_(MA104) sequencing information was used to construct plasmids for reverse genetics experiments. Full-length cDNAs of each genome segment were positioned within T7 plasmids in between an upstream T7 RNA polymerase promoter and a downstream hepatitis delta virus ribozyme [35,37]. In the presence of the T7 RNA polymerase, the OSU T7 plasmids produce full-length, positive-sense RNA with authentic 5' and 3' termini.

To confirm that each OSU T7 expression plasmid was functional, we generated 11 recombinant SA11/OSU (rSA11/OSU) monoreassortants. SA11 represents the prototype strain of simian RV [44–46]. Reverse genetics experiments were performed as previously described (Figure 1A) [35–37]. Briefly, 1 OSU T7 expression plasmid, 10 SA11 T7 expression plasmids, and pCMV-NP868R were transfected into BHK-T7 cells, which were subsequently overseeded with MA104 cells. Recombinant viruses generated in the transfected cells were amplified and their dsRNA profiles analyzed via gel electrophoresis. All transfection mixtures designed to produce monoreassortants resulted in the recovery of infectious virus (i.e., viral dsRNA) and induced a cytopathic effect within 5 days of infection. Thus, the genetic information obtained via Nanopore sequencing was functional for reverse genetics. By comparing the banding patterns to rSA11, we also determined the migration distance for each rOSU genome segment (Figure 1B, see red arrows). Functional differences between the SA11 and OSU genome segments and protein products may be investigated using the monoreassortants discussed here. Strikingly, we recovered recombinant virus that was composed of all 11 OSU genome segments (Figure 1B, see rOSU lane). Thus, we have developed a reverse genetics system for G5P[7] porcine RV that may be used for the production of vaccines targeting the most common cause of porcine RV infections. In contrast to current vaccines against porcine RV, which have been developed by serially passaging virulent strains in tissue culture, a costly and time-consuming process, the OSU reverse genetics system allows for the rapid generation of vaccine candidates and for the introduction of directed attenuating genetic mutations [41,42,47].

3.2. Recovery of OSU G5P[7] Porcine Rotavirus Encoding a Foreign Protein

RV can be modified to express fluorescent reporters, such as UnaG (green), mRuby (red), and TagBFP (blue), and other viral proteins, such as norovirus VP1 and SARS-COV-2 S1 [1–3,17–26]. These recombinant viruses are valuable for analyzing RV biology via fluorescence-based imaging and may be used to induce protective immunity responses against multiple pathogenic viruses. As a proof of concept, we explored the possibility of expressing UnaG from OSU genome segment 7. The porcine teschovirus 2A-like (2A) element was positioned at the 3' end of the NSP3 coding sequence, followed by the in-frame coding sequence of FLAG-tagged UnaG. Due to the activity of the 2A element [27–29], translation of the RNA produces two proteins: NSP3 with remnants of the 2A element and FLAG-tagged UnaG. We generated an additional expression plasmid with only the 2A element following NSP3 (Figure 2A). Using RV reverse genetics (Figure 1A), we recovered the rOSU-2A and rOSU-2A-UnaG viruses. For each, we observed a shift in the migration pattern of genome segment 7 to a larger molecular form (Figure 2B, see red arrows). These shifts are presumably due to the extra genetic material introduced into the segment 7 RNA by addition of the 2A and 2A-FLAG-UnaG coding sequences. We also noted similar dsRNA banding patterns for OSU-tc_(MA104) and rOSU, providing additional evidence for the utility of the OSU reverse genetics system.

To develop OSU G5P[7] porcine RV as a vector for expressing foreign protein, the recombinant virus must be genetically stable. This property is essential for scaling large quantities for vaccine production and for other applications. To examine genetic stability, we serially passaged rOSU-2A and rOSU-2A-UnaG at low multiplicity of infection. We observed no differences in the pattern of viral dsRNA recovered from the infected lysates during passage (Figure 2C, see red arrows). Notably, the 2A-FLAG-UnaG coding sequence is stable within the OSU background for at least five rounds of infection. This result is consistent with the genetic stability noted for other recombinant RV strains with modified segment 7 RNAs expressing UnaG (e.g., rSA11-2A-UnaG [19]). In contrast, insertion of foreign sequences longer that the 0.5-base UnaG sequence has been correlated with genetic instability, generating RV variants with segment 7 RNAs that retain the NSP3 ORF but lack all or portions of the foreign sequence [3,18–21]. Future studies will be required to determine if the introduction of longer foreign sequences into OSU segment 7 RNA likewise leads to instability.

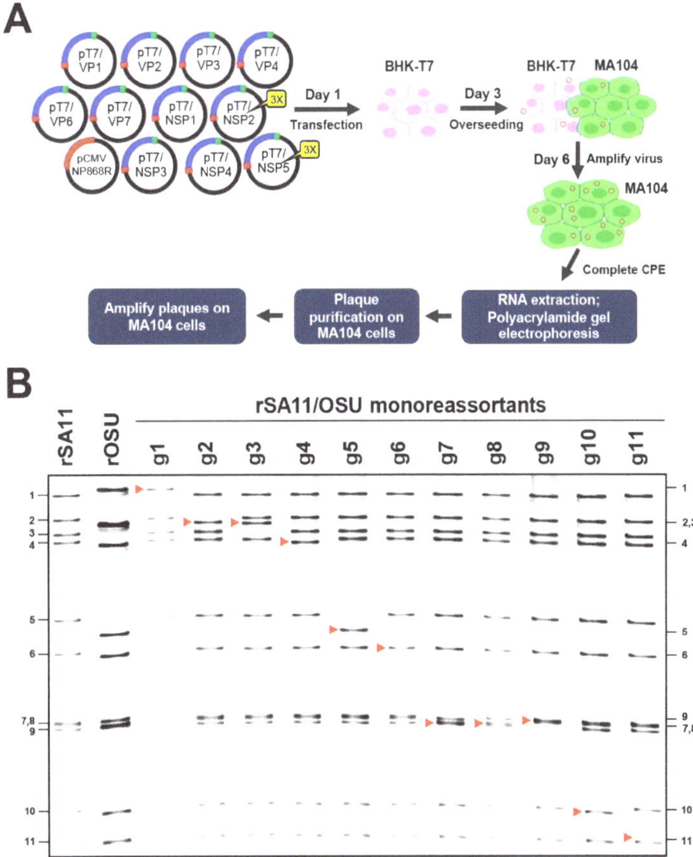

Figure 1. Production of OSU G5P[7] porcine rotavirus. (**A**) Rotavirus reverse genetics system [37]. Recombinant rotavirus was prepared by transfecting BHK-T7 cells with 11 T7 plasmids, which contain full-length cDNAs of rotavirus genome segments, and the CMV-NP868R plasmid, which encodes the African swine fever virus capping enzyme [3,36]. The BHK-T7 cells were overseeded 2 days post transfection with MA104 cells to facilitate the spread and amplification of recombinant rotavirus. At 3 days post overseeding, recombinant virus in the cells lysates was amplified on MA104 cells. The amplified virus was analyzed via RNA gel electrophoresis followed by plaque purification. Image was adapted from [37]. (**B**) Recovery of recombinant SA11/OSU monoreassortants via reverse genetics. Viral dsRNAs from rSA11, rOSU, and 11 rSA11/OSU monoreassortants were resolved via electrophoresis on an 8% polyacrylamide gel and stained with ethidium bromide. The migrations of rOSU gene segments in rSA11/OSU monoreassortants are indicated with red arrows. Genome segments 1–11 of rSA11 are indicated on the left side of the panel. Genome segments 1–11 of rOSU are indicated on the right side of the panel.

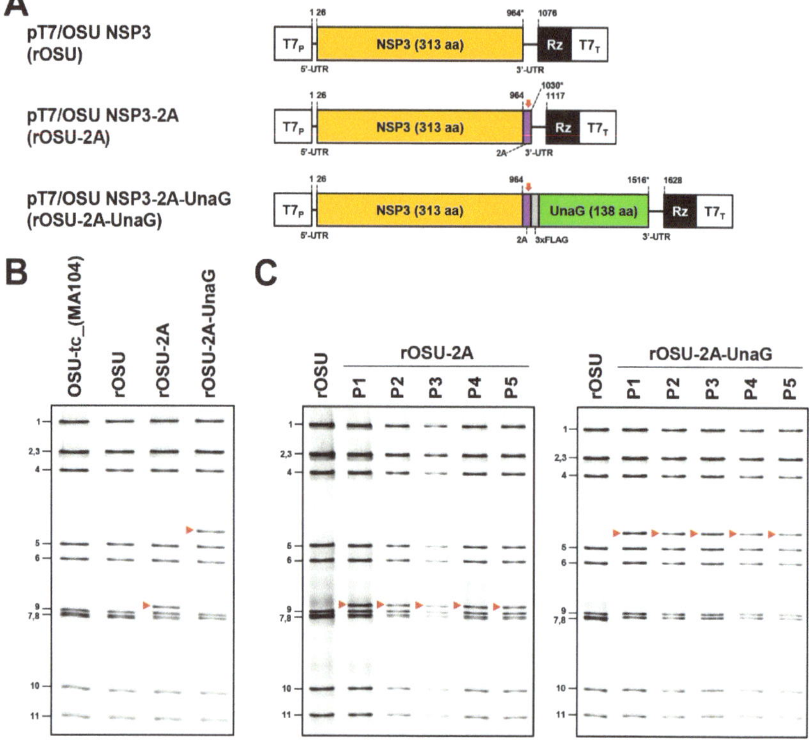

Figure 2. Production of recombinant OSU that encodes a foreign protein. (**A**) Modifications of rotavirus genome segment 7. The schematics indicate the nucleotide positions of the coding sequences for NSP3, the porcine teschovirus 2A element, 3X FLAG, and the fluorescent reporter UnaG (green). The red arrows indicate the positions of the 2A translational stop-restart elements, and the asterisks indicate the ends of the open reading frames. (**B**) Recovery of recombinant OSU-2A-UnaG via reverse genetics. Viral dsRNAs from OSU-tc_(MA104), rOSU, rOSU-2A, and rOSU-2A-UnaG were resolved via electrophoresis on an 8% polyacrylamide gel and stained with ethidium bromide. The migrations of modified genome segment 7 are indicated with red arrows. Genome segments 1–11 of rOSU are indicated on the left side of the panel. (**C**) Genetic stability. rOSU-2A and rOSU-2A-UnaG were serially passaged on MA104 cells. Viral dsRNAs from a total of five passages (P) were resolved via electrophoresis on an 8% polyacrylamide gel and stained with ethidium bromide. The migrations of modified genome segment 7 are indicated with a red arrow. Genome segments 1–11 of rOSU are indicated on the left side of the panels.

3.3. Growth Characteristics of rOSU G5P[7] Rotaviruses

The recombinant viruses generated in this work were derived from sequencing information gained for the laboratory-adapted strain. To determine if these viruses exhibit similar growth characteristics, we compared their growth kinetics and plaque morphologies. The analysis showed that rOSU was a well-growing virus, reaching peak titers of ~10^7 in MA104 cells. However, the peak titer reached by rOSU was ~0.5 log less than that reached by the OSU-tc_(MA104) virus. Moreover, single-step growth experiments indicated that rOSU grew slower than the OSU-tc_(MA104) virus. Plaque analysis also showed that the rOSU virus formed smaller plaques on MA104 cells than the OSU-tc_(MA104) virus (Figure 3A–D). These results suggest that sequence differences exist between rOSU and OSU-tc_(MA104) that impact virus growth. We conclude from this that the consensus sequence information generated for OSU-tc_(MA104) genome via Nanopore sequencing may not fully reflect the distribution and combinations of sequence variations in the

OSU-tc_(MA104) population associated with the fittest, best-growing viruses. Indeed, the OSU-tc_(MA104) population likely represents a quasi-species, of which rOSU may or may not be a single variant. A recent study suggests that serial passage of clonal RV isolates (vis-à-vis plaque isolates) results in the introduction of mutations that favor increased growth kinetics [48]. In a similar vein, serial passage of the rOSU virus may lead to a change in its phenotype that more closely resembles the OSU-tc_(MA104) virus. Such experiments may reveal the nature of nucleotide and amino acid changes in the RV genome correlated with growth characteristics.

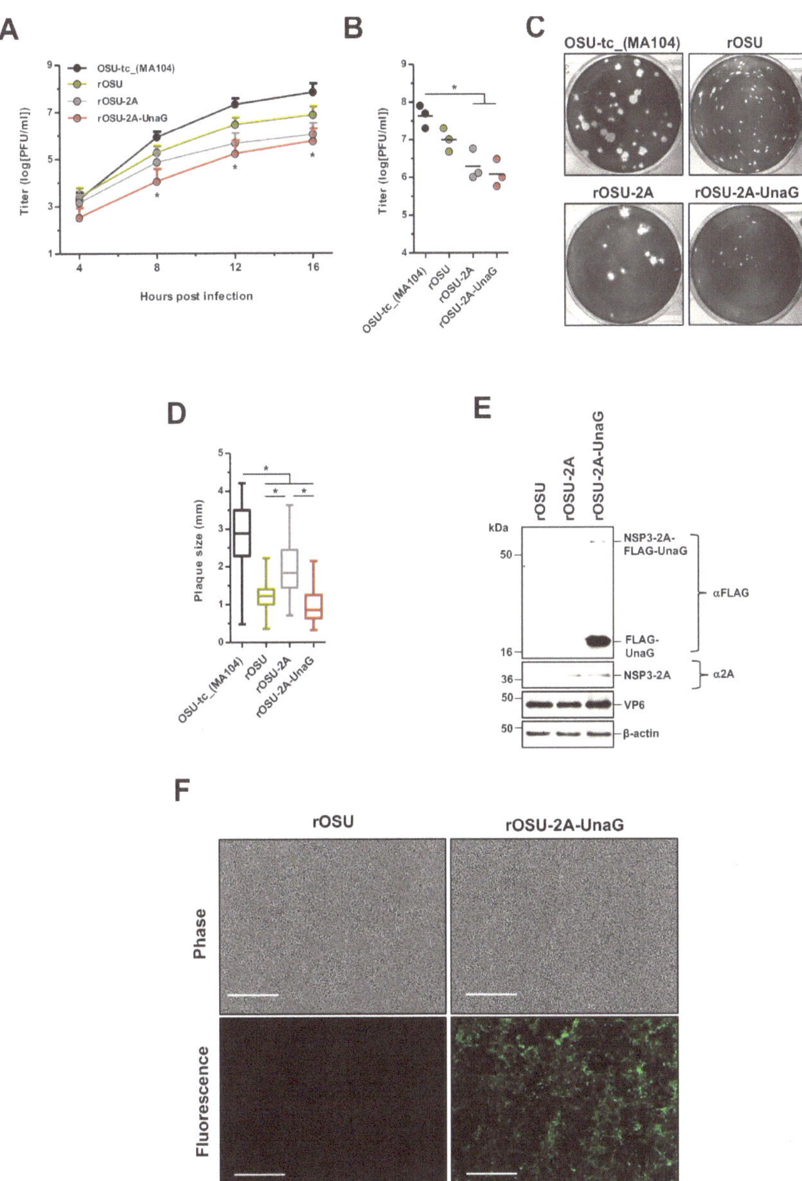

Figure 3. Characterization of recombinant OSU that expresses a foreign protein. (**A**,**B**) Production of infectious virus. MA104 cells were infected with the indicated viruses at an MOI of 5 PFU/cell.

In panel (**A**), titers were determined via plaque assay at the indicated times post infection. In panel (**B**), titers were determined via plaque assay upon complete cytopathic effect (typically 3–5 days). Error bars indicate the standard deviations (**A**); horizontal bars indicate the means (**B**). In panel (**A**), statistical analyses show the comparisons between OSU-tc_(MA104) and rOSU, rOSU-2A, and rOSU-2A-UnaG; *, $p < 0.05$ (n = 3 biological replicate). (**C,D**) Plaque morphologies and sizes. In panel (**C**), plaques on MA104 cells were detected via crystal violet staining. In panel (**D**), plaque diameters were measured using ImageJ software [38]. Fifty plaques were measured for each virus; *, $p < 0.05$ (n = 3 biological replicates). (**E,F**) Fluorescent reporter expression and activity of the 2A translational stop-restart element. In panel (**E**), MA104 cells were infected with the indicated viruses at an MOI of 5 PFU/cell. At 9 h post infection, infected cell lysates were prepared and analyzed via immunoblot assay. FLAG-UnaG and the read through product, NSP3-2A-FLAG-UnaG, were detected using anti-FLAG antibody, and NSP3-2A was detected using anti-2A antibody. Positions of molecular weight markers are indicated on the left side of panels (n = three biological replicates). In panel (**F**), MA104 were infected with the indicated viruses at an MOI of 0.05 PFU/cell. At 28 h post infection, the infected cells were imaged at $10\times$ magnification with an Incucyte live-cell analyzer (Satorius) using phase and green channels. Scale bars represent 400 μm (n = three biological replicates).

We next examined the growth characteristics of recombinant viruses that encode foreign protein. Compared to the rOSU, rOSU-2A-UnaG produced ~1 log-unit less infectious virus but generated similar plaque sizes (Figure 3A–D). This result is consistent with previous studies that show that the insertion of foreign genetic material into the segment 7 RNA is correlated with reduced virus growth [18–21].

3.4. Expression of Foreign Protein by rOSU G5P[7] Porcine Rotavirus

To develop OSU as a dual-vaccine platform, the vector must express a foreign protein. As such, the protein products made by rOSU-2A-UnaG in MA04-infected cells were probed via immunoblot assay using anti-FLAG antibody (Figure 3E). This assay revealed high levels of FLAG-UnaG (~18 kDa) expression, confirming that the rOSU-2A-UnaG virus directed expression of the UnaG foreign protein and contained a functional 2A element. We also detected minor amounts of the readthrough product, NSP3-2A-UnaG (~56 kDa); this likely derives from the failure of the 2A element to prevent peptide bond formation as the ribosome translates the PGP motif. Probing with an anti-2A antibody revealed a protein product that migrated at the expected molecular weight for NSP3 linked to the remnant residues of the 2A peptide (~38 kDa) (Figure 3E). Viral-protein VP6 and host-protein β-actin were detected under all infection conditions; however, NSP3-2A-UnaG, FLAG-UnaG, and NSP3-2A were not present in rOSU infected cell lysates. Finally, we used live cell imagining to determine whether the FLAG-UnaG expressed product of rOSU-2A-UnaG was functional. The results showed that in contrast to rOSU, rOSU-2A-UnaG produced high levels of fluorescent signal within live cells (Figure 3F). Thus, the rOSU-2A-UnaG virus expresses fluorescent UnaG protein, a feature not only enabling study of the replication and spread of the virus via live cell imagining but also suggesting that it may be possible to use recombinant OSU viruses as vectors in the development of combination vaccines.

4. Conclusions

The establishment of more effective porcine RV vaccines can be advanced through the application of reverse genetics technologies to create candidates that are superior in generating protective immunological responses. Such reverse genetics technologies allow for targeted genetic modifications and avoid the costly and time-consuming process of attenuating virulent strains through serially passage in vitro. Moreover, it may be possible to design combination vaccines that are capable of protecting individuals against RV and other pathogenic viruses. In this work, we developed a robust reverse genetics system for the G5P[7] OSU strain of porcine RV. Using information obtained by Nanopore sequencing the laboratory-adapted strain, we constructed 11 T7 expression plasmids that were sufficient for generating recombinant OSU (Figure 1). We leveraged the reverse genetics system to produce a recombinant virus that expressed UnaG within infected cells (Figures 2 and 3). Recombinant viruses that express fluorescent reporters are valuable tools for monitoring virus spread in infected animals and assessing immunological responses in pigs.

The development of the porcine OSU reverse genetics system also has possible application for understanding the biology of human RVs, given the close similarity of the genotype constellations of the OSU virus and the human Wa-like genogroup RVs (e.g., G1P[8], G3P[8], and G12P[8] RVs) [49]. Notably, the genotype of the genome segments for the nonstructural proteins and the core structural proteins of OSU are generally the same as found for the Wa-like viruses: R1-C1-M1-A1-N1-T1-E1-H1. Indeed, phylogenetic analyses have indicated that the human Wa-like viruses evolved from OSU-like porcine RVs [50]. Unlike many animal viruses, the NSP1 protein of the OSU virus reportedly relies on a mechanism similar to that of the NSP1 proteins of the Wa-like viruses to antagonize the interferon signaling system [51]. As a result of the ease of growing porcine RVs and their genetic similarity to Wa-like viruses, the possibility of generating human RV vaccines from porcine virus strains has been explored [52].

Author Contributions: Conceived and designed the experiments: A.J.S., C.A.A. and J.T.P. Performed the experiments: A.J.S. and C.A.A. Analyzed the data: A.J.S., C.A.A. and J.T.P. Wrote the paper: A.J.S., C.A.A. and J.T.P. All authors have read and agreed to the published version of the manuscript.

Funding: Research reported in this publication was supported by GIVax, Inc. JTP was also supported by Indiana University Start-Up Funding and the Lawrence M. Blatt Endowment.

Institutional Review Board Statement: Research reported in this study was approved by the Indiana University Institutional Biosafety Committee (IBC Protocol BL-879-07). The study did not involve animals or humans.

Informed Consent Statement: Not applicable.

Data Availability Statement: Data is contained within the article. Sequences used in the generation of recombinant OSU viruses are provided in NCBI GenBank {https://www.ncbi.nlm.nih.gov/genbank/, accessed on 29 January 2024} under accession numbers OP978238-OP978248, PP112343, and PP112344. Additional data related to this paper may be requested from the authors.

Acknowledgments: We thank members of the Indiana University virology community for their helpful comments and suggestions. Special thanks to Ulla Buchholz (Laboratory of Infectious Diseases, NIH, NIAID) for providing the BHK-T7 cells and to Taka Hoshino (Laboratory of Infectious Diseases, NIH, NIAID) for providing RVA/Pig-tc/USA/1975/OSU/G5P[7].

Conflicts of Interest: A.J.S., C.A.A., and J.T.P. are inventors of an Indiana University patent application related to the content of this work. J.T.P. has an interest in biotechnology companies developing vaccines using recombinant rotaviruses.

References

1. Kanai, Y.; Komoto, S.; Kawagishi, T.; Nouda, R.; Nagasawa, N.; Onishi, M.; Matsuura, Y.; Taniguchi, K.; Kobayashi, T. Entirely plasmid-based reverse genetics system for rotaviruses. *Proc. Natl. Acad. Sci. USA* **2017**, *114*, 2349–2354. [PubMed]
2. Komoto, S.; Fukuda, S.; Ide, T.; Ito, N.; Sugiyama, M.; Yoshikawa, T.; Murata, T.; Taniguchi, K. Generation of Recombinant Rotaviruses Expressing Fluorescent Proteins by Using an Optimized Reverse Genetics System. *J. Virol.* **2018**, *92*, e00588-18. [PubMed]
3. Philip, A.A.; Perry, J.L.; Eaton, H.E.; Shmulevitz, M.; Hyser, J.M.; Patton, J.T. Generation of Recombinant Rotavirus Expressing NSP3-UnaG Fusion Protein by a Simplified Reverse Genetics System. *J. Virol.* **2019**, *93*, e01616-19. [CrossRef] [PubMed]
4. Sanchez-Tacuba, L.; Feng, N.; Meade, N.J.; Mellits, K.H.; Jais, P.H.; Yasukawa, L.L.; Resch, T.K.; Jiang, B.; Lopez, S.; Ding, S.; et al. An Optimized Reverse Genetics System Suitable for Efficient Recovery of Simian, Human, and Murine-Like Rotaviruses. *J. Virol.* **2020**, *94*, e01294-20. [PubMed]
5. Komoto, S.; Fukuda, S.; Kugita, M.; Hatazawa, R.; Koyama, C.; Katayama, K.; Murata, T.; Taniguchi, K. Generation of Infectious Recombinant Human Rotaviruses from Just 11 Cloned cDNAs Encoding the Rotavirus Genome. *J. Virol.* **2019**, *93*, e02207-18. [CrossRef]
6. Kawagishi, T.; Nurdin, J.A.; Onishi, M.; Nouda, R.; Kanai, Y.; Tajima, T.; Ushijima, H.; Kobayashi, T. Reverse Genetics System for a Human Group A Rotavirus. *J. Virol.* **2020**, *94*, e00963-19. [CrossRef]

7. Hamajima, R.; Lusiany, T.; Minami, S.; Nouda, R.; Nurdin, J.A.; Yamasaki, M.; Kobayashi, N.; Kanai, Y.; Kobayashi, T. A reverse genetics system for human rotavirus G2P[4]. *J. Gen. Virol.* **2022**, *103*, 001816.

8. Philip, A.A.; Agbemabiese, C.A.; Yi, G.; Patton, J.T. T7 expression plasmids for producing a recombinant human G1P[8] rotavirus comprising RIX4414 sequences of the RV1 (Rotarix, GSK) vaccine strain. *Microbiol. Resour. Announc.* **2023**, *12*, e0060323.

9. Diebold, O.; Gonzalez, V.; Venditti, L.; Sharp, C.; Blake, R.A.; Tan, W.S.; Stevens, J.; Caddy, S.; Digard, P.; Borodavka, A.; et al. Using Species A Rotavirus Reverse Genetics to Engineer Chimeric Viruses Expressing SARS-CoV-2 Spike Epitopes. *J. Virol.* **2022**, *96*, e0048822.

10. Kanda, M.; Fukuda, S.; Hamada, N.; Nishiyama, S.; Masatani, T.; Fujii, Y.; Izumi, F.; Okajima, M.; Taniguchi, K.; Sugiyama, M.; et al. Establishment of a reverse genetics system for avian rotavirus A strain PO-13. *J. Gen. Virol.* **2022**, *103*, 001760. [CrossRef]

11. Papp, H.; Laszlo, B.; Jakab, F.; Ganesh, B.; De Grazia, S.; Matthijnssens, J.; Ciarlet, M.; Martella, V.; Banyai, K. Review of group A rotavirus strains reported in swine and cattle. *Vet. Microbiol.* **2013**, *165*, 190–199. [CrossRef]

12. Vlasova, A.N.; Amimo, J.O.; Saif, L.J. Porcine Rotaviruses: Epidemiology, Immune Responses and Control Strategies. *Viruses* **2017**, *9*, 48.

13. Saif, L.J.; Fernandez, F.M. Group A rotavirus veterinary vaccines. *J. Infect. Dis.* **1996**, *174* (Suppl. S1), S98–S106. [CrossRef]

14. Estes, M.K.; Kang, G.; Zeng, C.Q.; Crawford, S.E.; Ciarlet, M. Pathogenesis of rotavirus gastroenteritis. *Novartis Found. Symp.* **2001**, *238*, 82–96; discussion 96–100.

15. Estes, M.K.; Graham, D.Y.; Smith, E.M.; Gerba, C.P. Rotavirus stability and inactivation. *J. Gen. Virol.* **1979**, *43*, 403–409. [CrossRef]

16. Meng, Z.D.; Birch, C.; Heath, R.; Gust, I. Physicochemical stability and inactivation of human and simian rotaviruses. *Appl. Environ. Microbiol.* **1987**, *53*, 727–730. [CrossRef]

17. Philip, A.A.; Herrin, B.E.; Garcia, M.L.; Abad, A.T.; Katen, S.P.; Patton, J.T. Collection of Recombinant Rotaviruses Expressing Fluorescent Reporter Proteins. *Microbiol. Resour. Announc.* **2019**, *8*, e00523-19.

18. Philip, A.A.; Hu, S.; Dai, J.; Patton, J.T. Recombinant rotavirus expressing the glycosylated S1 protein of SARS-CoV-2. *J. Gen. Virol.* **2023**, *104*, 001899. [CrossRef] [PubMed]

19. Philip, A.A.; Patton, J.T. Expression of Separate Heterologous Proteins from the Rotavirus NSP3 Genome Segment Using a Translational 2A Stop-Restart Element. *J. Virol.* **2020**, *94*, e00959-20. [CrossRef] [PubMed]

20. Philip, A.A.; Patton, J.T. Rotavirus as an Expression Platform of Domains of the SARS-CoV-2 Spike Protein. *Vaccines* **2021**, *9*, 449. [CrossRef] [PubMed]

21. Philip, A.A.; Patton, J.T. Generation of Recombinant Rotaviruses Expressing Human Norovirus Capsid Proteins. *J. Virol.* **2022**, *96*, e0126222.

22. Kawagishi, T.; Sánchez-Tacuba, L.; Feng, N.; Costantini, V.P.; Tan, M.; Jiang, X.; Green, K.Y.; Vinjé, J.; Ding, S.; Greenberg, H.B. Mucosal and systemic neutralizing antibodies to norovirus induced in infant mice orally inoculated with recombinant rotaviruses. *Proc. Natl. Acad. Sci. USA* **2023**, *120*, e2214421120. [CrossRef]

23. Kanai, Y.; Kawagishi, T.; Nouda, R.; Onishi, M.; Pannacha, P.; Nurdin, J.A.; Nomura, K.; Matsuura, Y.; Kobayashi, T. Development of Stable Rotavirus Reporter Expression Systems. *J. Virol.* **2019**, *93*, e01774-18. [CrossRef]

24. Pannacha, P.; Kanai, Y.; Kawagishi, T.; Nouda, R.; Nurdin, J.A.; Yamasaki, M.; Nomura, K.; Lusiany, T.; Kobayashi, T. Generation of recombinant rotaviruses encoding a split NanoLuc peptide tag. *Biochem. Biophys. Res. Commun.* **2021**, *534*, 740–746. [CrossRef]

25. Wei, J.; Radcliffe, S.; Pirrone, A.; Lu, M.; Li, Y.; Cassaday, J.; Newhard, W.; Heidecker, G.J.; Rose Ii, W.A.; He, X.; et al. A Novel Rotavirus Reverse Genetics Platform Supports Flexible Insertion of Exogenous Genes and Enables Rapid Development of a High-Throughput Neutralization Assay. *Viruses* **2023**, *15*, 2034.

26. Kawamura, Y.; Komoto, S.; Fukuda, S.; Kugita, M.; Tang, S.; Patel, A.; Pieknik, J.R.; Nagao, S.; Taniguchi, K.; Krause, P.R.; et al. Development of recombinant rotavirus carrying herpes simplex virus 2 glycoprotein D gene based on reverse genetics technology. *Microbiol. Immunol.* **2024**, *68*, 56–64. [CrossRef]

27. Crawford, S.E.; Ramani, S.; Tate, J.E.; Parashar, U.D.; Svensson, L.; Hagbom, M.; Franco, M.A.; Greenberg, H.B.; O'Ryan, M.; Kang, G.; et al. Rotavirus infection. *Nat. Rev. Dis. Primers* **2017**, *3*, 17083.

28. de Felipe, P.; Luke, G.A.; Hughes, L.E.; Gani, D.; Halpin, C.; Ryan, M.D. E unum pluribus: Multiple proteins from a self-processing polyprotein. *Trends Biotechnol.* **2006**, *24*, 68–75.

29. Donnelly, M.L.L.; Hughes, L.E.; Luke, G.; Mendoza, H.; Ten Dam, E.; Gani, D.; Ryan, M.D. The 'cleavage' activities of foot-and-mouth disease virus 2A site-directed mutants and naturally occurring '2A-like' sequences. *J. Gen. Virol.* **2001**, *82*, 1027–1041. [CrossRef] [PubMed]

30. Langland, J.O.; Pettiford, S.; Jiang, B.; Jacobs, B.L. Products of the porcine group C rotavirus NSP3 gene bind specifically to double-stranded RNA and inhibit activation of the interferon-induced protein kinase PKR. *J. Virol.* **1994**, *68*, 3821–3829. [CrossRef] [PubMed]

31. Bohl, E.H.; Theil, K.W.; Saif, L.J. Isolation and serotyping of porcine rotaviruses and antigenic comparison with other rotaviruses. *J. Clin. Microbiol.* **1984**, *19*, 105–111. [CrossRef]

32. Theil, K.W.; Bohl, E.H.; Agnes, A.G. Cell culture propagation of porcine rotavirus (reovirus-like agent). *Am. J. Vet. Res.* **1977**, *38*, 1765–1768.

33. Guo, Y.; Wentworth, D.E.; Stucker, K.M.; Halpin, R.A.; Lam, H.C.; Marthaler, D.; Saif, L.J.; Vlasova, A.N. Amino Acid Substitutions in Positions 385 and 393 of the Hydrophobic Region of VP4 May Be Associated with Rotavirus Attenuation and Cell Culture Adaptation. *Viruses* **2020**, *12*, 408. [CrossRef]

34. Arnold, M.; Patton, J.T.; McDonald, S.M. Culturing, storage, and quantification of rotaviruses. *Curr. Protoc. Microbiol.* **2009**, *15*, 15C-3. [CrossRef]
35. Philip, A.A.; Dai, J.; Katen, S.P.; Patton, J.T. Simplified reverse genetics method to recover recombinant rotaviruses expressing reporter proteins. *J. Vis. Exp.* **2020**, *158*, e61039.
36. Eaton, H.E.; Kobayashi, T.; Dermody, T.S.; Johnston, R.N.; Jais, P.H.; Shmulevitz, M. African Swine Fever Virus NP868R Capping Enzyme Promotes Reovirus Rescue during Reverse Genetics by Promoting Reovirus Protein Expression, Virion Assembly, and RNA Incorporation into Infectious Virions. *J. Virol.* **2017**, *91*, e02416-16. [CrossRef]
37. Agbemabiese, C.A.; Philip, A.A.; Patton, J.T. Recovery of Recombinant Rotaviruses by Reverse Genetics. *Methods Mol. Biol.* **2024**, *2733*, 249–263. [PubMed]
38. Schneider, C.A.; Rasband, W.S.; Eliceiri, K.W. NIH Image to ImageJ: 25 years of image analysis. *Nat. Methods* **2012**, *9*, 671–675. [CrossRef] [PubMed]
39. Dewey, C.; Carman, S.; Pasma, T.; Josephson, G.; McEwen, B. Relationship between group A porcine rotavirus and management practices in swine herds in Ontario. *Can. Vet. J.* **2003**, *44*, 649–653. [PubMed]
40. Svensmark, B.; Nielsen, K.; Dalsgaard, K.; Willeberg, P. Epidemiological studies of piglet diarrhoea in intensively managed Danish sow herds. III. Rotavirus infection. *Acta Vet. Scand.* **1989**, *30*, 63–70. [CrossRef] [PubMed]
41. Kim, H.J.; Park, S.I.; Ha, T.P.; Jeong, Y.J.; Kim, H.H.; Kwon, H.J.; Kang, M.I.; Cho, K.O.; Park, S.J. Detection and genotyping of Korean porcine rotaviruses. *Vet. Microbiol.* **2010**, *144*, 274–286. [CrossRef]
42. Park, J.G.; Alfajaro, M.M.; Cho, E.H.; Kim, J.Y.; Soliman, M.; Baek, Y.B.; Park, C.H.; Lee, J.H.; Son, K.Y.; Cho, K.O.; et al. Development of a live attenuated trivalent porcine rotavirus A vaccine against disease caused by recent strains most prevalent in South Korea. *Vet. Res.* **2019**, *50*, 2. [CrossRef]
43. Bull, J.J. Evolutionary reversion of live viral vaccines: Can genetic engineering subdue it? *Virus Evol.* **2015**, *1*, vev005. [CrossRef] [PubMed]
44. Malherbe, H.H.; Strickland-Cholmley, M. Simian virus SA11 and the related O agent. *Arch. Gesamte Virusforsch.* **1967**, *22*, 235–245. [CrossRef] [PubMed]
45. Malherbe, H.; Harwin, R. Seven viruses isolated from the vervet monkey. *Br. J. Exp. Pathol.* **1957**, *38*, 539–541.
46. Malherbe, H.; Harwin, R. The cytopathic effects of vervet monkey viruses. *S. Afr. Med. J.* **1963**, *37*, 407–411.
47. Hanley, K.A. The double-edged sword: How evolution can make or break a live-attenuated virus vaccine. *Evolution* **2011**, *4*, 635–643. [CrossRef]
48. Kadoya, S.S.; Urayama, S.I.; Nunoura, T.; Hirai, M.; Takaki, Y.; Kitajima, M.; Nakagomi, T.; Nakagomi, O.; Okabe, S.; Nishimura, O.; et al. Bottleneck Size-Dependent Changes in the Genetic Diversity and Specific Growth Rate of a Rotavirus A Strain. *J. Virol.* **2020**, *94*, e02083-19. [CrossRef]
49. Matthijnssens, J.; Ciarlet, M.; McDonald, S.M.; Attoui, H.; Bányai, K.; Brister, J.R.; Buesa, J.; Esona, M.D.; Estes, M.K.; Gentsch, J.R.; et al. Uniformity of rotavirus strain nomenclature proposed by the Rotavirus Classification Working Group (RCWG). *Arch. Virol.* **2011**, *156*, 1397–1413. [CrossRef]
50. Matthijnssens, J.; Ciarlet, M.; Heiman, E.; Arijs, I.; Delbeke, T.; McDonald, S.M.; Palombo, E.A.; Iturriza-Gómara, M.; Maes, P.; Patton, J.T.; et al. Full genome-based classification of rotaviruses reveals a common origin between human Wa-Like and porcine rotavirus strains and human DS-1-like and bovine rotavirus strains. *J. Virol.* **2008**, *82*, 3204–3219. [CrossRef]
51. Morelli, M.; Dennis, A.F.; Patton, J.T. Putative E3 ubiquitin ligase of human rotavirus inhibits NF-κB activation by using molecular mimicry to target β-TrCP. *mBio* **2015**, *6*, e02490-14. [CrossRef]
52. Hoshino, Y.; Jones, R.W.; Ross, J.; Kapikian, A.Z. Porcine rotavirus strain Gottfried-based human rotavirus candidate vaccines: Construction and characterization. *Vaccine* **2005**, *23*, 3791–3799. [CrossRef]

Article

VP4 Mutation Boosts Replication of Recombinant Human/Simian Rotavirus in Cell Culture

Roman Valusenko-Mehrkens, Katja Schilling-Loeffler, Reimar Johne and Alexander Falkenhagen *

Department of Biological Safety, German Federal Institute for Risk Assessment, 10589 Berlin, Germany;
roman.valusenko-mehrkens@bfr.bund.de (R.V.-M.); katja.schilling-loeffler@bfr.bund.de (K.S.-L.);
reimar.johne@bfr.bund.de (R.J.)
* Correspondence: alexander.falkenhagen@bfr.bund.de; Tel.: +49-30-18412-24603

Abstract: Rotavirus A (RVA) is the leading cause of diarrhea requiring hospitalization in children and causes over 100,000 annual deaths in Sub-Saharan Africa. In order to generate next-generation vaccines against African RVA genotypes, a reverse genetics system based on a simian rotavirus strain was utilized here to exchange the antigenic capsid proteins VP4, VP7 and VP6 with those of African human rotavirus field strains. One VP4/VP7/VP6 (genotypes G9-P[6]-I2) triple-reassortant was successfully rescued, but it replicated poorly in the first cell culture passages. However, the viral titer was enhanced upon further passaging. Whole genome sequencing of the passaged virus revealed a single point mutation (A797G), resulting in an amino acid exchange (E263G) in VP4. After introducing this mutation into the VP4-encoding plasmid, a VP4 mono-reassortant as well as the VP4/VP7/VP6 triple-reassortant replicated to high titers already in the first cell culture passage. However, the introduction of the same mutation into the VP4 of other human RVA strains did not improve the rescue of those reassortants, indicating strain specificity. The results show that specific point mutations in VP4 can substantially improve the rescue and replication of recombinant RVA reassortants in cell culture, which may be useful for the development of novel vaccine strains.

Keywords: rotavirus; Sub-Saharan Africa; reverse genetics system; triple-reassortant; point mutation; next-generation sequencing; replication kinetics; cell culture

Citation: Valusenko-Mehrkens, R.;
Schilling-Loeffler, K.; Johne, R.;
Falkenhagen, A. VP4 Mutation
Boosts Replication of Recombinant
Human/Simian Rotavirus in Cell
Culture. *Viruses* **2024**, *16*, 565.
https://doi.org/10.3390/v16040565

Academic Editors: Ulrich
Desselberger and John T. Patton

Received: 23 February 2024
Revised: 22 March 2024
Accepted: 28 March 2024
Published: 5 April 2024

1. Introduction

Rotaviruses are double-stranded RNA viruses which belong to the family *Sedoreoviridae* [1] and infect wild animals, livestock as well as humans [2]. Human Rotavirus A (RVA) can cause severe gastroenteritis in infants and young children and is the leading cause of diarrhea requiring hospitalization in children under five years of age in low- and middle-income countries [3]. In the absence of symptomatic treatment, an infection can become life-threatening due to dehydration [4]. Based on recent data, RVA causes over 100,000 annual deaths in Sub-Saharan Africa alone [5].

The RVA capsid contains three concentric protein layers. The inner layer is formed by VP2, the middle layer by VP6 and the outer layer by VP7 and the VP4 spike protein [6]. Each viral spike is composed of three VP4 molecules that are anchored to the virus particle by the interaction of the VP4 base with VP6 and VP7 [7,8]. Upon proteolytic cleavage of the spikes, VP4 is divided into the N-terminal receptor-binding fragment VP8* and the C-terminal membrane-penetrating fragment VP5*.

The rotavirus genome consists of 11 double-stranded RNA segments each encoding one or two viral proteins [1]. A classification system for rotaviruses based on the complete nucleotide sequences of all eleven genome segments has been established, enabling the precise description of reassortant strains, in which the genotypes of the genome segments VP7-VP4-VP6-VP1-VP2-VP3-NSP1-NSP2-NSP3-NSP4-NSP5 are represented by Gx-P[x]-Ix-Rx-Cx-Mx-Ax-Nx-Tx-Ex-Hx, respectively [9]. Upon co-infection of the same host cell with two different RVA strains, the discrete genome segments can be shuffled, resulting in a novel rotavirus strain consisting of a mixture of segments from both parental strains [10]. This reassortment event greatly contributes to a high diversity in circulating RVA strains. Especially the RVA genome segments encoding VP7 and VP4, which have a very high genetic variability. To date, 42 different VP7 genotypes and 58 different VP4 genotypes have been described [11].

VP4 and VP7 contain the major antigenic epitopes that elicit neutralizing antibody responses [12], but VP6 has also been shown to induce protective immunity [13–15]. There are two approved vaccines that have mainly been used in Africa (RotaTeq and Rotarix) [16]. RotaTeq is a pentavalent, live-attenuated vaccine consisting of five reassortants containing VP4 (P[8]) or VP7 (G1, G2, G3 or G4) from human RVA strains in a bovine RVA backbone [17]. In contrast, Rotarix is a live-attenuated vaccine derived from only one human G1P[8] RVA isolate [18]. Although the effectiveness of both vaccines ranges from 85-98% in America, Europe and parts of Asia, they show a reduced efficacy and effectiveness in low-income countries in Africa (50–64%) [19,20]. There are several possible reasons proposed for the reduced vaccine effectiveness, including malnutrition, host genetic factors such as histo-blood group antigens (HBGAs), differences in gut microbiota or co-infections with other pathogens [20]. However, the high diversity and difference of RVA strains circulating in Sub-Saharan Africa also have to be considered as possible reasons for a lower vaccine efficacy [21,22]. Additional recently licensed rotavirus vaccines are Rotavac (monovalent human G9P[11] RVA, developed in New Delhi, India), RotaSIIL (pentavalent bovine reassortants with human RVA G1–G4 and G9, developed in Pune, India), Rotavin-M1 (monovalent human G1P[8] RVA, developed in Hanoi, Vietnam) and Lanzhou (monovalent lamb G10P[12] RVA, developed in Lanzhou, China). The introduction of vaccines led to changes in circulating rotavirus strains and genotypes [23,24]. Before the rotavirus vaccine was introduced in South Africa, G1P[8] was the most detected genotype. However, the proportion of G1 strains decreased and the proportion of non-G1P[8] strains increased after the vaccine's introduction [25], leading to a higher variety of circulating strains and an increase in uncommon genotype constellations.

Human RVA strains are difficult to adapt to replication in cell culture, which limits the possibilities to investigate reassortment and generate vaccine strains. Recently, entirely plasmid-based reverse genetics systems for RVA have been developed [26,27]. These reverse genetics systems are based on transfecting cell lines that constitutively express T7 RNA polymerase with plasmids encoding each rotavirus genome segment under the control of the T7 RNA polymerase promoter followed by infection of a cell line that is susceptible to rotavirus infection. The utilization of these reverse genetics systems enabled the generation of several human RVA strains including G1P[8] KU, G4P[8] Odelia, G2P[4] HN126 or G1P[8] CDC-9 [27–30]. The developed reverse genetics systems were also used to investigate the reassortment of diverse animal rotavirus genome segments [30–34]. Additionally, several studies investigated the reassortment of the genome segment encoding VP4 from human RVA strains in the backbone of the simian RVA strain SA11, resulting in the generation of SA11 reassortants with VP4 from Odelia, CDC-9, HN126 or clinical isolates (P[4] or P[8]) [28–30,35]. However, according to the mentioned studies, reassortants with VP4 from human RVA strains tend to replicate poorly in cell culture. Recently, two studies investigated whether different combinations of human RVA genome segments encoding VP4, VP7 and VP6 in an SA11 backbone improved replication [29,35]. While one study showed that combining human RVA P[8] VP4 with homologous G1 VP7 or with homologous VP7 and VP6 did not improve the rescue of SA11 reassortants, another study reported that the interaction of human RVA P[4] VP4 with homologous G2 VP7 contributed to efficient virus infectivity.

Previously, we also investigated the generation of SA11 reassortants containing VP4 and/or VP7 from three African human RVA strains that have never been adapted to cell culture: GR10924 (G9P[6]), Moz60a (G12P[8]) and Moz308 (G2P[4]) [36]. The strains were chosen because they represented common genotypes circulating in Africa and complete sequence data were available. We were able to rescue SA11 mono-reassortants with VP7 from all three human RVA strains as well as one slowly replicating SA11 mono-reassortant with P[6] VP4 from GR10924. However, the rescue of SA11 double-reassortants containing VP4 and VP7 from human RVAs was not possible. Recently, we have shown that restoring the natural interactions between VP4, VP7 and VP6 from the human RVA strain Wa improved the rescue of SA11/Wa reassortants [37].

In the current study, we investigated whether restoring the natural interactions between VP4, VP7 and VP6 from the three African RVA strains would enable us to generate viable reassortants containing their main antigens. Although this approach was of limited success, one triple-reassortant was generated that changed its phenotype and started to replicate to higher titers after initial cell culture passages. The sequencing of its whole genome identified a unique point mutation in VP4, which could be shown to substantially improve virus rescue and replication. The results may contribute to the improved generation of specific recombinant rotaviruses and may be useful for the development of novel vaccine strains containing human RVA P[6] VP4.

2. Materials and Methods

2.1. Cell Lines and Viruses

All cell culture reagents and media were obtained from Pan-Biotech (Aidenbach, Germany) unless indicated otherwise. Dulbecco's Modified Eagle's Medium and Minimal Essential Medium were supplemented with 10% fetal bovine serum (FBS), $1\times$ non-essential amino acids, 2 mM L-glutamine and 0.1 µg/mL gentamicin (hereafter referred to as DMEM and MEM, respectively). MA-104 cells were provided by the European Collection of Authenticated Cell Cultures (Salisbury, UK) and cultured in MEM. BSR-T7/5 cells were kindly provided by Dr. Karsten Tischer (Free University of Berlin, Berlin, Germany) and maintained in DMEM containing 1 mg/mL G418 (Biochrome, Berlin, Germany). All cells were incubated at 37 °C, 5% CO_2 and 85% RH. The virus strain RVA/Simian-tc/ZAF/SA11-L2/1958/G3P[2], referred to as SA11, was generated by using the plasmid-based reverse genetics system as described below.

2.2. Plasmids

The plasmids encoding the eleven SA11 genome segments, as well as the three helper plasmids pCAG-D1R, pCAG-D12L and pCAG-FAST-p10 encoding the vaccinia virus capping enzyme subunits D1R and D12L as well as the small fusion protein FAST were kindly provided by Takeshi Kobayashi [26] and obtained from Addgene (Watertown, MA, USA). The generation of the plasmids encoding VP4 and VP7 from the African human RVA strains RVA/Human-wt/ZAF/GR10924/1999/G9P[6], RVA/Human-wt/MOZ/0060a/2012/G12P[8] and RVA/Human-wt/MOZ/0308/2012/G2P[4] (referred to as GR10924, Moz60a and Moz308, respectively) has been described previously [36]. Expression cassettes containing a T7 RNA polymerase promoter, VP6 from human RVA strain GR10924, Moz60a or Moz308 (GenBank acc.-no. FJ183358.1, MG926762.1 or MG926729.1, respectively) [38,39], the hepatitis delta virus ribozyme and a T7 terminator were synthesized by Integrated DNA Technologies (IDT, Coralville, IA, USA) as dsDNA fragments. The promoter, hepatitis delta virus ribozyme and terminator sequence were identical to a plasmid described previously (Genbank: KT239165) [40]. The expression cassettes were cloned into pUC-IDT-Amp (IDT) using standard cloning techniques and sequence verified by Sanger sequencing (Eurofins Genomics GmbH, Ebersberg, Germany). Sequencing primers are available upon request. The VP4-encoding plasmid of GR10924 containing the mutation A797G was generated as described below. All plasmids were purified using the QIAfilter Plasmid Midi Kit (Qiagen GmbH, Hilden, Germany).

2.3. Plasmid-Based Reverse Genetics System

BSR-T7/5 cells were seeded in a 6-well plate (3.5×10^5 cells per well) and incubated for 24 h. At 90% confluency, the cells were co-transfected with the eleven plasmids encoding the individual rotavirus genome segments and the three helper plasmids (2250 ng for the

NSP2 and NSP5 encoding plasmids; 15 ng for the FAST-encoding plasmid and 750 ng for the remaining plasmids) using 30 µL of TransIT-LT1 transfection reagent (Mirus Bio, Madison, WI, USA). The transfected cells were incubated for 24 h before they were washed once with DMEM without FBS. Next, DMEM without FBS containing 0.5 µg/mL trypsin (Pan-Biotech) was added. After an additional 48 h of incubation, the transfected BSR-T7/5 cells were co-cultured with MA-104 cells (1×10^5 cells per well) in the presence of trypsin (2 µg/mL final concentration). After three days, the co-cultured cells including the culture media were frozen at $-20\ °C$ and thawed at room temperature. After low-speed centrifugation, clarified supernatants, referred to as freeze/thaw supernatants throughout the manuscript, were collected and used to infect MA-104 cells as described below.

2.4. Passaging of Reassortants

The reassortant viruses were essentially passaged as described previously [41,42]. In brief, for the first passage, trypsin (Pan-Biotech) was added to the entire (~2 mL) clarified freeze/thaw supernatants (final concentration: 20 µg/mL) from co-cultures of transfected BSR-T7/5 and MA-104 cells. These infection mixtures were incubated for one hour at $37\ °C$. Confluent MA-104 cells grown in a 6-well plate were washed twice with PBS and the infection mixtures were added. After an additional hour of incubation at $37\ °C$, the mixtures were removed from the cells, fresh MEM without FBS containing trypsin (final concentration: 2 µg/mL) were added, and the cells were incubated for seven days. For later passages, clarified freeze/thaw supernatants (~2 mL) were collected as described in Section 2.3 and 150 µL samples were taken for RNA analyses. The remaining clarified freeze/thaw supernatants were used to infect fresh MA-104 cells as described above.

2.5. RNA Extraction, qRT-PCR, RT-PCR and Sanger Sequencing

Viral RNA was extracted from freeze/thaw supernatants with the NUCLISENS easy-MAG system (bioMérieux, Marcy-l'Étoile, France) and digested with RNase-free DNase (Roche, Basel, Switzerland) according to the manufacturer's instructions before analyses by qRT-PCR or RT-PCR. The qRT-PCR was performed as described previously [43]. To determine the number of genome copy equivalents (GCEs)/mL culture supernatant, qRT-PCR analyses were performed with RNA isolated from culture supernatants and a pT7-NSP3SA11 plasmid standard with a known copy number. A quantification example is shown in Supplementary Figure S1. RT-PCR analyses were used to determine the presence of the expected virus genome segments and performed using the OneStep RT-PCR Kit (Qiagen, Hilden, Germany) with primers as listed in Supplementary Table S1, according to the manufacturer's instructions. After RT-PCR, 1 µL of $5\times$ DNA Loading Buffer, Blue (meridian Bioscience, Cincinnati, OH, USA) was added to the RNA, loaded onto a 2% agarose gel and separated at 100 V for 1 h. The gel was stained using ethidium bromide (Carl Roth, Karlsruhe, Germany) and visualized under UV light. For Sanger sequencing, PCR amplicons were cleaned up using the Monarch DNA Gel Extraction Kit (New England BioLabs, Ipswich, UK) and sent to Eurofins Genomics GmbH.

2.6. Whole Genome Sequencing and Sequence Analysis

Nucleic acid extracts (see Section 2.5) were used for preparing libraries with the KAPA RNA HyperPrep Kit (Roche Diagnostic, Mannheim, Germany) and the KAPA Unique Dual-Indexed Adapter Kit for Illumina® platforms (Roche Diagnostic) as previously described [44]. Resulting libraries were sequenced along with 119 libraries with 2×150 cycles using the NextSeq 500/550 Mid Output Kit v2.5 (Illumina, San Diego, CA, USA) on the NextSeq 500 Sequencer (Illumina). All of the sequence analysis was performed in Geneious Prime® 2023.2.1 (Biomatters Ltd., Auckland, New Zealand). Raw reads were trimmed using the BBDuk plugin. Segment sequences were assembled with the map to reference function, where eleven selected segment sequences were used as references (accession numbers: LC178570-LC178574, LC178564-LC178566, FJ183356, FJ183358 and FJ183360).

2.7. Site-Directed Mutagenesis

To introduce the point mutation A797G into the plasmids encoding VP4 from human RVA strains GR10924, Moz60a and Moz308, the Phusion Site-Directed Mutagenesis Kit (Thermo Fisher Scientific, Waltham, MA, USA) was used in accordance with the manufacturer's instructions. In brief, 5 ng of the targeted plasmid was amplified by PCR with two 5′-phosphorylated primers. The primers were designed to anneal back-to-back to the plasmid and the desired mutation was introduced into the forward primer. Primer sequences are shown in Supplementary Table S1. The PCR conditions were: Initial denaturation at 98 °C for 30 s (step 1); denaturation at 98 °C for 10 s (step 2); annealing at 65 °C for 20 s (step 3); extension at 72 °C for 150 s (step 4); final extension at 72 °C for 5 min (step 5). Step 2 to step 4 were repeated 25 times. After the digestion of parental methylated and hemimethylated DNA with FastDigest DpnI, the PCR product containing the mutation was circularized by ligation with T4 DNA Ligase. Finally, chemically competent One Shot® TOP10 E. coli (Thermo Fisher Scientific) was transformed as instructed by the manufacturer. For verification of the introduced mutation, the corresponding region of the plasmid was amplified by PCR using primers listed in Supplementary Table S1 and the PCR products were subjected to Sanger sequencing (Eurofins Genomics GmbH).

2.8. Replication Kinetics

Confluent MA-104 cells grown in 6-well plates were infected with cell culture supernatants containing viruses at 2×10^4 GCEs as described above. At the indicated time points, 500 μL samples were taken and the same volume of fresh media containing 2 μg/mL trypsin was added. Once all the samples were collected, viral RNA was extracted, digested with RNase-free DNase and analyzed by qRT-PCR as described above.

2.9. Sequence Analyses and Protein Structure Visualization

Sequences were constructed and analyzed with the SeqBuilder Pro software (Version 17.0.2; DNASTAR Inc., Madison, WI, USA). Alignments were performed using the MUSCLE method as implemented in MegAlign Pro (DNASTAR Inc.). Amino acid sequences were deduced using the SeqBuilder Pro software and the NCBI non-redundant protein sequences data base was screened using BLASTp search (https://blast.ncbi.nlm.nih.gov/Blast.cgi (accessed on 14 March 2024)). Protein structures were visualized and analyzed using Protean 3D (DNASTAR Inc.) or UCSF Chimera (University of California, San Francisco, CA, USA) [45] on the basis of the published atomic model of an infectious rotavirus particle (PDB 4v7q) [7].

2.10. Statistics

The data are presented as mean ± standard deviation. To determine statistical significance, a two-tailed unpaired *t*-test was used. Results with a *p*-value below 0.05, 0.01 or 0.001 were considered statistically significant and marked with one, two or three asterisks, respectively.

3. Results

3.1. Generation of Triple-Reassortants Carrying VP4, VP7 and VP6 from African Human Rotavirus A Strains

We aimed to perform plasmid-based reverse genetics for RVA using simian RVA strain SA11 as a backbone and replacing the genome segments encoding for VP4 (segment 4), VP7 (segment 9) and VP6 (segment 6) with the corresponding segments from African RVA strain Moz60a (rSA11/triple-Moz60a), Moz308 (rSA11/triple-Moz308) or GR10924 (rSA11/triple-GR10924). Rescue of recombinant simian RVA strain SA11 (rSA11) served as a positive control. BSR-T7/5 cells were transfected with the respective plasmids in duplicates and then co-cultured with MA-104 cells. Freeze/thaw supernatants from co-cultured cells were passaged on MA-104 cells and the inoculated cells were monitored for signs of an RVA-typical cytopathic effect (CPE). Figure 1a depicts an overview of the development of CPEs upon passaging in MA-104 cells.

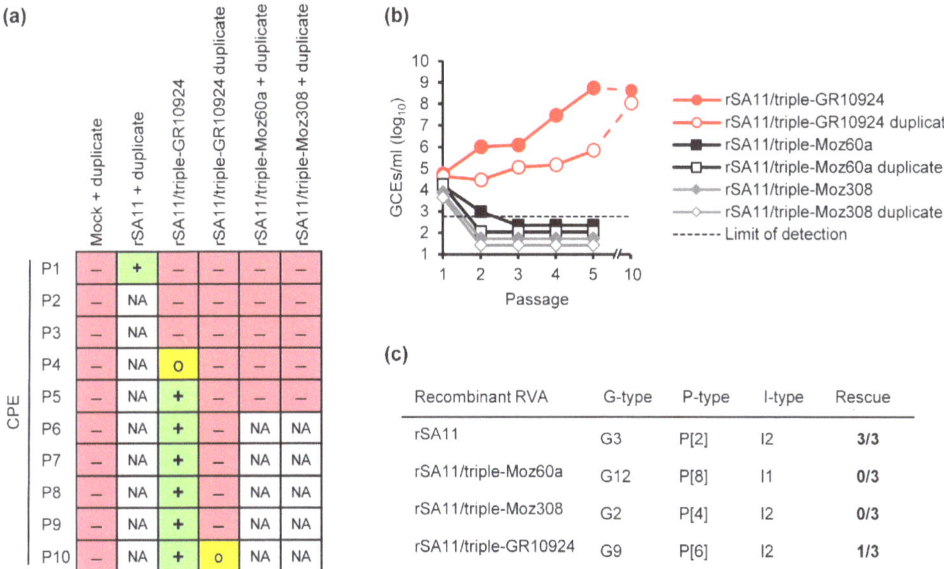

Recombinant RVA	G-type	P-type	I-type	Rescue
rSA11	G3	P[2]	I2	3/3
rSA11/triple-Moz60a	G12	P[8]	I1	0/3
rSA11/triple-Moz308	G2	P[4]	I2	0/3
rSA11/triple-GR10924	G9	P[6]	I2	1/3

Figure 1. Generation of SA11 triple-reassortants containing VP4, VP7 and VP6 from three African human RVA strains. (**a**) Overview of the developed cytopathic effect (CPE) upon passaging in MA-104 cells. (**b**) Analyses of the freeze–thaw supernatants by qRT-PCR after the indicated passages in MA-104 cells. (**c**) Overview of the VP7 (G-type), VP4 (P-type) and VP6 (I-type) genotypes and the number of successful rescue experiments. The first rescue experiment was performed in duplicates but counted as one experiment. Mock = Mock-infected cells; rSA11 = Recombinant SA11; rSA11/triple-GR10924, rSA11/triple-Moz60a and rSA11/triple-Moz308 = Recombinant rotaviruses carrying segment 4 (VP4), segment 9 (VP7) and segment 6 (VP6) from the indicated human RVA strain in the backbone of SA11; P1–10 = Passages 1–10; red minus = No CPE; yellow O = Mild CPE; green plus = Strong CPE; NA = Not analyzed; GCEs = Genome copy equivalents.

For rSA11 and the rSA11 duplicate, a clear cytopathic effect (CPE) was evident after the first passage, indicating that rescue was successful. However, no CPE was observed for any reassortant after passage 1. Cells infected with rSA11 were discarded to reduce the risk of cross-contamination, while all reassortants were passaged until passage 5. After the fifth passage, viral RNA was extracted from freeze/thaw supernatants of passage 1 to passage 5 and analyzed by qRT-PCR. No CPE was evident for rSA11/triple-Moz60a, rSA11/triple-Moz308 or their duplicates. In contrast, a CPE developed for rSA11/triple-GR10924 in the fourth passage and was clearly observable by passage 5, but no CPE was observed for the rSA11/triple-GR10924 duplicate by passage 5. Analyses by qRT-PCR showed that RVA RNA could be detected for rSA11/triple-GR10924 and the rSA11/triple-GR10924 duplicate after each passage, but the RNA titer of the reassortant that caused a CPE was higher after passages 2–5 (Figure 1b). For rSA11/triple-Moz60a, rSA11/triple-Moz308 and their duplicates, RVA RNA declined after the first passage and was not detectable anymore by passage 3 (Figure 1b), suggesting that the rescue of these triple-reassortants failed. Both rSA11/triple-GR10924 and the rSA11/triple-GR10924 duplicate were further passaged on MA-104 cells until passage 10, by which time the duplicate started to develop a CPE and the RNA titers increased (Figure 1a,b), indicating the successful rescue of rSA11/triple-GR10924 and the duplicate. Rescue experiments were repeated two additional times for each reassortant, but not in duplicates. While similar results were obtained for rSA11/triple-Moz60a and rSA11/triple-Moz308, we were unable to re-rescue the rSA11/triple-GR10924 reassortant (Supplementary Figure S2), suggesting that the rescue of this reassortant is possible but inefficient using the reverse genetics system.

Those experiments were stopped after passage 3 as the first rescue experiment showed that detection of RVA RNA corresponded with successful rescue at this passage number.

3.2. Next-Generation Sequencing Revealed Point Mutations in rSA11/triple-GR10924 and the Duplicate

After ten passages in MA-104 cells, the rSA11/triple-GR10924 and the rSA11/triple-GR10924 duplicate were analyzed by whole genome next-generation sequencing. For rSA11/triple-GR10924 and the duplicate, 2,268,340 and 2,986,102 reads were obtained, respectively. An overview of the average coverage, coverage range and the percentage of the open reading frame (ORF) sequenced is shown in Supplementary Table S2. The complete ORFs of each genome segment were covered, with the exception of genome segment 4 (VP4) from the rSA11/triple-GR10924 duplicate, where the first two nucleotides of the ORF could not be sequenced. For the rSA11/triple-GR10924 reassortant that replicated to a higher titer in early passages, only a single point mutation located in the ORF of genome segment 4 (VP4) was identified. The adenine at position 797 was substituted by a guanine, which led to an amino acid substitution of glutamic acid to glycine at position 263. For the duplicate, four mutations were located in ORFs: A1981G in genome segment 1 (VP1); C137T and G964T in genome segment 4 (VP4); C584T in genome segment 8 (NSP2). However, only the mutations detected in the VP4 ORF were non-synonymous, resulting in amino acid substitutions T43I and V322F. An overview of the identified mutations is depicted in Figure 2a.

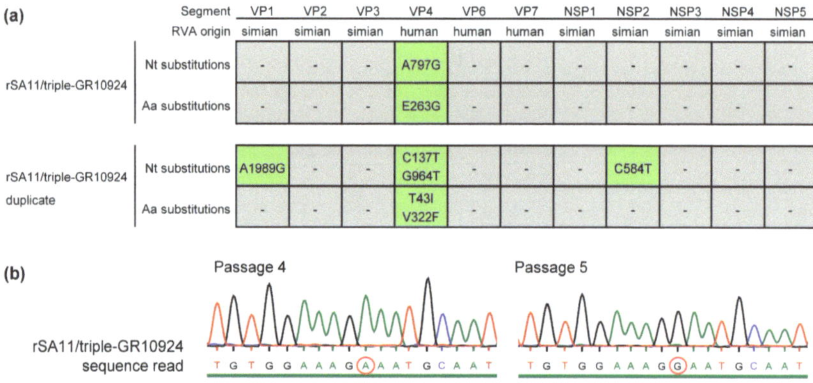

		VP1	VP2	VP3	VP4	VP6	VP7	NSP1	NSP2	NSP3	NSP4	NSP5
	Segment	VP1	VP2	VP3	VP4	VP6	VP7	NSP1	NSP2	NSP3	NSP4	NSP5
	RVA origin	simian	simian	simian	human	human	human	simian	simian	simian	simian	simian
rSA11/triple-GR10924	Nt substitutions	-	-	-	A797G	-	-	-	-	-	-	-
	Aa substitutions	-	-	-	E263G	-	-	-	-	-	-	-
rSA11/triple-GR10924 duplicate	Nt substitutions	A1989G	-	-	C137T G964T	-	-	-	C584T	-	-	-
	Aa substitutions	-	-	-	T43I V322F	-	-	-	-	-	-	-

Figure 2. Sequence analyses of rSA11/triple-GR10924 and the duplicate. (**a**) Nucleotide (Nt) and amino acid (Aa) substitutions in the open reading frames of the eleven rotavirus genome segments identified by next-generation sequencing after passage 10. (**b**) Sanger sequencing analyses of the VP4-encoding genome segment from the rSA11/triple-GR10924 reassortant that replicated to a higher titer in early passages. Respective sequencing chromatograms of passage 4 and 5 are shown. The red circle marks nucleotide position 797 in the VP4-encoding genome segment from human RVA strain GR10924. The green line below the chromatograms indicates that the probability for a wrong base call was equal to or less than 1 in 1000. rSA11/triple-GR10924 = Recombinant rotavirus carrying segment 4 (VP4), segment 9 (VP7) and segment 6 (VP6) from human RVA strain GR10924 in the backbone of SA11.

Additionally, partial untranslated region (UTR) sequences were obtained. The Supplementary Tables S3 and S4 show the identified UTR sequences for the genome segments encoding structural and non-structural proteins, respectively. No nucleotide substitutions were detected in the UTRs of rSA11/triple-GR10924, but one nucleotide substitution located in the 3′UTR of genome segment 10 encoding NSP4 (T705C) was identified for the duplicate.

The point mutation in the rSA11/triple-GR10924 reassortant that replicated to a higher titer in early passages was further characterized. First, the presence of the mutation in passage 10 virus was confirmed by Sanger sequencing. Analysis of passages 1–5 by Sanger sequencing showed that the virus with the mutation A797G (E263G) in VP4 became predominant in the fifth passage. Figure 2b shows VP4 sequencing chromatograms of the corresponding region from passage 4 and passage 5 virus. Sequencing results from passages 1–3 are shown in Supplementary Figure S3.

3.3. Introduction of Mutation A797G into the VP4-Encoding Plasmid of GR10924 Improves Rescue of Reassortants

To determine whether VP4-E263G affects the rescue of reassortants, the A797G mutation was introduced into the VP4-encoding plasmid of GR10924 by site-directed mutagenesis. Rescue experiments to generate rSA11/triple-GR10924 without and with VP4-E263G (rSA11/triple-GR10924$_{E263G}$) were performed as described above. An overview of the appearance of CPEs during each passage is shown in Figure 3a. After the first passage, a CPE appeared in rSA11/triple-GR10924$_{E263G}$. In contrast, rSA11/triple-GR10924 without the mutation did not develop a CPE by the end of passage 4. Analyses of freeze/thaw supernatants collected at the end of every passage by qRT-PCR revealed that rSA11/triple-GR10924$_{E263G}$ already replicated to a high titer in the first passage, while rSA11/triple-GR10924 without the mutation could not be rescued again (Figure 3b).

We have previously been able to generate a recombinant rotavirus containing genome segment 4 (VP4) from human RVA strain GR10924 in the backbone of SA11 (rSA11/mono-GR10924) using a similar reverse genetics approach, but poor replication was observed in MA-104 cells [36]. To analyze whether VP4-E263G also improved the rescue of this mono-reassortant, the generation of rSA11/mono-GR10924 without and with VP4-E263G (rSA11/mono-GR10924$_{E263G}$) was attempted. While a CPE was only evident in passage 4 for rSA11/mono-GR10924 without the mutation, a CPE could already be detected in passage 1 for rSA11/mono-GR10924$_{E263G}$ (Figure 3a). The rSA11/mono-GR10924$_{E263G}$ reassortant also reached higher RNA titers than the reassortant without the mutation after passages 1–4 (Figure 3b). All rescue experiments were repeated with similar results for the rSA11/triple-GR10924, rSA11/triple-GR10924$_{E263G}$ and rSA11/mono-GR10924$_{E263G}$, but the second rescue attempt of the rSA11/mono-GR10924 reassortant without the mutation was not successful (Supplementary Figure S4), suggesting that the rescue of this reassortant was also not efficient using the reverse genetics system employed here.

In order to confirm the identity of the rescued reassortants, RVA RNA from the generated reassortants was analyzed by RT-PCR using specific primer pairs for the genome segments encoding VP4 and VP7 from human RVA GR10924 as well as for the segments encoding VP4, VP7 and VP2 from simian RVA SA11. Analyses of the resulting PCR products by agarose gel electrophoresis confirmed that the expected genome fragments were present for each reassortant (Figure 3c). To confirm that the expected VP6-encoding genome segment was present in the rescued viruses, a primer pair that could bind to the VP6-encoding segment from SA11 and GR10924 was used in RT-PCR analyses and the PCR products were analyzed by Sanger sequencing (Figure 3d).

Having shown that the amino acid substitution E263G improved the rescue of reassortants containing VP4 from GR10924, we were interested in examining whether the mutation also improved the rescue of recombinant rotaviruses with VP4-encoding segments from the other African human RVA strains in the backbone of SA11. The mutation was introduced into the VP4-encoding plasmids of Moz60a and Moz308. However, rescue of Moz60a or Moz308 triple-reassortants with VP4-E263G (rSA11/triple-Moz60a$_{E263G}$ or rSA11/triple-Moz308$_{E263G}$, respectively) as well as mono-reassortants with VP4-E263G (rSA11/mono-Moz60a$_{E263G}$ or rSA11/mono-Moz308$_{E263G}$, respectively) was not successful, as indicated by the absence of a CPE (Supplementary Figure S5a) and RVA RNA (Supplementary Figure S5b) after four passages in MA-104 cells.

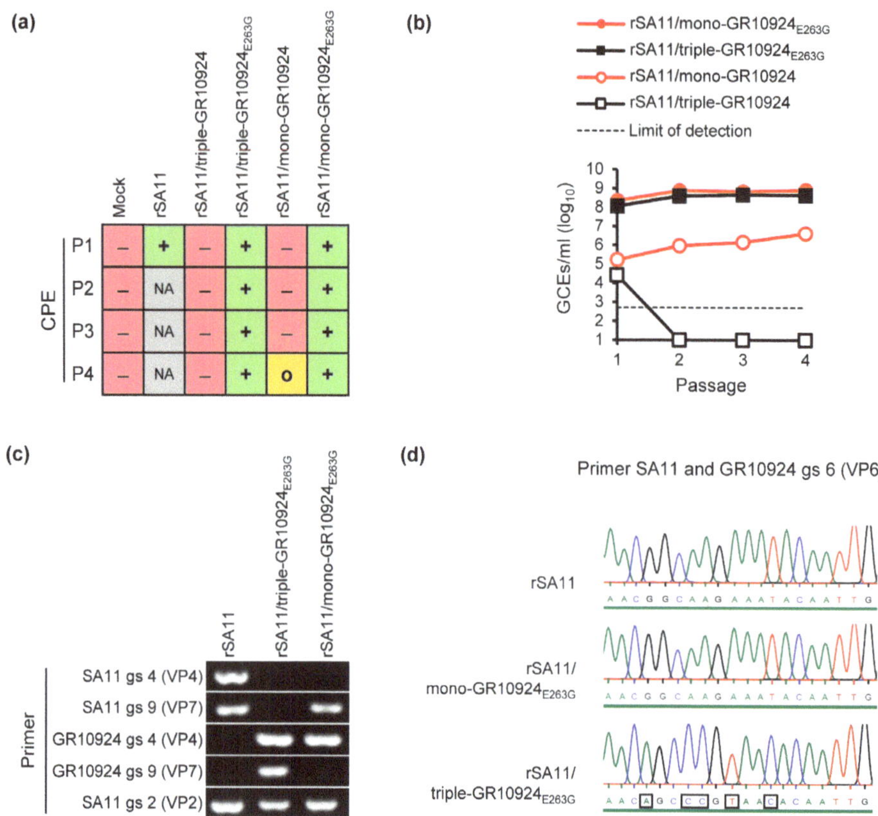

Figure 3. Rescue of SA11 mono- and triple-reassortants containing VP4 from human RVA strain GR10924 with and without the mutation A797G in the VP4-encoding genome segment. (**a**) Cytopathic effect (CPE) upon passage in MA-104 cells. (**b**) Determined number of genome copy equivalents (GCEs)/mL in freeze–thaw supernatant after each passage. (**c**) Detection of VP4- and VP7-encoding genome segments from rSA11 and rescued reassortants via RT-PCR using strain- and genome segment-specific primer pairs followed by agarose gel electrophoresis analysis. (**d**) Detection of VP6-encoding genome segments from rSA11 and rescued reassortants via RT-PCR using VP6-specific primer pairs followed by Sanger sequencing. The black squares mark nucleotide differences between the VP6-encoding genome segment from SA11 and GR10924. The green line below the chromatograms indicates that the probability for a wrong base call was equal to or less than 1 in 1000. rSA11 = Recombinant SA11; rSA11/triple-GR10924 = Recombinant rotavirus carrying segment 4 (VP4), segment 9 (VP7) and segment 6 (VP6) from human RVA strain GR10924 in the backbone of SA11; rSA11/triple-GR10924$_{E263G}$ = rSA11/triple-GR10924 with VP4-E263G; rSA11/mono-GR10924 = Recombinant rotavirus carrying segment 4 (VP4) from human RVA strain GR10924 in the backbone of SA11; rSA11/mono-GR10924$_{E263G}$ = rSA11/mono-GR10924 with VP4-E263G; gs = genome segment; P1–4 = Passages 1–4; red minus = No CPE; yellow O = Mild CPE; green plus = Strong CPE; NA = Not analyzed; GCEs = Genome copy equivalents.

3.4. VP4-E263G Improves Replication of Reassortants

In order to determine the growth kinetics of the generated reassortants and to investigate whether VP4-E263G improves replication, MA-104 cells were infected with rSA11 and reassortants containing VP4 with and without the mutation. As we were unable to rescue rSA11/triple-GR10924 without any mutation in the VP4-encoding genome segment, the rSA11/triple-GR10924 duplicate that contained two other amino acid substitutions in VP4 (see Section 3.2) was used in this experiment. MA-104 cells were infected with rSA11 and the respective reassortants using an equal number of GCEs, and cell culture supernatants were collected at indicated time points post-infection, and the number of GCEs/mL was determined by qRT-PCR (Figure 4). While higher mean titers were observed for the rSA11/mono-GR10924$_{E263G}$ in comparison to the rSA11/triple-GR10924 without the mutation throughout the experiment, the individual titers varied considerably on day 1 and day 2 post-infection. Titers became more consistent by day 3 post-infection. At that time point, the titer of the rSA11/triple-GR10924$_{E263G}$ was 2.5 log$_{10}$ higher than the titer of rSA11/triple-GR10924 without the mutation ($p < 0.01$). Similarly, the titer of rSA11/mono-GR10924$_{E263G}$ was 1.3 log$_{10}$ higher than the titer of rSA11/mono-GR10924 without the mutation ($p < 0.05$) on day 3 post-infection.

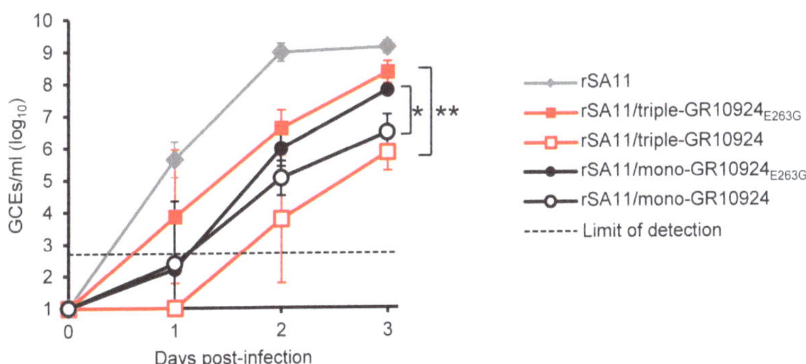

Figure 4. Replication kinetics in MA-104 cells. Cells were infected with 2×10^4 genome copy equivalents (GCEs) corresponding to 0.04 GCEs/cell and the number of GCEs in culture supernatants was determined by qRT-PCR at the indicated time points post-infection. Data are means ± standard deviation from three independent experiments. rSA11 = Recombinant SA11; rSA11/triple-GR10924 = Recombinant rotavirus carrying segment 4 (VP4), segment 9 (VP7) and segment 6 (VP6) from human RVA strain GR10924 in the backbone of SA11; rSA11/triple-GR10924$_{E263G}$ = rSA11/triple-GR10924 with VP4-E263G; rSA11/mono-GR10924 = Recombinant rotavirus carrying segment 4 (VP4) from human RVA strain GR10924 in the backbone of SA11; rSA11/mono-GR10924$_{E263G}$ = rSA11/mono-GR10924 with VP4-E263G; ** $p < 0.01$ for rSA11/triple-GR10924$_{E263G}$ versus rSA11/triple-GR10924 on day 3. * $p < 0.05$ for rSA11/mono-GR10924$_{E263G}$ versus rSA11/mono-GR10924 on day 3.

3.5. Searching for the Presence of E263G, T43I and V322F Exchanges in Reported RVA Field Strain Sequences

The NCBI non-redundant protein sequences data base was screened by BLASTp search using the deduced complete amino acid sequence of the wildtype GR10924 VP4. It was found that no sequence containing the E263G exchange was present in the 100 most closely related hits, indicating that the mutation is not common in field strains. To search more specifically for the mutation, a BLASTp search was performed with a short amino acid sequence (residues 252–272 of GR10924 mVP4) including the E263G exchange. This search resulted in 3/100 hits which contained the E263G exchange. This included two human P[6] strains from Mali (AB938246) [46] and India (EU753965) [47], as well as one human P[8] strain from China [48]. In all cases, the amino acid exchange resulted from an A797G mutation in the VP4 gene. A BLASTp search was also performed with short amino acid residues 33–53 and 312–332 of GR10924 VP4 including the T43I and V322F exchanges, respectively. This search only identified one human P[6] strain with T43I from South Korea (KF650088) [49] and no hit for V322F.

4. Discussion

Cell culture isolation of human RVA strains is difficult and often not successful, leading to a lack of cell culture isolates for many important human RVA genotypes and antigenetic variants. The utilization of the recently established, plasmid-based reverse genetics system facilitates the generation of recombinant rotaviruses, which can also contain antigens of several human RVA genotypes. Using that system, we have previously generated diverse reassortants with VP4, VP7 and VP6 in various combinations from human and non-human rotavirus strains [31–34,36,37]. However, the generation of reassortants containing VP4 of human wildtype RVA was often not successful or resulted in only slowly replicating viruses. Here, we adapted our latest insights to rotavirus strains that were identified in Sub-Saharan Africa. By restoring the natural human RVA VP4, VP7 and VP6 interaction, we hoped to improve the rescue of these strains. However, we were only able to rescue one of those reassortants, rSA11/triple-GR10924. The rescue of rSA11/triple-GR10924 was inconsistent using our reverse genetics approach and replication to higher titers was linked to the development of mutations in VP4, of which the A797G mutation identified by NGS occurred at an early passage (P5) and was confirmed by reverse genetics. The results indicate that single-point mutations in the VP4 gene of human wildtype RVA can substantially improve cell culture replication. In addition, the availability of well-replicating reassortants with human RVA P[6] VP4 may be useful for basic and applied research.

We have previously rescued an SA11 mono-reassortant carrying VP4 from human RVA strain GR10924, but the rescue of SA11 double-reassortants containing both VP4 and VP7 from GR10924 was unsuccessful [36]. We have also shown that combining the genome segments encoding VP4, VP7 and VP6 from the cell culture-adapted human RVA strain Wa in the backbone of SA11 improved the rescue of reassortants [37]. Here, we generated SA11 containing VP4 and VP7 from GR10924 by including VP6 from GR10924. However, the rescue of SA11 triple-reassortants containing VP4, VP7 and VP6 from two other human RVA strains of African origin was not successful. Kanai et al. also tested the generation of SA11 reassortants carrying combinations of VP4, VP7 and VP6 from clinical isolate U14 in the backbone of SA11 [35]. While a poorly replicating mono-reassortant with VP4 from U14 was generated, rescue attempts of an SA11 double-reassortant with VP4 and VP7 from U14 or SA11 triple-reassortant with VP4, VP7 and VP6 from U14 were unsuccessful. In contrast, Hamajima et al. showed that the interplay of VP4 and VP7 from human RVA strain HN126 was important for the generation of SA11 reassortants [29]. Taken together, these results confirm that the generation of reassortants highly depends on the interaction of the capsid proteins, but that their interplay is complex and simply restoring the natural VP4, VP7 and VP6 interaction is not sufficient in all cases. Other factors could be the varying ability of VP4 from different wildtype human RVA strains to mediate entry into target cells, as shown by exchanging the receptor-binding fragment of VP4 with that of a cell culture-adapted strain [31]. In addition, interaction of VP6 with VP2 in the mature virus particles and VP6 interaction with NSP4 during virus assembly [50] may interfere with the generation of viable reassortants.

The two rescued rSA11/triple-GR10924 reassortants developed non-synonymous mutations in VP4 within ten passages in MA-104 cells, suggesting that there was selective pressure on VP4. Indeed, most amino acid residue substitutions seem to be detected in VP4 upon long-term passaging of human RVA strains in cell culture [51]. However, the occurrence of a single amino acid sequence exchange in a very early passage number that coincided with a steep increase in titer was intriguing. In comparison, when human RVA strain CDC-9 was grown in MA-104 cells to passage 11 or 12, no nucleotide sequence changes from the original virus in stool were detected [52]. Interestingly, the single-point mutation that caused a titer increase was not detectable by Sanger sequencing at the end of passage 4, but it was predominant at the end of passage 5. As we were using nearly the entire clarified freeze/thaw culture supernatants for passaging, plaque-purifying the virus from early passages could not be performed, but should be considered for future rescue experiments to identify minor virus sub-populations. One possible explanation for the sudden dominance of the reassortant with the mutation in VP4 after passage 5 may be that only a small proportion of the virus without the mutation was able to infect MA-104 cells going from one passage to the next, while the virus with the mutation was much more efficient at infecting MA-104 cells, resulting in the selection of the virus with the mutation. By introducing the mutation identified here into the VP4-encoding plasmid of GR10294 and performing rescue experiments as well as replication kinetics analyses, we could confirm that this mutation substantially improved rescue and replication in MA-104 cells. This improvement was only found for VP4 from GR10924, but it did not improve virus rescue when introduced into VP4 of strains Moz60a and Moz308, indicating strain specificity.

The mutation at VP4 position 263 was not present in other human RVA strains that have been continuously passaged in cell culture, e.g., CDC-9, Wa, M or the Rotarix precurser vaccine strain 89–12 [52–54]. However, mutations in close proximity (K262R, N267D and R268T) were identified in a study comparing the cell culture adaptation of three human strains in two different cell lines [51]. In addition, we could identify three human wildtype RVA strains which contained the E263G exchange [46–48] by screening of the NCBI sequence database. In each case, E263G was caused by an A797G mutation also found in passaged rSA11/triple-GR10924. This indicates that this mutation may be rare in field strains, but as it can be found in some of them it seems to also support replication in humans. In contrast, we could only identify one field strain containing the T43I exchange [49] and no strain that contained V322F, which indicates that these mutations are very rare in field strains.

It is unclear how the amino acid substitution VP4-E263G improved replication. Glutamic acid is a comparatively large, bulky and rigid amino acid that is negatively charged. Meanwhile, glycine is the smallest amino acid, which is very flexible and does not have a charge. There could be multiple explanations for the observed enhancement in replication. VP4 is proteolytically cleaved by trypsin into VP8* and VP5*. The amino acid residue E263 is downstream of the trypsin cleavage site (residues 231–248 of VP4 from human RVA strain GR10924) and located at the N-terminal region of VP5* (Figure 5a). As E263G is distant from the cleavage site, an effect on proteolytic cleavage seems unlikely, but trypsin cleavage analysis of VP4-E263G would be required to exclude this possibility with certainty.

VP5* plays a role in the perforation of the cellular membrane, but Dowling et al. have shown that VP5* deletion mutants containing residues 265 to 474 or 265 to 404 of VP4 still retained cell permeabilization capabilities [55], suggesting that VP4-E263G is at least not directly affecting cell permeabilization. However, membrane penetration requires the VP4 spike to undergo conformational changes from an upright to a folded-back structure following attachment mediated by VP8* [6,56]. Jenni et al. have reported that mutations in VP5* had a stabilizing effect on the upright conformation of VP4, which resulted in increased infectivity of human RVA CDC-9 [57]. E263G could also lead to changes in the VP4 conformation. Protein structure analysis of the single VP4 molecule (Figure 5b) predicted that a hydrogen bond is formed between glutamic acid at position 264 and arginine at position 369 of VP4 from RRV, which corresponds to glutamic acid at position 263 and arginine at position 368 of VP4 from GR10924, respectively.

Additionally, the rotavirus spike is formed by three VP4 subunits (VP4A, VP4B and VP4C). In VP4A and VP4B, the amino acid residue corresponding to E263 from GR10924 is distant from VP7 or VP6. However, E263 in VP4A is in contact with VP4B and vice versa (Figure S6a). Interestingly, the mutation T42I identified in the rSA11/triple-GR10924 duplicate is close to the same VP4A/VP4B contact site (Figure S6a), which could alter the VP4 subunit interaction with each other. The V322F mutation identified in the duplicate is distant from other VP4 subunits, VP7 or VP6 in VP4A and VP4B (Figure S6a). In VP4C, the N-terminal tip of VP5* lays in a gap formed by two VP7 trimers. In this conformation, it has previously been reported that residue 267 of RRV VP4 (corresponding to residue 266 of GR10924 VP4) is in contact with the RRV VP7 loop containing residue 200 [7]. The residue T42 was not resolved in the VP4C structure, but V322 was in proximity to the N-terminal region of VP4B (Figure S6b). It is of note that, aside from affecting the virion structure and entry, mutations could also impact other VP4 functions. For example, it has recently been shown that VP4 interacts with actin and facilitates the viroplasm assembly process [58,59].

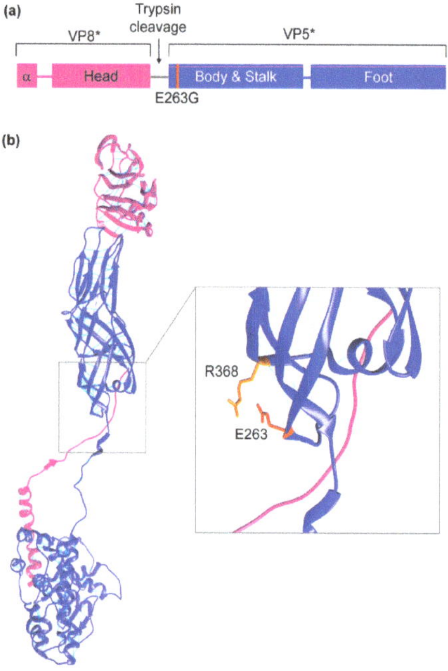

Figure 5. Location of E263 in VP4. (**a**) Schematic of VP4. VP4 is cleaved by trypsin into VP8* and VP5*. The location of the trypsin cleavage site is indicated. VP8* is composed of an α-helix at the N-terminus followed by the head region. The N-terminal helix interacts with the foot region of VP5* and the head region contains the putative receptor-binding site. VP5* contains the body and stalk region at the N-terminus and the foot region at the C-terminus. The location of E263G is shown in red. (**b**) Three-dimensional structure of VP4 on the basis of the atomic model of an infectious rhesus rotavirus (RRV) particle (PDB 4v7q, chain BX). VP8* is colored in magenta and VP5* in blue. Predicted hydrogen bonds are in cyan. The location of E263 in VP4 from GR10924 corresponding to E264 in VP4 from RRV is highlighted in red. R368 in VP4 from GR10924 corresponding to R369 in VP4 from RRV is shown in orange.

Reverse-engineered rotaviruses could serve as next-generation rotavirus vaccine strains and recent studies have already explored the generation of recombinant rotaviruses that express foreign immunogens to use as multivalent vaccine vectors [60–63]. The rSA11/triple-GR10924$_{E263G}$ reassortant generated in our study replicated to high titers in cell culture and could have potential as a vaccine candidate as there are currently no approved vaccines that contain human P[6] VP4, although the rotavirus vaccine candidate RV3-BB (monovalent human G3P[6] RVA, Parkville, Australia) is being tested in in Blantyre, Malawi [64,65]. VP4 contains multiple antigenic epitopes that induce neutralizing antibody responses. Known VP5* antigenic epitopes [66] were mapped to the three-dimensional structure of VP4 from RRV, showing that the amino acid residue corresponding to E263 in VP4 from GR10924 is distant from these epitopes (Supplementary Figure S7). While the mutation identified in VP4 was outside of known VP4 antigenic epitopes, immunization and neutralization studies will have to be conducted to verify that this reassortant can induce cross-neutralizing antibodies. The triple-reassortant could also be used to further investigate factors that influence reassortment, e.g., compatibility with other human RVA VP7 genotypes or other human RVA structural and non-structural proteins.

In conclusion, we have successfully generated a triple-reassortant of an African human RVA strain, where we included all major antigens into one virus by integrating the genome segments encoding human RVA P[6] VP4, G9 VP7 and I2 VP6 into the backbone of the simian RVA strain SA11. Additionally, we identified a mutation in human RVA P[6] VP4 that substantially improved replication in cell culture by a yet unknown mechanism, indicating that single-point mutations in human wildtype RVA VP4 genes can substantially improve cell culture replication. In the future, the use of this triple-reassortant as a potential next-generation vaccine strain should be investigated.

Supplementary Materials: The following supporting information can be downloaded at: https://www.mdpi.com/article/10.3390/v16040565/s1, Supplementary Information. Figure S1: Example of an RT-qPCR analysis of an SA11 NSP3-encoding plasmid standard for the determination of genome copy equivalents (GCEs)/ml culture supernatant. Figure S2: Repetition experiments for the generation of SA11 triple-reassortants containing VP4, VP7 and VP6 from three African human RVA strains. Figure S3: Sanger sequencing analyses of the VP4-encoding genome segment from the rSA11/triple-GR10924 reassortant that replicated to a higher titer in early passages. Figure S4: Repetition experiment for the rescue of SA11 mono- and triple-reassortants containing VP4 from human RVA strain GR10924 with and without the mutation A797G in VP4. Figure S5: Attempts to rescue of SA11 mono- and triple-reassortants containing VP4 from human RVA strains Moz60a or Moz308 with the mutation A797G in VP4. Figure S6: Mapping of E263, T42 and V322 to the VP4 trimeric spike. Figure S7: Mapping of known VP5* antigenic epitopes to the three-dimensional structure of VP4 from rhesus rotavirus (PDB 4v7q, chain BX). Table S1: Primer pairs used for RT-PCR analyses, Sanger sequencing and mutagenesis. Table S2: Average coverage, coverage range, and % ORF sequenced of rSA11/triple-GR10924 and rSA11/triple-GR10924-duplicate after next generation sequencing of virus present in the supernatant of the 10th passage. Table S3: UTRs of the genome segments encoding structural viral proteins from rSA11/triple-GR10924 and rSA11/triple-GR10924-duplicate. Table S4: UTRs of the genome segments encoding non-structural viral proteins from rSA11/triple-GR10924 and rSA11/triple-GR10924-duplicate.

Author Contributions: Conceptualization, R.J. and A.F.; Data curation, R.V.-M., K.S.-L. and A.F.; Funding acquisition, R.J.; Investigation, R.V.-M., K.S.-L. and A.F.; Methodology, R.V.-M., K.S.-L. and A.F.; Writing—original draft, R.V.-M.; Writing—review and editing, R.V.-M., K.S.-L., R.J. and A.F. All authors have read and agreed to the published version of the manuscript.

Funding: This study was supported by the German Research Foundation (DFG), grant numbers JO369/5-1 and JO369/5-2.

Institutional Review Board Statement: Not applicable.

Informed Consent Statement: Not applicable.

Data Availability Statement: All data are contained in the manuscript or are available upon reasonable request from the corresponding author.

Acknowledgments: We would like to thank Silke Apelt, Stefanie Prosetzky and Anja Schlosser for their excellent technical assistance. Development of UCSF Chimera by the Resource for Biocomputing, Visualization and Informatics at the University of California, San Francisco, was supported by NIH P41-GM103311.

Conflicts of Interest: The authors declare no conflicts of interest.

References

1. Matthijnssens, J.; Attoui, H.; Banyai, K.; Brussaard, C.P.D.; Danthi, P.; Del Vas, M.; Dermody, T.S.; Duncan, R.; Fang, Q.; Johne, R.; et al. ICTV Virus Taxonomy Profile: Sedoreoviridae 2022. *J. Gen. Virol.* **2022**, *103*, 001782. [CrossRef] [PubMed]
2. Doro, R.; Farkas, S.L.; Martella, V.; Banyai, K. Zoonotic transmission of rotavirus: Surveillance and control. *Expert. Rev. Anti Infect. Ther.* **2015**, *13*, 1337–1350. [CrossRef] [PubMed]
3. Cohen, A.L.; Platts-Mills, J.A.; Nakamura, T.; Operario, D.J.; Antoni, S.; Mwenda, J.M.; Weldegebriel, G.; Rey-Benito, G.; de Oliveira, L.H.; Ortiz, C.; et al. Aetiology and incidence of diarrhoea requiring hospitalisation in children under 5 years of age in 28 low-income and middle-income countries: Findings from the Global Pediatric Diarrhea Surveillance network. *BMJ Glob. Health* **2022**, *7*, e009548. [CrossRef] [PubMed]

4. Crawford, S.E.; Ramani, S.; Tate, J.E.; Parashar, U.D.; Svensson, L.; Hagbom, M.; Franco, M.A.; Greenberg, H.B.; O'Ryan, M.; Kang, G.; et al. Rotavirus infection. *Nat. Rev. Dis. Primers* **2017**, *3*, 17083. [CrossRef] [PubMed]
5. Troeger, C.; Khalil, I.A.; Rao, P.C.; Cao, S.; Blacker, B.F.; Ahmed, T.; Armah, G.; Bines, J.E.; Brewer, T.G.; Colombara, D.V.; et al. Rotavirus Vaccination and the Global Burden of Rotavirus Diarrhea Among Children Younger Than 5 Years. *JAMA Pediatr.* **2018**, *172*, 958–965. [CrossRef]
6. Desselberger, U. Rotaviruses. *Virus Res.* **2014**, *190*, 75–96. [CrossRef]
7. Settembre, E.C.; Chen, J.Z.; Dormitzer, P.R.; Grigorieff, N.; Harrison, S.C. Atomic model of an infectious rotavirus particle. *EMBO J.* **2011**, *30*, 408–416. [CrossRef] [PubMed]
8. Baker, M.; Prasad, B.V. Rotavirus cell entry. *Curr. Top. Microbiol. Immunol.* **2010**, *343*, 121–148. [CrossRef]
9. Matthijnssens, J.; Ciarlet, M.; Heiman, E.; Arijs, I.; Delbeke, T.; McDonald, S.M.; Palombo, E.A.; Iturriza-Gomara, M.; Maes, P.; Patton, J.T.; et al. Full genome-based classification of rotaviruses reveals a common origin between human Wa-Like and porcine rotavirus strains and human DS-1-like and bovine rotavirus strains. *J. Virol.* **2008**, *82*, 3204–3219. [CrossRef]
10. McDonald, S.M.; Nelson, M.I.; Turner, P.E.; Patton, J.T. Reassortment in segmented RNA viruses: Mechanisms and outcomes. *Nat. Rev. Microbiol.* **2016**, *14*, 448–460. [CrossRef]
11. RCWG Rotavirus Classification Working Group. List of Accepted Genotypes. Laboratory of Viral Metagenomics. Available online: https://rega.kuleuven.be/cev/viralmetagenomics/virus-classification/rcwg (accessed on 17 November 2023).
12. Desselberger, U.; Huppertz, H.I. Immune responses to rotavirus infection and vaccination and associated correlates of protection. *J. Infect. Dis.* **2011**, *203*, 188–195. [CrossRef] [PubMed]
13. Feng, N.; Lawton, J.A.; Gilbert, J.; Kuklin, N.; Vo, P.; Prasad, B.V.; Greenberg, H.B. Inhibition of rotavirus replication by a non-neutralizing, rotavirus VP6-specific IgA mAb. *J. Clin. Investig.* **2002**, *109*, 1203–1213. [CrossRef]
14. Burns, J.W.; Siadat-Pajouh, M.; Krishnaney, A.A.; Greenberg, H.B. Protective effect of rotavirus VP6-specific IgA monoclonal antibodies that lack neutralizing activity. *Science* **1996**, *272*, 104–107. [CrossRef] [PubMed]
15. Caddy, S.L.; Vaysburd, M.; Wing, M.; Foss, S.; Andersen, J.T.; O'Connell, K.; Mayes, K.; Higginson, K.; Iturriza-Gomara, M.; Desselberger, U.; et al. Intracellular neutralisation of rotavirus by VP6-specific IgG. *PLoS Pathog.* **2020**, *16*, e1008732. [CrossRef] [PubMed]
16. Mwenda, J.M.; Parashar, U.D.; Cohen, A.L.; Tate, J.E. Impact of rotavirus vaccines in Sub-Saharan African countries. *Vaccine* **2018**, *36*, 7119–7123. [CrossRef] [PubMed]
17. Heaton, P.M.; Goveia, M.G.; Miller, J.M.; Offit, P.; Clark, H.F. Development of a pentavalent rotavirus vaccine against prevalent serotypes of rotavirus gastroenteritis. *J. Infect. Dis.* **2005**, *192* (Suppl. 1), S17–S21. [CrossRef] [PubMed]
18. Ward, R.L.; Bernstein, D.I. Rotarix: A rotavirus vaccine for the world. *Clin. Infect. Dis.* **2009**, *48*, 222–228. [CrossRef] [PubMed]
19. Jonesteller, C.L.; Burnett, E.; Yen, C.; Tate, J.E.; Parashar, U.D. Effectiveness of Rotavirus Vaccination: A Systematic Review of the First Decade of Global Postlicensure Data, 2006–2016. *Clin. Infect. Dis.* **2017**, *65*, 840–850. [CrossRef] [PubMed]
20. Desselberger, U. Differences of Rotavirus Vaccine Effectiveness by Country: Likely Causes and Contributing Factors. *Pathogens* **2017**, *6*, 65. [CrossRef] [PubMed]
21. Todd, S.; Page, N.A.; Duncan Steele, A.; Peenze, I.; Cunliffe, N.A. Rotavirus strain types circulating in Africa: Review of studies published during 1997–2006. *J. Infect. Dis.* **2010**, *202* (Suppl. 1), S34–S42. [CrossRef]
22. Patton, J.T. Rotavirus diversity and evolution in the post-vaccine world. *Discov. Med.* **2012**, *13*, 85–97.
23. Bonura, F.; Mangiaracina, L.; Filizzolo, C.; Bonura, C.; Martella, V.; Ciarlet, M.; Giammanco, G.M.; De Grazia, S. Impact of Vaccination on Rotavirus Genotype Diversity: A Nearly Two-Decade-Long Epidemiological Study before and after Rotavirus Vaccine Introduction in Sicily, Italy. *Pathogens* **2022**, *11*, 424. [CrossRef]
24. Mwangi, P.N.; Page, N.A.; Seheri, M.L.; Mphahlele, M.J.; Nadan, S.; Esona, M.D.; Kumwenda, B.; Kamng'ona, A.W.; Donato, C.M.; Steele, D.A.; et al. Evolutionary changes between pre- and post-vaccine South African group A G2P[4] rotavirus strains, 2003–2017. *Microb. Genom.* **2022**, *8*, 000809. [CrossRef]
25. Page, N.A.; Seheri, L.M.; Groome, M.J.; Moyes, J.; Walaza, S.; Mphahlele, J.; Kahn, K.; Kapongo, C.N.; Zar, H.J.; Tempia, S.; et al. Temporal association of rotavirus vaccination and genotype circulation in South Africa: Observations from 2002 to 2014. *Vaccine* **2018**, *36*, 7231–7237. [CrossRef]
26. Kanai, Y.; Komoto, S.; Kawagishi, T.; Nouda, R.; Nagasawa, N.; Onishi, M.; Matsuura, Y.; Taniguchi, K.; Kobayashi, T. Entirely plasmid-based reverse genetics system for rotaviruses. *Proc. Natl. Acad. Sci. USA* **2017**, *114*, 2349–2354. [CrossRef]
27. Komoto, S.; Fukuda, S.; Kugita, M.; Hatazawa, R.; Koyama, C.; Katayama, K.; Murata, T.; Taniguchi, K. Generation of Infectious Recombinant Human Rotaviruses from Just 11 Cloned cDNAs Encoding the Rotavirus Genome. *J. Virol.* **2019**, *93*, e02207-18. [CrossRef]
28. Kawagishi, T.; Nurdin, J.A.; Onishi, M.; Nouda, R.; Kanai, Y.; Tajima, T.; Ushijima, H.; Kobayashi, T. Reverse Genetics System for a Human Group A Rotavirus. *J. Virol.* **2020**, *94*, e00963-19. [CrossRef]
29. Hamajima, R.; Lusiany, T.; Minami, S.; Nouda, R.; Nurdin, J.A.; Yamasaki, M.; Kobayashi, N.; Kanai, Y.; Kobayashi, T. A reverse genetics system for human rotavirus G2P[4]. *J. Gen. Virol.* **2022**, *103*, 001816. [CrossRef]
30. Sanchez-Tacuba, L.; Feng, N.; Meade, N.J.; Mellits, K.H.; Jais, P.H.; Yasukawa, L.L.; Resch, T.K.; Jiang, B.; Lopez, S.; Ding, S.; et al. An Optimized Reverse Genetics System Suitable for Efficient Recovery of Simian, Human, and Murine-Like Rotaviruses. *J. Virol.* **2020**, *94*, e01294-20. [CrossRef]

31. Falkenhagen, A.; Huyzers, M.; van Dijk, A.A.; Johne, R. Rescue of Infectious Rotavirus Reassortants by a Reverse Genetics System Is Restricted by the Receptor-Binding Region of VP4. *Viruses* **2021**, *13*, 363. [CrossRef]

32. Falkenhagen, A.; Patzina-Mehling, C.; Ruckner, A.; Vahlenkamp, T.W.; Johne, R. Generation of simian rotavirus reassortants with diverse VP4 genes using reverse genetics. *J. Gen. Virol.* **2019**, *100*, 1595–1604. [CrossRef]

33. Patzina-Mehling, C.; Falkenhagen, A.; Trojnar, E.; Gadicherla, A.K.; Johne, R. Potential of avian and mammalian species A rotaviruses to reassort as explored by plasmid only-based reverse genetics. *Virus Res.* **2020**, *286*, 198027. [CrossRef]

34. Falkenhagen, A.; Tausch, S.H.; Labutin, A.; Grutzke, J.; Heckel, G.; Ulrich, R.G.; Johne, R. Genetic and biological characteristics of species A rotaviruses detected in common shrews suggest a distinct evolutionary trajectory. *Virus Evol.* **2022**, *8*, veac004. [CrossRef]

35. Kanai, Y.; Onishi, M.; Kawagishi, T.; Pannacha, P.; Nurdin, J.A.; Nouda, R.; Yamasaki, M.; Lusiany, T.; Khamrin, P.; Okitsu, S.; et al. Reverse Genetics Approach for Developing Rotavirus Vaccine Candidates Carrying VP4 and VP7 Genes Cloned from Clinical Isolates of Human Rotavirus. *J. Virol.* **2020**, *95*, e01374-20. [CrossRef]

36. Falkenhagen, A.; Patzina-Mehling, C.; Gadicherla, A.K.; Strydom, A.; O'Neill, H.G.; Johne, R. Generation of Simian Rotavirus Reassortants with VP4- and VP7-Encoding Genome Segments from Human Strains Circulating in Africa Using Reverse Genetics. *Viruses* **2020**, *12*, 201. [CrossRef]

37. Valusenko-Mehrkens, R.; Gadicherla, A.K.; Johne, R.; Falkenhagen, A. Strain-Specific Interactions between the Viral Capsid Proteins VP4, VP7 and VP6 Influence Rescue of Rotavirus Reassortants by Reverse Genetics. *Int. J. Mol. Sci.* **2023**, *24*, 5670. [CrossRef]

38. Potgieter, A.C.; Page, N.A.; Liebenberg, J.; Wright, I.M.; Landt, O.; van Dijk, A.A. Improved strategies for sequence-independent amplification and sequencing of viral double-stranded RNA genomes. *J. Gen. Virol.* **2009**, *90*, 1423–1432. [CrossRef]

39. Strydom, A.; Joao, E.D.; Motanyane, L.; Nyaga, M.M.; Christiaan Potgieter, A.; Cuamba, A.; Mandomando, I.; Cassocera, M.; de Deus, N.; O'Neill, H.G. Whole genome analyses of DS-1-like Rotavirus A strains detected in children with acute diarrhoea in southern Mozambique suggest several reassortment events. *Infect. Genet. Evol.* **2019**, *69*, 68–75. [CrossRef]

40. Trojnar, E.; Sachsenroder, J.; Twardziok, S.; Reetz, J.; Otto, P.H.; Johne, R. Identification of an avian group A rotavirus containing a novel VP4 gene with a close relationship to those of mammalian rotaviruses. *J. Gen. Virol.* **2013**, *94*, 136–142. [CrossRef]

41. Chen, J.Z.; Settembre, E.C.; Aoki, S.T.; Zhang, X.; Bellamy, A.R.; Dormitzer, P.R.; Harrison, S.C.; Grigorieff, N. Molecular interactions in rotavirus assembly and uncoating seen by high-resolution cryo-EM. *Proc. Natl. Acad. Sci. USA* **2009**, *106*, 10644–10648. [CrossRef]

42. Uprety, T.; Wang, D.; Li, F. Recent advances in rotavirus reverse genetics and its utilization in basic research and vaccine development. *Arch. Virol.* **2021**, *166*, 2369–2386. [CrossRef]

43. Otto, P.H.; Rosenhain, S.; Elschner, M.C.; Hotzel, H.; Machnowska, P.; Trojnar, E.; Hoffmann, K.; Johne, R. Detection of rotavirus species A, B and C in domestic mammalian animals with diarrhoea and genotyping of bovine species A rotavirus strains. *Vet. Microbiol.* **2015**, *179*, 168–176. [CrossRef]

44. Johne, R.; Schilling-Loeffler, K.; Ulrich, R.G.; Tausch, S.H. Whole Genome Sequence Analysis of a Prototype Strain of the Novel Putative Rotavirus Species L. *Viruses* **2022**, *14*, 462. [CrossRef]

45. Pettersen, E.F.; Goddard, T.D.; Huang, C.C.; Couch, G.S.; Greenblatt, D.M.; Meng, E.C.; Ferrin, T.E. UCSF Chimera—A visualization system for exploratory research and analysis. *J. Comput. Chem.* **2004**, *25*, 1605–1612. [CrossRef]

46. Nakagomi, T.; Do, L.P.; Agbemabiese, C.A.; Kaneko, M.; Gauchan, P.; Doan, Y.H.; Jere, K.C.; Steele, A.D.; Iturriza-Gomara, M.; Nakagomi, O.; et al. Whole-genome characterisation of G12P[6] rotavirus strains possessing two distinct genotype constellations co-circulating in Blantyre, Malawi, 2008. *Arch. Virol.* **2017**, *162*, 213–226. [CrossRef]

47. Mukherjee, A.; Dutta, D.; Ghosh, S.; Bagchi, P.; Chattopadhyay, S.; Nagashima, S.; Kobayashi, N.; Dutta, P.; Krishnan, T.; Naik, T.N.; et al. Full genomic analysis of a human group A rotavirus G9P[6] strain from Eastern India provides evidence for porcine-to-human interspecies transmission. *Arch. Virol.* **2009**, *154*, 733–746. [CrossRef]

48. Wang, Y.H.; Pang, B.B.; Ghosh, S.; Zhou, X.; Shintani, T.; Urushibara, N.; Song, Y.W.; He, M.Y.; Liu, M.Q.; Tang, W.F.; et al. Molecular epidemiology and genetic evolution of the whole genome of G3P[8] human rotavirus in Wuhan, China, from 2000 through 2013. *PLoS ONE* **2014**, *9*, e88850. [CrossRef]

49. Mun, S.K.; Cho, H.G.; Lee, H.K.; Park, S.H.; Park, P.H.; Yoon, M.H.; Jeong, H.S.; Lim, Y.H. High incidence of group A rotaviruses G4P[6] strains among children in Gyeonggi province of South Korea, from 2009 to 2012. *Infect. Genet. Evol.* **2016**, *44*, 351–355. [CrossRef]

50. Meyer, J.C.; Bergmann, C.C.; Bellamy, A.R. Interaction of rotavirus cores with the nonstructural glycoprotein NS28. *Virology* **1989**, *171*, 98–107. [CrossRef]

51. Tsugawa, T.; Tsutsumi, H. Genomic changes detected after serial passages in cell culture of virulent human G1P[8] rotaviruses. *Infect. Genet. Evol.* **2016**, *45*, 6–10. [CrossRef]

52. Resch, T.K.; Wang, Y.; Moon, S.; Jiang, B. Serial Passaging of the Human Rotavirus CDC-9 Strain in Cell Culture Leads to Attenuation: Characterization from In Vitro and In Vivo Studies. *J. Virol.* **2020**, *94*, e00889-20. [CrossRef]

53. Guo, Y.; Wentworth, D.E.; Stucker, K.M.; Halpin, R.A.; Lam, H.C.; Marthaler, D.; Saif, L.J.; Vlasova, A.N. Amino Acid Substitutions in Positions 385 and 393 of the Hydrophobic Region of VP4 May Be Associated with Rotavirus Attenuation and Cell Culture Adaptation. *Viruses* **2020**, *12*, 408. [CrossRef]

54. Ward, R.L.; Kirkwood, C.D.; Sander, D.S.; Smith, V.E.; Shao, M.; Bean, J.A.; Sack, D.A.; Bernstein, D.I. Reductions in cross-neutralizing antibody responses in infants after attenuation of the human rotavirus vaccine candidate 89-12. *J. Infect. Dis.* **2006**, *194*, 1729–1736. [CrossRef]
55. Dowling, W.; Denisova, E.; LaMonica, R.; Mackow, E.R. Selective membrane permeabilization by the rotavirus VP5* protein is abrogated by mutations in an internal hydrophobic domain. *J. Virol.* **2000**, *74*, 6368–6376. [CrossRef]
56. Rodriguez, J.M.; Chichon, F.J.; Martin-Forero, E.; Gonzalez-Camacho, F.; Carrascosa, J.L.; Caston, J.R.; Luque, D. New insights into rotavirus entry machinery: Stabilization of rotavirus spike conformation is independent of trypsin cleavage. *PLoS Pathog.* **2014**, *10*, e1004157. [CrossRef]
57. Jenni, S.; Li, Z.; Wang, Y.; Bessey, T.; Salgado, E.N.; Schmidt, A.G.; Greenberg, H.B.; Jiang, B.; Harrison, S.C. Rotavirus VP4 Epitope of a Broadly Neutralizing Human Antibody Defined by Its Structure Bound with an Attenuated-Strain Virion. *J. Virol.* **2022**, *96*, e00627-22. [CrossRef]
58. Vetter, J.; Papa, G.; Seyffert, M.; Gunasekera, K.; De Lorenzo, G.; Wiesendanger, M.; Reymond, J.L.; Fraefel, C.; Burrone, O.R.; Eichwald, C. Rotavirus Spike Protein VP4 Mediates Viroplasm Assembly by Association to Actin Filaments. *J. Virol.* **2022**, *96*, e01074-22. [CrossRef]
59. Trejo-Cerro, O.; Eichwald, C.; Schraner, E.M.; Silva-Ayala, D.; Lopez, S.; Arias, C.F. Actin-Dependent Nonlytic Rotavirus Exit and Infectious Virus Morphogenetic Pathway in Nonpolarized Cells. *J. Virol.* **2018**, *92*, e02076-17. [CrossRef]
60. Philip, A.A.; Hu, S.; Dai, J.; Patton, J.T. Recombinant rotavirus expressing the glycosylated S1 protein of SARS-CoV-2. *J. Gen. Virol.* **2023**, *104*, 001899. [CrossRef]
61. Philip, A.A.; Patton, J.T. Rotavirus as an Expression Platform of Domains of the SARS-CoV-2 Spike Protein. *Vaccines* **2021**, *9*, 449. [CrossRef]
62. Kawamura, Y.; Komoto, S.; Fukuda, S.; Kugita, M.; Tang, S.; Patel, A.; Pieknik, J.R.; Nagao, S.; Taniguchi, K.; Krause, P.R.; et al. Development of recombinant rotavirus carrying herpes simplex virus 2 glycoprotein D gene based on reverse genetics technology. *Microbiol. Immunol.* **2024**, *68*, 56–64. [CrossRef]
63. Kawagishi, T.; Sanchez-Tacuba, L.; Feng, N.; Costantini, V.P.; Tan, M.; Jiang, X.; Green, K.Y.; Vinje, J.; Ding, S.; Greenberg, H.B. Mucosal and systemic neutralizing antibodies to norovirus induced in infant mice orally inoculated with recombinant rotaviruses. *Proc. Natl. Acad. Sci. USA* **2023**, *120*, e2214421120. [CrossRef]
64. Desselberger, U. Potential of plasmid only based reverse genetics of rotavirus for the development of next-generation vaccines. *Curr. Opin. Virol.* **2020**, *44*, 1–6. [CrossRef]
65. Witte, D.; Handley, A.; Jere, K.C.; Bogandovic-Sakran, N.; Mpakiza, A.; Turner, A.; Pavlic, D.; Boniface, K.; Mandolo, J.; Ong, D.S.; et al. Neonatal rotavirus vaccine (RV3-BB) immunogenicity and safety in a neonatal and infant administration schedule in Malawi: A randomised, double-blind, four-arm parallel group dose-ranging study. *Lancet Infect. Dis.* **2022**, *22*, 668–678. [CrossRef]
66. McDonald, S.M.; Matthijnssens, J.; McAllen, J.K.; Hine, E.; Overton, L.; Wang, S.; Lemey, P.; Zeller, M.; Van Ranst, M.; Spiro, D.J.; et al. Evolutionary dynamics of human rotaviruses: Balancing reassortment with preferred genome constellations. *PLoS Pathog.* **2009**, *5*, e1000634. [CrossRef]

viruses

MDPI

Article

Reverse Genetics of Murine Rotavirus: A Comparative Analysis of the Wild-Type and Cell-Culture-Adapted Murine Rotavirus VP4 in Replication and Virulence in Neonatal Mice

Takahiro Kawagishi [1,2,3,4,*,†], Liliana Sánchez-Tacuba [2,3,4], Ningguo Feng [2,3,4], Harry B. Greenberg [2,3,4,*] and Siyuan Ding [1,*]

1 Department of Molecular Microbiology, Washington University School of Medicine, St. Louis, MO 63110, USA
2 Department of Medicine, Division of Gastroenterology and Hepatology, Stanford University School of Medicine, Stanford, CA 94305, USA
3 Department of Microbiology and Immunology, Stanford University School of Medicine, Stanford, CA 94305, USA
4 VA Palo Alto Health Care System, Department of Veterans Affairs, Palo Alto, CA 94304, USA
* Correspondence: tkawagis@biken.osaka-u.ac.jp (T.K.); hbgreen@stanford.edu (H.B.G.); siyuan.ding@wustl.edu (S.D.)
† Present Address: Department of Virology, Research Institute for Microbial Diseases, Osaka University, Osaka 565-0871, Japan.

Citation: Kawagishi, T.; Sánchez-Tacuba, L.; Feng, N.; Greenberg, H.B.; Ding, S. Reverse Genetics of Murine Rotavirus: A Comparative Analysis of the Wild-Type and Cell-Culture-Adapted Murine Rotavirus VP4 in Replication and Virulence in Neonatal Mice. *Viruses* 2024, *16*, 767. https://doi.org/10.3390/v16050767

Academic Editor: Ulrich Desselberger

Received: 9 April 2024
Revised: 3 May 2024
Accepted: 5 May 2024
Published: 12 May 2024

Abstract: Small-animal models and reverse genetics systems are powerful tools for investigating the molecular mechanisms underlying viral replication, virulence, and interaction with the host immune response in vivo. Rotavirus (RV) causes acute gastroenteritis in many young animals and infants worldwide. Murine RV replicates efficiently in the intestines of inoculated suckling pups, causing diarrhea, and spreads efficiently to uninoculated littermates. Because RVs derived from human and other non-mouse animal species do not replicate efficiently in mice, murine RVs are uniquely useful in probing the viral and host determinants of efficient replication and pathogenesis in a species-matched mouse model. Previously, we established an optimized reverse genetics protocol for RV and successfully generated a murine-like RV rD6/2-2g strain that replicates well in both cultured cell lines and in the intestines of inoculated pups. However, rD6/2-2g possesses three out of eleven gene segments derived from simian RV strains, and these three heterologous segments may attenuate viral pathogenicity in vivo. Here, we rescued the first recombinant RV with all 11 gene segments of murine RV origin. Using this virus as a genetic background, we generated a panel of recombinant murine RVs with either N-terminal VP8* or C-terminal VP5* regions chimerized between a cell-culture-adapted murine ETD strain and a non-tissue-culture-adapted murine EW strain and compared the diarrhea rate and fecal RV shedding in pups. The recombinant viruses with VP5* domains derived from the murine EW strain showed slightly more fecal shedding than those with VP5* domains from the ETD strain. The newly characterized full-genome murine RV will be a useful tool for dissecting virus–host interactions and for studying the mechanism of pathogenesis in neonatal mice.

Keywords: rotavirus; reverse genetics; small-animal model

1. Introduction

Rotavirus (RV) is the most common causative agent of severe acute diarrhea in infants and small animals worldwide [1]. While RV has been isolated from many mammalian and avian species, RV infection relatively infrequently demonstrates cross-species transmission and persistence in a heterologous host species, a phenomenon known as host-range restriction (HRR) [2–10]. For instance, the genome sequences of RV strains isolated from humans generally belong to groups of human RV strains isolated before. RVs from other animal species are occasionally isolated from humans but rarely spread or persist in the human population [11]. HRR has been exploited to generate two live-attenuated RV vaccines currently used worldwide (i.e., RotaTeq (Merck) and RotaSiil (Serum Institute of India)), using bovine RV strains as a genetic backbone [12].

Since HRR contributes to natural attenuation, the molecular basis of RV HRR has been studied in animal infection models, especially in mice [2,3,6,10,13,14]. Homologous murine RV strains replicate and spread efficiently in the mouse model; however, heterologous RV strains (e.g., simian, bovine, porcine, and human RVs) do not. The RV genome consists of 11 segmented double-stranded RNAs that encode six structural proteins (VP1 to VP4, VP6, and VP7) and six nonstructural proteins (NSP1 to NSP6). Previous studies, including research by our group and others, have demonstrated that gene segments encoding VP3, VP4, VP7, NSP1, NSP2, NSP3, and NSP4 proteins can all be associated with RV HRR [5,7,9,10,13,14].

A natural mono-reassortant RV D6/2 strain was isolated by plaque assays from an intestinal homogenate of a suckling mouse co-infected with wild-type murine EDIM-EW and the tissue-culture-adapted simian RRV strain [5]. Unlike wild-type murine RV, D6/2 replicates in cultured cell lines while still efficiently causing diarrhea in inoculated pups [5,10]. Of note, 10 of the 11 gene segments of D6/2 are derived from the EDIM-EW strain; for the exception is gene segment 4, which encodes the cell attachment protein VP4 (Table 1). It has been shown that cell-culture-adapted murine RV strains do not cause diarrhea as efficiently as the wild-type murine RV [6,15–17]. Therefore, D6/2 provides a unique opportunity to further interrogate the viral determinants of HRR.

Table 1. Gene constellation of the D6/2 and recombinant viruses.

Gene Segment (Viral Protein)	D6/2	rD6/2-2g	rEW/ETD-VP4
Gene segment 1 (VP1)	EW	SA11	EW
Gene segment 2 (VP2)	EW	EW	EW
Gene segment 3 (VP3)	EW	EW	EW
Gene segment 4 (VP4)	RRV	RRV	ETD
Gene segment 5 (NSP1)	EW	EW	EW
Gene segment 6 (VP6)	EW	EW	EW
Gene segment 7 (NSP3)	EW	EW	EW
Gene segment 8 (NSP2)	EW	EW	EW
Gene segment 9 (VP7)	EW	EW	EW
Gene segment 10 (NSP4)	EW	SA11	EW
Gene segment 11 (NSP5/6)	EW	EW	EW

Since the first plasmid-based reverse genetics was developed for simian SA11 strain [18], reverse genetics has been established for several other animal RV strains [19–26]. We previously used D6/2 as a genetic backbone and rescued a recombinant murine-like RV by introducing two more gene segments from the simian SA11 strain: gene segments 1 and 10, which encode viral polymerase VP1 and viral enterotoxin NSP4, respectively (Table 1) [21]. The new recombinant virus D6/2 with the two additional genes derived from simian RV (so-called rD6/2-2g) replicated in the intestine of inoculated pups and transmitted to their uninoculated littermates. Using rD6/2-2g as the backbone, we have since demonstrated that murine RV NSP1, an interferon antagonist, plays a critical role in viral replication in vivo [27]. In addition, by comparing VP4s from heterologous RV strains in an isogenic rD6/2-2g background, we demonstrated that VP4s from heterologous RV strains contribute to HRR to varying degrees [14]. Furthermore, we rescued an rD6/2-2g RV expressing a bioluminescent reporter, Nano-Luciferase, to characterize systemic dissemination of RV in vivo in a non-invasive manner [28].

Despite the utility of rD6/2-2g, this recombinant murine-like RV is a compromise between rescue efficiency and in vivo virulence. It still harbors three gene segments (gene segments 1, 4, and 10) from heterologous simian RV strains (Table 1). Considering that these three gene segments may be potentially associated with HRR in mice, it is desirable to create an RV that grows well in culture but has all 11 gene segments derived from a murine strain to be used in a mouse model. Toward that objective, we attempted to replace the three heterologous gene segments in rD6/2-2g with those from a murine-origin RV strain.

The obvious primary challenge for this objective is the RV structural protein VP4. It is the spike protein present on the surface of RV virions and is required for cell attachment and entry in host cells. VP4 is cleaved by trypsin into two distinct domains (Figure 1A). The N-terminal VP8* domain forms the head structure of the virion spike and engages in attachment to the target cell surface. The β-barrel domain in VP5* forms the body of the spike, and the C-terminal region of the VP5* domain functions as the foot of the spike by interacting with VP7 and VP6 proteins in the virion [29–31]. Here, we used reverse genetics and improved the virulence of rD6/2-2g by rescuing a series of recombinant RVs with all 11 gene segments from a murine RV strain. We also generated murine RV VP4 chimeric viruses between the cell-culture-adapted ETD_822 strain and the wild-type murine EW strain and compared diarrheal diseases in the suckling mouse model. The data suggest that murine RV reverse genetics offers a new tool to study molecular mechanisms of RV replication, virulence, and spread in the homologous murine model.

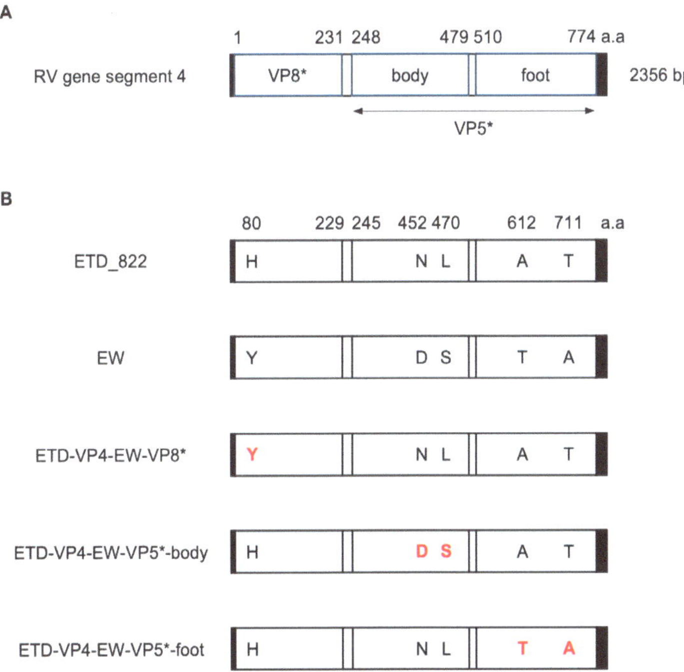

Figure 1. Schematic presentation of the murine RV VP4 gene. (**A**) Schematic presentation of RV gene segment 4. The 5′ and 3′ UTRs are shown as black boxes. VP8* and the body and foot regions of the VP5* domain in the VP4 gene are shown in light blue boxes. The numbers above the box indicate the amino acid positions. (**B**) Schematic presentation of the murine RV ETD_822 and EW strains and the VP4 chimeric viruses generated in this study. The five amino acids that differ between the ETD_822 and EW strains are highlighted in red inside the blue boxes. The number above the box indicates the amino acid positions.

2. Materials and Methods

2.1. Cells and Viruses

Monkey kidney MA104 cells (ATCC CRL-2378.1) were grown in Medium 199 (Gibco) supplemented with 10% fetal bovine serum (FBS), 2 mM L-glutamine, 100 I.U./mL penicillin, and 100 μg/mL streptomycin. Baby hamster kidney cells constitutively expressing T7 RNA polymerase (BHK-T7 cells) were kindly gifted by Dr. Buchholz at the NIH [32] and cultured in Dulbecco-modified essential medium (DMEM) (CORNING) supplemented with 10% FBS, L-glutamine (2 mM), penicillin (100 I.U./mL), and streptomycin (100 μg/mL). The cells were cultured in the presence of Geneticin (1mg/mL) every other passage to maintain the clone expressing T7 RNA polymerase. The natural reassortant D6/2 was generated previously [5] and was propagated in MA104 cells. Briefly, the virus was activated by 5 μg/mL of trypsin (Sigma-Aldrich, St. Louis, MO, USA) at 37 °C for 15 min and propagated in MA104 cells in serum-free Medium 199 (SFM) with 0.5 μg/mL of trypsin.

2.2. Plasmid Construction

To construct rescue plasmids for D6/2 gene segments 1 (VP1) and 10 (NSP4), viral dsRNAs were extracted from freeze-thawed stock of D6/2 with Trizol Reagent (Thermo Fisher Scientific, Waltham, MA, USA). Viral gene segments were determined as previously described [33]. Briefly, a self-anchoring primer was ligated to the 3′ termini of viral dsRNAs with T4 RNA Ligase (New England Biolabs, Ipswich, MA, USA), and then viral cDNAs were synthesized with High-Capacity cDNA Reverse Transcription Kit (Applied Biosystems, Waltham, MA, USA). Gene segments 1 and 10 were amplified by PrimerSTAR HS DNA Polymerase (Takara Bio, San Jose, CA, USA) and used to replace the cDNA of SA11 NSP4 in the pT7-SA11-NSP4 plasmid by the NEBuilder HiFi DNA Assembly Master Mix (New England Biolabs, Ipswich, MA, USA). Nine rescue plasmids for D6/2 (pT7-D6/2-VP2, -VP3, -VP4, -VP6, -VP7, -NSP1, -NSP2, -NSP3, and -NSP5) and the rescue plasmid encoding ETD_822-VP4 (pT7-ETD_822-VP4) were previously constructed [14,21]. To generate the rescue plasmid for EW VP4 (pT7-EW-VP4), the cDNA (GenBank accession number: U08429) was synthesized and cloned between the T7 promoter and HDV ribozyme sequences in the pT7-SA11-NSP4 plasmid by replacing the cDNA of SA11-NSP4 with that of EW-VP4. To generate VP4 chimeric plasmids between the ETD_822 and EW strains (pT7-ETD-VP4-EW-VP8*, pT7-ETD-VP4-EW-VP5*-body, and pT7-ETD-VP4-EW-VP5*-foot), sequences for nucleotides 1 to 1000, 1001 to 1500, and 1500 to 2331 in pT7-ETD_822-VP4 were replaced with those for EW VP4.

2.3. Reverse Genetics

Recombinant viruses were generated using an optimized reverse genetics protocol, as reported previously [21]. Briefly, we mixed 11 rescue plasmids and one helper plasmid (0.4 μg for nine of the rescue plasmids (excluding for NSP2 and NSP5), 1.2 μg of the two rescue plasmids for NSP2 and NSP5, and 0.8 μg of C3P3-G1) in OPTI-MEM I Reduced-Serum Medium (Thermo Fisher Scientific, Waltham, MA, USA). The mixture of the plasmids was transfected into BHK-T7 cells using TransIT-LT1 (Mirus, Madison, WI, USA). The next day, the medium was replaced with serum-free DMEM and cultured overnight. Then, MA104 cells were added to the BHK-T7 cells and cultured in the presence of 0.5 μg/mL of trypsin. To generate the VP4 chimeric viruses, we replaced the rescue plasmid for VP4 with the appropriate plasmids. Rescued viruses were amplified in MA104 cells, and the VP4 sequence of the viruses was confirmed by DNA sequencing before use.

2.4. Focus-Forming Unit Assays

MA104 cells were seeded on 96-well plates and cultured for 2 to 3 days. Virus samples were activated with 5 µg of trypsin, serially diluted with SFM, and inoculated into MA104 cells. The cells were fixed with 10% formalin (Fisherbrand Waltham, MA, USA) 14 h after inoculation, permeabilized with PBS with 0.05% Triton X-100, and stained with rabbit anti-RV DLP and HRP-conjugated anti-rabbit IgG polyclonal antibody (Sigma-Aldrich, St. Louis, MO, USA). RV antigen was visualized with the AEC Substrate Kit and peroxidase (Vector Laboratories, Newark, CA, USA). The number of foci was counted under a microscope, and the virus titer was expressed as FFU/mL.

2.5. Mouse Infection

129sv mice were purchased from Taconic Biosciences Inc. and maintained at the animal facility in the Veterinary Medical Unit of the Palo Alto VA Health Care System. Five-day-old 129sv pups were orally inoculated by gastric lavage with 1×10^3 FFU of recombinant viruses or 1×10^3 DD_{50} of the wild-type EW strain. Mice were monitored to collect stool samples by gentle abdominal pressure for 12 days. Stool samples were collected in 40 µL of PBS (+) (CORNING) and stored at -80 °C until use. The animal experiment protocol was approved by the Stanford Institutional Animal Care Committee.

2.6. ELISA

The relative quantity of RV fecal shedding was assessed by sandwich ELISA, as previously described, using guinea pig anti-RV TLP antiserum and rabbit anti-RV DLP antiserum generated in the Greenberg lab [34]. Briefly, ELISA plates (E&K Scientific Products, Swedesboro, NJ, USA, cat. #EK-25061) were coated with guinea pig anti-RV TLP antiserum and blocked with PBS supplemented with 2% BSA. After washing the plate with PBS containing 0.05% Tween 20, 70 µL of PBS containing 2% BSA and 2 µL of the fecal samples were added to the plate and incubated at 4 °C overnight. The RV antigen in the stool samples was detected by rabbit anti-RV DLP antiserum, HRP conjugated anti-rabbit IgG (Sigma-Aldrich, St. Louis, MO, USA, cat. #A0545), and peroxidase substrate (SeraCare, Milford, MA, USA). The signal intensity at 450nm was measured with the ELx800 microplate reader (BIO-TEK, Shoreline, WA, USA).

2.7. Statistical Analysis

Fecal shedding curves by RVs were analyzed by two-way ANOVAs with the Tukey multiple comparison test using GraphPad Prism 8.

3. Results

3.1. Generation of Recombinant Murine RVs

In a previous study, we synthesized all 11 rescue plasmids from the D6/2 strain. However, we were unable to rescue a completely recombinant D6/2 strain after multiple trials [21]. Therefore, we used recombinant D6/2 with gene segments 1 and 10, which encode VP1 and NSP4, from the simian SA11 strain as an alternative approach [21]. The amino acid sequence identity of VP1 and NSP4 between the murine EW and the simian SA11 strains showed 86.2% and 62.3% homology, respectively. To generate a recombinant virus with a gene constellation closer to a fully murine RV, we reconstructed rescue plasmids for gene segments 1 and 10 from the D6/2 strain. To our surprise, we obtained recombinant D6/2 (rD6/2) with the new rescue plasmids, despite there being no difference in the cDNA sequences of gene segments 1 and 10 compared with those in the previous failed rescue plasmids.

We next attempted to further optimize gene segment 4 in rD6/2, which is derived from the simian RRV strain. Wild-type murine RV strains (including the EW strain) propagated in mouse intestines do not efficiently infect immortalized cell lines. Previous studies that compared the nucleotide sequences of murine RV before and after adaptation to cultured cells reported that gene segment 4 is one of the determinants for effective viral replication in cultured cell lines [17]. It suggests that murine RV from mouse intestines poorly replicates in the cell line possibly due to a partial restriction at the attachment and entry process. Therefore, we used the rescue plasmid for gene segment 4 of the cell-culture-adapted murine ETD_822 strain, and we generated a recombinant virus with 10 gene segments from murine EW and gene segment 4 from the ETD_822 strain (rEW/ETD-VP4) (Table 1).

We compared the nucleotide sequence of VP4 between the EW and ETD_822 strains to better understand the difference in VP4 in the murine RV strain used in this study. Sequence alignment shows that compared with the wild-type EW strain, ETD_822 has only five non-synonymous amino acid substitutions (Y80H, D452N, S470L, T612A, and A711T) in VP4. Y80H is the only amino acid difference found in the VP8* domain, and the VP5* domain has two amino acid differences in either the body (D452N and S470L) or the C-terminal foot (T612A and A711T) regions (Figure 1B). It is known that the cell-culture-adapted EDIM murine RV strains are attenuated in suckling mice in terms of diarrheal dose and duration of shedding while having acquired the ability to replicate in cultured cell lines [6,15–17]. We speculated that some amino acids are strongly associated with the adaptation to cultured cell lines, but not all amino acids are necessary for efficient replication in cell lines. To test whether we could rescue a recombinant RV with a VP4 protein closer to the more virulent, non-cell-culture-adapted progenitor EW strain, we constructed three VP4 chimeric plasmids between the ETD_822 and EW strains. These plasmids harbor nucleotide sequences of the VP8*, VP5*-body, or VP5*-foot domains from the EW strain in ETD_822 VP4 (Figure 1B). Of note, we successfully rescued all three chimeric viruses, namely rEW/ETD-VP4-EW-VP8*, rEW/ETD-VP4-EW-VP5*-body, and rEW/ETD-VP4-EW-VP5*-foot. These data suggest that amino acid differences in these three regions in ETD_822 are not individually involved in the adaptation to cultured cell lines.

3.2. Comparative Analysis of Diarrhea Rate by Recombinant Murine RVs in a Suckling Mouse Model

To assess the capacity of the rescued viruses to induce diarrhea, we inoculated 5-day-old 129sv pups with 1×10^3 FFU of the recombinant murine RVs and 1×10^3 DD$_{50}$ of the highly virulent non-cell-culture-adapted murine RV EW strain as a control. We monitored the mice for 12 days to compare the percentage and duration of diarrhea occurrence. The wild-type, non-cell-culture-adapted murine EW strain caused 100% diarrhea in all inoculated pups from 2 to 9 days post-inoculation (Figure 2A). Compared with EW, rD6/2 was slightly attenuated and did not cause diarrhea in all the pups (Figure 2B), consistent with the previous literature [10]. The new rEW/ETD-VP4 virus had a similar disease phenotype in that it caused diarrhea but not in all pups over time (Figure 2C), suggesting that ETD VP4 is not more virulent than RRV VP4. The three VP4 chimeric viruses (rEW/ETD-VP4-EW-VP8*, rEW/ETD-VP4-EW-VP5*-body, and rEW/ETD-VP4-EW-VP5*-foot) caused diarrhea in inoculated pups but did not show a diarrhea phenotype as robust as that by murine EW (Figure 2D–F). The data suggest that VP8* or the body or foot regions of VP5* from the EW strain did not individually increase the diarrheal rates compared with the parental virus with ETD-VP4 (rEW/ETD-VP4) in the suckling mouse model.

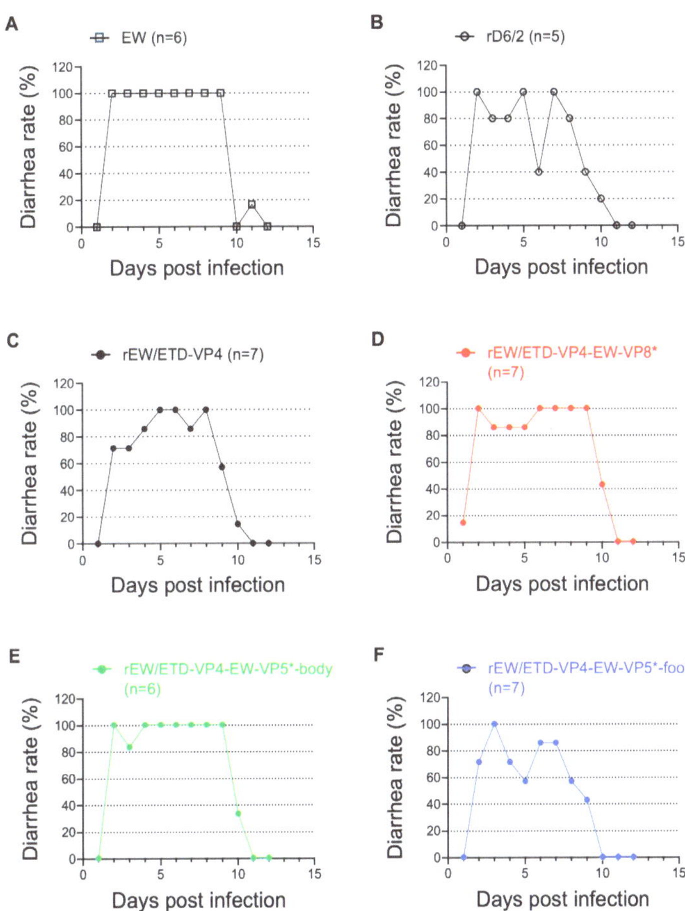

Figure 2. Percentage of diarrhea caused by the wild-type murine RV and recombinant murine RVs. Five-day-old 129sv pups were inoculated with (**A**) 1×10^3 DD$_{50}$ of EW, or 1×10^3 FFU of (**B**) rD6/2, (**C**) rEW/ETD-VP4, (**D**) rEW/ETD-VP4-EW-VP8*, (**E**) rEW/ETD-VP4-ETD-VP5*-body, or (**F**) rEW/ETD-VP4-ETD-VP5*-foot. The infected mice were monitored for diarrheal stool for 12 days by gentle abdominal pressure.

3.3. Comparative Analysis of Fecal RV Shedding by Recombinant Murine RVs in a Suckling Mouse Model

Next, we compared the amount of fecal RV shedding among the various VP4 constructs. Wild-type murine RV caused a curve with a single peak of more than 2.0 at OD$_{450}$ at 4 days post-inoculation, demonstrating robust replication in the mouse intestine (Figure 3A). In contrast, the fecal RV shedding curve from the rD6/2-inoculated pups showed two peaks on days 2 and 6 post-inoculation, and the OD values did not reach as high as those of the EW strain (Figure 3B). The other four viruses that had the murine RV VP4 gene demonstrated three peaks on day 2, from day 5 to day 7, and from day 10 to 11 days post-inoculation, and none of these viruses reached the high levels of fecal shedding as seen with the wild-type EW strain (Figure 3C–F). Statistical analysis of the fecal RV shedding between the recombinant viruses and the EW strain confirmed that none of the recombinant viruses were shed to the same level as that by the wild-type murine EW strain (Table 2). We also found that, compared with rEW/ETD-VP4, two of the three VP4 chimeras, i.e., rEW/ETD-VP4-EW-VP5*-body and rEW/ETD-VP4-EW-VP5*-foot, caused more fecal RV shedding, whereas rEW/ETD-VP4-EW-VP8* did not (Table 2). These results suggest that, among the five different amino acids in VP4 between the EW and ETD_822 strains, amino acids in the VP5* region are positively associated with efficient replication in the mouse intestine.

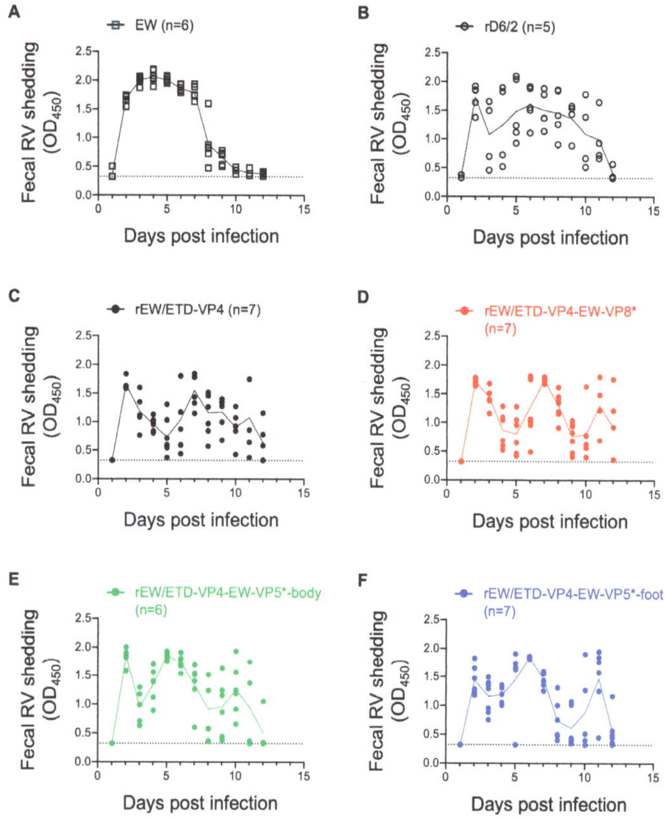

Figure 3. Fecal RV shedding by wild-type murine RV and recombinant murine RVs. Five-day-old 129sv pups were inoculated with the same doses and viruses as in Figure 2. (**A**) 1×10^3 DD$_{50}$ of EW, or 1×10^3 FFU of (**B**) rD6/2, (**C**) rEW/ETD-VP4, (**D**) rEW/ETD-VP4-EW-VP8*, (**E**) rEW/ETD-VP4-ETD-VP5*-body, or (**F**) rEW/ETD-VP4-ETD-VP5*-foot. The amount of RV in the stool samples was determined by ELISA. Each dot shows data from one pup and the line shows the average score. The dotted lines indicate the score of the limit of detection determined from the stool of uninfected pups.

Table 2. Summary of the statistical analysis of fecal RV shedding [1,2].

Virus	Fecal RV Shedding (Compared with EW)	Fecal RV Shedding (Compared with rEW/ETD-VP4)
EW	n.a.	***
rEW/ETD-VP4	***	n.a.
rEW/ETD-VP4-EW-VP8*	**	n.s.
rEW/ETD-VP4-EW-VP5*-body	**	***
rEW/ETD-VP4-EW-VP5*-foot	**	*

[1] Fecal RV shedding curves were compared with either EW or rEW/ETD-VP4 by two-way ANOVA with the Tukey multiple comparison test. [2] Statistical significance is indicated as n.s.: not significant; * $p < 0.05$; ** $p < 0.01$; *** $p < 0.001$; n.a.: not applicable.

4. Discussion

In this study, we leveraged an optimized reverse genetics system to improve the virulence of the murine RV rD6/2-2g strain by exchanging the remaining three gene segments from heterologous simian SA11 or RRV strains with its homologous murine counterparts and rescued a recombinant RV with 11 gene segments all derived from a murine RV strain (rEW/ETD-VP4). We previously attempted to rescue rD6/2 with rescue plasmids of gene segments 1 and 10 constructed by DNA synthesis. After constructing the plasmids, we performed reverse genetics with different clones and repeated this multiple times; however, none of the rescue experiments were successful. In the current study, we constructed the plasmids again by cloning the gene segments from the original D6/2 stock. Of note, the new plasmid sequences of the T7 promoter, RV cDNA, hepatitis delta virus ribozyme, and T7 terminator, although identical to the original plasmids, led to the successful rescue of rD6/2. It suggests that clonal differences might affect the rescue efficiency in reverse genetics. It is uncertain whether there is a difference in some other parts of the plasmid, and, if that is the case, whether this affects the reverse genetics results. Whole-plasmid sequencing of the plasmids would be helpful to examine whether there is any difference between the clones. It would be important to test multiple clones prepared separately when some rescue plasmids do not work, even if the plasmid has the correct sequence.

We replaced three gene segments, which encode the RNA-dependent RNA polymerase VP1 (encoded by gene segment 1), the cell attachment protein VP4 (encoded by gene segment 4), and the viral enterotoxin NSP4 (encoded by gene segment 10). Among these gene segments, gene segment 4 has been implicated in RV HRR; however, the contribution of gene segments 1 and 10 to HRR is less clear. VP1 interaction with VP2 is critical for transcription and genome replication [35]. Group A RVs have 28 VP1 genotypes and 24 VP2 genotypes (Rotavirus Classification Working Group: RCWG updated on April 3rd 2023 (https://rega.kuleuven.be/cev/viralmetagenomics/virus-classification/rcwg)) [36,37], and it is reported that the combination of VP1 and VP2 genotypes changes the VP1 polymerase activity in some cases [38]. RV NSP4 is an enterotoxin that increases host calcium levels in the cytoplasm and activates calcium-ion-dependent chloride channels, and it is directly involved in causing diarrhea [39]. In light of the sequence differences between the EW and SA11 strains, we preferred using gene segments 1 and 10 originating from murine RV to specifically focus on studying viral replication, virulence, and spread of murine RV in a mouse model.

In our previous study, we compared the role of VP8* and VP5* from heterologous RV strains in virus replication and diarrhea in a suckling mouse model. We generated VP8* and VP5* chimeric viruses between homologous ETD and heterologous bovine UK strains on an rD6/2-2g background [14]. The results showed that, in the case of comparison between homologous and heterologous VP4s, both VP8* and VP5* from ETD contributed to increased diarrhea in the suckling mouse model [14]. In the present study, we evaluated the role of VP8* and VP5* from murine RV strains in a murine RV backbone. This is important because we are now testing VP4 in a genetic backbone identical to the homologous murine RV backbone, as opposed to the murine-like condition used in the previous study. Despite the different genetic background, we came to the same conclusion that ETD VP4 is not more virulent than RRV VP4, suggesting that when ETD VP4 is not available, RRV VP4 can serve as a robust surrogate for in vivo studies. To delineate the contributions of VP8* versus VP5*, we generated VP4 chimeric viruses between a non-tissue-culture-adapted EW strain and a tissue-culture-adapted ETD_822 strain and compared the role of VP8* and VP5* in a homologous murine RV strain. Of interest, VP4 chimeric viruses with VP5* body or foot regions, but not VP8*, slightly increased the amount of RV shedding in the feces compared with a control virus with ETD-VP4 (Figure 3C,E,F and Table 2). It is possible that the rEW/ETD-VP4-EW-VP5*-body and the rEW/ETD-VP4-EW-VP5*-foot replicate better than rEW/ETD-VP4 in MA104 cells. Previous studies on host factors involved with RV entry demonstrated that the VP8* domain of VP4 attaches to cell-surface glycans (e.g., sialic acid and histo-blood group antigens), while the VP5* domain interacts with other coreceptors (e.g., integrins and heat-shock cognate protein 70). Subsequently, VP5* likely plays a role in membrane penetration at a post-attachment step [31]. Our current results suggest that the difference in VP4 between non-tissue-culture-adapted EW and cell-culture-adapted ETD_822 occurs after the initial virion attachment step with cell-surface glycans. Of note, none of the recombinant viruses caused the same severe diarrheal diseases as EDIM-EW did (Figure 2). These data suggest that multiple mutations in VP4 or other viral proteins are required for robust replication in the mouse intestine.

One can imagine that there are multiple avenues available to leverage this powerful murine RV system to identify and study the molecular factors that modulate the severity of diarrhea and viral replication. For example, it would be interesting to further passage these recombinant viruses in mouse intestines, determine the nucleotide differences by next-generation sequencing, and introduce the mutations into the rescue plasmids to pinpoint the precise amino acids important for more robust replication in the mouse intestine without losing the ability of the virus to replicate in cultured cells. It would also be of interest to test these viruses in an adult mouse model to see if different results are obtained to those in the neonatal mouse system. Finally, although human enteroid cultures have proven a great tool for modeling primary human intestinal epithelial cells and for studying RV infection [40–42], such a system is lacking for the murine enteroids, which would be useful for teasing apart the stage of entry affected by VP8* and/or VP5* mutations. In conclusion, we have developed a reverse genetics for murine RV. This system will provide a useful tool for understanding the biology of RV in mouse models.

Author Contributions: T.K., L.S.-T., H.B.G. and S.D. designed the research; T.K., L.S.-T. and N.F. performed the research; T.K. analyzed the data; and T.K., S.D. and H.B.G. wrote the paper. All authors have read and agreed to the published version of the manuscript.

Funding: This work is supported by the Stanford Maternal and Child Health Research Institute Postdoctoral Support FY2020 (T.K.), the R01 AI125249 and VA Merit Grant (GRH0022) (H.B.G.), and the U19 AI116484 and R01 AI150796 grants (S.D.).

Institutional Review Board Statement: The animal study protocol was approved by the Institutional Review Board of the Stanford Institutional Animal Care Committee (GRH1862, 5 October 2022).

Informed Consent Statement: Not applicable.

Data Availability Statement: Data related to this paper may be requested from the authors.

Acknowledgments: We thank the members of the Greenberg lab and the Ding lab for helpful discussions, Linda Jacob for her secretarial work, and Taufeeq Ahmed for his technical assistance with mouse colony management.

Conflicts of Interest: The authors declare no conflicts of interest in this project.

References

1. Crawford, S.E.; Ding, S.; Greenberg, H.B.; Estes, M.K. Rotaviruses. In *Fields Virology*, 7th ed.; Lippincott Williams & Wilkins: Philadelphia, PA, USA, 2022; Volume 3, pp. 362–413.
2. Greenberg, H.B.; Vo, P.T.; Jones, R. Cultivation and characterization of three strains of murine rotavirus. *J. Virol.* **1986**, *57*, 585–590. [CrossRef] [PubMed]
3. Ramig, R.F. The effects of host age, virus dose, and virus strain on heterologous rotavirus infection of suckling mice. *Microb. Pathog.* **1988**, *4*, 189–202. [CrossRef] [PubMed]
4. Conner, M.E.; Estes, M.K.; Graham, D.Y. Rabbit model of rotavirus infection. *J. Virol.* **1988**, *62*, 1625–1633. [CrossRef] [PubMed]
5. Broome, R.L.; Vo, P.T.; Ward, R.L.; Clark, H.F.; Greenberg, H.B. Murine rotavirus genes encoding outer capsid proteins VP4 and VP7 are not major determinants of host range restriction and virulence. *J. Virol.* **1993**, *67*, 2448–2455. [CrossRef] [PubMed]
6. Feng, N.; Burns, J.W.; Bracy, L.; Greenberg, H.B. Comparison of mucosal and systemic humoral immune responses and subsequent protection in mice orally inoculated with a homologous or a heterologous rotavirus. *J. Virol.* **1994**, *68*, 7766–7773. [CrossRef] [PubMed]
7. Hoshino, Y.; Saif, L.J.; Kang, S.Y.; Sereno, M.M.; Chen, W.K.; Kapikian, A.Z. Identification of group A rotavirus genes associated with virulence of a porcine rotavirus and host range restriction of a human rotavirus in the gnotobiotic piglet model. *Virology* **1995**, *209*, 274–280. [CrossRef]

8. Bridger, J.C.; Dhaliwal, W.; Adamson, M.J.; Howard, C.R. Determinants of rotavirus host range restriction--a heterologous bovine NSP1 gene does not affect replication kinetics in the pig. *Virology* **1998**, *245*, 47–52. [CrossRef] [PubMed]

9. Ciarlet, M.; Estes, M.K.; Barone, C.; Ramig, R.F.; Conner, M.E. Analysis of host range restriction determinants in the rabbit model: Comparison of homologous and heterologous rotavirus infections. *J. Virol.* **1998**, *72*, 2341–2351. [CrossRef]

10. Feng, N.; Yasukawa, L.L.; Sen, A.; Greenberg, H.B. Permissive replication of homologous murine rotavirus in the mouse intestine is primarily regulated by VP4 and NSP1. *J. Virol.* **2013**, *87*, 8307–8316. [CrossRef]

11. Matthijnssens, J.; Van Ranst, M. Genotype constellation and evolution of group A rotaviruses infecting humans. *Curr. Opin. Virol.* **2012**, *2*, 426–433. [CrossRef]

12. Glass, R.I.; Tate, J.E.; Jiang, B.; Parashar, U. The Rotavirus Vaccine Story: From Discovery to the Eventual Control of Rotavirus Disease. *J. Infect. Dis.* **2021**, *224*, S331–S342. [CrossRef]

13. Offit, P.A.; Blavat, G.; Greenberg, H.B.; Clark, H.F. Molecular basis of rotavirus virulence: Role of gene segment 4. *J. Virol.* **1986**, *57*, 46–49. [CrossRef] [PubMed]

14. Sánchez-Tacuba, L.; Kawagishi, T.; Feng, N.; Jiang, B.; Ding, S.; Greenberg, H.B. The Role of the VP4 Attachment Protein in Rotavirus Host Range Restriction in an In Vivo Suckling Mouse Model. *J. Virol.* **2022**, *96*, e0055022. [CrossRef] [PubMed]

15. Ward, R.L.; McNeal, M.M.; Sheridan, J.F. Development of an adult mouse model for studies on protection against rotavirus. *J. Virol.* **1990**, *64*, 5070–5075. [CrossRef]

16. Burns, J.W.; Krishnaney, A.A.; Vo, P.T.; Rouse, R.V.; Anderson, L.J.; Greenberg, H.B. Analyses of homologous rotavirus infection in the mouse model. *Virology* **1995**, *207*, 143–153. [CrossRef] [PubMed]

17. Tsugawa, T.; Tatsumi, M.; Tsutsumi, H. Virulence-associated genome mutations of murine rotavirus identified by alternating serial passages in mice and cell cultures. *J. Virol.* **2014**, *88*, 5543–5558. [CrossRef]

18. Kanai, Y.; Komoto, S.; Kawagishi, T.; Nouda, R.; Nagasawa, N.; Onishi, M.; Matsuura, Y.; Taniguchi, K.; Kobayashi, T. Entirely plasmid-based reverse genetics system for rotaviruses. *Proc. Natl. Acad. Sci. USA* **2017**, *114*, 2349–2354. [CrossRef] [PubMed]

19. Komoto, S.; Fukuda, S.; Kugita, M.; Hatazawa, R.; Koyama, C.; Katayama, K.; Murata, T.; Taniguchi, K. Generation of Infectious Recombinant Human Rotaviruses from Just 11 Cloned cDNAs Encoding the Rotavirus Genome. *J. Virol.* **2019**, *93*, e02207-18. [CrossRef] [PubMed]

20. Kawagishi, T.; Nurdin, J.A.; Onishi, M.; Nouda, R.; Kanai, Y.; Tajima, T.; Ushijima, H.; Kobayashi, T. Reverse Genetics System for a Human Group A Rotavirus. *J. Virol.* **2020**, *94*, e00963-19. [CrossRef]

21. Sánchez-Tacuba, L.; Feng, N.; Meade, N.J.; Mellits, K.H.; Jaïs, P.H.; Yasukawa, L.L.; Resch, T.K.; Jiang, B.; López, S.; Ding, S.; et al. An Optimized Reverse Genetics System Suitable for Efficient Recovery of Simian, Human, and Murine-Like Rotaviruses. *J. Virol.* **2020**, *94*, e01294-20. [CrossRef]

22. Kanda, M.; Fukuda, S.; Hamada, N.; Nishiyama, S.; Masatani, T.; Fujii, Y.; Izumi, F.; Okajima, M.; Taniguchi, K.; Sugiyama, M.; et al. Establishment of a reverse genetics system for avian rotavirus A strain PO-13. *J. Gen. Virol.* **2022**, *103*, 001760. [CrossRef] [PubMed]

23. Diebold, O.; Gonzalez, V.; Venditti, L.; Sharp, C.; Blake, R.A.; Tan, W.S.; Stevens, J.; Caddy, S.; Digard, P.; Borodavka, A.; et al. Using Species a Rotavirus Reverse Genetics to Engineer Chimeric Viruses Expressing SARS-CoV-2 Spike Epitopes. *J. Virol.* **2022**, *96*, e0048822. [CrossRef] [PubMed]

24. Hamajima, R.; Lusiany, T.; Minami, S.; Nouda, R.; Nurdin, J.A.; Yamasaki, M.; Kobayashi, N.; Kanai, Y.; Kobayashi, T. A reverse genetics system for human rotavirus G2P[4]. *J. Gen. Virol.* **2022**, *103*, 001816. [CrossRef]

25. Philip, A.A.; Agbemabiese, C.A.; Yi, G.; Patton, J.T. T7 expression plasmids for producing a recombinant human G1P[8] rotavirus comprising RIX4414 sequences of the RV1 (Rotarix, GSK) vaccine strain. *Microbiol. Resour. Announc.* **2023**, *12*, e0060323. [CrossRef] [PubMed]

26. Snyder, A.J.; Agbemabiese, C.A.; Patton, J.T. Production of OSU G5P[7] Porcine Rotavirus Expressing a Fluorescent Reporter via Reverse Genetics. *Viruses* **2024**, *16*, 411. [CrossRef] [PubMed]

27. Hou, G.; Zeng, Q.; Matthijnssens, J.; Greenberg, H.B.; Ding, S. Rotavirus NSP1 Contributes to Intestinal Viral Replication, Pathogenesis, and Transmission. *mBio* **2021**, *12*, e0320821. [CrossRef] [PubMed]

28. Zhu, Y.; Sánchez-Tacuba, L.; Hou, G.; Kawagishi, T.; Feng, N.; Greenberg, H.B.; Ding, S. A recombinant murine-like rotavirus with Nano-Luciferase expression reveals tissue tropism, replication dynamics, and virus transmission. *Front. Immunol.* **2022**, *13*, 911024. [CrossRef] [PubMed]

29. Dormitzer, P.R.; Nason, E.B.; Prasad, B.V.; Harrison, S.C. Structural rearrangements in the membrane penetration protein of a non-enveloped virus. *Nature* **2004**, *430*, 1053–1058. [CrossRef] [PubMed]

30. Settembre, E.C.; Chen, J.Z.; Dormitzer, P.R.; Grigorieff, N.; Harrison, S.C. Atomic model of an infectious rotavirus particle. *EMBO J.* **2011**, *30*, 408–416. [CrossRef]

31. Herrmann, T.; Torres, R.; Salgado, E.N.; Berciu, C.; Stoddard, D.; Nicastro, D.; Jenni, S.; Harrison, S.C. Functional refolding of the penetration protein on a non-enveloped virus. *Nature* **2021**, *590*, 666–670. [CrossRef]

32. Buchholz, U.J.; Finke, S.; Conzelmann, K.K. Generation of bovine respiratory syncytial virus (BRSV) from cDNA: BRSV NS2 is not essential for virus replication in tissue culture, and the human RSV leader region acts as a functional BRSV genome promoter. *J. Virol.* **1999**, *73*, 251–259. [CrossRef] [PubMed]

33. Maan, S.; Rao, S.; Maan, N.S.; Anthony, S.J.; Attoui, H.; Samuel, A.R.; Mertens, P.P. Rapid cDNA synthesis and sequencing techniques for the genetic study of bluetongue and other dsRNA viruses. *J. Virol. Methods* **2007**, *143*, 132–139. [CrossRef] [PubMed]

34. Kawagishi, T.; Sánchez-Tacuba, L.; Feng, N.; Costantini, V.P.; Tan, M.; Jiang, X.; Green, K.Y.; Vinjé, J.; Ding, S.; Greenberg, H.B. Mucosal and systemic neutralizing antibodies to norovirus induced in infant mice orally inoculated with recombinant rotaviruses. *Proc. Natl. Acad. Sci. USA* **2023**, *120*, e2214421120. [CrossRef] [PubMed]

35. Jenni, S.; Salgado, E.N.; Herrmann, T.; Li, Z.; Grant, T.; Grigorieff, N.; Trapani, S.; Estrozi, L.F.; Harrison, S.C. In situ Structure of Rotavirus VP1 RNA-Dependent RNA Polymerase. *J. Mol. Biol.* **2019**, *431*, 3124–3138. [CrossRef]

36. Matthijnssens, J.; Ciarlet, M.; Heiman, E.; Arijs, I.; Delbeke, T.; McDonald, S.M.; Palombo, E.A.; Iturriza-Gómara, M.; Maes, P.; Patton, J.T.; et al. Full genome-based classification of rotaviruses reveals a common origin between human Wa-Like and porcine rotavirus strains and human DS-1-like and bovine rotavirus strains. *J. Virol.* **2008**, *82*, 3204–3219. [CrossRef] [PubMed]

37. Matthijnssens, J.; Ciarlet, M.; Rahman, M.; Attoui, H.; Bányai, K.; Estes, M.K.; Gentsch, J.R.; Iturriza-Gómara, M.; Kirkwood, C.D.; Martella, V.; et al. Recommendations for the classification of group A rotaviruses using all 11 genomic RNA segments. *Arch. Virol.* **2008**, *153*, 1621–1629. [CrossRef] [PubMed]

38. Steger, C.L.; Boudreaux, C.E.; LaConte, L.E.; Pease, J.B.; McDonald, S.M. Group A Rotavirus VP1 Polymerase and VP2 Core Shell Proteins: Intergenotypic Sequence Variation and In Vitro Functional Compatibility. *J. Virol.* **2019**, *93*, e01642-18. [CrossRef] [PubMed]

39. Crawford, S.E.; Ramani, S.; Tate, J.E.; Parashar, U.D.; Svensson, L.; Hagbom, M.; Franco, M.A.; Greenberg, H.B.; O'Ryan, M.; Kang, G.; et al. Rotavirus infection. *Nat. Rev. Dis. Primers* **2017**, *3*, 17083. [CrossRef] [PubMed]

40. Saxena, K.; Blutt, S.E.; Ettayebi, K.; Zeng, X.L.; Broughman, J.R.; Crawford, S.E.; Karandikar, U.C.; Sastri, N.P.; Conner, M.E.; Opekun, A.R.; et al. Human Intestinal Enteroids: A New Model To Study Human Rotavirus Infection, Host Restriction, and Pathophysiology. *J. Virol.* **2016**, *90*, 43–56. [CrossRef]

41. Nolan, L.S.; Baldridge, M.T. Advances in understanding interferon-mediated immune responses to enteric viruses in intestinal organoids. *Front. Immunol.* **2022**, *13*, 943334. [CrossRef]

42. Adeniyi-Ipadeola, G.; Nwanosike, H.; Ramani, S. Human intestinal organoids as models to study enteric bacteria and viruses. *Curr. Opin. Microbiol.* **2023**, *75*, 102362. [CrossRef] [PubMed]

Article

Generation of Recombinant Authentic Live Attenuated Human Rotavirus Vaccine Strain RIX4414 (Rotarix®) from Cloned cDNAs Using Reverse Genetics

Saori Fukuda [1], Masanori Kugita [2], Kanako Kumamoto [2], Yuki Akari [1,3], Yuki Higashimoto [4,5], Shizuko Nagao [2], Takayuki Murata [1,6], Tetsushi Yoshikawa [5,6], Koki Taniguchi [1] and Satoshi Komoto [1,3,6,*]

[1] Department of Virology, Fujita Health University School of Medicine, Toyoake 470-1192, Aichi, Japan; saorif@fujita-hu.ac.jp (S.F.); m24d9001@oita-u.ac.jp (Y.A.); tmurata@fujita-hu.ac.jp (T.M.); kokitani@fujita-hu.ac.jp (K.T.)

[2] Education and Research Facility of Animal Models for Human Diseases, Fujita Health University, Toyoake 470-1192, Aichi, Japan; m-kugi@fujita-hu.ac.jp (M.K.); kumamoto@fujita-hu.ac.jp (K.K.); shizun@fujita-hu.ac.jp (S.N.)

[3] Division of One Health, Research Center for GLOBAL and LOCAL Infectious Diseases (RCGLID), Oita University, Yufu 879-5593, Oita, Japan

[4] Department of Clinical Microbiology, Fujita Health University School of Medical Sciences, Toyoake 470-1192, Aichi, Japan; yhigashi@fujita-hu.ac.jp

[5] Department of Pediatrics, Fujita Health University School of Medicine, Toyoake 470-1192, Aichi, Japan; tetsushi@fujita-hu.ac.jp

[6] Center for Infectious Disease Research, Research Promotion Headquarters, Fujita Health University, Toyoake 470-1192, Aichi, Japan

* Correspondence: satoshik@oita-u.ac.jp or satoshik@fujita-hu.ac.jp

Citation: Fukuda, S.; Kugita, M.; Kumamoto, K.; Akari, Y.; Higashimoto, Y.; Nagao, S.; Murata, T.; Yoshikawa, T.; Taniguchi, K.; Komoto, S. Generation of Recombinant Authentic Live Attenuated Human Rotavirus Vaccine Strain RIX4414 (Rotarix®) from Cloned cDNAs Using Reverse Genetics. *Viruses* **2024**, *16*, 1198. https://doi.org/10.3390/v16081198

Academic Editors: Ulrich Desselberger and John T. Patton

Received: 7 July 2024
Revised: 23 July 2024
Accepted: 24 July 2024
Published: 25 July 2024

Abstract: The live attenuated human rotavirus vaccine strain RIX4414 (Rotarix®) is used worldwide to prevent severe rotavirus-induced diarrhea in infants. This strain was attenuated through the cell culture passaging of its predecessor, human strain 89-12, which resulted in multiple genomic mutations. However, the specific molecular reasons underlying its attenuation have remained elusive, primarily due to the absence of a suitable reverse genetics system enabling precise genetic manipulations. Therefore, we first completed the sequencing of its genome and then developed a reverse genetics system for the authentic RIX4414 virus. Our experimental results demonstrate that the rescued recombinant RIX4414 virus exhibits biological characteristics similar to those of the parental RIX4414 virus, both in vitro and in vivo. This novel reverse genetics system provides a powerful tool for investigating the molecular basis of RIX4414 attenuation and may facilitate the rational design of safer and more effective human rotavirus vaccines.

Keywords: human rotavirus; live attenuated rotavirus vaccine; Rotarix®; reverse genetics

1. Introduction

Group A rotavirus (RVA), a member of *Sedoreoviridae*, is a primary cause of severe gastroenteritis in young children worldwide, accounting for approximately 128,500–215,000 deaths annually in children aged under 5 years of age [1,2]. The virion contains an 11-segment double-stranded RNA (dsRNA) genome encoding six structural proteins (VP1-VP4, VP6, and VP7) and six non-structural proteins (NSP1-NSP6) [3]. Each dsRNA segment possesses one or two protein-coding sequences flanked by untranslated regions (UTRs).

Currently, there are no specific antiviral treatments for RVA gastroenteritis, and prevention through vaccination remains the most effective approach. Live oral human RVA (HuRVA) vaccines have been developed through the serial passaging of clinical HuRVAs in cell cultures and were found safe in neonates and infants. Rotarix® (GlaxoSmithKline), one of the most widely used live attenuated vaccines globally, is licensed in >100 countries and universally recommended for all infants in many of these [4–6]. This vaccine is based on the

RIX4414 (G1P[8]) strain, which was derived from the HuRVA strain 89-12 (G1P[8]), isolated from a child with diarrhea in the United States during the 1988–1989 rotavirus season. To develop RIX4414, strain 89-12 was passaged 33 times in primary African green monkey cells and further passaged in monkey Vero cells to acquire attenuating mutations in its genome [7]. Infants orally administered the serially passaged strain 89-12 (strain RIX4414) did not develop diarrhea, confirming its safety. However, the genetic basis for the attenuation of vaccine strain RIX4414 remains unknown. Elucidating these mechanisms at the molecular level will enhance our knowledge of RVA pathology and may contribute to the development of new vaccines.

The advances in reverse genetics have allowed RVA genomes to be artificially manipulated, significantly enhancing our ability to research these viruses. However, HuRVA strains typically do not grow as robustly in cell culture (~10^6 plaque-forming units (PFU)/mL) as various animal RVA strains, such as simian SA11 virus (~10^8 PFU/mL), making the development of reverse genetics systems more challenging for HuRVAs than for animal RVAs [8–10]. Despite these difficulties, reverse genetics systems have been developed for several HuRVA strains. Strain KU (G1P[8]) [11] was the first HuRVA to be successfully manipulated by reverse genetics to generate an infectious virus [8]. Subsequently, systems for a few other HuRVAs, including CDC-9 (G1P[8]) [9], HN126 (G2P[4]) [12], and Odelia (G4P[8]) [13], have been described. However, none of these HuRVA strains represent licensed, clinically validated attenuated vaccine strains. Given this, a reverse genetics platform based on authentic vaccine strain RIX4414 would be highly valuable, enabling targeted genetic manipulations to unravel the attenuation mechanisms of the live HuRVAs vaccines.

Very recently, the rescue of the recombinant RIX4414-like virus by reverse genetics was reported [14]. This RIX4414-like virus was mostly based on RIX4414 but incorporated segments of the 5′- and 3′-UTRs from the wild-type HuRVA strain Wa (G1P[8]) [15], a modification required because of incomplete sequence data for certain UTRs of strain RIX4414 in the GenBank/EMBL/DDBJ data libraries. Additionally, the RIX4414-like virus VP2 protein included residues shared with several wild-type HuRVAs (Wa (G1P[8]), KU (G1P[8]), and Odelia (G4P[8])) instead of those found in RIX4414 VP2, because the originally attempted rescue T7 plasmid carrying the VP2 gene of RIX4414 did not function in reverse genetics [14]. The UTR regions and VP2 core shell protein are involved in all stages of RVA replication and virion assembly [16–21]. Therefore, a more reliable and complete understanding of the molecular mechanisms of attenuation in vaccine strain RIX4414 necessitates a more authentic reverse genetics system.

In this study, we determined the complete sequences of all 11 dsRNA segments of RIX4414 using next-generation sequencing (NGS) and successfully developed a reverse genetics system that accurately represents the authentic RIX4414 virus. This system provides a valuable genetic platform for elucidating the attenuated pathogenic mechanisms of RIX4414.

2. Materials and Methods

2.1. Cells and Viruses

A baby hamster kidney cell line stably expressing the T7 RNA polymerase (BHK/T7-9) [22] was cultured in Dulbecco's modified Eagle medium (DMEM; Nacalai, Kyoto, Japan) supplemented with 5% fetal calf serum (FCS; Gibco, Tokyo, Japan) (complete medium) in the presence of 600 ng/mL hygromycin (Invitrogen, Tokyo, Japan). Monkey kidney cell lines, MA104 and CV-1, were cultured in complete medium. HuRVA vaccine strain RIX4414 (G1-P[8]-I1-R1-C1-M1-A1-N1-T1-E1-H1) was obtained directly from a vial of Rotarix® (GlaxoSmithKline, Tokyo, Japan). Strain RIX4414 and recombinant simian RVA strain SA11-L2 (G3-P[2]-I2-R2-C5-M5-A5-N5-T5-E2-H5) (rSA11-L2) [23] were propagated as described previously [24]. Briefly, RIX4414 and rSA11-L2 viruses were pretreated with trypsin (type IX, from porcine pancreas; 10 μg/mL) (Sigma-Aldrich, Tokyo, Japan) and then propagated in MA104 cells in Eagle's minimum essential medium (MEM; Nissui, Tokyo, Japan) without FCS (incomplete medium) but containing trypsin (1 μg/mL).

2.2. cDNA Library Construction, Illumina MiSeq Sequencing, and Sequence Analysis of RIX4414 Virus

Construction of a cDNA library and Illumina MiSeq sequencing for RIX4414 virus were conducted as described previously [25,26]. Viral genomic dsRNAs were directly extracted from a suspension of RIX4414 virus in a vial of Rotarix®, without any passaging in our laboratory, using a QIAamp Viral RNA Mini Kit (Qiagen, Tokyo, Japan). A 200-bp fragment cDNA library ligated with bar-coded adapters was prepared using an NEBNext Ultra RNA Library Prep Kit for Illumina v1.2 (New England Biolabs, Tokyo, Japan) according to the manufacturer's instructions. The cDNA library was isolated using Agencourt AMPure XP magnetic beads (Beckman Coulter, Tokyo, Japan). After assessing the quality and quantity of the purified cDNA library, nucleotide sequencing was performed five times on an Illumina MiSeq sequencer (Illumina, Tokyo, Japan) using a MiSeq Reagent Kit v2 (Illumina) to generate 151 paired-end reads. Analysis of the MiSeq sequencing data was performed using a CLC Genomics Workbench v8.0.1 (CLC Bio, Tokyo, Japan). Contigs were assembled from the yielded sequence reads (trimmed) by de novo assembly. Using the assembled contigs as query sequences and the Basic Local Alignment Search Tool (BLAST, https://blast.ncbi.nlm.nih.gov/Blast.cgi) for searching the non-redundant nucleotide database of the National Center for Biotechnology Information (NCBI; https://www.ncbi.nlm.nih.gov/, accessed on 1 September 2018), it was determined which contigs represented the full-length nucleotide sequence for each segment of RIX4414 virus including the typical RVA-segment endings. The determined sequences have been deposited in GenBank/EMBL/DDBJ, and the accession numbers for the nucleotide sequences of the VP1-VP4, VP6, VP7, and NSP1-NSP5 genes of strain RIX4414 are LC822556-LC822566, respectively.

2.3. Construction of Rescue T7 Plasmids Carrying All 11 dsRNA Segments of RIX4414 Virus

To develop a reverse genetics system for the RIX4414 virus, we newly constructed 11 rescue T7 plasmids for transcription of each mRNA of the 11 dsRNA segments of RIX4414 virus. For this, full-length cDNA fragments matching the 11 dsRNA segments were biochemically synthesized by Eurofins Genomics (Tokyo, Japan) or GENEWIZ (Tokyo, Japan) based on the full-length genomic sequences of the RIX4414 virus determined in this study, and each was individually cloned into a pUC57-derived pUC57R vector that carries the antigenomic hepatitis delta virus (HDV) ribozyme and T7 RNA polymerase terminator sequences [23]. In each of the constructed T7 plasmids, a cDNA copy of a full-length dsRNA segment is flanked by the T7 RNA polymerase promoter and HDV ribozyme sequences [27], and followed by the T7 RNA polymerase terminator sequence. A rescue T7 plasmid containing a signature mutation to destroy the unique HindIII restriction enzyme site in the VP3 gene at position 1039 (by a synonymous mutation) was also constructed using artificial synthesis by GENEWIZ. The 12 rescue T7 plasmids containing the genome of strain RIX4414 were pT7/VP1RIX, pT7/VP2RIX, pT7/VP3RIX, pT7/VP3RIX-ΔHindIII, pT7/VP4RIX, pT7/VP6RIX, pT7/VP7RIX, pT7/NSP1RIX, pT7/NSP2RIX, pT7/NSP3RIX, pT7/NSP4RIX, and pT7/NSP5RIX.

2.4. Reverse Genetics System

The protocol was basically as described previously [8,23]. For strain RIX4414, the above-described rescue T7 plasmids were employed. As a comparison, 11 rescue T7 plasmids encoding the genome of simian laboratory strain SA11-L2 [28] were also employed, namely, pT7/VP1SA11, pT7/VP2SA11, pT7/VP3SA11, pT7/VP4SA11-ΔPstI, pT7/VP6SA11, pT7/VP7SA11, pT7/NSP1SA11, pT7/NSP2SA11, pT7/NSP3SA11, pT7/NSP4SA11, and pT7/NSP5SA11 [23]. Briefly, the protocol was as follows. Monolayers of BHK/T7-9 cells in 6-well plates (Falcon, Bedford, MA, USA) were cotransfected with 11 T7 plasmids, representing the cloned cDNAs of 11 RVA dsRNA segments, in the following different quantities:

pT7/VP1RIX (0.75 µg), pT7/VP1SA11 (0.75 µg), pT7/VP2RIX (0.75 µg), pT7/VP2SA11 (0.75 µg), pT7/VP3RIX-ΔHindIII (0.75 µg), pT7/VP3SA11 (0.75 µg), pT7/VP4RIX (0.75 µg), pT7/VP4SA11-ΔPstI (0.75 µg), pT7/VP6RIX (0.75 µg), pT7/VP6SA11 (0.75 µg), pT7/VP7RIX (0.75 µg), pT7/VP7SA11 (0.75 µg), pT7/NSP1RIX (0.75 µg), pT7/NSP1SA11 (0.75 µg), pT7/NSP2RIX (2.25 µg), pT7/NSP2SA11 (2.25 µg), pT7/NSP3RIX (0.75 µg), pT7/NSP3SA11 (0.75 µg), pT7/NSP4RIX (0.75 µg), pT7/NSP4SA11 (0.75 µg), pT7/NSP5RIX (2.25 µg), and/or pT7/NSP5SA11 (2.25 µg). To generate recombinant SA11-L2 x RIX4414 single segment-reassortant viruses, the rescue T7 plasmid encoding the segment of SA11-L2 virus, which was to be replaced, was exchanged with a rescue T7 plasmid representing the corresponding dsRNA segment of RIX4414 virus; this process was carried out for each of the 11 individual dsRNA segments. Following a 1-day incubation, the transfected BHK/T7-9 cells were washed with incomplete medium and then cocultured with overlaid CV-1 cells (5×10^4 cells/mL) for 3 days in incomplete medium containing trypsin (0.3 or 0.9 µg/mL). After this, the cultures were subjected to two cycles of freezing and thawing and then treated with trypsin (10 µg/mL) for RVA activation, followed by inoculation onto MA104 cells in a roller-tube culture [29]; after one additional passage in such a culture and a 1-day incubation, recombinant RVAs were rescued and subsequently plaque purified in CV-1 cells, as previously described [30].

2.5. PAGE Analysis of Viral Genomic dsRNAs

Viral genomic dsRNAs were extracted from cell cultures using a QIAamp Viral RNA Mini Kit (Qiagen). The extracted viral genomic dsRNAs were subjected to polyacrylamide gel electrophoresis (PAGE) analysis: they were run in a 10% polyacrylamide gel for 16 h at 20 mA at room temperature, followed by silver staining [24] to visualize the genomic dsRNA migration profiles.

2.6. Multiple-Step Virus Growth

Monolayers of MA104 cells in 12-well plates (Thermo Fisher Scientific, Rochester, NY, USA) were infected in triplicate with trypsin-pretreated RVAs at an MOI of 0.01, washed twice with incomplete medium, and then incubated in incomplete medium containing trypsin (1 µg/mL) over various time periods. The infected cells were frozen and thawed twice before the measurement of viral titers by plaque assay.

2.7. Plaque Assay

The plaque assays were conducted as described previously [31]. Briefly, confluent monolayers of CV-1 cells in 6-well plates (Falcon) were infected with trypsin-pretreated RVAs, washed twice with incomplete medium, and then cultured with trypsin (1 µg/mL) in primary overlay medium (0.7% agarose). After 2 or 3 days, the cells were stained with secondary overlay medium containing 0.005% neutral red (Sigma-Aldrich) and 0.7% agarose. Plaque sizes were determined by measuring the mean diameters of 25 plaques in 2 independent assays.

2.8. Mouse Experiment

The mouse experiment was basically performed as previously described [32–35]. Briefly, pregnant BALB/cCrSlc mice (16 days of gestation) were purchased from Japan SLC Inc, Shizuoka, Japan. The mice were housed individually, and suckling mice were born on day 19.5 of gestation on average. To assess the rate and score of diarrhea, 5-day-old suckling mice were orally administered 50 µL of cell culture supernatant containing RIX4414, rRIX4414, or rSA11-L2 (1.0×10^5 PFU/mouse), or cell culture medium without RVA (mock), and were then monitored daily over a period of 5 days for diarrhea following gentle abdominal palpation. Mock inoculations were performed using MEM without additives. Rating of diarrhea (diarrhea score) was performed based on the following scale: 0, no diarrhea (normal stool or no stool); 1, soft orange stool; 2, soft mucous stool; and 3, liquid stool [32,33]. In this study, the 'diarrhea' status was defined as diarrhea score: ≥ 1. To compare the pathological changes in intestinal tissues after RVA infection and confirm RVA infection in the small intestines, suckling mice at 2–4 days post-infection were subjected to analysis of histopathology and RVA antigen expression of the small intestines. The animal experiment protocol was approved by the Fujita Health University Animal Care and Use Committee (Approval No.: APU19037-MD2).

2.9. Histopathology and Immunochemistry of Small Intestines

Under anesthesia, the small intestines of RVA-inoculated suckling mice were harvested and fixed in 4% paraformaldehyde phosphate-buffered solution (FUJIFILM Wako Chemicals, Osaka, Japan) for 1 day, followed by preparation of paraffin blocks, hematoxylin and eosin (HE) staining, and immunohistochemistry. In immunohistochemistry, a major RVA antigen was detected using BOND RX stainer (Leica, Tokyo, Japan). Deparaffinized and rehydrated sections were used for detection of RVA VP6 protein by using a mouse monoclonal antibody recognizing VP6 protein (YO-156 antibody [36]). HE-stained sections and immunostained sections were used for observation of RVA-induced lesions and VP6 protein expression, respectively, using a BX51 optical microscope (Olympus, Tokyo, Japan).

2.10. Statistics

Virus titers were evaluated by means of a two-way ANOVA with Sidak's post-test. Statistical analyses were completed using GraphPad Prism 7 (GraphPad Software 7, Boston, MA). p values of < 0.05 were considered statistically significant.

3. Results

3.1. Sequence Determination of the Full RIX4414 Genome

Sequence information for the $5'$- and/or $3'$-UTRs of multiple segments is absent in the available sequence reports for the authentic RIX4414 virus ("Rotarix-A41CB052A") in the GenBank/EMBL/DDBJ data libraries (JN849113, JN849114, and KX954616-KX954624), reflecting a focus on the gene coding sequences by Zeller et al. [37,38]. The reported lengths of the sequences for VP1-VP4, VP6, VP7, and NSP1-NSP5 of Rotarix-A41CB052A virus are 3267, 2673, 2508, 2359, 1194, 978, 1461, 954, 933, 528, and 594 nucleotides, respectively. In this study, we determined the complete nucleotide sequences of all 11 dsRNA segments of vaccine RIX4414 virus in a vial of Rotarix® by performing deep sequencing with Illumina MiSeq. This approach allowed the determination of the complete nucleotide sequences of all 11 gene segments of the authentic RIX4414 virus. The lengths in nucleotides of the VP1-VP4, VP6, VP7, and NSP1-NSP5 dsRNA segments of the RIX4414 were found to be 3302, 2717, 2591, 2359, 1356, 1062, 1566, 1059, 1074, 750, and 664, respectively. All the determined genetic sequences of the RIX4414 were identical to those of the Rotarix-A41CB052A, the exception being the VP4 gene that showed a one nucleotide difference (the residues are R and G at nucleotide position 1112 for Rotarix-A41CB052A and RIX4414, respectively). The genotype constellation of RIX4414 was identified as G1-P[8]-I1-R1-C1-M1-A1-N1-T1-E1-H1.

3.2. Construction of 11 Rescue T7 Plasmids for Live Attenuated Vaccine RIX4414 Virus

Our strategy for constructing 11 rescue T7 plasmids for the RIX4414 virus was essentially based on the approach used for the wild-type HuRVA strain KU, which established the first HuRVA reverse genetics system [8]. To generate recombinant authentic RIX4414, 11 T7 rescue plasmids were created, each designed to express mRNA corresponding to one of the full-length dsRNA segments as determined by our sequencing analysis. These segment sequences were synthesized biochemically and cloned into individual T7-driven plasmids, flanked by the T7 RNA polymerase promoter and HDV ribozyme sequences (Figure 1A).

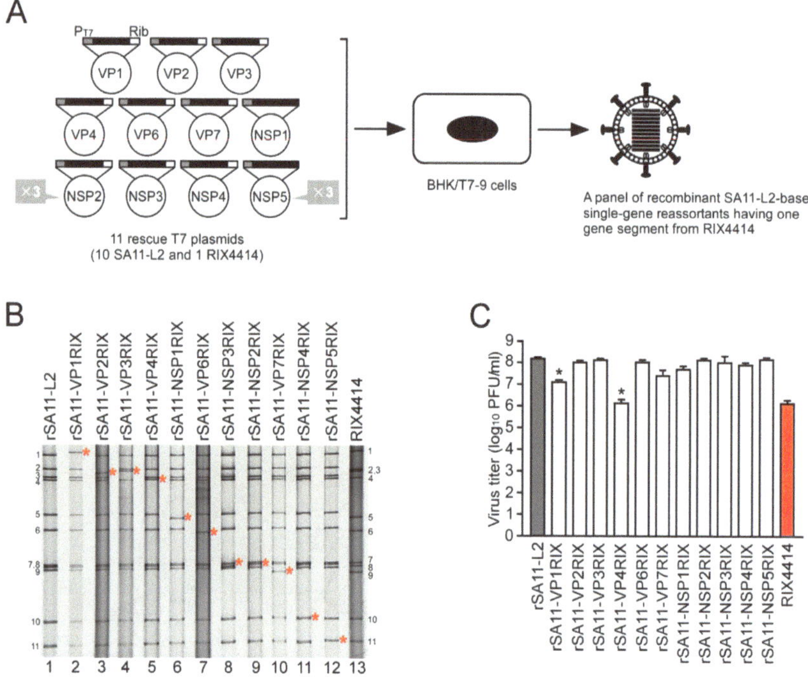

Figure 1. Generation of a panel of recombinant SA11-L2-based single-segment reassortants having one gene segment from RIX4414. (**A**) Schematic presentation of an 11-plasmid reverse genetics system to generate SA11-L2-based single-segment reassortants. The 11 rescue T7 plasmids include the full-length segment of cDNA of each dsRNA segment of RVA, flanked by the T7 RNA polymerase promoter (PT7) and the HDV ribozyme (Rib). To generate SA11-L2-based single-segment reassortants having one gene segment from RIX4414, BHK/T7-9 cells were cotransfected with the 11 rescue T7 plasmids (10 for SA11-L2 plus one for RIX4414) with 3-fold increased amounts of the two plasmids carrying the NSP2 and NSP5 genes. A panel of recombinant SA11-L2-based single-segment reassortants having one segment from RIX4414 were rescued from the cultures of the transfected BHK/T7-9 cells. (**B**) PAGE analysis of recombinant SA11-L2-based single-segment reassortants with each of the 11 segments from RIX4414. Lanes 1 and 13, dsRNAs from rSA11-L2 (lane 1) and RIX4414 (lane 13); lanes 2–12, dsRNAs from rescued rSA11-VP1RIX (lane 2), rSA11-VP2RIX (lane 3), rSA11-VP3RIX (lane 4), rSA11-VP4RIX (lane 5), rSA11-NSP1RIX (lane 6), rSA11-VP6RIX (lane 7), rSA11-NSP3RIX (lane 8), rSA11-NSP2RIX (lane 9), rSA11-VP7RIX (lane 10), rSA11-NSP4RIX (lane 11), and rSA11-NSP5RIX (lane 12). Red asterisks indicate the positions of the cDNA-derived RIX4414 segments. The numbers on the left and right indicate the orders of the genomic dsRNA segments of rSA11-L2 and RIX4414, respectively. (**C**) Infectivity of recombinant SA11-L2-based single-segment reassortants with each of the 11 segments from RIX4414. MA104 cells were infected with RVAs at an MOI of 0.01 and then incubated for 36 h. The viral titers in the cultures were determined by plaque assay. The data shown are the mean viral titers and standard deviations (SDs) for three independent cell cultures. Asterisks indicate significant differences between rSA11-L2 and recombinant single-segment reassortants; *, $p < 0.05$ (as calculated by two-way ANOVA with Sidak's post-test).

3.3. Generation of a Panel of Recombinant SA11-L2-Based Single-Segment Reassortant Viruses Carrying One RIX4414-Derived Segment

To individually access the functionality of each of the newly constructed 11 rescue T7 plasmids encoding the RIX4414 virus genome, we generated a panel of recombinant SA11-L2 × RIX4414 single-segment reassortant viruses. These viruses utilized a simian SA11-L2 genetic backbone, consisting of 10 T7 plasmids that each encoded a different

SA11-L2 dsRNA-segment, and one T7 plasmid encoding the RIX4414 version of the missing dsRNA-segment. These 11 T7 plasmids were co-transfected into BHK/T7-9 cells using the 11 plasmid-based reverse genetics system (Figure 1A) [8,23]. For each of the 11 individual RIX4414 T7 plasmids, this method successfully produced recombinant SA11-L2-based single-segment reassortant viruses, thereby demonstrating the functionality of all the plasmids in reverse genetics.

In PAGE analysis, the viral genomic dsRNAs extracted from the 11 rescued recombinant single-segment reassortants displayed migration patterns where each respective RIX4414 segment was aligned with the corresponding segment in the RIX4414 virus (Figure 1B). Furthermore, the nucleotide sequence analysis of the extracted viral genomic dsRNAs validated the authenticity of each manipulated genomic segment. Thus, these results collectively confirm the functionality in reverse genetics of all 11 rescue T7 plasmids for RIX4414 that we constructed.

To estimate the growth potential of the rescued 11 recombinant SA11-L2-based single-segment reassortants, viral titers of rSA11-L2, the 11 recombinant SA11-L2 x RIX4414 single-segment reassortants, and RIX4414 were determined at 36 h after infection of MA104 cells at an MOI of 0.01 PFU/cell (Figure 1C). Nine single-segment reassortants, rSA11-VP2RIX, rSA11-VP3RIX, rSA11-VP6RIX, rSA11-VP7RIX, rSA11-NSP1RIX, rSA11-NSP2RIX, rSA11-NSP3RIX, rSA11-NSP4RIX, and rSA11-NSP5RIX, exhibited virus growth similar to rSA11-L2. On the other hand, two reassortants, rSA11-VP1RIX and rSA11-VP4RIX, exhibited impaired growth, with titers that were ~10-fold and ~100-fold lower, respectively. This is consistent with previous evidence that the HuRVA-derived spike VP4 proteins are associated with reduced viral growth in cell culture [8,9,12,13,24,28,39,40]. The involvement of HuRVA-derived RNA-dependent RNA polymerase VP1 proteins as a determinant of viral growth in cell culture was reported for HuRVA strains KU (G1P[8]), CDC-9 (G1P[8]), and Odelia (G4P[8]), but not for HuRVA strain HN126 (G2P[4]) [9,12,13,39], suggesting a possible sub-optimal protein interaction between the exchanged HuRVA-derived VP1 protein and the other existing structural and/or non-structural proteins of simian SA11-L2 virus in a strain-specific fashion.

3.4. Generation of Recombinant Authentic Live Attenuated Vaccine Strain RIX4414 (rRIX4414) from Cloned cDNAs

We proceeded using the 11 rescue T7 plasmids that encode the RIX4414 genome to generate a replicative, authentic RIX4414 virus. Notably, for marking purposes, instead of pT7/VP3RIX plasmid, we used a pT7/VP3RIX-ΔHindIII plasmid, in which a unique HindIII site in the VP3 gene was eliminated via a silent G-to-A mutation as described in the Materials and Methods section. The 11 plasmids were co-transfected into BHK/T7-9 cells using a modified 11-plasmid reverse genetics system for HuRVAs [8]. While no striking cytopathic effect (CPE) was observed in the first passage of the virus in MA104 cells using lysates of the cocultures of transfected BHK/T7-9 cells and overlaid CV-1 cells, a typical RVA CPE appeared in the MA104 cells during the second virus passage. This indicated the successful generation of recombinant authentic RIX4414 virus, named rRIX4414, entirely from cloned cDNAs.

To validate the genetic integrity of rRIX4414, we conducted PAGE, sequence, and restriction analyses. PAGE analysis of the viral genomic dsRNAs extracted from the rescued virus showed that rRIX4414 exhibited an RNA migration pattern indistinguishable from that of the parental RIX4414 virus (Figure 2A). To confirm that the rescued virus was generated from the cloned cDNAs, we confirmed the absence of a unique HindIII site, a marker genetic mutation introduced into the VP3 gene of rRIX4414 (Figure 2B, upper panel). The sequence analysis demonstrated that the VP3 gene from rRIX4414 had the introduced mutation at nucleotide position 1039, whereas the VP3 gene amplified from the parental RIX4414 virus did not (Figure 2B, lower panel). Furthermore, RT-PCR products derived from the VP3 gene of rRIX4414 were resistant to HindIII digestion, unlike those from the parental RIX4414 strain (Figure 2C). Whole-genomic sequencing using Illumina

MiSeq also confirmed the presence of the expected G-to-A mutation in the VP3 gene and verified the absence of additional mutations across all 11 segments of the rescued rRIX4414 virus. These results confirm that the recombinant authentic RIX4414 virus was successfully generated via reverse genetics.

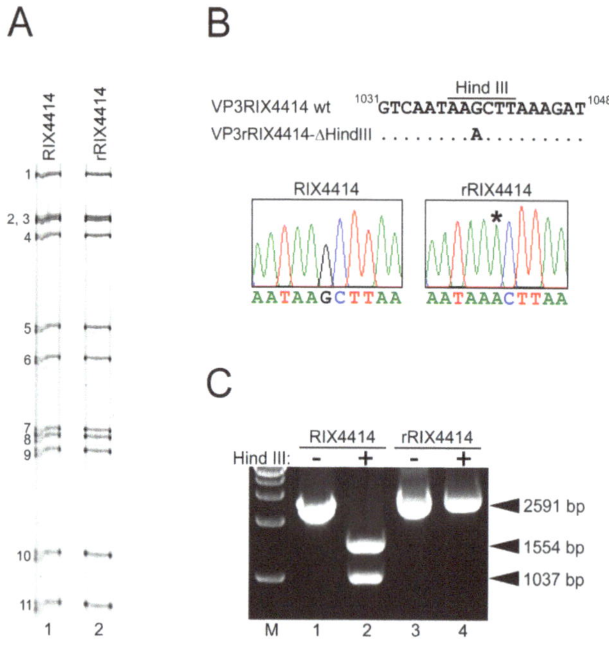

Figure 2. Generation of recombinant authentic rRIX4414 virus entirely from cloned cDNAs. (**A**) PAGE of viral genomic dsRNAs extracted from the parental RIX4414 and rescued rRIX4414. Lane 1, dsRNAs from the parental RIX4414; lane 2, dsRNAs from rescued rRIX4414. The numbers on the left indicate the order of the genomic dsRNA segments of RIX4414. (**B**) Rescued rRIX4414 contains a signature mutation (synonymous mutation) in its VP3 gene, because a nucleotide substitution (G-to-A at nucleotide position 1039) was introduced to abolish a unique HindIII site. The VP3 genes of RIX4414 and rRIX4414 were amplified by RT-PCR using specific primers, and sequencing electrograms show that indeed rRIX4414 possesses a G-to-A mutation at nucleotide position 1039. An asterisk indicates the G-to-A mutation introduced into the VP3 gene of rRIX4414. (**C**) Confirmation of the expected susceptibility to HindIII digestion. The 2591-bp VP3 gene RT-PCR products obtained for RIX4414 (lanes 1 and 2) and rRIX4414 (lanes 3 and 4) were treated with HindIII (+) or not (−), and separated in a 1% agarose gel. M, 1-kb DNA ladder marker.

3.5. Characterization of rRIX4414 Virus in Cultured Cells

To evaluate whether the rescued rRIX4414 virus possesses the replication characteristics of the parental RIX4414 virus, multiple-step growth curves for RIX4414 and rRIX4414 were determined after infection of MA104 cells at an MOI of 0.01 PFU/cell. The growth curves showed that the replication of rRIX4414 was virtually identical to that of the parental RIX4414 (Figure 3A). We also compared the plaque sizes in CV-1 cells for these viruses by measuring the mean diameters of 25 plaques each in two independent assay repeats (examples in Figure 3B); this revealed that the rRIX4414 virus produced plaques of virtually the same size (diameter, 2.01 ± 0.46 mm) as those produced by the parental RIX4414 (diameter, 2.05 ± 0.42 mm). These results demonstrate that the replication characteristics of rRIX4414 in cultured cells are indistinguishable from those of the parental RIX4414.

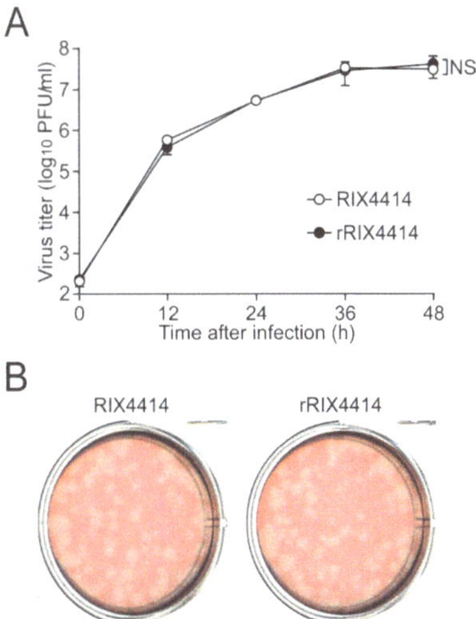

Figure 3. Growth properties of rRIX4414 virus in cultured cells. (**A**) Multiple-step growth curves for RIX4414 and rRIX4414. MA104 cells were infected with RIX4414 or rRIX4414 at an MOI of 0.01 and then incubated for various times (0, 12, 24, 36, and 48 h). The viral titers in the cultures were determined by plaque assay. The data shown are the mean viral titers and SDs from three independent cell cultures. NS, $p > 0.05$ (as calculated by two-way ANOVA with Sidak's post-test). (**B**) Plaque formation by RIX4414 and rRIX4414. RIX4414 or rRIX4414 was directly plated onto CV-1 cells to form plaques. The experiment was repeated three times with similar results, and representative results are shown.

3.6. Characterization of rRIX4414 Virus in Suckling Mice

To estimate whether the rescued rRIX4414 virus also possesses the biological characteristics of the parental RIX4414 virus in vivo, 5-day-old suckling mice were orally infected with RIX4414 or rRIX4414. To the best of our knowledge, there was no prior information about the pathogenicity of the vaccine strain RIX4414 in animal models including suckling mice, and therefore rSA11-L2 was included as a positive control known to induce diarrhea in suckling mice [32,33,35]. The RVA-inoculated mice were observed daily to check the possible onset of diarrhea according to previously established criteria [32,33]. Diarrhea was observed in the mice infected with RIX4414 and rRIX4414 solely on day 2 after infection, while those infected with rSA11-L2 experienced diarrhea on days 1–4 after infection (Table 1). In the mice infected with rSA11-L2, the average diarrhea score during days 1–5 was +1.26, whereas the average diarrhea scores in the same period were only +0.03 and +0.11 in the mice infected with RIX4414 and rRIX4414, respectively (Figure 4A). This indicates a reduced pathogenicity of RIX4414 and rRIX4414 relative to SA11-L2 in suckling mice.

Table 1. Pathogenicity of RIX4414, rRIX4414, and rSA11-L2 viruses in 5-day-old suckling mice.

Days after Infection	RIX4414	rRIX4414	rSA11-L2	Mock
0	0% (0/23) *	0% (0/28)	0% (0/21)	0% (0/11)
1	0% (0/23)	0% (0/28)	66.7% (12/18)	0% (0/11)
2	13.0% (3/23)	35.7% (10/28)	58.8% (10/17)	0% (0/11)
3	0% (0/21)	0% (0/26)	73.3% (11/15)	0% (0/10)
4	0% (0/19)	0% (0/24)	23.1% (3/13)	0% (0/9)
5	0% (0/17)	0% (0/21)	0% (0/11)	0% (0/9)

* Rate of diarrhea (No. of diarrhea/inoculated).

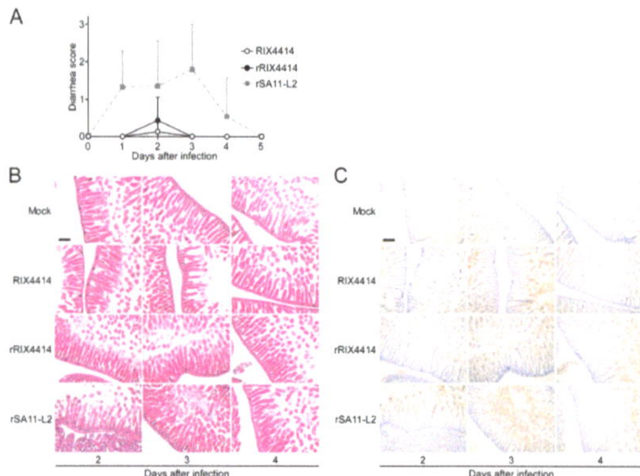

Figure 4. Biological properties of rRIX4414 virus in suckling mice. Five-day-old suckling mice were orally administered 50 µL of cell culture supernatants containing RIX4414, rRIX4414, rSA11-L2, or a mock infection. (**A**) Induction of diarrhea after infection with RIX4414, rRIX4414, or rSA11-L2. Diarrhea scores for individual mice were monitored daily. The data presented are the mean diarrhea scores and SDs for 9–28 pups. (**B**) Cytological changes in the small intestines. Small intestines were removed daily from day 2 to 4 after infection. Paraffin-embedded sections were stained with HE reagent. Two pups from each RVA-infected group were sacrificed for HE staining with similar results, and representative results are shown. From the mock-infection group, only one pup was sacrificed for HE staining. The bar represents 200 µm. (**C**) Histological analysis of rotaviral antigen expression in the small intestines. Small intestines of the infected mice were prepared as described in (**B**). Sections were stained by immunohistochemistry using a monoclonal antibody recognizing RVA VP6 antigen. VP6 protein expression in villus enterocytes was detected. The bar represents 200 µm.

We examined the pathological changes in the small intestine of infected mice on days 2–4 after infection by using HE staining of thin section preparations. Unlike the mock-infected mice, which showed no histological changes, significant epithelial vacuolization and villus shortening were observed in the small intestines of RVA-infected mice (Figure 4B). Notably, epithelial vacuolar degradation, a hallmark of RVA infection, was only observed on day 2 in the mice infected with RIX4414 or rRIX4414, but persisted throughout days 2 to 4 after infection in the rSA11-L2-infected mice.

To confirm the RVA infection in the small intestine, immunohistochemistry was performed using a mouse monoclonal antibody recognizing the VP6 protein, a major RVA antigen. On days 2–4 after RVA administration, the VP6 protein was detected in epithelial cells in the upper part of the villi of RVA-infected mice and not in mock-infected mice (Figure 4C). A visual estimation of the numbers of small intestine villus epithelial cells that expressed VP6 could not determine clear differences between rRIX4414-, RIX4414-, or rSA11-L2-infected mice.

The important conclusion from the combined results is that the biological characteristics of rRIX4414 in suckling mice are comparable to those of the parental RIX4414.

4. Discussion and Conclusions

We successfully established an authentic reverse genetics system for the HuRVA vaccine strain RIX4414. The recombinant strain, rRIX4414, exhibited properties both in vitro and in vivo that were indistinguishable from those of the original RIX4414 strain.

One of the limitations of this study is the comparison of the pathogenicity of rRIX4414 and RIX4414 with rSA11-L2 in suckling mice. Both HuRVA viruses, rRIX4414 and RIX4414, belong to the G1P[8] genotype, while the simian rSA11-L2 virus belongs to the G3P[2] genotype. This distinction is relevant for RVA infection because the P[8]-VP4 spike proteins recognize fucosylated histo-blood group antigens such as H1 antigen, while the P[2]-VP4 spike proteins recognize sialic acid present in gangliosides [41–46]. Therefore, to confirm the attenuated phenotype of rRIX4414, it would have been preferable to compare the vaccine strains with a human strain of the same P[8] genotype (e.g., the virulent Wa strain (G1P[8])). In any case, it was observed that the biological characteristics of rRIX4414 in suckling mice are comparable to those of the parental RIX4414 and that they were very mild compared to a more virulent RVA.

One future application of the recombinant rRIX4414 system is the further improvement of HuRVA vaccines. Currently, four live attenuated HuRVA vaccines are licensed and WHO-prequalified: Rotarix® (G1P[8]), RotaTeq® (five reassortant viruses; G1, G2, G3, G4, and P[8]), ROTAVAC® (G9P[11]), and ROTASIIL® (five reassortant viruses; G1, G2, G3, G4, and G5). Although these vaccines generally provide cross-protection against heterotypic HuRVA strains, outbreaks caused by strains heterotypic to the vaccines still occasionally occur [47–51], and wider vaccine immunogenicity would be advantageous. Additionally, despite several attempts, the attenuating mutations in the genome of RIX4414 have not yet been identified [52,53], and knowing them could benefit the development of other RVA vaccines. Furthermore, while the Rotarix® vaccine is safe and well-tolerated [54], its vaccine strain RIX4414 is still capable of reversion to increased virulence due to mutations [55–58]. Hence, the identification and modification of these risk factors by using our established RIX4414 reverse genetics system may yield a safer vaccine strain. In short, we envision that the rRIX4414 reverse genetics platform will not only help uncover why RIX4414 is attenuated but will also serve as a base for rationally enhancing vaccine safety and immunogenicity.

Another long-term goal of the rRIX4414 platform is to develop an RVA-based enteric delivery vector. Since the advent of reverse genetics systems for RVA, it has been speculated that recombinant RVA could be used as a vector to deliver foreign genes to intestinal cells. Given the safety profile of Rotarix®, which is a vaccine licensed by the WHO, the rRIX4414 system is ideally positioned for exploring such possibilities.

Author Contributions: Conceptualization, S.F., S.N., T.M., T.Y., K.T., and S.K.; data curation, S.F., M.K., K.K., Y.H., and S.K.; funding acquisition, S.F., T.Y., and S.K.; investigation, S.F., M.K., K.K., Y.A., Y.H., and S.K.; methodology, S.F., K.T., and S.K.; writing—original draft preparation, S.F.; writing—review and editing, S.F., T.Y., and S.K. All authors have read and agreed to the published version of the manuscript.

Funding: This study was supported in part by AMED (22fk0108121h0603, 23fk0108669h0401, and 24fk0108669h0402 to S.K.), JSPS KAKENHI (21K08498 and 24K10235 to S.F., 18H02784 to T.Y., and 21K07057 and 24K11806 to S.K.), RCGLID, Oita University (2023B07 to S.F.), GSK Japan Research Grant 2021 (S.F.), the Mochida Memorial Foundation for Medical and Pharmaceutical Research (S.F. and S.K.), the Takeda Science Foundation (S.F. and S.K.), and the Public Foundation of Vaccination Research Center (T.Y.).

Institutional Review Board Statement: The suckling mice were handled ethically according to the Regulations for the Management of Laboratory Animals of Fujita Health University. The experimental protocol for the ethical use of these animals was approved by the Animal Care and Use Committee of Fujita Health University (Permit No.: APU19037-MD2).

Informed Consent Statement: Not applicable.

Data Availability Statement: The nucleotide sequence data obtained in this study have been deposited in the DDBJ and EMBL/GenBank data libraries. The accession numbers for the nucleotide sequences of the VP1-VP4, VP6, VP7, and NSP1-NSP5 genes of strain RIX4414 are LC822556-LC822566, respectively.

Acknowledgments: We wish to thank Johannes M. Dijkstra (Office of Research Administration, Fujita Health University) for his English editing support.

Conflicts of Interest: We have no conflicts of interest to declare.

References

1. Tate, J.E.; Burton, A.H.; Boschi-Pinto, C.; Parashar, U.D.; World Health Organization–Coordinated Global Rotavirus Surveillance Network. Global, regional, and national estimates of rotavirus mortality in children <5 years of age, 2000–2013. *Clin. Infect. Dis.* **2016**, *62*, S96–S105. [CrossRef] [PubMed]
2. Troeger, C.; Khalil, I.A.; Rao, P.C.; Cao, S.; Blacker, B.F.; Ahmed, T.; Armah, G.; Bines, J.E.; Brewer, T.G.; Colombara, D.V.; et al. Rotavirus vaccination and the global burden of rotavirus diarrhea among children younger than 5 years. *JAMA Pediatr.* **2018**, *172*, 958–965. [CrossRef]
3. Estes, M.K.; Greenberg, H.B. Rotaviruses. In *Fields Virology*, 6th ed.; Knipe, D.M., Howley, P.M., Cohen, J.I., Griffin, D.E., Lamb, R.A., Martin, M.A., Racaniello, V.R., Roizman, B., Eds.; Wolters Kluwer Health/Lippincott Williams & Wilkins: Philadelphia, PA, USA, 2013; pp. 1347–1401.
4. Bernstein, D.I.; Ward, R.L. Rotarix: Development of a live attenuated monovalent human rotavirus vaccine. *Pediatr. Ann.* **2006**, *35*, 38–43. [CrossRef]
5. Burnett, E.; Parashar, U.D.; Tate, J.E. Global impact of rotavirus vaccination on diarrhea hospitalizations and deaths among children <5 years old: 2006–2019. *J. Infect. Dis.* **2020**, *222*, 1731–1739.
6. Ward, R.L.; Bernstein, D.I. Rotarix: A rotavirus vaccine for the world. *Clin. Infect. Dis.* **2009**, *48*, 222–228. [CrossRef]
7. Bernstein, D.I.; Smith, V.E.; Sherwood, J.R.; Schiff, G.M.; Sander, D.S.; DeFeudis, D.; Spriggs, D.R.; Ward, R.L. Safety and immunogenicity of live, attenuated human rotavirus vaccine 89-12. *Vaccine* **1998**, *16*, 381–387. [CrossRef]
8. Komoto, S.; Fukuda, S.; Kugita, M.; Hatazawa, R.; Koyama, C.; Katayama, K.; Murata, T.; Taniguchi, K. Generation of infectious recombinant human rotaviruses from just 11 cloned cDNAs encoding the rotavirus genome. *J. Virol.* **2019**, *93*, e02207-18. [CrossRef] [PubMed]
9. Sánchez-Tacuba, L.; Feng, N.; Meade, N.J.; Mellits, K.H.; Jaïs, P.H.; Yasukawa, L.L.; Resch, T.K.; Jiang, B.; López, S.; Ding, S.; et al. An optimized reverse genetics system suitable for efficient recovery of simian, human, and murine-like rotaviruses. *J. Virol.* **2020**, *94*, e01294-20. [CrossRef] [PubMed]
10. Ward, R.L.; Knowlton, D.R.; Pierce, M.J. Efficiency of human rotavirus propagation in cell culture. *J. Clin. Microbiol.* **1984**, *19*, 748–753. [CrossRef]
11. Urasawa, S.; Urasawa, T.; Taniguchi, K.; Chiba, S. Serotype determination of human rotavirus isolates and antibody prevalence in pediatric population in Hokkaido, Japan. *Arch. Virol.* **1984**, *81*, 1–12. [CrossRef]
12. Hamajima, R.; Lusiany, T.; Minami, S.; Nouda, R.; Nurdin, J.A.; Yamasaki, M.; Kobayashi, N.; Kanai, Y.; Kobayashi, T. A reverse genetics system for human rotavirus G2P[4]. *J. Gen. Virol.* **2022**, *103*, 001816. [CrossRef] [PubMed]
13. Kawagishi, T.; Nurdin, J.A.; Onishi, M.; Nouda, R.; Kanai, Y.; Tajima, T.; Ushijima, H.; Kobayashi, T. Reverse genetics system for a human group A rotavirus. *J. Virol.* **2020**, *94*, e00963-19. [CrossRef] [PubMed]
14. Philip, A.A.; Agbemabiese, C.A.; Yi, G.; Patton, J.T. T7 expression plasmids for producing a recombinant human G1P[8] rotavirus comprising RIX4414 sequences of the RV1 (Rotarix, GSK) vaccine strain. *Microbiol. Resour. Announc.* **2023**, *12*, e0060323. [CrossRef] [PubMed]
15. Wyatt, R.G.; James, W.D.; Bohl, E.H.; Theil, K.W.; Saif, L.J.; Kalica, A.R.; Greenberg, H.B.; Kapikian, A.Z.; Chanock, R.M. Human rotavirus type 2: Cultivation in vitro. *Science* **1980**, *207*, 189–191. [CrossRef] [PubMed]
16. Anderson, M.L.; Sullivan, O.M.; Nichols, S.L.; Kaylor, L.; Kelly, D.F.; McDonald Esstman, S. Rotavirus core shell protein sites that regulate intra-particle polymerase activity. *J. Virol.* **2023**, *97*, e0086023. [CrossRef]
17. Chen, D.; Patton, J.T. Rotavirus RNA replication requires a single-stranded 3′ end for efficient minus-strand synthesis. *J. Virol.* **1998**, *72*, 7387–7396. [CrossRef] [PubMed]
18. De Lorenzo, G.; Drikic, M.; Papa, G.; Eichwald, C.; Burrone, O.R.; Arnoldi, F. An inhibitory motif on the 5′UTR of several rotavirus genome segments affects protein expression and reverse genetics strategies. *PLoS ONE* **2016**, *11*, e0166719. [CrossRef] [PubMed]
19. Patton, J.T. Rotavirus VP1 alone specifically binds to the 3′ end of viral mRNA, but the interaction is not sufficient to initiate minus-strand synthesis. *J. Virol.* **1996**, *70*, 7940–7947. [CrossRef] [PubMed]
20. Vetter, J.; Papa, G.; Tobler, K.; Rodriguez, J.M.; Kley, M.; Myers, M.; Wiesendanger, M.; Schraner, E.M.; Luque, D.; Burrone, O.R.; et al. The recruitment of TRiC chaperonin in rotavirus viroplasms correlates with virus replication. *mBio* **2024**, *15*, e0049924. [CrossRef]
21. Wentz, M.J.; Patton, J.T.; Ramig, R.F. The 3′-terminal consensus sequence of rotavirus mRNA is the minimal promoter of negative-strand RNA synthesis. *J. Virol.* **1996**, *70*, 7833–7841. [CrossRef]
22. Ito, N.; Takayama-Ito, M.; Yamada, K.; Hosokawa, J.; Sugiyama, M.; Minamoto, N. Improved recovery of rabies virus from cloned cDNA using a vaccinia virus-free reverse genetics system. *Microbiol. Immunol.* **2003**, *47*, 613–617. [CrossRef]
23. Komoto, S.; Fukuda, S.; Ide, T.; Ito, N.; Sugiyama, M.; Yoshikawa, T.; Murata, T.; Taniguchi, K. Generation of recombinant rotaviruses expressing fluorescent proteins by using an optimized reverse genetics system. *J. Virol.* **2018**, *92*, e00588-18. [CrossRef] [PubMed]
24. Komoto, S.; Sasaki, J.; Taniguchi, K. Reverse genetics system for introduction of site-specific mutations into the double-stranded RNA genome of infectious rotavirus. *Proc. Natl. Acad. Sci. USA* **2006**, *103*, 4646–4651. [CrossRef]
25. Dennis, F.E.; Fujii, Y.; Haga, K.; Damanka, S.; Lartey, B.; Agbemabiese, C.A.; Ohta, N.; Armah, G.E.; Katayama, K. Identification of novel Ghanaian G8P[6] human-bovine reassortant rotavirus strain by next generation sequencing. *PLoS ONE* **2014**, *9*, e100699. [CrossRef]

26. Komoto, S.; Adah, M.I.; Ide, T.; Yoshikawa, T.; Taniguchi, K. Whole genomic analysis of human and bovine G8P[1] rotavirus strains isolated in Nigeria provides evidence for direct bovine-to-human interspecies transmission. *Infect. Genet. Evol.* **2016**, *43*, 424–433. [CrossRef] [PubMed]

27. Johne, R.; Reetz, J.; Kaufer, B.B.; Trojnar, E. Generation of an avian-mammalian rotavirus reassortant by using a helper virus-dependent reverse genetics system. *J. Virol.* **2016**, *90*, 1439–1443. [CrossRef] [PubMed]

28. Taniguchi, K.; Nishikawa, K.; Kobayashi, N.; Urasawa, T.; Wu, H.; Gorziglia, M.; Urasawa, S. Differences in plaque size and VP4 sequence found in SA11 virus clones having simian authentic VP4. *Virology* **1994**, *198*, 325–330. [CrossRef]

29. Urasawa, T.; Urasawa, S.; Taniguchi, K. Sequential passages of human rotavirus in MA-104 cells. *Microbiol. Immunol.* **1981**, *25*, 1025–1035. [CrossRef]

30. Taniguchi, K.; Morita, Y.; Urasawa, T.; Urasawa, S. Cross-reactive neutralization epitopes on VP3 of human rotavirus: Analysis with monoclonal antibodies and antigenic variants. *J. Virol.* **1987**, *61*, 1726–1730. [CrossRef]

31. Urasawa, S.; Urasawa, T.; Taniguchi, K. Three human rotavirus serotypes demonstrated by plaque neutralization of isolated strains. *Infect. Immun.* **1982**, *38*, 781–784. [CrossRef]

32. Fukuda, S.; Kugita, M.; Higashimoto, Y.; Shiogama, K.; Tsujikawa, H.; Moriguchi, K.; Ito, N.; Sugiyama, M.; Nagao, S.; Murata, T.; et al. Rotavirus incapable of NSP6 expression can cause diarrhea in suckling mice. *J. Gen. Virol.* **2022**, *103*, 001745. [CrossRef] [PubMed]

33. Kawahara, T.; Makizaki, Y.; Oikawa, Y.; Tanaka, Y.; Maeda, A.; Shimakawa, M.; Komoto, S.; Moriguchi, K.; Ohno, H.; Taniguchi, K. Oral administration of Bifidobacterium bifidum G9-1 alleviates rotavirus gastroenteritis through regulation of intestinal homeostasis by inducing mucosal protective factors. *PLoS ONE* **2017**, *12*, e0173979. [CrossRef] [PubMed]

34. Kawamura, Y.; Komoto, S.; Fukuda, S.; Kugita, M.; Tang, S.; Patel, A.; Pieknik, J.R.; Nagao, S.; Taniguchi, K.; Krause, P.R.; et al. Development of recombinant rotavirus carrying herpes simplex virus 2 glycoprotein D gene based on reverse genetics technology. *Microbiol. Immunol.* **2024**, *68*, 56–64. [CrossRef] [PubMed]

35. Offit, P.A.; Clark, H.F.; Kornstein, M.J.; Plotkin, S.A. A murine model for oral infection with a primate rotavirus (simian SA11). *J. Virol.* **1984**, *51*, 233–236. [CrossRef] [PubMed]

36. Taniguchi, K.; Urasawa, T.; Urasawa, S.; Yasuhara, T. Production of subgroup-specific monoclonal antibodies against human rotaviruses and their application to an enzyme-linked immunosorbent assay for subgroup determination. *J. Med. Virol.* **1984**, *14*, 115–125. [CrossRef] [PubMed]

37. Zeller, M.; Patton, J.T.; Heylen, E.; De Coster, S.; Ciarlet, M.; Van Ranst, M.; Matthijnssens, J. Genetic analyses reveal differences in the VP7 and VP4 antigenic epitopes between human rotaviruses circulating in Belgium and rotaviruses in Rotarix and RotaTeq. *J. Clin. Microbiol.* **2012**, *50*, 966–976. [CrossRef] [PubMed]

38. Zeller, M.; Heylen, E.; Tamim, S.; McAllen, J.K.; Kirkness, E.F.; Akopov, A.; De Coster, S.; Van Ranst, M.; Matthijnssens, J. Comparative analysis of the Rotarix™ vaccine strain and G1P[8] rotaviruses detected before and after vaccine introduction in Belgium. *PeerJ* **2017**, *5*, e2733. [CrossRef] [PubMed]

39. Fukuda, S.; Hatazawa, R.; Kawamura, Y.; Yoshikawa, T.; Murata, T.; Taniguchi, K.; Komoto, S. Rapid generation of rotavirus single-gene reassortants by means of eleven plasmid-only based reverse genetics. *J. Gen. Virol.* **2020**, *101*, 806–815. [CrossRef]

40. Greenberg, H.B.; Flores, J.; Kalica, A.R.; Wyatt, R.G.; Jones, R. Gene coding assignments for growth restriction, neutralization and subgroup specificities of the Wa and DS-1 strains of human rotavirus. *J. Gen. Virol.* **1983**, *64*, 313–320. [CrossRef]

41. Barbé, L.; Le Moullac-Vaidye, B.; Echasserieau, K.; Bernardeau, K.; Carton, T.; Bovin, N.; Nordgren, J.; Svensson, L.; Ruvoën-Clouet, N.; Le Pendu, J. Histo-blood group antigen-binding specificities of human rotaviruses are associated with gastroenteritis but not with in vitro infection. *Sci. Rep.* **2018**, *8*, 12961. [CrossRef]

42. Böhm, R.; Fleming, F.E.; Maggioni, A.; Dang, V.T.; Holloway, G.; Coulson, B.S.; von Itzstein, M.; Haselhorst, T. Revisiting the role of histo-blood group antigens in rotavirus host-cell invasion. *Nat. Commun.* **2015**, *6*, 5907. [CrossRef]

43. Delorme, C.; Brüssow, H.; Sidoti, J.; Roche, N.; Karlsson, K.A.; Neeser, J.R.; Teneberg, S. Glycosphingolipid binding specificities of rotavirus: Identification of a sialic acid-binding epitope. *J. Virol.* **2001**, *75*, 2276–2287. [CrossRef]

44. Hu, L.; Crawford, S.E.; Czako, R.; Cortes-Penfield, N.W.; Smith, D.F.; Le Pendu, J.; Estes, M.K.; Prasad, B.V. Cell attachment protein VP8* of a human rotavirus specifically interacts with A-type histo-blood group antigen. *Nature* **2012**, *485*, 256–259. [CrossRef]

45. Le Pendu, J.; Nyström, K.; Ruvoën-Clouet, N. Host-pathogen co-evolution and glycan interactions. *Curr. Opin. Virol.* **2014**, *7*, 88–94. [CrossRef] [PubMed]

46. Yu, X.; Coulson, B.S.; Fleming, F.E.; Dyason, J.C.; von Itzstein, M.; Blanchard, H. Novel structural insights into rotavirus recognition of ganglioside glycan receptors. *J. Mol. Biol.* **2011**, *413*, 929–939. [CrossRef]

47. Burke, R.M.; Tate, J.E.; Barin, N.; Bock, C.; Bowen, M.D.; Chang, D.; Gautam, R.; Han, G.; Holguin, J.; Huynh, T.; et al. Three rotavirus outbreaks in the postvaccine era—California, 2017. *MMWR Morb. Mortal. Wkly. Rep.* **2018**, *67*, 470–472. [CrossRef]

48. Cates, J.E.; Amin, A.B.; Tate, J.E.; Lopman, B.; Parashar, U. Do rotavirus strains affect vaccine effectiveness? A systematic review and meta-analysis. *Pediatr. Infect. Dis. J.* **2021**, *40*, 1135–1143. [CrossRef] [PubMed]

49. Mwanga, M.J.; Verani, J.R.; Omore, R.; Tate, J.E.; Parashar, U.D.; Murunga, N.; Gicheru, E.; Breiman, R.F.; Nokes, D.J.; Agoti, C.N. Multiple introductions and predominance of rotavirus group A genotype G3P[8] in Kilifi, coastal Kenya, 4 years after nationwide vaccine introduction. *Pathogens* **2020**, *9*, 981. [CrossRef] [PubMed]

50. Pitzer, V.E.; Bilcke, J.; Heylen, E.; Crawford, F.W.; Callens, M.; De Smet, F.; Van Ranst, M.; Zeller, M.; Matthijnssens, J. Did large-scale vaccination drive changes in the circulating rotavirus population in Belgium? *Sci. Rep.* **2015**, *5*, 18585. [CrossRef]

51. Roczo-Farkas, S.; Kirkwood, C.D.; Cowley, D.; Barnes, G.L.; Bishop, R.F.; Bogdanovic-Sakran, N.; Boniface, K.; Donato, C.M.; Bines, J.E. The impact of rotavirus vaccines on genotype diversity: A comprehensive analysis of 2 decades of Australian surveillance data. *J. Infect. Dis.* **2018**, *218*, 546–554. [CrossRef]

52. Ward, R.L.; Mason, B.B.; Bernstein, D.I.; Sander, D.S.; Smith, V.E.; Zandle, G.A.; Rappaport, R.S. Attenuation of a human rotavirus vaccine candidate did not correlate with mutations in the NSP4 protein gene. *J. Virol.* **1997**, *71*, 6267–6270. [CrossRef] [PubMed]

53. Ward, R.L.; Kirkwood, C.D.; Sander, D.S.; Smith, V.E.; Shao, M.; Bean, J.A.; Sack, D.A.; Bernstein, D.I. Reductions in cross-neutralizing antibody responses in infants after attenuation of the human rotavirus vaccine candidate 89-12. *J. Infect. Dis.* **2006**, *194*, 1729–1736. [CrossRef] [PubMed]

54. Vesikari, T.; Karvonen, A.; Korhonen, T.; Espo, M.; Lebacq, E.; Forster, J.; Zepp, F.; Delem, A.; De Vos, B. Safety and immunogenicity of RIX4414 live attenuated human rotavirus vaccine in adults, toddlers and previously uninfected infants. *Vaccine* **2004**, *22*, 2836–2842. [CrossRef] [PubMed]

55. Boom, J.A.; Sahni, L.C.; Payne, D.C.; Gautam, R.; Lyde, F.; Mijatovic-Rustempasic, S.; Bowen, M.D.; Tate, J.E.; Rench, M.A.; Gentsch, J.R.; et al. Symptomatic infection and detection of vaccine and vaccine-reassortant rotavirus strains in 5 children: A case series. *J. Infect. Dis.* **2012**, *206*, 1275–1279. [CrossRef] [PubMed]

56. Gower, C.M.; Dunning, J.; Nawaz, S.; Allen, D.; Ramsay, M.E.; Ladhani, S. Vaccine-derived rotavirus strains in infants in England. *Arch. Dis. Child.* **2020**, *105*, 553–557. [CrossRef] [PubMed]

57. Kaplon, J.; Cros, G.; Ambert-Balay, K.; Leruez-Ville, M.; Chomton, M.; Fremy, C.; Pothier, P.; Blanche, S. Rotavirus vaccine virus shedding, viremia and clearance in infants with severe combined immune deficiency. *Pediatr. Infect. Dis. J.* **2015**, *34*, 326–328. [CrossRef]

58. Simsek, C.; Bloemen, M.; Jansen, D.; Descheemaeker, P.; Reynders, M.; Van Ranst, M.; Matthijnssens, J. Rotavirus vaccine-derived cases in Belgium: Evidence for reversion of attenuating mutations and alternative causes of gastroenteritis. *Vaccine* **2022**, *40*, 5114–5125. [CrossRef]

Article

Chimeric Viruses Enable Study of Antibody Responses to Human Rotaviruses in Mice

Sarah Woodyear [1], Tawny L. Chandler [1], Takahiro Kawagishi [2], Tom M. Lonergan [1], Vanshika A. Patel [1], Caitlin A. Williams [3], Sallie R. Permar [3], Siyuan Ding [2] and Sarah L. Caddy [1,*]

[1] Baker Institute for Animal Health, Cornell University, Ithaca, NY 14850, USA; sarahwoodyear@cornell.edu (S.W.)

[2] Department of Molecular Microbiology, Washington University in St. Louis, St. Louis, MO 63101, USA

[3] Department of Pediatrics, Weill Cornell Medicine, New York, NY 10001, USA

* Correspondence: sarahcaddy@cornell.edu

Abstract: The leading cause of gastroenteritis in children under the age of five is rotavirus infection, accounting for 37% of diarrhoeal deaths in infants and young children globally. Oral rotavirus vaccines have been widely incorporated into national immunisation programs, but whilst these vaccines have excellent efficacy in high-income countries, they protect less than 50% of vaccinated individuals in low- and middle-income countries. In order to facilitate the development of improved vaccine strategies, a greater understanding of the immune response to existing vaccines is urgently needed. However, the use of mouse models to study immune responses to human rotavirus strains is currently limited as rotaviruses are highly species-specific and replication of human rotaviruses is minimal in mice. To enable characterisation of immune responses to human rotavirus in mice, we have generated chimeric viruses that combat the issue of rotavirus host range restriction. Using reverse genetics, the rotavirus outer capsid proteins (VP4 and VP7) from either human or murine rotavirus strains were encoded in a murine rotavirus backbone. Neonatal mice were infected with chimeric viruses and monitored daily for development of diarrhoea. Stool samples were collected to quantify viral shedding, and antibody responses were comprehensively evaluated. We demonstrated that chimeric rotaviruses were able to efficiently replicate in mice. Moreover, the chimeric rotavirus containing human rotavirus outer capsid proteins elicited a robust antibody response to human rotavirus antigens, whilst the control chimeric murine rotavirus did not. This chimeric human rotavirus therefore provides a new strategy for studying human-rotavirus-specific immunity to the outer capsid, and could be used to investigate factors causing variability in rotavirus vaccine efficacy. This small animal platform therefore has the potential to test the efficacy of new vaccines and antibody-based therapeutics.

Keywords: rotavirus; antibody; reverse genetics; vaccine

Citation: Woodyear, S.; Chandler, T.L.; Kawagishi, T.; Lonergan, T.M.; Patel, V.A.; Williams, C.A.; Permar, S.R.; Ding, S.; Caddy, S.L. Chimeric Viruses Enable Study of Antibody Responses to Human Rotaviruses in Mice. *Viruses* **2024**, *16*, 1145. https://doi.org/10.3390/v16071145

Academic Editor: Mamta Chawla Sarkar

Received: 14 May 2024
Revised: 10 July 2024
Accepted: 11 July 2024
Published: 16 July 2024

1. Introduction

Rotavirus vaccines have been highly successful at reducing the burden of rotavirus-induced gastroenteritis in children in high-income countries. Two oral vaccines were first approved in 2006 and 2008, developed from a human rotavirus strain (Rotarix, GSK, Rixensart, Belgium) and a human–bovine rotavirus reassortant virus (Rotateq, Merck & Co., Rahway, NJ, USA) [1,2]. A total of four live attenuated rotavirus vaccines are now pre-qualified with the World Health Organization, and a further two are country-specific [3]. However, live attenuated vaccines have failed to protect infants in low- to middle-income countries, with vaccine efficacy often lower than 50% [4,5], leading to higher rates of gastroenteritis deaths in unprotected infants. Consequently, there is a pressing need for the development of improved vaccines.

A major hurdle for vaccine development has been the absence of suitable pre-clinical models to test new vaccine candidates. Rotavirus strains are highly species-specific, meaning that human rotavirus strains are primarily pathogenic in humans, and murine rotavirus strains only cause disease in mice [6,7]. This hinders the study of disease pathology and immune responses to human rotavirus strains in mice, as human rotavirus strains only replicate to low titres and cause minimal disease in this species.

In this study, we aimed to develop a new strategy to study the immune response to human rotaviruses in a mouse model. To achieve this, we generated a chimeric rotavirus strain that can replicate well in mice and contains key immunogenic proteins from a human rotavirus strain. This approach builds on previous work that showed reassortment of certain murine rotavirus genes with those from a non-murine rotavirus strain could still permit virus replication in mice [6].

The advent of rotavirus reverse genetics has provided the field with the capacity to rapidly generate chimeric rotaviruses with relative ease [8,9]. Rotaviruses have a triple-layered structure with an outer capsid composed of two proteins, VP4 and VP7, which become a major target of the adaptive immune response [10]. Chimeric rotaviruses have previously been produced using reverse genetics with a murine backbone and human rotavirus VP4 protein [11], and we have now extended this by successfully generating a chimeric rotavirus with VP4 and VP7 from a human rotavirus strain. As a control, we also generated a chimeric virus encoding the outer capsid proteins of a heterologous murine rotavirus strain. We demonstrated that both viruses could infect neonatal mice and replicate to comparable titres.

Antibody responses to human rotavirus proteins in chimeric rotaviruses have not previously been studied in mice. We aimed to determine the magnitude and specificity of the antibody response to human rotavirus outer capsid proteins using our small animal model. We observed strong germinal centre formation in the draining lymph nodes of mice infected with both chimeric viruses, which correlated with antibody production. Neutralisation and ELISpot assays were used to clearly demonstrate that the chimeric virus with human rotavirus outer capsid proteins induced antibodies with human rotavirus specificity. This novel approach to studying human rotaviruses in a small animal model will be valuable for pre-clinical evaluation of vaccine efficacy and therapeutics targeting the outer capsid of human rotaviruses.

2. Materials and Methods

2.1. Cells and Viruses

MA104 African green monkey kidney cells, provided by Dr. John Parker (Baker Institute for Animal Health, Cornell University, Ithaca, NY, USA), were grown in Dulbecco's Minimum Essential Media (DMEM) supplemented with 10% heat-inactivated fetal bovine serum, 100 IU/mL penicillin, and 100 µg/mL streptomycin (complete DMEM). Reagent details are listed in Table 1.

The pre-existing rotavirus strains used in this study were the Rotarix vaccine strain (G1P [8]) and the primate strain SA11 (G3P [2]). Two rotavirus chimeric strains were developed using a published plasmid-based reverse genetics system [12]. Using the reassortant virus rD6/2-2g as the backbone, chimeric viruses were rescued with a human CDC-9 strain VP4 and VP7 (human outer capsid proteins), or a murine strain (ETD) VP4 and VP7, validated by sequencing. All viruses were propagated in MA104 cells following activation with 10 µg/mL TPCK-treated trypsin at 37 °C for 30 min. Prior to in vivo infection, viruses were diluted to the appropriate titre in sterile PBS without calcium chloride and magnesium chloride.

Table 1. Reagent details.

Reagent	Source	Identifier
	Antibodies and Dyes	
Alexa Fluor 488 Anti-Sheep IgG (Donkey)	Invitrogen, Carlsbad, CA, USA	CAT#A-11015
Anti-Mouse IgG Biotin (Goat)	Mabtech, Cincinnati, OH, USA	CAT#3825-6
Anti-Mouse IgG HRP (Goat)	Sigma Aldrich, St. Louis, MO, USA	CAT#A0168
CD45 (BUV395 Rat Anti-Mouse)	BD Biosciences, San Jose, CA, USA	CAT#564279; Clone: 30-F11
CD45R (PerCP/Cyanine 5.5 Anti-Mouse/Human CD45R/B220)	BioLegend, San Diego, CA, USA	CAT#103236; Clone: RA3-6B2
CD95 (APC Anti-Mouse (Fas))	BioLegend, San Diego, CA, USA	CAT#152603; Clone: SA367H8
Clarity Western ECL Substrate	Bio-Rad, Hercules, CA, USA	CAT#170-5060S
Fc Block (TruStain FcX Anti-Mouse CD16/32)	BioLegend, San Diego, CA, USA	CAT#101320; Clone: 93
Fixable Viability Dye eFluor 780	Invitrogen, Carlsbad, CA, USA	CAT#65-0865
GL7 Antigen (Pacific Blue Anti-Mouse/Human)	BioLegend, San Diego, CA, USA	CAT#144614; Clone: GL7
Hoechst 33342	Invitrogen, Carlsbad, CA, USA	CAT#H3570
Rotavirus Polyclonal Antibody (Sheep)	Invitrogen, Carlsbad, CA, USA	CAT#PA1-85845
Streptavidin–ALP	Mabtech, Cincinnati, OH, USA	CAT#3310-10
	Chemicals	
BCIP/NBT-plus for ALP	Mabtech, Cincinnati, OH, USA	CAT#3650-10
Dulbecco's Modified Eagle's Medium (DMEM)	Corning, Corning, NY, USA	CAT#10-013
Fetal Bovine Serum (FBS)	Corning, Corning, NY, USA	CAT#28622001
Laemmli Sample Buffer (4×)	Bio-Rad, Hercules, CA, USA	CAT#1610747
Pierce Protease Inhibitor Mini Tablets	Thermo Fisher Scientific, Waltham, MA, USA	CAT#A32953
Red Blood Cell Lysis Buffer (10×)	BioLegend, San Diego, CA, USA	CAT#420301
RIPA Lysis and Extraction Buffer	Thermo Fisher Scientific, Waltham, MA, USA	CAT#89900
RPMI 1640	Corning, Corning, NY, USA	CAT#10-040-CV
TPCK-Treated Trypsin	Worthington Biochemical, Lakewood, NJ, USA	CAT#LS003740
3,3′,5,5′-Tetramethylbenzidine (TMB) Membrane Peroxidase Substrate Plus	Avantor, Radnor, PA, USA	CAT#K830
	Commercial Assays	
BCA Protein Assay Kit	Thermo Fisher Scientific, Waltham, MA, USA	CAT#23227
Luna Universal One-Step RT-qPCR Kit	New England Biolabs, Ipswich, MA, USA	CAT#E3006
Monarch Total RNA Miniprep Kit	New England Biolabs, Ipswich, MA, USA	CAT#T2010S
Neon Transfection System 100 µL Kit	Invitrogen, Carlsbad, CA, USA	CAT#MPK10025

2.2. Virus Quantification by Fluorescent Focus Assay (FFA)

A fluorescent focus assay was used to determine viral titre (in fluorescent focus units, FFU) as previously described [13]. Briefly, MA104 cells were infected with the virus for 16 h, then cells were fixed with 1:1 methanol–acetone at $-20\ ^{\circ}$C for 20 min. After blocking with PBS-2%FBS for 20 min at room temperature, 10 µg/mL rotavirus polyclonal antibody (sheep) diluted in PBS-2%FBS was added for 1 h at room temperature. After three washes with PBS supplemented with 0.1% Tween-20 (PBS-T), 4 µg/mL Alexa Fluor 488 Donkey Anti-Sheep IgG and Hoechst 33,342 diluted in PBS-2%FBS were added to each well, and the plate was incubated for 1 h at room temperature. All plates were coated with PBS and kept at 4–6 °C prior to imaging. Quantification of rotavirus-infected cells was achieved using the BioTek Cytation 7 Cell Imaging Multimode Reader and Gen5 Image Prime (v3.13) software.

2.3. Rotavirus Infection of Mice

129S6/SvEvTac mice (Taconic Biosciences, Germantown, NY, USA) were maintained by an in-house breeding colony housed at the Baker Institute for Animal Health. All mouse work was approved by the Cornell University Institutional Animal Care and Use Committee (IACUC), Protocol 2022-0152. Seven-day-old pups were infected with 1×10^4 FFU of virus by oral gavage. Pups were monitored for the development of diarrhoea, and scored positive if mucus or liquid stool was observed. Stool samples were collected once daily post-infection from each litter, then pooled and diluted 1:10 in PBS. Diluted stool was centrifuged at $8000 \times g$ for 5 min to remove debris, and the supernatant was stored at -80 °C. Blood samples were collected from the lateral saphenous vein at 4, 6, and 8 weeks old, and by terminal cardiac puncture at 10 weeks of age. All blood samples were centrifuged at $6000 \times g$ for 5 min, and sera were stored short-term at 4 °C. In separate experiments, 14-days-post-infection mice were humanely culled with terminal cardiac puncture samples collected, and the Peyer's patches (PPs), mesenteric lymph nodes (MLNs), and spleens were harvested.

2.4. Quantification of Virus Shedding by RT-qPCR

RNA extraction from a clarified stool suspension was achieved using the Monarch Total RNA Miniprep Kit according to manufacturer's instructions. Each sample was eluted in a total volume of 50 µL nuclease-free water, followed by denaturation of dsRNA at 95 °C for 5 min. RT-qPCR was performed using the Luna Universal One-Step RT-qPCR Kit per the manufacturer's instructions, with 5 µL of RNA in a total reaction volume of 20 µL using NSP5 forward primer CTGCTTCAAACGATCCACTCAC at 400 nM, NSP5 reverse primer TGAATCCATAGACACGCC at 400 nM, and NSP5 TaqMan probe FAM-TCAAATGCAGTTAAGACAAATGCAGACGCT-TAMRA at 200 nM. The reaction was carried out on a QuantStudio 3 thermocycler (Applied Biosystems, Waltham, MA, USA) under the cycling conditions of 55 °C for 10 min, 95 °C for 1 min, and 40 cycles of 95 °C for 10 s and 60 °C for 30 s. A 10-fold serial dilution of SA11 total RNA was included on each plate to quantify rotavirus genome copies per mL of stool supernatant using QuantStudio Design & Analysis Software (v1.5.1). A lower limit of quantification of 100 genome copy numbers was set and assigned to samples with no detectable virus.

2.5. ELISAs

ELISAs were performed to detect IgG-specific anti-rotavirus antibodies using an in-house method as previously described [13]. MA104 cells were infected with rotavirus to produce infected cell lysate, or phosphate-buffered saline (PBS) to produce control cell lysate. Cells were collected and resuspended in Radio-Immunoprecipitation Assay (RIPA) buffer supplemented with protease inhibitors. The Bicinchoninic Acid Kit for protein determination was used to measure protein concentration following the manufacturer's instructions, and lysate stocks were diluted to 1 mg/mL in PBS.

Plates were washed three times using PBS-T between each step. High-binding 96-well plates (Greiner Bio-One, Monroe, NC, USA) were coated with 5 µg/mL rotavirus-specific polyclonal antibody (sheep) in PBS and incubated at 4–6 °C for 16 h. Plates were then blocked with 5% milk–PBS-T at room temperature for 1 h. The purified cell culture lysates (Rotarix- and SA11 virus-infected lysate or mock-infected control lysate) were diluted to 10 µg/mL in PBS and incubated at 37 °C for two hours. Sera were diluted 1:200 in 5% milk–PBS-T and added in duplicate and incubated at 37 °C for 2 h. Positive and negative control sera from known infected and uninfected mice were included on each plate. The anti-mouse IgG HRP secondary antibody was diluted 1:1000 in 5% milk–PBS-T and added before plates were incubated at 37 °C for 1 h. To detect the bound antibody, 3,3',5,5'-tetramethylbenzidine (TMB) was incubated at room temperature for 10 min. The reaction was stopped with 1M sulfuric acid (H_2SO_4), and the optical density (OD) was read at 450 nm using the BioTek Cytation 7 Cell Imaging Multimode Reader. The OD was normalised by subtracting the OD of the mock-infected control cell lysate well from the OD of the virus-infected cell lysate well.

2.6. Western Blot

Rotarix- and SA11-infected cell lysates were denatured in $4\times$ Laemli buffer at 95 °C for 5 min, then separated on a 4–15% Mini-PROTEAN TGX gel (Bio-Rad, CAT#456-1083). Proteins were transferred to a polyvinylidene difluoride (PVDF) membrane by a Trans-blot Turbo Transfer System (Bio-Rad, Hercules, CA, USA). The membrane was blocked with 5% milk–PBS-T at room temperature for 1 h, then incubated with sera from mice infected with either human outer capsid rotavirus or murine outer capsid rotavirus diluted 1:250 in 5% milk–PBS-T for 12 h at 4–6 °C. Following three washes in PBS-T, the membrane was incubated with anti-mouse IgG HRP diluted 1:500 in 5% milk–PBS-T for 1 h at room temperature. Blots were washed three times in PBS-T and visualised using clarity Western ECL substrate and a ChemiDoc MP imaging system.

2.7. Extracellular Neutralisation Assay

MA104 cells were seeded in a 96-well black-sided plate (Corning 3340) at 2×10^4 per well in complete DMEM and incubated at 37 °C for 4 h to allow cells to adhere. One 1:10 dilution of serum in serum-free media (SFM) was incubated with trypsin-activated rotavirus at 37 °C for 1 h. The serum–virus mixture was then added in triplicate to seeded cells. After 1 h at 37 °C, 50 µL complete DMEM was added to each well and the plate was incubated at 37 °C for 16 h. Rotavirus neutralisation was quantified by FFA.

2.8. Intracellular Neutralisation Assay

A previously published intracellular neutralisation assay protocol was applied to sera from mice infected with the human or mouse outer capsid chimeric viruses [9]. Using a Neon Transfection System Kit, sera were diluted 1:3 in PBS and mixed with 2×10^5 MA104 cells suspended in Resuspension Buffer R. Sera were then electroporated with two pulses at 1400 V and a 20-pulse width using the Neon® Transfection System (Thermo Fisher Scientific, Waltham, MA, USA). Electroporated cells were resuspended in complete DMEM and plated onto a 96-well black-sided plate in duplicate. After incubation at 37 °C for 4 h, wells were washed once with PBS, and trypsin-activated rotavirus in SFM was added to each well. Infection and virus quantification then proceeded as for the extracellular neutralisation assay, described above.

2.9. Isolation of Single Cells from Lymph Nodes and Spleens

Peyer's patches (PPs), mesenteric lymph nodes (MLNs), and spleens were harvested from infected mice 14 days post-infection or control uninfected mice. The tissues were homogenised through 70 µm mesh cell strainers to obtain single-cell suspensions, and then washed with RPMI supplemented with 2%FBS (RPMI-2%FBS) by centrifugation at $300\times g$ for 5 min. The PP and MLN cell pellets were resuspended in 1 mL cold staining buffer (PBS-1%FBS), then filtered through 70 µm mesh, and washed and resuspended in 100 µL staining buffer. Spleen cell pellets were resuspended in red blood cell lysis buffer and incubated at room temperature for 3 min. These were then washed with PBS by centrifugation and resuspended in RPMI-2%FBS.

2.10. Flow Cytometry

PP and MLN single-cell suspensions were incubated with Fc Block (1:100) in staining buffer for 30 min at 4 °C, then cells were incubated with viability dye and fluorescently conjugated antibodies targeting B220, GL7 Antigen, CD95, and CD45 in staining buffer for 30 min at 4 °C. Single-color controls were included using PP cells or compensation beads. Cells were washed twice with staining buffer by centrifugation and resuspended in 100 µL

4%PFA-PBS at 4 °C for 15 min to fix. Cells were then washed twice by centrifugation, and the cell pellets were resuspended in 300 µL staining buffer. Cells were analysed using a BD LSRFortessa X-20 (BD Biosciences, San Jose, CA, USA) and BD FACSDiva Software (v9.0), and data analysis was performed in FlowJo (v10.9.0).

2.11. ELISpot Assay

Spleen single-cell suspensions were analysed by ELISpot to identify rotavirus-specific B cell responses. PVDF-based membrane plates (MSIP white, Mabtech, Cincinnati, OH, USA) were activated with 35% ethanol, followed by five washes with sterile water. All further washing steps were performed five times with sterile PBS. Plates were coated with 2.7×10^4 FFU Rotarix per well and incubated at 4 °C for 16 h. Wells were washed to remove excess antigen, and incubated with RPMI supplemented with 10% FBS, 100 IU/mL penicillin, and 100 µg/mL streptomycin for 30 min at room temperature. The media was removed, and 3×10^5 spleen cells were incubated in triplicate at 37 °C for 16 h. The plate was then washed, and 1 µg/mL anti-mouse IgG biotin in PBS-0.5%FBS was incubated for 2 h at room temperature. After washing, 1:1000 streptavidin–ALP diluted in PBS-0.5%FBS was incubated for 1 h at room temperature. The plate was washed a final time before BCIP/NBT-plus for ALP was added and allowed to develop until distinct spots emerged. Colour development was then stopped by rinsing the plate with water, and plates were left to dry before imaging on the upright microscope of a BioTek Cytation 7 Cell Imaging Multimode Reader.

2.12. Statistics

Statistical analysis was performed using GraphPad Prism (v10.2.0). Immune response outcomes for two groups were analysed by unpaired two-tailed t-tests. Dependent outcomes reported for three or more groups were analysed by one-way ANOVA and pair-wise comparisons reported with Tukey's adjustment for multiple comparisons. Outcomes reported over time were analysed by repeated measures ANOVA and pair-wise comparisons reported with Bonferroni's correction for multiple comparisons. Statistical differences were considered significant at p-values < 0.05 for all comparisons. Error bars indicate the standard error of the mean.

3. Results

3.1. Construction and Characterisation of Chimeric Rotaviruses Expressing Murine or Human Strain Outer Capsid Proteins

We previously generated a chimeric virus named rD6/2-2g, which has a murine rotavirus backbone of the non-tissue-culture-adapted wild-type murine EW strain with gene segment 4 (VP4) of the simian rotavirus RRV strain. Gene segments 1 (VP1) and 10 (NSP4) of simian rotavirus strain SA11 were introduced to enhance rescue by reverse genetics. In addition, we generated a murine reassortant virus with gene segment 4 (encoding VP4) from the human rotavirus strain CDC9 (G1P [8]) that replicates in mice [11]. To generate a murine-like virus with an entirely human outer capsid composition, we inserted both gene segment 4 and gene segment 9 (encoding VP7) of CDC-9 into the murine-like rD6/2-2g backbone. As a control, we also produced a chimeric virus encoding the VP4 of the cell-culture-adapted murine ETD strain. This generated two different chimeric viruses that were identical except one encoded a human rotavirus outer capsid and the other a murine rotavirus outer capsid, as depicted in Figure 1.

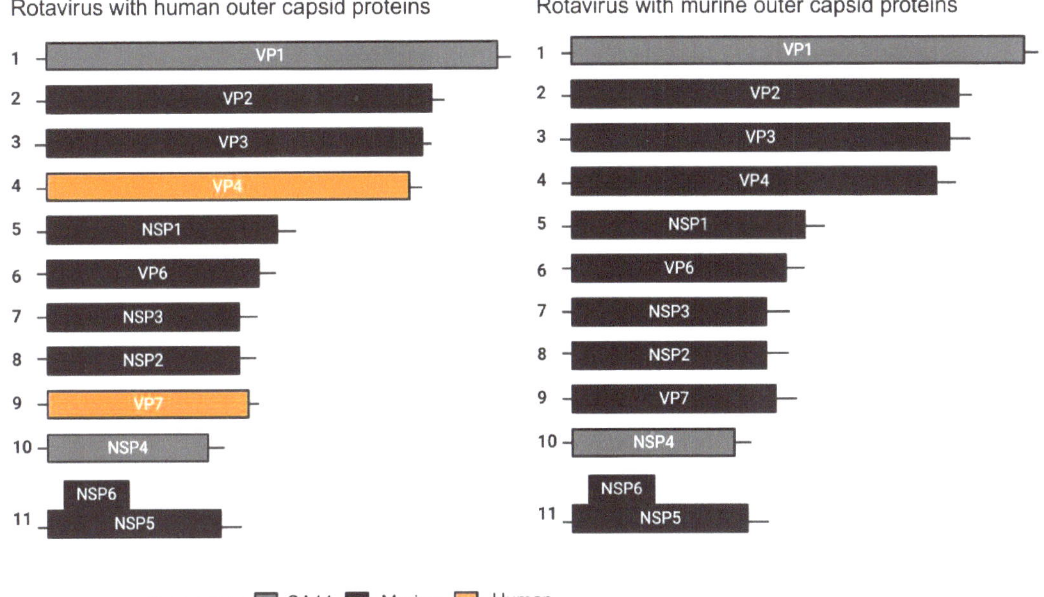

Figure 1. Schematic diagram of segmented dsRNA genome of chimeric rotaviruses. Reverse genetics was used to generate chimeric viruses encoding either human or murine outer capsid proteins.

3.2. Chimeric Rotaviruses with Human Outer Capsid Proteins Replicate in Mice

To examine the ability of chimeric rotaviruses to replicate and cause disease in mice, two separate litters of five pups (mouse outer capsid rotavirus) or six pups (human outer capsid rotavirus) were infected with virus by oral gavage at seven days old. For comparison with an entirely human rotavirus strain, we used Rotarix (human vaccine strain) to infect an additional litter of five pups at the same viral titre. Rotarix has an amino acid identity to CDC-9 of 98.3% for VP4 and 94.5% for VP7, so serological cross-reactivity was expected. All pups in each litter were monitored daily for the development of diarrhoea, and stool samples were collected for quantification by qPCR.

As expected, Rotarix did not robustly replicate and was detectable at only low copy numbers in pup stool (Figure 2A). In contrast, both chimeric rotaviruses showed strong evidence of viral amplification from day 2 onwards. Viral shedding was detectable for at least seven days. As shown in Figure 2B, diarrhoea was only detected in the litter of pups infected with the chimeric virus expressing the murine rotavirus outer capsid proteins. The chimeric virus with human rotavirus outer capsid proteins was significantly attenuated in comparison. Given that the viral loads were comparable, this suggests that the outer capsid proteins are important in the pathogenesis of diarrhoea in this model, but the mechanism remains unclear.

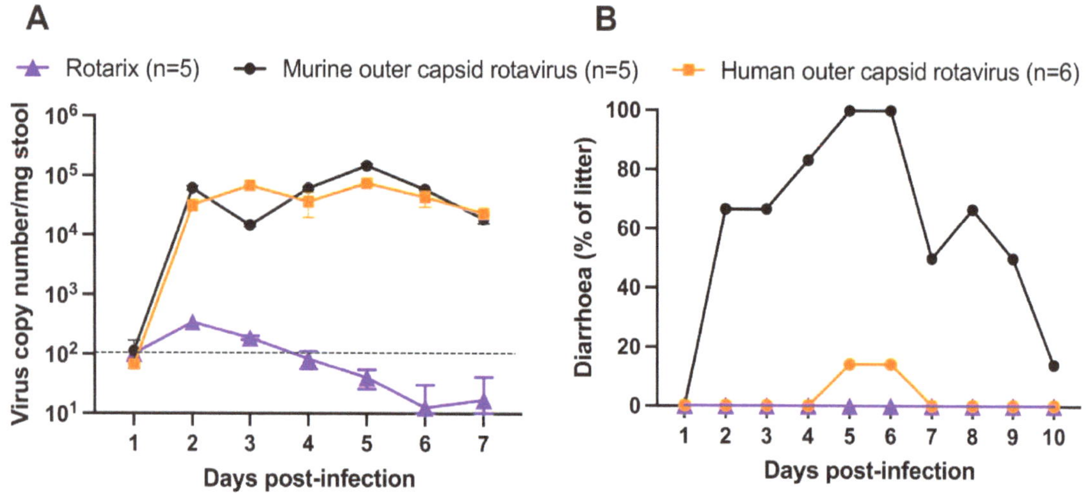

Figure 2. Virus shedding and clinical disease induced by oral infection of neonatal mice with chimeric rotaviruses compared to a human rotavirus strain. (**A**) Viral load detected in stool by qPCR (dotted line for lower limit of quantification). (**B**) Diarrhoea observed in neonatal mice infected at seven days old. Numbers in brackets in the key indicate the number of pups per litter.

3.3. Human Rotavirus-Specific Antibody Responses Are Generated in Mice Infected with Chimeric Rotaviruses

Antibody responses following the infection of three litters of seven-day-old mice with different rotaviruses were analysed to determine the immunogenicity of the chimeric strains. Serum samples were collected from mice infected with chimeric rotaviruses at 2 weeks post-infection and compared with samples from mice infected with Rotarix. An in-house sandwich ELISA based on the SA11 primate rotavirus strain (therefore distinct from both the human and murine strains) was used to show that rotavirus-specific IgGs were readily detected in mice infected with the chimeric rotaviruses, but no antibody response was evident in mice infected with Rotarix (Figure 3B).

Next, we wanted to determine if there were any differences between longitudinal antibody responses in mice infected with either chimeric virus, so we infected two further litters of six pups each. We demonstrated that antibody responses were detected for over two months following infection (Figure 3C). Despite heterogeneity in the antibody responses within litters, there were no significant differences between IgG titres in mice infected with chimeric viruses expressing the murine or human outer capsid proteins from weeks 6 to 10.

To determine whether a sandwich ELISA could differentiate between the antibody responses of the two litters if a homologous antigen to one of the strains was used, we generated a lysate of cells infected with the Rotarix strain. We analysed samples collected at the 10-week timepoint and, as shown in Figure 3E, there was no significant difference between IgG responses detected using the Rotarix-based ELISA. This likely reflects the fact that the infected cell lysate used in the sandwich ELISA contains all rotavirus proteins, and the chimeric viruses are identical except for VP4 and VP7. This supports previous work that has shown this type of ELISA predominantly detects antibodies specific for the inner capsid protein VP6 [14], an immunodominant antigen that is identical between both chimeric virus strains. To confirm this, we performed a Western blot with the Rotarix- and SA11-infected cell lysates and probed with sera from mice infected with either the murine outer capsid rotavirus or human outer capsid rotavirus (Figure 3D). A distinct band of ~45 kD, corresponding to VP6, was detected by antibodies from both infected mice at a comparable level for either lysate.

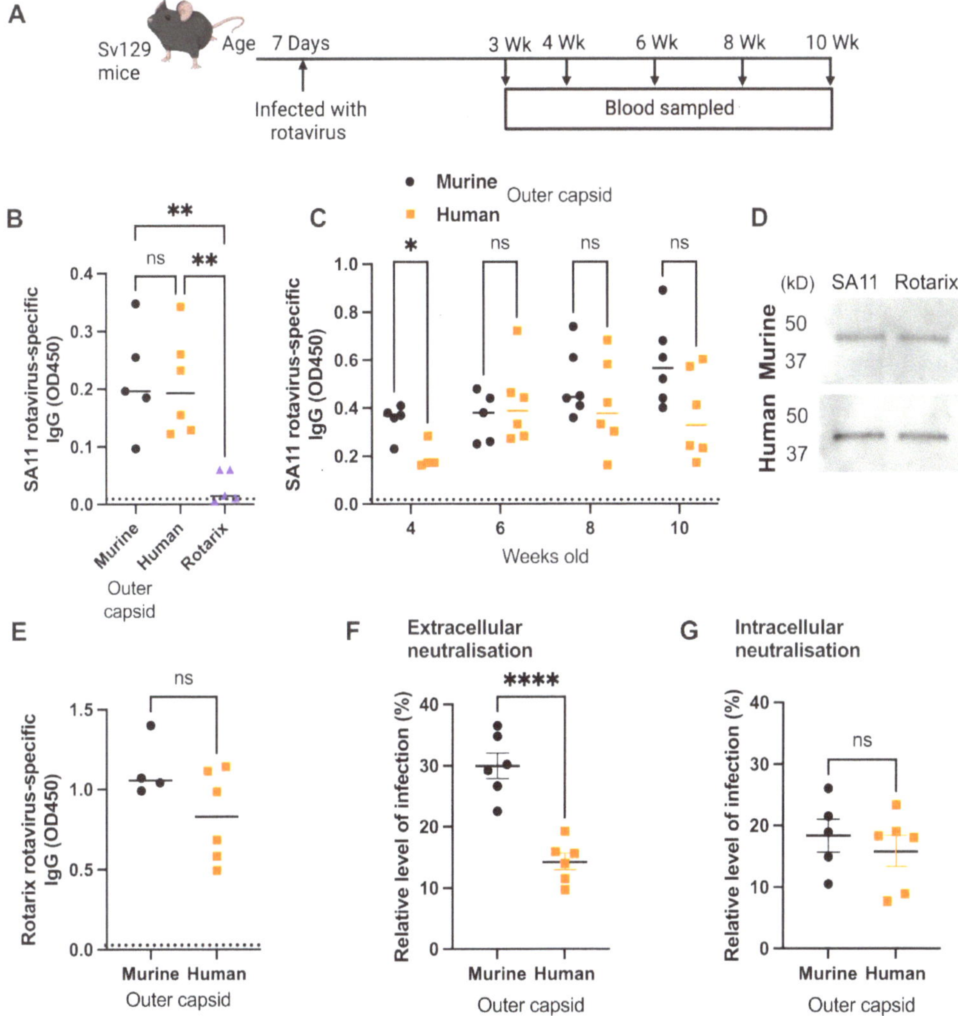

Figure 3. Analysis of serum antibody responses in mice infected with chimeric rotaviruses. Mice were infected with chimeric rotaviruses (murine outer capsid as black circles, human outer capsid as orange squares) or Rotarix control (shown as purple triangles) at seven days old and serum samples were collected for antibody analysis at the timepoints shown in the schematic diagram (**A**). For all graphs, each point corresponds to an individual mouse; note that some samples were not available for all assays due to limited sample volumes. (**B**) Analysis of serum from 21-day-old mice by sandwich ELISA with primate lysate. (**C**) Longitudinal samples analysed by sandwich ELISA with primate rotavirus. (**D**) Western blot of SA11- and Rotarix-infected cell lysates, probed with sera from mice infected with either murine outer capsid rotavirus or human outer capsid rotavirus. (**E**) Samples from 10-week-old mice analysed by sandwich ELISA with human rotavirus (Rotarix strain). (**F**) Extracellular neutralisation of human rotavirus by serum samples from 10-week-old mice as quantified by fluorescent focus assay. (**G**) Intracellular neutralisation of human rotavirus by serum samples from 10-week-old mice as quantified by fluorescent focus assay. Dashed horizontal lines in Figure (**A**–**C**) represent the positive threshold based on the OD450 of serum from uninfected control mice. Statistical significance was determined by one-way ANOVA (**B**), repeated measures ANOVA (**C**), or unpaired two-tailed t-tests (* $p < 0.05$; ** $p < 0.01$; **** $p < 0.0001$). Tukey's adjusted pair-wise comparisons (**B**) and Bonferroni-corrected pair-wise comparisons (**C**) are shown.

To evaluate whether antibodies specifically targeting the human rotavirus capsid were induced by the human chimeric strain, we next performed serum neutralisation assays. Sera from 10-week-old mice were incubated with Rotarix for 1 h, and then the resulting complexes were added to MA104 cells and infection was allowed to proceed overnight (Figure 3F). Whereas sera from mice infected with entirely murine chimeric virus could neutralise to a mean of 30.0% relative to the no-sera control, sera from mice infected with virus containing human outer capsid proteins neutralised to 14.2% ($p < 0.0001$). This provides clear evidence that a human-outer-capsid-specific antibody response was induced by this chimeric virus.

To verify that this functional response was specific for the outer capsid and not for other rotavirus proteins, we also performed an intracellular neutralisation assay. This assay evaluates the activity of VP6-specific antibodies inside cells, achieved when serum antibodies are electroporated into the cytoplasm of MA104 cells [13]. As the VP6 sequence was identical for both chimeric viruses, we hypothesised that intracellular neutralisation by sera from infected mice would be very similar. As shown in Figure 3G, there was indeed no significant difference in intracellular neutralisation induced by the antibodies raised to both chimeric viruses.

3.4. Chimeric Viruses Induced Human-Rotavirus-Specific B Cell Responses in Mice

To complement and extend the results obtained from the serum antibody analysis of mice infected with chimeric viruses, we also characterised B cell responses to infection. Two litters of mice were infected with chimeric viruses, and 14 days post-infection, cells from the spleen and draining lymph nodes (PPs and MLNs) were analysed. A third litter of mice was infected with an equal titre of the Rotarix vaccine strain to verify the inability of an entirely human rotavirus strain to induce a detectable B cell response in mice. Two age-matched uninfected mice were also included as controls. Flow cytometry on 50,000 cells was used to identify the germinal centre B cells present in the PPs and MLNs of each litter of mice (Figure 4A,B). No germinal centre formation was observed in pups infected with Rotarix, in accordance with Figure 3A, and in line with the lack of virus replication measured in Figure 2A. In contrast, germinal centre formation in draining lymph nodes was readily observed in both litters of mice infected with the two chimeric rotavirus strains. Interestingly, there was a small but significant difference ($p = 0.0128$) between the number of germinal centre B cells measured in the MLNs. This indicates that the chimeric strain with the human rotavirus outer capsid induces fewer germinal centre B cells than the entirely murine strain.

Whilst germinal centre B cell quantification clearly shows a strong B cell response, this approach does not identify antigen specificity of the B cells. To investigate this, we performed ELISpot assays, using Rotarix as the antigen coated onto wells of an ELISpot plate. Splenocytes from three mice from each litter, plus from two uninfected control mice, were incubated in the antigen-coated wells overnight, then staining for IgG production was completed the following day. Figure 4C,D show that mice infected with chimeric virus encoding the human outer capsid generated B cells that produced significantly more human-rotavirus-specific antibodies than the other viruses. This therefore confirms that a human-rotavirus-specific B cell response can be readily induced and detected by infection with a chimeric virus.

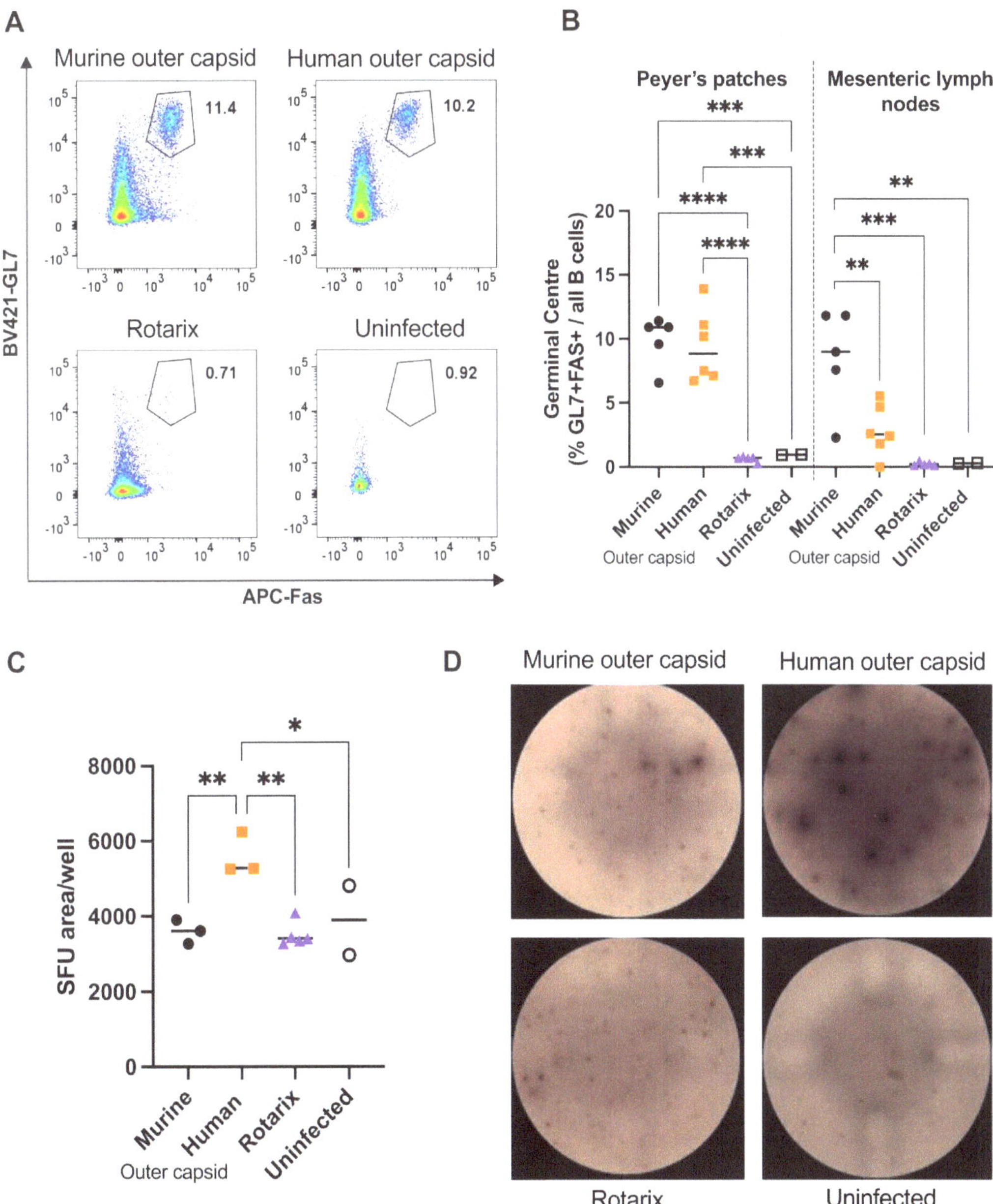

Figure 4. Analysis of B cell responses in mice infected with chimeric (murine outer capsid as black circles, human outer capsid as orange squares) or human Rotarix control (shown as purple triangles). Seven-day-old mice were infected with the panel of viruses, and B cell analysis was performed 14 days later. (**A**) Representative flow plots of germinal centre (GL7 + FAS+) B cells identified in Peyer's patches (PPs) by flow cytometry. Numbers represent the percentage of total B220 + cells. (**B**) Quantification of germinal centres in PPs and MLNs by flow cytometry. (**C**) Quantification of spot forming unit (SFU) area by B cell ELISpot for Rotarix-specific B cells. (**D**) Representative images of B cell ELISpot wells. Statistical significance was determined by one-way ANOVA. Tukey's adjusted pair-wise comparisons are shown for $p < 0.05$ (* $p < 0.05$; ** $p < 0.01$; *** $p < 0.001$; **** $p < 0.0001$).

4. Discussion

The species-specificity of rotaviruses has made development of a small animal model to study human rotavirus strains a significant challenge. This has hampered efforts to characterise the immune responses to human rotavirus by experimental infections, and means that a robust pre-clinical system to analyse the efficacy of vaccine candidates and therapeutics has been lacking. To address this, we have successfully generated a novel approach to permit replication of rotaviruses encoding human rotavirus proteins in mice, and shown that these mice generate robust antibody responses to human rotavirus proteins. This was achieved using reverse genetics to create a chimeric rotavirus that encodes the outer capsid proteins of a human rotavirus. We verified that whereas human rotavirus stains replicate poorly in mice and are not immunogenic, a chimeric rotavirus replicates to a high titre and enables characterisation and quantification of human-rotavirus-outer-capsid-specific antibody responses. These viruses could also facilitate future analysis of additional immune responses to human outer capsid proteins, including B cell memory induction and VP4- and VP7-specific T cells.

A common alternative approach to studying non-murine viruses in mouse models is to use immunocompromised mice [15,16]. This has been used to permit replication of non-murine rotaviruses in mouse models, e.g., STAT1 knockout mice [6,17], but this does not provide a comprehensive overview of the interactions between the innate and adaptive immune responses. Therefore, use of our chimeric virus system has an advantage over the use of immunocompromised mice, as an immune response that more closely resembles the complexity and functionality seen in humans is induced. This generates a more representative picture of the multifaceted immune response to the outer capsid epitopes that occurs in a human rotavirus infection. It is acknowledged that the use of humanised mice could be a means of advancing this model further, but use of the widely available wild-type 129S6/SvEvTac mouse strain makes our approach more accessible.

We propose a number of different situations where this chimeric rotavirus could be valuable for advancing our understanding of how to control human rotavirus infections. Firstly, the ability to study antibody responses to human rotavirus outer capsid proteins could be valuable for investigating a number of factors that have been associated with reduced rotavirus vaccine efficacy in low- and middle-income countries. For example, maternal antibodies have been correlated with a reduced ability of infants to seroconvert following vaccination in a number of vaccine clinical trials, yet the target of these interfering maternal antibodies is unclear [18,19]. Infection of female mice with one strain, and then infection of their pups with another would determine whether antibodies targeting the outer capsid protein are responsible for interference. A second situation where these viruses could be useful is in pre-clinical vaccine trials. Whereas immune responses to mice vaccinated with new strategies targeting human rotaviruses, e.g., recently described rotavirus VP8 * mRNA vaccines [20], can be readily studied in mice, the ability of these immune responses to protect against human rotavirus infection is not possible using standard strains. A chimeric virus infection would provide a solution for this. Finally, chimeric viruses could be used to test new therapeutic strategies targeting the human outer capsid protein. This would be especially useful for testing monoclonal antibodies specific for human rotavirus strains [21,22].

One potential limitation of our model is the inability of the human chimeric virus to recapitulate the gastrointestinal disease seen in natural species-specific rotavirus infections. Current rotavirus vaccines do not induce sterilising immunity, but instead reduce the severity of clinical signs following infection. The ideal model system would therefore induce gastroenteritis in mice in order to enable testing of vaccine candidates or therapeutics that aim to reduce the severity of clinical disease. Whilst the viral replication in mice infected with rotaviruses with murine or human outer capsid proteins was similar, there was an interesting decrease in pathogenicity when switching from mouse to human. We found that the incidence of diarrhoea in pups infected with the rotavirus with a human capsid was significantly reduced. It is known that the pathogenesis of diarrhoea in rotavirus infection is multi-factorial, with reduced epithelial absorption, NSP4 enterotoxins, and activation of the nervous system all reported to be involved [23]. As the only difference between our two chimeric viruses was the outer capsid protein, this eliminates a possible role for NSP4, and instead suggests that the interaction of the outer capsid proteins and the murine intestinal tract is important. We hypothesise that differences in outer capsid proteins alter the region of the intestines where the virus preferentially binds and replicates. This could be further explored by immunohistochemistry of the entire intestinal tract post-infection.

An additional limitation of our outer capsid chimera approach is that this model does not enable study of a human-rotavirus-VP6-specific immune response. This is an issue, as the middle capsid protein VP6 is highly immunogenic and known to be the target of many rotavirus-specific antibodies in humans [24,25]. Furthermore, VP6-specific antibodies have been shown to be protective in mouse models [13,26]. The absence of human-rotavirus-specific VP6 in our chimeric virus system means the repertoire of antibodies induced by chimeric viruses will not fully recapitulate that induced by natural human rotavirus strains. Similarly, any T cell responses to human strain VP6 will not be evident. One possible solution to this issue could be to generate a chimeric rotavirus with human VP4, VP7, and VP6 on a murine backbone. However, this is predicted to be technically challenging, as VP6 plays a crucial role in the structure and function of rotaviruses, and therefore a human strain VP6 may not be compatible with a murine rotavirus backbone.

5. Conclusions

In conclusion, we have shown that using reverse genetics to manipulate rotaviruses can be an effective strategy to study human rotaviruses in an immunocompetent pre-clinical model. This approach has the potential to facilitate future vaccine and therapeutic development against a childhood pathogen whose global disease burden is still unacceptably high.

Author Contributions: Conceptualization, C.A.W., S.R.P. and S.L.C.; funding acquisition, C.A.W., S.R.P. and S.L.C.; investigation, S.W., T.L.C., T.K., T.M.L., and V.A.P.; methodology, S.D.; resources, S.D. and S.R.P.; writing—original draft, S.W. and S.L.C.; writing—review and editing, T.L.C., T.K. and S.D. All authors have read and agreed to the published version of the manuscript.

Funding: This work was supported by a Multi-Investigator Seed Grant (MISG) from Cornell University, New York, USA, and a Wellcome Trust Clinical Research Career Development Fellowship to S.L.C. (211138/A/18/Z).

Institutional Review Board Statement: All mouse work was approved by the Cornell University Institutional Animal Care and Use Committee (IACUC), Protocol 2022-0152. Date of approval September 2022.

Informed Consent Statement: Not applicable.

Data Availability Statement: All the relevant data are provided in this paper.

Acknowledgments: We thank the staff at the Baker Institute for Animal Health for helping to care for our mice, in particular Julie Reynolds, and John Parker for kindly gifting MA104 cells.

Conflicts of Interest: The authors declare no conflicts of interest. The funders had no role in the design of the study, collection, analyses, interpretation of data, manuscript writing, or in the decision to publish the results.

References

1. Ruiz-Palacios, G.M.; Pérez-Schael, I.; Velázquez, F.R.; Abate, H.; Breuer, T.; Clemens, S.C.; Cheuvart, B.; Espinoza, F.; Gillard, P.; Innis, B.L.; et al. Safety and Efficacy of an Attenuated Vaccine against Severe Rotavirus Gastroenteritis. *N. Engl. J. Med.* **2006**, *354*, 11–22. [CrossRef] [PubMed]
2. Vesikari, T.; Matson, D.O.; Dennehy, P.; Van Damme, P.; Santosham, M.; Rodriguez, Z.; Dallas, M.J.; Heyse, J.F.; Goveia, M.G.; Black, S.B.; et al. Safety and Efficacy of a Pentavalent Human–Bovine (WC3) Reassortant Rotavirus Vaccine. *N. Engl. J. Med.* **2006**, *354*, 23–33. [CrossRef] [PubMed]

3. Cates, J.E.; Tate, J.E.; Parashar, U. Rotavirus vaccines: Progress and new developments. *Expert Opin. Biol. Ther.* **2022**, *22*, 423–432. [CrossRef] [PubMed]
4. Desselberger, U. Differences of Rotavirus Vaccine Effectiveness by Country: Likely Causes and Contributing Factors. *Pathogens* **2017**, *6*, 65. [CrossRef] [PubMed]
5. Caddy, S.; Papa, G.; Borodavka, A.; Desselberger, U. Rotavirus research: 2014–2020. *Virus Res.* **2021**, *304*, 198499. [CrossRef] [PubMed]
6. Feng, N.; Yasukawa, L.L.; Sen, A.; Greenberg, H.B. Permissive Replication of Homologous Murine Rotavirus in the Mouse Intestine Is Primarily Regulated by VP4 and NSP1. *J. Virol.* **2013**, *87*, 8307–8316. [CrossRef]
7. Cook, N. The zoonotic potential of rotavirus. *J. Infect.* **2004**, *48*, 289–302. [CrossRef] [PubMed]
8. Kanai, Y.; Komoto, S.; Kawagishi, T.; Nouda, R.; Nagasawa, N.; Onishi, M.; Matsuura, Y.; Taniguchi, K.; Kobayashi, T. Entirely plasmid-based reverse genetics system for rotaviruses. *Proc. Natl. Acad. Sci. USA* **2017**, *114*, 2349–2354. [CrossRef] [PubMed]
9. Kanai, Y.; Kobayashi, T. Rotavirus reverse genetics systems: Development and application. *Virus Res.* **2021**, *295*, 198296. [CrossRef]
10. Desselberger, U. Rotaviruses. *Virus Res.* **2014**, *190*, 75–96. [CrossRef]
11. Sánchez-Tacuba, L.; Kawagishi, T.; Feng, N.; Jiang, B.; Ding, S.; Greenberg, H.B. The Role of the VP4 Attachment Protein in Rotavirus Host Range Restriction in an In Vivo Suckling Mouse Model. *J. Virol.* **2022**, *96*, e00550-22. [CrossRef] [PubMed]
12. Sánchez-Tacuba, L.; Feng, N.; Meade, N.J.; Mellits, K.H.; Jaïs, P.H.; Yasukawa, L.L.; Resch, T.K.; Jiang, B.; López, S.; Ding, S.; et al. An Optimized Reverse Genetics System Suitable for Efficient Recovery of Simian, Human, and Murine-Like Rotaviruses. *J. Virol.* **2020**, *94*, e01294-20. [CrossRef] [PubMed]
13. Caddy, S.L.; Vaysburd, M.; Wing, M.; Foss, S.; Andersen, J.T.; O'Connell, K.; Mayes, K.; Higginson, K.; Iturriza-Gómara, M.; Desselberger, U.; et al. Intracellular neutralisation of rotavirus by VP6-specific IgG. *PLoS Pathog.* **2020**, *16*, e1008732. [CrossRef] [PubMed]
14. Ishida, S.; Feng, N.; Gilbert, J.M.; Tang, B.; Greenberg, H.B. Immune Responses to Individual Rotavirus Proteins following Heterologous and Homologous Rotavirus Infection in Mice. *J. Infect. Dis.* **1997**, *175*, 1317–1323. [CrossRef] [PubMed]
15. Winkler, C.W.; Peterson, K.E. Using immunocompromised mice to identify mechanisms of Zika virus transmission and pathogenesis. *Immunology* **2018**, *153*, 443–454. [CrossRef] [PubMed]
16. Kutle, I.; Dittrich, A.; Wirth, D. Mouse Models for Human Herpesviruses. *Pathogens* **2023**, *12*, 953. [CrossRef] [PubMed]
17. Vancott, J.L.; McNeal, M.M.; Choi, A.H.C.; Ward, R.L. The Role of Interferons in Rotavirus Infections and Protection. *J. Interferon Cytokine Res.* **2003**, *23*, 163–170. [CrossRef] [PubMed]
18. Otero, C.E.; Langel, S.N.; Blasi, M.; Permar, S.R. Maternal antibody interference contributes to reduced rotavirus vaccine efficacy in developing countries. *PLoS Pathog.* **2020**, *16*, e1009010. [CrossRef]
19. Mwila, K.; Chilengi, R.; Simuyandi, M.; Permar, S.R.; Becker-Dreps, S. Contribution of Maternal Immunity to Decreased Rotavirus Vaccine Performance in Low- and Middle-Income Countries. *Clin. Vaccine Immunol.* **2017**, *24*. [CrossRef]
20. Roier, S.; Mangala Prasad, V.; McNeal, M.M.; Lee, K.K.; Petsch, B.; Rauch, S. mRNA-based VP8* nanoparticle vaccines against rotavirus are highly immunogenic in rodents. *NPJ Vaccines* **2023**, *8*, 190. [CrossRef]
21. Langel, S.; Steppe, J.; Chang, J.; Travieso, T.; Webster, H.; Otero, C.; Williamson, L.; Crowe, J.; Greenberg, H.; Wu, H.; et al. Protective Transfer: Maternal Passive Immunization with a Rotavirus-Neutralizing Dimeric IgA Protects against Rotavirus Disease in Suckling Neonates. *bioRxiv* **2021**. [CrossRef]
22. Zha, M.; Yang, J.; Zhou, L.; Wang, H.; Pan, X.; Deng, Z.; Yang, Y.; Li, W.; Wang, B.; Li, M. Preparation of mouse anti-human rotavirus VP7 monoclonal antibody and its protective effect on rotavirus infection. *Exp. Ther. Med.* **2019**, *18*, 1384–1390. [CrossRef] [PubMed]
23. Crawford, S.E.; Ramani, S.; Tate, J.E.; Parashar, U.D.; Svensson, L.; Hagbom, M.; Franco, M.A.; Greenberg, H.B.; O'Ryan, M.; Kang, G.; et al. Rotavirus infection. *Nat. Rev. Dis. Primers* **2017**, *3*, 17083. [CrossRef]
24. Svensson, L.; Sheshberadaran, H.; Vene, S.; Norrby, E.; Grandien, M.; Wadell, G. Serum Antibody Responses to Individual Viral Polypeptides in Human Rotavirus Infections. *J. Gen. Virol.* **1987**, *68*, 643–651. [CrossRef] [PubMed]
25. Johansen, K.; Granqvist, L.; Karlén, K.; Stintzing, G.; Uhnoo, I.; Svensson, L. Serum IgA immune response to individual rotavirus polypeptides in young children with rotavirus infection. *Arch. Virol.* **1994**, *138*, 247–259. [CrossRef]
26. Burns, J.W.; Siadat-Pajouh, M.; Krishnaney, A.A.; Greenberg, H.B. Protective Effect of Rotavirus VP6-Specific IgA Monoclonal Antibodies That Lack Neutralizing Activity. *Science* **1996**, *272*, 104–107. [CrossRef]

 *viruses*

MDPI

Article

Towards the Development of a Minigenome Assay for Species A Rotaviruses

Ola Diebold [†], Shu Zhou, Colin Peter Sharp, Blanka Tesla, Hou Wei Chook, Paul Digard * and Eleanor R. Gaunt *

Virology Division, Roslin Institute, University of Edinburgh, Easter Bush Campus, Midlothian EH25 9RG, UK
* Correspondence: paul.digard@roslin.ed.ac.uk (P.D.); elly.gaunt@ed.ac.uk (E.R.G.)
[†] Current address: Pandemic Sciences Institute, Centre for Human Genetics, Nuffield Department of Medicine, University of Oxford, Roosevelt Dr, Oxford OX3 7BN, UK.

Abstract: RNA virus polymerases carry out multiple functions necessary for successful genome replication and transcription. A key tool for molecular studies of viral RNA-dependent RNA polymerases (RdRps) is a 'minigenome' or 'minireplicon' assay, in which viral RdRps are reconstituted in cells in the absence of full virus infection. Typically, plasmids expressing the viral polymerase protein(s) and other co-factors are co-transfected, along with a plasmid expressing an RNA encoding a fluorescent or luminescent reporter gene flanked by viral untranslated regions containing *cis*-acting elements required for viral RdRp recognition. This reconstitutes the viral transcription/replication machinery and allows the viral RdRp activity to be measured as a correlate of the reporter protein signal. Here, we report on the development of a 'first-generation' plasmid-based minigenome assay for species A rotavirus using a firefly luciferase reporter gene.

Keywords: rotavirus; minigenome; RNA-dependent RNA polymerase; reporter assay

Citation: Diebold, O.; Zhou, S.; Sharp, C.P.; Tesla, B.; Chook, H.W.; Digard, P.; Gaunt, E.R. Towards the Development of a Minigenome Assay for Species A Rotaviruses. *Viruses* **2024**, *16*, 1396. https://doi.org/10.3390/v16091396

Academic Editors: Ulrich Desselberger, John T. Patton and Feng Li

Received: 24 April 2024
Revised: 12 August 2024
Accepted: 26 August 2024
Published: 31 August 2024

1. Introduction

Rotaviruses (RVs) are segmented, double-stranded RNA (dsRNA) viruses that cause acute gastroenteritis in infants, young children and livestock worldwide [1]. Currently, there are 11 distinct RV species (A–L, no E) with species A being the most predominant, accounting for over 90% of infections in humans and animals [2–6].

The RV virion is a non-enveloped triple-layered particle (TLP) containing 11 segments of dsRNA as its genome [7,8]. The core shell is formed by 60 asymmetric dimers of the viral protein 2 (VP2) and is surrounded by an intermediate layer of VP6 forming the transcriptionally active, non-infectious double-layered particle (DLP) [8,9]. The polymerase complex, composed of the RNA-dependent RNA polymerase (RdRp) (VP1) and the RNA-capping enzyme (VP3), is anchored at the five-fold axes through simultaneous interactions with multiple subdomains of VP2 [10,11]. The dsRNA segments are thought to be organised within the core in a way that each genome segment interacts with one specific polymerase complex [9,12].

The RdRp has a cage-like structure, with four tunnels leading to a catalytic core (residues 333–778) enclosed between the N-terminal (residues 1–332) and C-terminal domains (residues 779–1088) [13]. During transcription, the RdRp synthesises the capped, non-polyadenylated, positive-sense RNA ((+)RNA) transcripts from the minus strand of the genomic dsRNA, which are extruded out of the DLP into the cytoplasm [14]. The (+)RNA functions as mRNA for viral protein translation and as templates for the synthesis of new dsRNA genomes [15]. The cap-binding site of the N-terminal domain of VP1 splits the dsRNA genome through its interaction with the 5' conserved m7GpppGGC residue of (+)RNA present in all the RV segments [16,17]. After a short part of the helix is unwound, the unpaired negative sense RNA ((−)RNA) traverses towards the active site of the RdRp and immediately pairs with complementary NTPs within the core that form a backbone of the nascent RNA [18]. The dsRNA genome is pushed along by the newly synthesised nascent RNA backbone until it reaches the C-terminal domain of VP1, where the coding strand reanneals with the template and reforms the dsRNA genome [13]. The presence of distinct exit tunnels ensures that the nascent RNA is released into the cytoplasm while the $(-)$RNA is reused in subsequent rounds of $(+)$RNA synthesis [13,19].

In RVs, highly conserved *cis*-acting elements that enhance $(-)$RNA synthesis were also shown to be present at the 5′-end of $(+)$RNA, which sometimes extended into the coding region [17,20–23]. Previous studies using in vitro replication systems showed that the complementary base pairing of 5′ and 3′ regions of each segment is predicted to facilitate RNA circularisation by forming panhandle structures where the 3′-GACC conserved terminal sequence extends as a single-stranded tail [17,24,25]. The RdRp specifically recognises the conserved consensus sequences at the 3′-end of $(+)$RNA to initiate $(-)$RNA synthesis during genome replication [16]. This interaction is catalytically inactive and requires the N-terminal domain of VP2, which leads to conformational changes in the priming loop within the catalytic core of VP1, stabilising the initiating nucleotide in the priming site of RdRp [26–29]. This correct alignment results in the formation of the first phosphodiester bond of the $(-)$RNA product [30]. Simultaneously, the priming loop retracts, allowing elongation of the dsRNA product out of the polymerase [13]. The ratio of VP1:VP2 required to achieve this replicase activity was shown to be 1:10, the same ratio that forms the vertices of the core [17,31]. Studies showed that assembly of VP2 into cores was required for RNA replication and encapsidation of VP1 and VP3, demonstrating its direct role in core assembly and the packaging of newly made dsRNA products [16,29].

The above understanding is derived from in vitro biochemistry experiments, which typically require laborious protein purification and which cannot easily interrogate host interactions or rapidly test mutant viral polypeptides. While reverse genetics systems for RVs now exist [32–35], they do not easily allow separate interrogation of viral transcription. For other Baltimore groups, the study of the RdRp function of plus-strand viruses is readily accessible due to minigenome assays developed in the 1990s [36], with the development of assays for the study of minus-strand virus RdRps following shortly thereafter [37–39]. Thus, the lack of a minigenome system for dsRNA viruses is a major roadblock to the study of the RdRps of dsRNA viruses.

In this study, we aimed to establish a plasmid-based minigenome assay to recapitulate viral transcription and replication, using a luciferase reporter construct flanked by viral UTRs. We found that the luciferase signal could be generated by expressing the subset of viral structural proteins required for DLP formation, if the ratios of the VP1:VP2 constructs were optimised. Mutations of the conserved residues in the catalytic core of the VP1 RdRp only modestly reduced the reporter activity but negated the virus rescue, suggesting that the reporter signal may also be amplified by viral proteins other than the VP1 RdRp. This new 'first generation' plasmid-based minigenome system has established a potential model for measuring the polymerase activity in vitro, with several avenues for how to improve this system now possible.

2. Materials and Methods

Reporter segment construction. The constructs were designed to encode the firefly luciferase gene in either a positive or negative orientation, flanked by 5′- and 3′-UTRs under a bacteriophage T7 RNA promoter (T7P), containing sequences for antigenomic hepatitis delta virus (HDV) ribozyme and T7 transcription terminator (T7T) sequences at the 3′-end. These constructs were synthesised by Invitrogen GeneArt on pMA (ampicillin resistance) vectors. The plasmids were amplified by transformation into chemically competent *E. coli* DH5α and purified using the QIAGEN® Plasmid Midi Kit (QIAGEN, Manchester, UK) according to the manufacturer's protocol. The inserts in each plasmid were verified by Sanger sequencing (GATC Biotech or Genewiz, Germany) using the primers listed in Table 1. The sequence results were analysed in SSE v1.4 software [40].

Table 1. Sequences of primers used in this study.

Target gene	Sequence (5′ to 3′)	Use
Fluc in pMA plasmid	TAATACGACTCACTATAGGG TCGTCCACTCGGATGGCTA	Sequence 5′- and 3′-plasmids containing Fluc gene
VP1 plasmid	GGAAGGAGAGATGTACCAGGA	Sequence mutations of GDD motif in VP1 plasmid

Cell lines. BSR-T7 cells, a derivative of baby hamster kidney fibroblasts (BHK-21 cells), constitutively expressing T7 RNA polymerase, were cultured in complete cell culture medium consisting of Glasgow's Minimal Essential Medium (GMEM) (Gibco) supplemented with 1% tryptose phosphate broth (TPB) (Gibco), heat inactivated 10% foetal bovine serum (FBS) (Gibco) and 1% penicillin–streptomycin (Gibco). The cells were a kind gift from the laboratory of Prof. Massimo Palmarini (MRC-University of Glasgow Centre for Virus Research, UK). The cells were passaged twice weekly and maintained at 37 °C, 5% CO_2. At every fifth passage, the G-418 selection drug (1 mg/mL) (Scientific Laboratory Supplies) was added to the cell media.

Site-directed mutagenesis. Site-directed mutagenesis was performed on the RF VP1 plasmid using the QuikChange II Site-Directed Mutagenesis Kit (Agilent Technologies, Cheadle, UK) according to the manufacturer's instructions but using half-volume reactions. The thermal cycling parameters were as follows: 2 min denaturing at 95 °C, followed by 18 cycles of 30 s denaturing at 95 °C, 1 min primer annealing at 55 °C, 6 min elongation at 68 °C, with final 10 min elongation at 68 °C. The PCR products were digested using 1 μL of DpnI restriction enzyme to remove parental methylated DNA before transformation into competent *E. coli* cells. The products were visualised using gel electrophoresis. Successful mutagenesis was confirmed by Sanger sequencing.

Virus rescue. RV RF strain viruses and derivatives thereof were recovered using our previously described protocol [32]. In summary, BSR-T7 cells in 6-well plates were co-transfected with 11 plasmids corresponding to each RV genome segment (2.5 μg for plasmids encoding NSP2 and NSP5; 0.8 μg for the remaining plasmids) using 16 μL Lipofectamine 2000 (Invitrogen) per transfection. After 24 h incubation, MA104 cells (1×10^5 cells/well) were added to the transfected BSR-T7 cells and co-cultured for 4 days in FBS-free Dulbecco's Modified Eagle Medium (DMEM) (Sigma-Aldrich, Gillingham, UK) supplemented with 0.5 μg/mL porcine pancreatic trypsin type IX (Sigma-Aldrich). The co-cultured cells were then lysed three times by freeze/thaw and the lysates were incubated with trypsin at a final concentration of 10 μg/mL for 30 min to activate the virus. The lysates were then transferred to fresh MA104 cells in T25 flasks with 0.5 μg/mL porcine pancreatic trypsin type IX for up to 7 days and the viruses were harvested. Mock preparations with the mutated segment omitted were generated for use as negative controls throughout. All the rescue experiments were performed three times for each virus. The viruses were titred by plaque assays, and the presence of mutations in the VP1 gene segment was confirmed by Sanger sequencing (GATC Biotech or Genewiz, Germany).

Plaque assay. Plaque assays for RVs were performed using adapted methods [41,42]. Confluent monolayers of MA104 cells in 6-well plates were washed with FBS-free DMEM and infected with 800 μL of ten-fold serially diluted virus for 1 h at 37 °C 5% CO_2. Following virus adsorption, 2 mL/well overlay medium was added (1:1 ratio of 2.4% Avicel (FMC Biopolymer, Philadelphia, USA) and FBS-free DMEM supplemented with 0.5 μg/mL porcine pancreatic trypsin type IX) and incubated for 4 days. The cells were then fixed for 1 h with 1 mL/well of 10% neutral buffered formalin (CellPath, Newtown, UK) and stained for 1 h with 0.1% Toluidine blue (Sigma-Aldrich) dissolved in H_2O.

Luciferase assay. At 70–80% confluency, BSR-T7 cells in a 24-well plate were transfected with plasmid DNA using Lipofectamine 2000 reagent (Invitrogen, Loughborough, UK) according to the manufacturer's protocol. Plasmid DNA and Lipofectamine reagent (1 μL of Lipofectamine for 1 μg of DNA) were separately diluted in 50 μL Opti-MEM. After a 5 min incubation at room temperature, the diluted mixes were combined and incubated for a further 25 min. During this time, the complete cell culture medium was changed to Opti-MEM (200 µL/well) and transfection mix was added to cells dropwise, which were then incubated for 48 h at 37 °C, 5% CO_2. Following incubation, the supernatant was removed and the cells were lysed in 150 µL of Active Lysis Buffer (Promega, Chilworth, UK). The luciferase activity was analysed using Luciferase Assay Reagent (Promega) and the signal was read on a Cytation 3 plate reader (BioTek, Vermont, USA). A positive control plasmid expressing the Fluc gene driven by the cytomegalovirus (CMV) immediate early promoter, pVR1255, was used as a positive control throughout and is referred to as '+ve'.

Statistical analysis. GraphPad Prism v9 was used for all the statistical analyses. Data are presented as the mean and standard error of the mean from at least three independent experiments with technical duplicates unless otherwise stated. The *p* values were determined by ratio-paired *t*-test and were considered statistically significant at <0.05.

3. Results

Minigenome reporter construct design. The highly conserved *cis*-acting signals in the 5′- and 3′-UTRs of all the RV segments are recognised by the viral RdRp for (+)RNA synthesis and dsRNA replication [16,17]. To set up a plasmid-based minigenome assay, synthetic DNA constructs were initially designed to express the firefly luciferase (Fluc) gene in either a positive or negative orientation flanked by the 5′- and 3′-UTRs from the RF strain NSP1 gene (Figure 1A; positive and negative sense reporters were named 5′-reporter and 3′-reporter respectively). NSP1 was selected by analogy with the influenza A virus minigenome assay, in which the open reading frame of NS1, also a broadly acting interferon pathway antagonist, is replaced by a reporter gene [43]. The viral UTRs were flanked by T7P and HDV ribozyme sequences, as used successfully in the development of the RV reverse genetics system [33]. Thus, transcription of the resulting vector would generate full length viral (+) single-stranded RNA transcripts containing native viral 5′ and 3′ termini [44]. If these RNAs were transcribed and/or replicated by the viral RdRp, the luciferase levels in the transfected cells would be expected to increase.

To determine the quantity of the reporter plasmids required that could be transfected while generating only minimal background signal, increasing amounts were transfected into BSR-T7 cells (Figure 1B). The positive control ('+ve') luciferase-expressing plasmid produced a strong luciferase signal that increased with the plasmid dose. In contrast, the RV reporter genes gave very low levels of signal, not significantly above the background of no luciferase gene at doses of 100 ng and lower (Figure 1B). Both reporters gave levels of signal that were above background at 200 ng, but this was still around 30 RLU or lower, so this was chosen as the amount of reporter construct to take forward for further assay development.

Reporter expression by RV polymerase. During RV reverse genetics, 11 plasmids (1 per genome segment) are co-transfected into BSR-T7 cells for successful virus rescue, meaning that viral polymerase is functional and is able to copy both transcript polarities, thereby generating the complete viral genome. We therefore predicted that co-transfection of either of our luciferase reporters with the full complement of reverse genetics plasmids would result in an amplification of the luciferase signal above the background. To test this, a dose-dependent titration was performed for all 11 plasmids in the presence of 200 ng of the 5′- and 3′-reporters to determine the minimal amount of RV plasmids needed to generate the strongest luciferase signal (Figure 2A; as in reverse genetics [32,34], NSP2 and NSP5 plasmids were used in 3.125X amounts relative to the other nine plasmids). The pVR1255-positive control plasmid produced a strong luciferase signal (Figure 2B,C). Throughout, both 5′- and 3′-reporter plasmids expressed alone gave similar levels of background to those seen in Figure 1B, and so the cognate background reporter signals were subtracted from all the readings from samples transfected with RV plasmids. The 5′-reporter yielded a 2–3 $\log_{10}$ increase in luminescence above the background between 25 and 200 ng of the 11 RG plasmids, with apparent saturation at 200 ng and a decrease at 400 ng (Figure 2B). The same trend was observed for the 3′-reporter, but the overall luminescence signals were around one $\log_{10}$ lower than for the 5′-reporter. The 5′-reporter was therefore considered to be more suitable than the 3′-reporter for the minigenome assay, with 100–200 ng (312.5–625 ng for NSP2 and NSP5) of each reverse genetics plasmid being the optimal amount.

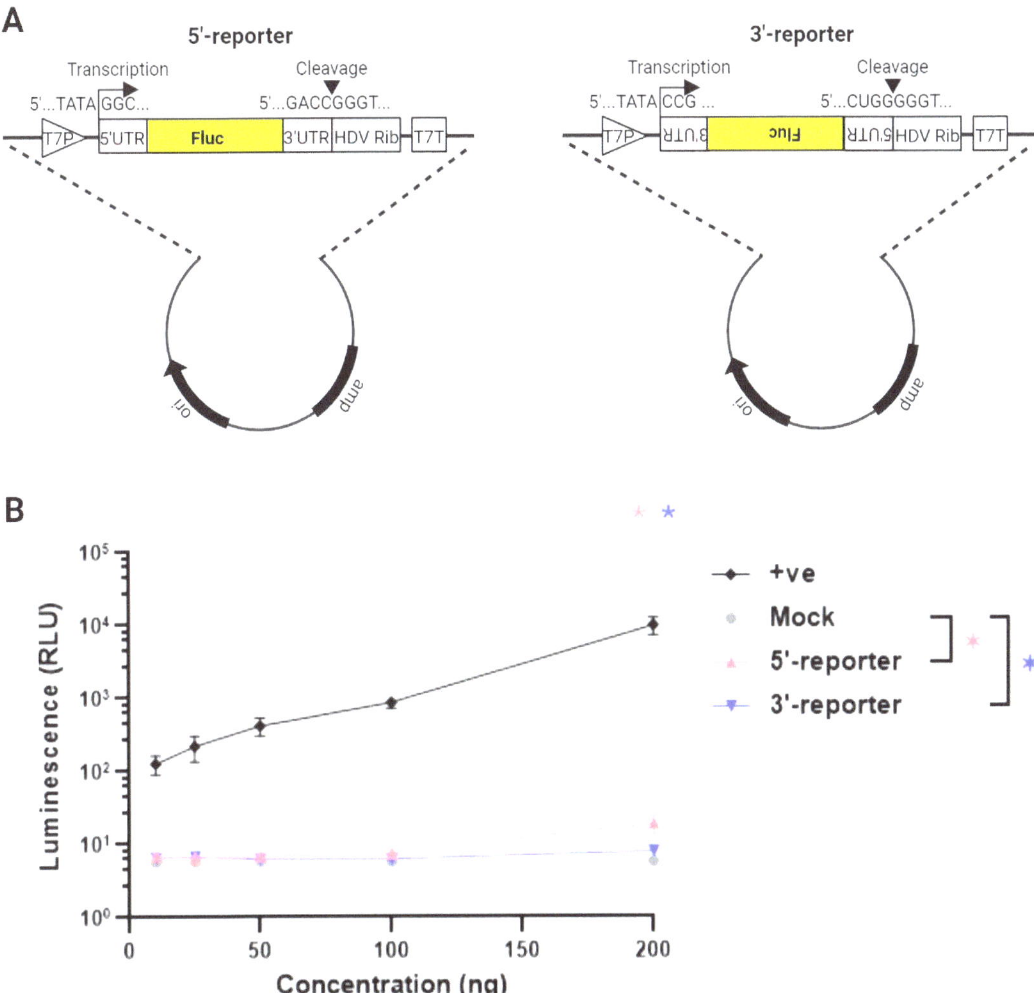

Figure 1. Minigenome reporter gene construct design. (**A**) Schematic of the reporter gene construct for the minigenome assay. Plasmids under a T7 promoter (T7P) encoding the Fluc gene in either the positive (5′-reporter) or negative sense (3′-reporter) flanked by the 5′- and 3′-UTRs. The plasmids included HDV ribozyme and T7 terminator sequences (T7T). Pink boxes, UTRs; yellow box, firefly luciferase ORF; blue box, HDV ribozyme sequence. (**B**) Dose-dependent titration of the 5′- and 3′-reporter plasmids. The pVR1255 plasmid expressing the Fluc gene was used as a positive control (denoted as '+ve'). The mock sample contained transfection reagent only. Data are the mean ± SEM from four independent experiments.

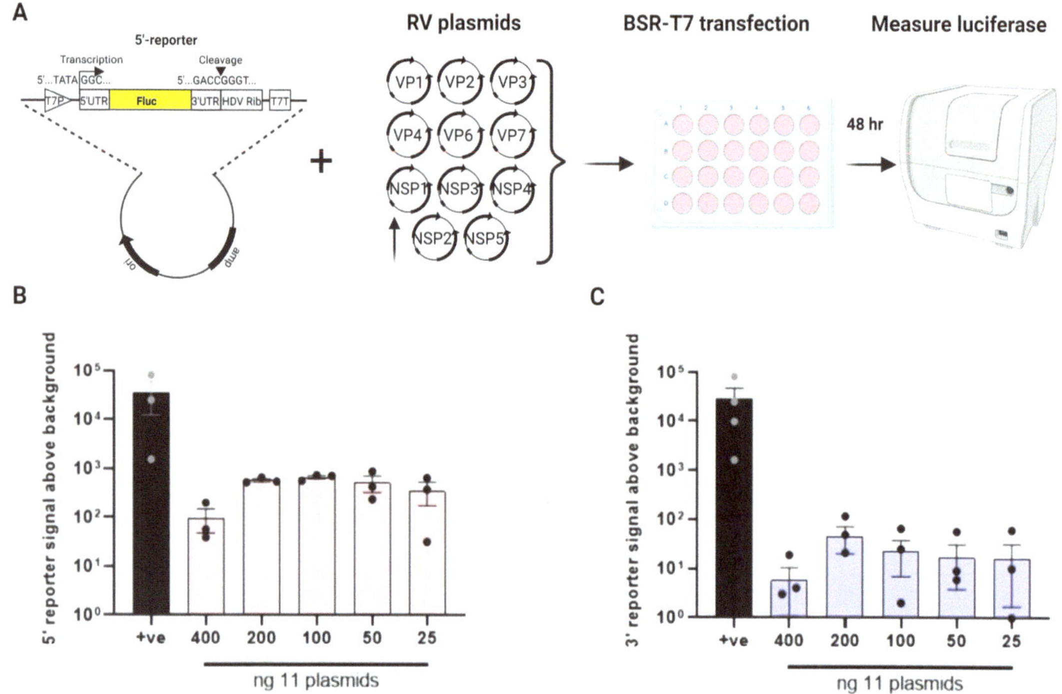

Figure 2. Reporter expression by rotavirus polymerase. (**A**) Schematic of the proposed minigenome assay. RV plasmids encoding each bovine RF stain gene were co-transfected with T7 reporter plasmids expressing the Fluc gene in either the positive (5′-reporter) or negative sense (3′-reporter). Luciferase activity was measured after 48 h post transfection. Dose-dependent titration of 11 RV plasmids with 200 ng of the 5′-reporter in (**B**) and with 200 ng of the 3′-reporter in (**C**), with the RLU values for the reporter-only signal ('background') subtracted. As in reverse genetics, the amount of plasmids expressing NSP2 and NSP5 genes was increased to scale.

Generation of an inactive RdRp. The increase in the luciferase signal in the presence of all the RV polypeptides was suggestive but not conclusive evidence of viral polymerase activity. Therefore, for further assay validation, we sought to generate a viral RdRp VP1 mutant lacking polymerase activity. The RV RdRp resembles a 'right-handed' architecture made up of the N-terminal domain, the core and the C-terminal domain, where the core is further split into the 'palm, finger and thumb' subdomains [19,27]. Ogden et al. (2012) showed that the conserved aspartate residues within the 'GDD' motif in the palm subdomain (Figure 3A) were critical for RNA synthesis [26]. These conserved aspartate residues, D631 and D632, were mutated to alanine by site-directed mutagenesis, creating the mutants D631A and D632A, respectively (Figure 3B). To test for successful inhibition of viral RdRp activity, rescues of the VP1 mutants were attempted using reverse genetics, and as expected, no virus was recovered in the presence of the mutated polymerase (Figure 3C). This confirmed that the mutations rendered the virus replication incompetent. The same VP1 mutants were therefore tested in the minigenome assay, with 11 RV gene segments (100 ng each except 312.5 ng for NSP2 and NSP5), co-transfected into BSR-T7 with 200 ng of the 5′-reporter. As before, a pVR1255-positive control produced a strong luciferase signal (Figure 3D). Unexpectedly, however, only a modest decrease in signal was observed for the D631A mutant and only the D632A mutant yielded a significant reduction. Thus, while the entire GDD motif appears to be essential for viral rescue, single amino acid mutations were tolerated in the minigenome assay. Combining D631A and D632A mutations emphasised the reduction in luciferase signal, but this was still above the background (Figure 3D). The

VP1 mutants possibly retained some polymerase activity, or some of the luciferase signal was generated by other viral proteins.

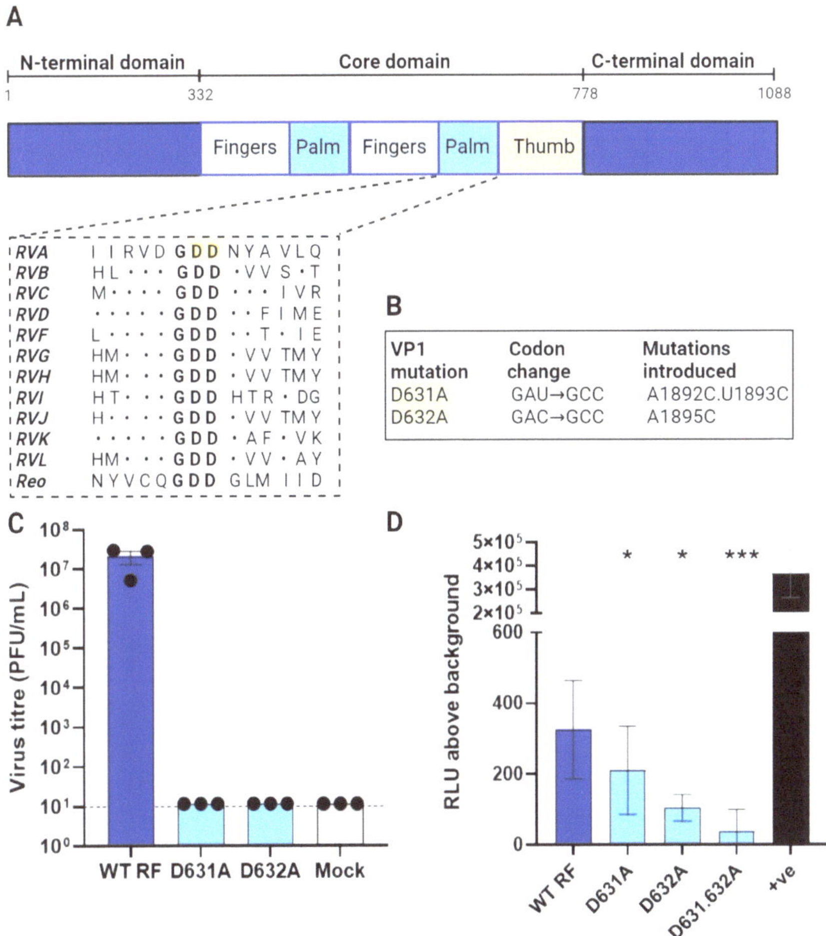

Figure 3. Generation of inactive polymerase. (**A**) VP1 shown as a linear schematic and coloured according to the domain organisation, with amino acid numbers labelled above. Adapted from [27], with alignments for RVH-RVL updated/added based on their reference sequences (KT962027.1, NC_026825.2, NC_055268.1, OQ934016.1, OM101015.1). The N-terminal and C-terminal domains (deep blue) flank the core domain containing the fingers (pale blue), palm (mid blue) and thumb (orange). Sequence-based alignment of RdRps across RV species and the related reovirus showing the conserved catalytic 'GDD' site. Dots indicate amino acid conservation. Highlighted in yellow are the two conserved aspartic acid residues targeted for mutagenesis. (**B**) Mutagenesis strategy for evolutionarily conserved aspartic acid residues in the VP1 catalytic domain. (**C**) Viral titres of WT RF and of VP1 mutants. (**D**) Minigenome assay for VP1 mutants. In all cases, all 11 RG plasmids were transfected (with 3.125×X amounts of NSP2 and NSP5 plasmids) along with 200 ng 5′-reporter. pVR1255 plasmid expressing Fluc gene was used as a positive control (denoted as '+ve'). The RLU values for the reporter-only signal ('background') were subtracted. * $p < 0.05$; *** $p < 0.001$.

Exploring the minimal requirements for the RV minigenome assay. Although 11 plasmids are required for viral rescue, we considered the possibility that not all 11 segments may be needed for the minigenome assay. During RV replication, upon cell entry, the loss of the outer protein layer of VP4 and VP7 triggers a conformational switch that induces polymerase activity in the now double-layered virus particle (DLP) [45]. The DLP contains a core comprising VP1, VP2 and VP3, surrounded by a shell of VP6. We therefore considered the possibility that reconstitution of the DLP alone may be sufficient for polymerase activity and whether transfection of constructs delivering only these four proteins with the 5′-reporter was sufficient to yield a luciferase signal. However, transfecting equimolar amounts of the VP1-3 and VP6 plasmids resulted in a dramatic loss of signal relative to the 11-plasmid system that was barely above the background (Figure 4A). Patton et al. (1997) showed that a molar ratio of 1 VP1 to 11 VP2, similar to that found in virion cores, produced the highest level of dsRNA synthesis in the cell-free system [46]. Therefore, we also tested whether adjusting the VP1:VP2 ratio to 1:11 would improve the signal. Indeed, when the amount of VP2 plasmid was increased 11-fold, the signal increased significantly and was only slightly lower than that of the 11-plasmid system.

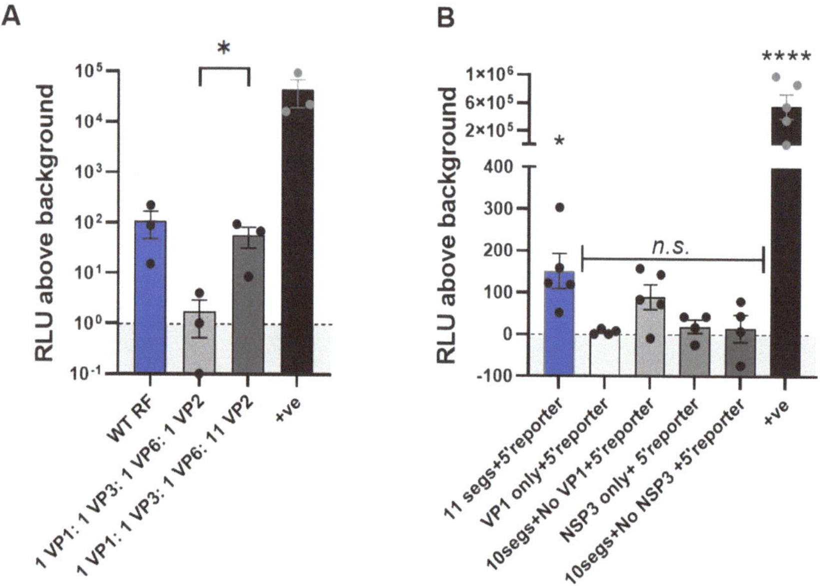

Figure 4. Measuring RdRp activity. Luciferase activity following co-transfection of RV plasmids with 200 ng of 5′-reporter. WT RF denotes co-transfection of all 11 RG plasmids (with 3.125× amounts of NSP2 and NSP5 plasmids) with the reporter. The pVR1255 plasmid expressing the Fluc gene was used as a positive control (denoted as '+ve'). The RLU values for the reporter-only signal ('background') were subtracted. (**A**) To test whether the number of RG plasmids expressed could be reduced, 4 plasmids corresponding to VP1, VP2, VP3 and VP6 were co-transfected with the 5′-reporter ('4 plasmids + 5′ rep). To test whether the 4-plasmid system could be improved upon, the amount of the VP2 plasmid was increased 11-fold ('VP1:VP2 ratio'). (**B**) To test the dependency of the system on VP1 and NSP3, polymerase assays were undertaken either by co-transfecting reporter with VP1 or NSP3 alone, or with ten segments minus either VP1 or NSP3. * $p < 0.05$; **** $p < 0.0001$; *n.s*, not significant.

We next sought to determine whether VP1 RdRp alone could support amplification of the reporter signal, and whether NSP3 would serve to enhance the translation. To test this, each of these constructs was expressed either alone with the luciferase reporter or reporter was expressed with constructs representing the remaining ten segments in the absence of VP1 or NSP3. However, in none of these cases was the signal significantly above the background (Figure 4B).

4. Discussion

Minigenome assays have been used to study the in-cell polymerase activity of a range of RNA viruses with single-stranded genomes, including influenza A virus [46], respiratory syncytial virus [47] and poliovirus [48]. As far as we are aware, in-cell reconstitution of viral polymerase activity has not been achieved previously for a virus with a double-stranded RNA genome, although polymerase activity has been assayed in a cell-free system for bluetongue virus [49]. Minigenome assays have yielded significant breakthroughs in the understanding of viral polymerase functions and domains, and this development represents an opportunity for such research questions to be applied to RVs.

Through this work, we unexpectedly found that mutating the highly conserved catalytic GDD motif of RV VP1 only modestly reduced the luciferase signal (Figure 3D). This may suggest that a significant part of the luciferase signal generated in the polymerase assay is attributable to the activity of viral protein(s) other than VP1. When either VP1 or NSP3 were excluded from the system but the remaining ten viral proteins were co-expressed with the reporter, the luciferase signal was not significantly above the background (Figure 4B), although some signal was observed in the 'no VP1' transfection. As exclusion of NSP3, or expression of NSP3 or VP1 alone, did not increase the reporter signal above the background, this suggests that if the reporter signal is indeed generated by viral proteins other than the VP1 polymerase, the way this arises is complicated by interactions between multiple viral proteins. The minor reduction in the luciferase signal brought about by the D631A and D632A mutations in VP1 is likely insufficient to explain why the same mutations abrogated virus production. Possibly, this domain of VP1 has multiple functions required to complete a virus lifecycle beyond its well-characterised role in dsRNA replication, such as genome packaging [19,26].

We have demonstrated that the reporter signal is generated when only components of the DLP are expressed, with no absolute requirement for non-structural proteins. NSP2 and NSP5 are together necessary and sufficient for the formation of viroplasms (or viroplasm-like structures), which form a sequestered environment for the accumulation of viral proteins and DLPs [45]. NSP3 replaces poly-A-binding protein in ribosomal complexes to bind viral non-polyadenylated transcripts, thereby enhancing their translation (and also reducing the translation of cellular polyadenylated transcripts) [50]; nevertheless, in its absence, viral polymerase activity was apparent (Figure 4A), demonstrating that translation of viral proteins occurs readily in NSP3's absence. To improve the dynamic range of this assay, co-expression of this subset of non-structural proteins could be explored. Co-expression of capping enzymes such as that of African swine fever virus, shown to be efficient in RV reverse genetics [51], may also augment the efficiency of this minigenome assay. Alternatively, it may be possible to reduce the number of plasmids required; we have shown that a signal can be generated when expressing only VP1, 2, 3 and 6, but it may be possible to reduce this further by removing VP 3 and/or 6; expression of VP1 alone did not yield a reporter signal significantly above the background (Figure 4B).

Further improvements to this system might include the co-expression of multiple viral gene segments from the same plasmid. This would reduce the number of plasmids being co-transfected and so theoretically increase the number of cells receiving the full complement of viral proteins required for viral polymerase activity to occur. There may be a trade-off due to the possibility of larger plasmids transfecting with poorer efficiency, but we have found that analogous multi-segment plasmids improve the minigenome assay efficiency in the influenza A virus system (unpublished data).

The polymerase assay system reported here was established using reverse genetics plasmids and so all viral gene segments were under a T7 promoter, necessitating the use of BSR-T7 cells. It is likely that higher transfection efficiencies would be achieved in HEK293T cells (infectable with RV in our hands), which are used for other minigenome assays, including that of influenza A virus. Testing this would require cloning of the constructs used into a different backbone so that viral genes are expressed under a mammalian promoter such as CMV. Alternatively, the T7 polymerase would need to be expressed in HEK293T cells.

Here, we have established a 'first generation' minigenome assay for the RF strain, which is a widely used and well-characterised lab strain of RV. The generalisability of this approach to other RV strains should be tested in the future using the analogous approach of generating a reporter construct encoded by flanking NSP1 UTRs of the cognate strain, analogous to strategies for maximising influenza A virus gene expression [52]. It is possible that the use of UTRs from other viral segments would yield a higher translational efficiency, which was not examined here. Reporter constructs comprising the UTRs from heterologous RV strains should also be explored as this would negate the requirement for strain-specific reporter constructs.

Author Contributions: Conceptualisation, P.D. and E.R.G.; methodology, O.D., S.Z., P.D. and E.R.G.; validation, O.D., S.Z. and E.R.G.; formal analysis, O.D. and E.R.G.; investigation, O.D., S.Z., C.P.S., B.T., H.W.C. and E.R.G.; resources, P.D. and E.R.G.; data curation, O.D. and E.R.G.; writing—original draft preparation, O.D. and E.R.G.; writing—reviewing and editing, ED, H.W.C., P.D. and E.R.G.; visualisation, O.D., P.D. and E.R.G.; supervision, C.P.S., P.D. and E.R.G.; project administration, E.R.G.; funding acquisition, P.D. and E.R.G. All authors have read and agreed to the published version of the manuscript.

Funding: P.D., E.R.G. and C.P.S. are supported by a BBSRC Institute Strategic Programme grant (BBS/E/RL/230002C). E.R.G., C.P.S. and B.T. are also supported by a Wellcome Trust/Royal Society Sir Henry Dale Fellowship (211222_Z_18_Z). The funders had no role in the study design, data collection and analysis, decision to publish, or preparation of the manuscript. For the purpose of open access, the authors have applied a CC 775 BY public copyright licence to any Author Accepted Manuscript version arising from this submission.

Data Availability Statement: All data presented in this manuscript are available at the discretion of the corresponding authors.

Acknowledgments: We want to thank the lab members, central support unit (CSU) and technical staff for their support and assistance with this project. We are grateful to Valeria Lulla and Ulrich Desselberger (both University of Cambridge) for the helpful discussions.

Conflicts of Interest: The authors declare no conflict of interest.

References

1. Troeger, C.; Khalil, I.A.; Rao, P.C.; Cao, S.; Blacker, B.F.; Ahmed, T.; Armah, G.; Bines, J.E.; Brewer, T.G.; Colombara, D.V.; et al. Rotavirus Vaccination and the Global Burden of Rotavirus Diarrhea among Children Younger than 5 Years. *JAMA Pediatr.* **2018**, *172*, 958–965. [CrossRef]
2. Bányai, K.; Kemenesi, G.; Budinski, I.; Földes, F.; Zana, B.; Marton, S.; Varga-Kugler, R.; Oldal, M.; Kurucz, K.; Jakab, F. Candidate new rotavirus species in Schreiber's bats, Serbia. *Infect. Genet. Evol.* **2017**, *48*, 19–26. [CrossRef]
3. Johne, R.; Schilling-Loeffler, K.; Ulrich, R.G.; Tausch, S.H. Whole Genome Sequence Analysis of a Prototype Strain of the Novel Putative Rotavirus Species L. *Viruses* **2022**, *14*, 462. [CrossRef]
4. Johne, R.; Tausch, S.H.; Ulrich, R.G.; Schilling-Loeffler, K. Genome analysis of the novel putative rotavirus species K. *Virus Res.* **2023**, *334*, 199171. [CrossRef] [PubMed]
5. Matthijnssens, J.; Otto, P.H.; Ciarlet, M.; Desselberger, U.; Van Ranst, M.; Johne, R. VP6-sequence-based cutoff values as a criterion for rotavirus species demarcation. *Arch. Virol.* **2012**, *157*, 1177–1182. [CrossRef] [PubMed]
6. Mihalov-Kovács, E.; Gellért, Á.; Marton, S.; Farkas, S.L.; Fehér, E.; Oldal, M.; Jakab, F.; Martella, V.; Bányai, K. Candidate new rotavirus species in sheltered dogs, Hungary. *Emerg. Infect. Dis.* **2015**, *21*, 660–663. [CrossRef] [PubMed]
7. Li, Z.; Baker, M.L.; Jiang, W.; Estes, M.K.; Prasad, B.V. Rotavirus architecture at subnanometer resolution. *J. Virol.* **2009**, *83*, 1754–1766. [CrossRef]
8. Settembre, E.C.; Chen, J.Z.; Dormitzer, P.R.; Grigorieff, N.; Harrison, S.C. Atomic model of an infectious rotavirus particle. *Embo J.* **2011**, *30*, 408–416. [CrossRef]
9. McClain, B.; Settembre, E.; Temple, B.R.; Bellamy, A.R.; Harrison, S.C. X-ray crystal structure of the rotavirus inner capsid particle at 3.8 A resolution. *J. Mol. Biol.* **2010**, *397*, 587–599. [CrossRef]
10. Estrozi, L.F.; Settembre, E.C.; Goret, G.; McClain, B.; Zhang, X.; Chen, J.Z.; Grigorieff, N.; Harrison, S.C. Location of the dsRNA-dependent polymerase, VP1, in rotavirus particles. *J. Mol. Biol.* **2013**, *425*, 124–132. [CrossRef]
11. Prasad, B.V.V.; Rothnagel, R.; Zeng, C.Q.Y.; Jakana, J.; Lawton, J.A.; Chiu, W.; Estes, M.K. Visualization of ordered genomic RNA and localization of transcriptional complexes in rotavirus. *Nature* **1996**, *382*, 471–473. [CrossRef]

12. Guglielmi, K.M.; McDonald, S.M.; Patton, J.T. Mechanism of intraparticle synthesis of the rotavirus double-stranded RNA genome. *J. Biol. Chem.* **2010**, *285*, 18123–18128. [CrossRef] [PubMed]

13. Jenni, S.; Salgado, E.N.; Herrmann, T.; Li, Z.; Grant, T.; Grigorieff, N.; Trapani, S.; Estrozi, L.F.; Harrison, S.C. In situ Structure of Rotavirus VP1 RNA-Dependent RNA Polymerase. *J. Mol. Biol.* **2019**, *431*, 3124–3138. [CrossRef] [PubMed]

14. Crawford, S.E.; Ding, S.D.; Greenberg, H.B.; Estes, M.K. Rotaviruses. In *Fields Virology: RNA Viruses*; Wolters Kluwer: Alphen aan den Rijn, The Netherlands, 2023.

15. Periz, J.; Celma, C.; Jing, B.; Pinkney, J.N.; Roy, P.; Kapanidis, A.N. Rotavirus mRNAS are released by transcript-specific channels in the double-layered viral capsid. *Proc. Natl. Acad. Sci. USA* **2013**, *110*, 12042–12047. [CrossRef] [PubMed]

16. Tortorici, M.A.; Broering, T.J.; Nibert, M.L.; Patton, J.T. Template recognition and formation of initiation complexes by the replicase of a segmented double-stranded RNA virus. *J. Biol. Chem.* **2003**, *278*, 32673–32682. [CrossRef]

17. Tortorici, M.A.; Shapiro, B.A.; Patton, J.T. A base-specific recognition signal in the $5'$ consensus sequence of rotavirus plus-strand RNAs promotes replication of the double-stranded RNA genome segments. *RNA* **2006**, *12*, 133–146. [CrossRef]

18. Ding, K.; Celma, C.C.; Zhang, X.; Chang, T.; Shen, W.; Atanasov, I.; Roy, P.; Zhou, Z.H. In situ structures of rotavirus polymerase in action and mechanism of mRNA transcription and release. *Nat. Commun.* **2019**, *10*, 2216. [CrossRef]

19. Lu, X.; McDonald, S.M.; Tortorici, M.A.; Tao, Y.J.; Vasquez-Del Carpio, R.; Nibert, M.L.; Patton, J.T.; Harrison, S.C. Mechanism for coordinated RNA packaging and genome replication by rotavirus polymerase VP1. *Structure* **2008**, *16*, 1678–1688. [CrossRef]

20. Barro, M.; Mandiola, P.; Chen, D.; Patton, J.T.; Spencer, E. Identification of sequences in rotavirus mRNAs important for minus strand synthesis using antisense oligonucleotides. *Virology* **2001**, *288*, 71–80. [CrossRef]

21. Chen, D.; Barros, M.; Spencer, E.; Patton, J.T. Features of the $3'$-consensus sequence of rotavirus mRNAs critical to minus strand synthesis. *Virology* **2001**, *282*, 221–229. [CrossRef]

22. Navarro, A.; Trask, S.D.; Patton, J.T. Generation of genetically stable recombinant rotaviruses containing novel genome rearrangements and heterologous sequences by reverse genetics. *J. Virol.* **2013**, *87*, 6211–6220. [CrossRef] [PubMed]

23. Patton, J.T.; Chnaiderman, J.; Spencer, E. Open reading frame in rotavirus mRNA specifically promotes synthesis of double-stranded RNA: Template size also affects replication efficiency. *Virology* **1999**, *264*, 167–180. [CrossRef]

24. Chen, D.; Patton, J.T. Rotavirus RNA replication requires a single-stranded $3'$ end for efficient minus-strand synthesis. *J. Virol.* **1998**, *72*, 7387–7396. [CrossRef]

25. Li, W.; Manktelow, E.; von Kirchbach, J.C.; Gog, J.R.; Desselberger, U.; Lever, A.M. Genomic analysis of codon, sequence and structural conservation with selective biochemical-structure mapping reveals highly conserved and dynamic structures in rotavirus RNAs with potential cis-acting functions. *Nucleic Acids Res.* **2010**, *38*, 7718–7735. [CrossRef] [PubMed]

26. Ogden, K.M.; Ramanathan, H.N.; Patton, J.T. Mutational analysis of residues involved in nucleotide and divalent cation stabilization in the rotavirus RNA-dependent RNA polymerase catalytic pocket. *Virology* **2012**, *431*, 12–20. [CrossRef] [PubMed]

27. Steger, C.; Brown, M.; Sullivan, O.; Boudreaux, C.; Cohen, C.; LaConte, L.; McDonald, S. In Vitro Double-Stranded RNA Synthesis by Rotavirus Polymerase Mutants with Lesions at Core Shell Contact Sites. *J. Virol.* **2019**, *93*, 10–1128. [CrossRef]

28. Tao, Y.; Farsetta, D.L.; Nibert, M.L.; Harrison, S.C. RNA synthesis in a cage--structural studies of reovirus polymerase lambda3. *Cell* **2002**, *111*, 733–745. [CrossRef]

29. Zeng, C.Q.; Estes, M.K.; Charpilienne, A.; Cohen, J. The N terminus of rotavirus VP2 is necessary for encapsidation of VP1 and VP3. *J. Virol.* **1998**, *72*, 201–208. [CrossRef]

30. Gridley, C.L.; Patton, J.T. Regulation of rotavirus polymerase activity by inner capsid proteins. *Curr. Opin. Virol.* **2014**, *9*, 31–38. [CrossRef]

31. Patton, J.T.; Jones, M.T.; Kalbach, A.N.; He, Y.W.; Xiaobo, J. Rotavirus RNA polymerase requires the core shell protein to synthesize the double-stranded RNA genome. *J. Virol.* **1997**, *71*, 9618–9626. [CrossRef]

32. Diebold, O.; Gonzalez, V.; Venditti, L.; Sharp, C.; Blake, R.A.; Tan, W.S.; Stevens, J.; Caddy, S.; Digard, P.; Borodavka, A.; et al. Using Species a Rotavirus Reverse Genetics to Engineer Chimeric Viruses Expressing SARS-CoV-2 Spike Epitopes. *J. Virol.* **2022**, *96*, e0048822. [CrossRef] [PubMed]

33. Kanai, Y.; Komoto, S.; Kawagishi, T.; Nouda, R.; Nagasawa, N.; Onishi, M.; Matsuura, Y.; Taniguchi, K.; Kobayashi, T. Entirely plasmid-based reverse genetics system for rotaviruses. *Proc. Natl. Acad. Sci. USA* **2017**, *114*, 2349. [CrossRef]

34. Komoto, S.; Fukuda, S.; Ide, T.; Ito, N.; Sugiyama, M.; Yoshikawa, T.; Murata, T.; Taniguchi, K. Generation of Recombinant Rotaviruses Expressing Fluorescent Proteins by Using an Optimized Reverse Genetics System. *J. Virol.* **2018**, *92*, e00588-18. [CrossRef] [PubMed]

35. Komoto, S.; Kanai, Y.; Fukuda, S.; Kugita, M.; Kawagishi, T.; Ito, N.; Sugiyama, M.; Matsuura, Y.; Kobayashi, T.; Taniguchi, K. Reverse Genetics System Demonstrates that Rotavirus Nonstructural Protein NSP6 Is Not Essential for Viral Replication in Cell Culture. *J. Virol.* **2017**, *91*, e00695-17. [CrossRef]

36. Percy, N.; Barclay, W.S.; Sullivan, M.; Almond, J.W. A poliovirus replicon containing the chloramphenicol acetyltransferase gene can be used to study the replication and encapsidation of poliovirus RNA. *J. Virol.* **1992**, *66*, 5040–5046. [CrossRef]

37. Sidhu, M.S.; Chan, J.; Kaelin, K.; Spielhofer, P.; Radecke, F.; Schneider, H.; Masurekar, M.; Dowling, P.C.; Billeter, M.A.; Udem, S.A. Rescue of Synthetic Measles Virus Minireplicons: Measles Genomic Termini Direct Efficient Expression and Propagation of a Reporter Gene. *Virology* **1995**, *208*, 800–807. [CrossRef] [PubMed]

38. Groseth, A.; Feldmann, H.; Theriault, S.; Mehmetoglu, G.; Flick, R. RNA Polymerase I-Driven Minigenome System for Ebola Viruses. *J. Virol.* **2005**, *79*, 4425–4433. [CrossRef]

39. Lutz, A.; Dyall, J.; Olivo, P.D.; Pekosz, A. Virus-inducible reporter genes as a tool for detecting and quantifying influenza A virus replication. *J. Virol. Methods* **2005**, *126*, 13–20. [CrossRef]

40. Simmonds, P. SSE: A nucleotide and amino acid sequence analysis platform. *BMC Res. Notes* **2012**, *5*, 50. [CrossRef]

41. Matrosovich, M.; Matrosovich, T.; Garten, W.; Klenk, H.-D. New low-viscosity overlay medium for viral plaque assays. *Virol. J.* **2006**, *3*, 63. [CrossRef]

42. Arnold, M.; Patton, J.T.; McDonald, S.M. Culturing, Storage, and Quantification of Rotaviruses. *Curr. Protoc. Microbiol.* **2009**, *15*, 15C-3. [CrossRef] [PubMed]

43. Wise, H.M.; Foeglein, A.; Sun, J.; Dalton Rosa, M.; Patel, S.; Howard, W.; Anderson, E.C.; Barclay, W.S.; Digard, P. A Complicated Message: Identification of a Novel PB1-Related Protein Translated from Influenza A Virus Segment 2 mRNA. *J. Virol.* **2009**, *83*, 8021–8031. [CrossRef]

44. Roner, M.R.; Joklik, W.K. Reovirus reverse genetics: Incorporation of the CAT gene into the reovirus genome. *Proc. Natl. Acad. Sci. USA* **2001**, *98*, 8036–8041. [CrossRef] [PubMed]

45. Desselberger, U. Rotaviruses. *Virus Res.* **2014**, *190*, 75–96. [CrossRef] [PubMed]

46. te Velthuis, A.; Long, J.; Barclay, W. Assays to Measure the Activity of Influenza Virus Polymerase. *Influenza Virus Methods Protoc.* **2018**, *1836*, 343–374.

47. Noton, S.L.; Cowton, V.M.; Zack, C.R.; McGivern, D.R.; Fearns, R. Evidence that the polymerase of respiratory syncytial virus initiates RNA replication in a nontemplated fashion. *Proc. Natl. Acad. Sci. USA* **2010**, *107*, 10226–10231. [CrossRef]

48. Paul, A.V.; Rieder, E.; Kim, D.W.; Boom, J.H.v.; Wimmer, E. Identification of an RNA Hairpin in Poliovirus RNA That Serves as the Primary Template in the In Vitro Uridylylation of VPg. *J. Virol.* **2000**, *74*, 10359–10370. [CrossRef]

49. Wehrfritz, J.M.; Boyce, M.; Mirza, S.; Roy, P. Reconstitution of bluetongue virus polymerase activity from isolated domains based on a three-dimensional structural model. *Biopolymers* **2007**, *86*, 83–94. [CrossRef]

50. Vende, P.; Piron, M.; Castagné, N.; Poncet, D. Efficient Translation of Rotavirus mRNA Requires Simultaneous Interaction of NSP3 with the Eukaryotic Translation Initiation Factor eIF4G and the mRNA 3′ End. *J. Virol.* **2000**, *74*, 7064–7071. [CrossRef]

51. Sánchez-Tacuba, L.; Feng, N.; Meade, N.J.; Mellits, K.H.; Jaïs, P.H.; Yasukawa, L.L.; Resch, T.K.; Jiang, B.; López, S.; Ding, S.; et al. An Optimized Reverse Genetics System Suitable for Efficient Recovery of Simian, Human, and Murine-Like Rotaviruses. *J. Virol.* **2020**, *94*, 10–1128. [CrossRef]

52. Zheng, H.; Palese, P.; García-Sastre, A. Nonconserved nucleotides at the 3′ and 5′ ends of an influenza A virus RNA play an important role in viral RNA replication. *Virology* **1996**, *217*, 242–251. [CrossRef] [PubMed]

Review

The Role of the Host Cytoskeleton in the Formation and Dynamics of Rotavirus Viroplasms

Janine Vetter [†], Melissa Lee [†] and Catherine Eichwald [*]

Institute of Virology, University of Zurich, 8057 Zurich, Switzerland; janine.vetter@uzh.ch (J.V.); melissa.lee2@uzh.ch (M.L.)
* Correspondence: ceichwald@vetvir.uzh.ch
† These authors contributed equally to this work.

Abstract: Rotavirus (RV) replicates within viroplasms, membraneless electron-dense globular cytosolic inclusions with liquid–liquid phase properties. In these structures occur the virus transcription, replication, and packaging of the virus genome in newly assembled double-layered particles. The viroplasms are composed of virus proteins (NSP2, NSP5, NSP4, VP1, VP2, VP3, and VP6), single- and double-stranded virus RNAs, and host components such as microtubules, perilipin-1, and chaperonins. The formation, coalescence, maintenance, and perinuclear localization of viroplasms rely on their association with the cytoskeleton. A stabilized microtubule network involving microtubules and kinesin Eg5 and dynein molecular motors is associated with NSP5, NSP2, and VP2, facilitating dynamic processes such as viroplasm coalescence and perinuclear localization. Key post-translation modifications, particularly phosphorylation events of RV proteins NSP5 and NSP2, play pivotal roles in orchestrating these interactions. Actin filaments also contribute, triggering the formation of the viroplasms through the association of soluble cytosolic VP4 with actin and the molecular motor myosin. This review explores the evolving understanding of RV replication, emphasizing the host requirements essential for viroplasm formation and highlighting their dynamic interplay within the host cell.

Keywords: rotavirus; viroplasm; cytoskeleton; microtubule; actin; molecular motors; lipid droplets; NSP5; NSP2; VP2; VP4

Citation: Vetter, J.; Lee, M.; Eichwald, C. The Role of the Host Cytoskeleton in the Formation and Dynamics of Rotavirus Viroplasms. *Viruses* **2024**, *16*, 668. https://doi.org/10.3390/v16050668

Academic Editors: Ulrich Desselberger and John T. Patton

Received: 24 March 2024
Revised: 22 April 2024
Accepted: 23 April 2024
Published: 25 April 2024

1. Rotavirus

Rotavirus (RV) was initially observed in 1963 via electron microscopy of feces samples of young monkeys and mice presenting diarrhea [1]. In humans, the virus was first described in 1973 in the duodenal mucosa of infants with acute nonbacterial gastroenteritis [2]. Fifty years later, RV infections are the leading cause of severe gastroenteritis and dehydration in infants and young animals [3]. In 2008, before worldwide RV vaccine programs, RV gastroenteritis led to 435′000 deaths worldwide, mainly in developing countries, and high-cost hospitalization in developed countries [4]. The introduction of vaccine programs reduced the disease burden by 85% in developed countries [5]; however, developing countries show a much more modest reduction in disease burden [6].

RV is a nonenveloped virus belonging to the order of *Reovirales* within the family *Sedoreoviridae*, where it forms the genus *Rotavirus* [7,8]. The genus *Rotavirus* currently entails nine different RV species, A-D and F-J, distinguished by serological criteria, host range, and sequence analysis [9]. The strains designated as RV species E were nonrecoverable from long-term storage, and no sequence information is available to support its existence, being consequently removed from the RV species list by the ICTV in 2019 [10]. Moreover, recent reports indicate the identification and sequencing of RV species K and L [11,12]. However, the ICTV has not yet approved them.

The mature RV virion is a nonenveloped, icosahedral (T = 13), triple-layered particle (TLP) [13] of about 100 nm in diameter [14]. The virion encapsidates one copy of each of the eleven double-stranded RNA (dsRNA) genome segments. Each genome segment encodes for one protein, six structural proteins (VP1, VP2, VP3, VP4, VP6, and VP7), which are incorporated into the mature virion, and five (or six) nonstructural proteins (NSP1, NSP2, NSP3, NSP4, NSP5, and, in certain strains, also NSP6) [15].

The spike protein, VP4, is cleaved in the intestine tract by a trypsin-like protease in two main products, VP8* (28kDa, amino acids 1-247) and VP5* (60 kDa, amino acids 248–776), that remain noncovalently associated with the infectious particle allowing the initiation of the RV entry [16,17]. In this context, VP8* initiates RV cell entry by attaching to various cellular glycans [18], among them terminal sialic acids [19,20] and histo-blood group antigens (HBGAs) [21]. After cell binding, RV favors entry to the cell via clathrin-mediated endocytosis, although some strains also use clathrin-independent pathways [22,23].

A common step in RV entry is the localization of the virus particles into early endosomes, where they are exposed to environmental changes, such as an acidic pH, low calcium concentrations, or other lysosomal components. Those factors seem to be involved in the entrance of the virus into the cytosol [24], but they appear to be strain-dependent, as some strains seem to profit from maturing endosomes, while others use late endosomes [25]. It is thought that VP5* is involved in forming pore-like structures in the endosomes, eventually allowing endosomal escape [26,27].

In this process, the outer layer of the virion is detached, and the double-layered particles (DLPs) are released in the cytosol. These DLPs become transcriptionally active [28], releasing into the cytosol capped, nonpolyadenylated (+)ssRNAs [29] for direct translation of the virus proteins required to (i) block the innate immune response of the host (NSP1 and VP3) and (ii) build viroplasms (NSP5, NSP2, and VP2) [30–33]. Moreover, increased levels of virus mRNA transcripts inhibit the translation of the host polyadenylated mRNAs [34].

Specific RV proteins accumulate within specialized cytosolic inclusions called viroplasms, where newly synthesized RV genome segments are packaged in newly formed cores, followed by the addition of a middle coat layer to form DLPs. Subsequently, DLPs exit the viroplasms via a not fully understood pathway. The current proposed mechanism involves the association of VP6 with NSP4 embedded in the membrane of the endoplasmic reticulum (ER) [35,36]. Simultaneously, the spike protein VP4 localizes between the viroplasm and the ER, associating with NSP4 [37]. These associations with NSP4 bring VP6 and VP4 in proximity, leading to the formation of a transiently enveloped DLP (eDLP) in the lumen of the ER [38]. In fact, eDLP reconstructions appear as DLPs with 60 trimeric VP4 spikes, which connect the particles to the transient envelope where VP7 and NSP4 are not discernible in images [38].

The assembly is completed in a poorly understood process by incorporating the outer-layer protein VP7, which is present in the ER [37,38]. The fully formed TLPs are then released either via cell lysis [39] or in an actin-dependent process from the cell surface [40–42].

2. Viroplasms

2.1. Spatial and Temporal Organization

Viroplasms are membraneless globular electron-dense cytosolic inclusions (Figure 1a). These structures are responsible for virus genome replication and the generation of new rotavirus virions. So far, only RVA has been shown experimentally to induce viroplasm formation; their formation in other species remains to be demonstrated. The viroplasms comprise NSP5, NSP4, NSP2, VP6, VP1, VP2, VP3, virus single- and double-stranded RNA, and host components such as tubulin, perilipin, the host proteasome, and cellular chaperonins [28,43–48]. Finally, viroplasms are adjacent to the ER, enriched in VP4 and VP7 [43,49].

Viroplasms are detected as early as 2 h postinfection (hpi), and their size steadily increases [50,51]. Interestingly, the number of viroplasms per cell decreases at 6 hpi, indicating coalescence between the structures and their dynamic nature [50,52]. Indeed, it has recently been shown that viroplasms are liquid-like inclusions [53]. Furthermore, during the infection, viroplasms move toward the perinuclear region of the cell in a process that is dependent on the microtubule (MT) cytoskeleton [52].

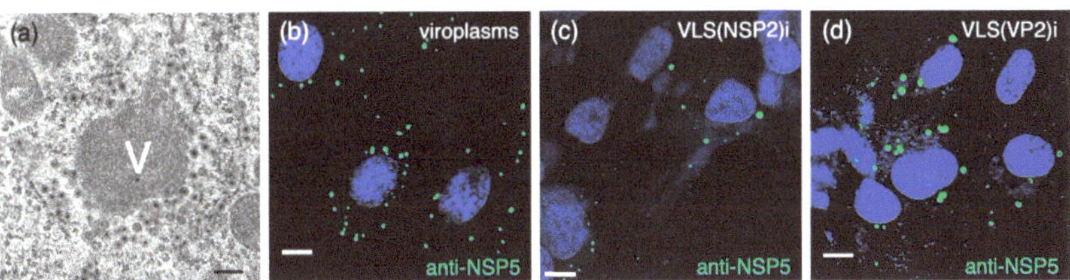

Figure 1. Comparison of viroplasms and VLSs. (**a**) Electron micrograph of viroplasm (V) at 6 hpi. The electron dense viroplasm structure is surrounded by the endoplasmic reticulum membrane filled with TLPs at diverse stages of maturation. Scale bar is 200 nm. Immunofluorescence of images of viroplasm (**b**), VLS (NSP2)i (**c**), and VLS (VP2)i (**d**) immunostained with anti-NSP5 (green). Nuclei are stained with DAPI (blue). Scale bar is 10 μm for **b–d**.

Remarkably, the expression of NSP5 in the presence of NSP2 or VP2 induces the formation of viroplasm-like structures (VLS) [54]. VLSs closely resemble the morphology of RV viroplasms (Figure 1b–d). Due to their simplicity, VLSs are a valuable model for studying viroplasms within a host using an in vivo approach, as they share characteristics such as coalescence and perinuclear condensation [50,54].

2.2. Replication Steps within Viroplasms

The exact mechanism of genome packaging and assembly of virions within the viroplasms is unclear. The current model assumes that pregenomic (+)ssRNAs are organized sequence-specifically through the assistance of NSP2 [55] in the viroplasms and packaged in the assembling core while simultaneously being replicated into dsRNA [56]. The filled cores move towards the periphery of the viroplasms, which are rich in VP6 for converting the cores into DLPs. These DLPs would produce (i) more (+)ssRNA [57] or (ii) migrate to the ER to become mature TLPs [58].

2.3. NSP5

Inhibition of NSP5 expression via RNA interference completely abolishes viroplasm formation and synthesis of genomic dsRNA as well as progeny virus, revealing the essential role of NSP5 in viroplasm formation [59]. While the complete structure of NSP5 remains unknown, it has been shown to form dimers and oligomers through its C-terminal region [60,61]. In this context, NSP5 has been shown to be an intrinsically disordered protein [60], which is consistent with a high propensity to phase separation of the viroplasms [53]. NSP5 is a hyperphosphorylated protein in infected cells, achieved by phosphorylation by cellular kinases, such as casein kinase (CK1alpha), and regulated via autoregulation and interaction with NSP2 and VP2 [62–67]. The phosphorylation cascade is critically dependent on the presence of a serine at position 67 [62]. A recombinant RV (rRV) harboring a point mutation in NSP5 in serine 67 to alanine (S67A) shows aberrant viroplasms. This observation suggests that hyperphosphorylation of NSP5 is crucial for viroplasm morphology [68]. Additionally, the study highlights the significance of the NSP5 tail region in the phosphorylation cascade and viroplasm formation [63,68]. Despite the initial notion that autokinase activity could be described, no kinase activity could be attributed to NSP5, which, in addition to being the primary driver of viroplasm formation, displays ATPase activity [61,63,69].

2.4. NSP2

Another critical protein in viroplasm formation is NSP2. Similarly to NSP5, inhibition of NSP2 expression also leads to impairment of viroplasm formation [70]. NSP2 self-assembles into donut-shaped octamers, as denoted with crystallographic and cryogenic electron microscopy analyses for species A, B, and C [71–74]. These multimers can interact with the RNA-dependent RNA polymerase VP1 and viral RNA [75,76]. Furthermore, it has been demonstrated that NSP2 is an RNA chaperone, capable of binding to RNA transcripts and consequently controlling their interaction and unfolding [77]. NSP2 has been linked to several enzymatic activities, among them a nucleoside diphosphate kinase-like activity [78], RNA-helix-destabilizing activities [78], and nucleoside triphosphatase (NTPase) activity [71]. NSP2 plays a direct role in viroplasm coalescence events [50]. NSP2 is found to be dispersed in the cytosol (dNSP2), and its phosphorylated version is exclusively found in the viroplasms (vNSP2) [79,80]. The phosphorylation of NSP2 occurs uniquely in S313 by CK1 alpha. Notably, the phosphorylation of NSP2 has been implicated in viroplasm formation, as evidenced by the delayed formation of viroplasms observed in a rRV harboring an NSP2 S313D phosphomimetic mutant [81]. Additionally, studies using a mutant NSP2 harboring a lysine-to-glutamic acid change in the C-terminal region revealed the importance of a flexible tail in viroplasm biogenesis and coalescence properties [82].

2.5. VP2

An often-overlooked protein in the context of viroplasms is VP2. Silencing of VP2 expression in infected cells reduces the number of viroplasms per cell [83]. VP2, primarily studied as the main structural core protein, is also an inducer of VLS formation when co-expressed with NSP5 [44]. It has been shown that VLS formation is critically dependent on the presence of the three amino acids, L124, V865, and I878, with residues highly conserved in VP2 of RV species A-H [84]. Previous studies have demonstrated that NSP2 [64] and VP2 [44,84] trigger the hyperphosphorylation cascade of NSP5. Additionally, VP2 has been implicated in modifying viroplasm perinuclear localization [52].

3. Host-Cell Cytoskeleton

3.1. Microtubules

MTs are a significant component of the cytoskeletal network in eukaryotic cells, forming a dynamic network of polymeric filaments distributed throughout the cytoplasm. MTs play pivotal roles in numerous cellular processes, such as cell division, intracellular transport, motility, and organelle positioning. MTs are hollowed-out tubes formed from α-tubulin and β-tubulin ($\alpha\beta$-tubulin) heterodimers that are polarized and typically oriented toward the cell periphery [85]. The polarity, a crucial requirement for MT function, results from the head-to-tail polymerization of tubulin dimers with α-tubulin at the minus end and β-tubulin at the plus end [86]. Notably, individual filaments can reach up to 5000 μm persistence length in vitro, much longer than actin filaments, which can only reach persistence lengths of 15–20 μm [87].

An exciting feature of tubulins is their ability to undergo various reversible post-translational modifications (PTMs), such as acetylation, phosphorylation, polyglycylation, polyglutamylation, (de)tyrosination, and palmitoylation [88,89]. Most PTMs occur in the carboxy-terminal tails of tubulin, with the notable exception of acetylation [89]. Acetylation mainly occurs after the assembly of MTs and is associated with stabilizing the MT structure [89]. In addition, acetylation can improve the binding and transport of molecular motors, such as kinesin-1 or dynein [90,91]. Another way to regulate MT functions is with nonmotor MT-associated proteins (MAPs), classified as MT-stabilizers, destabilizers, or plus-end tracking proteins [92,93]. MAPs also play a major role in MT bundling, a process that further regulates the stability of MT filaments [94,95].

The MT cytoskeleton is exploited by numerous viruses throughout almost all stages of the viral life cycle [96], including internalization [97], viral factory formation [98], assembly [99], and virus release [100].

3.2. MT-Dependent Molecular Motors

Two main classes of molecular motors specialize in transport along the MT network, corresponding to kinesin and dynein motors. Kinesin motors move towards the MT plus-end in what is known as anterograde transport. The diverse cargoes can either associate directly with the heavy chain or bind to specific regions in the C-terminus of the light chain [101]. In contrast, the molecular motor dynein moves towards the MT minus end, performing retrograde transport [102]. The cargo can bind to dynein in numerous ways, allowing for a wide range of client proteins [103,104]. Viruses, as cargoes, exploit cytoplasmic dynein to facilitate their transport within the cell [101].

3.3. Actin

Actin is the most abundant protein in many eukaryotic cells. Accordingly, several viruses subvert the actin cytoskeleton to spread and move over long distances [105].

Actin is expressed as a globular monomer known as G-actin [106]. When it polymerizes, it forms F-actin, filamentous structures that can form spontaneously in physiological conditions. Actin fibers play a fundamental role in many cellular processes, including motility, morphogenesis, cytokinesis, or endocytosis [107]. Actin-bundling proteins can crosslink actin filaments into actin bundles, which are the main components of the actin network [108]. When smaller filaments are organized into microvilli in the plasma membrane protrusions and tightly packed into arrays, the filaments are referred to as brush borders [109]. Within the cells, the force of actin is produced by myosin molecular motors that move along the long actin domains, referred to as stress fibers [107,110]. These stress fibers are often anchored to focal adhesions corresponding to complex structures responsible for crucial scaffolding interactions with actin [111].

3.4. Actin-Dependent Molecular Motors

Over forty classes of myosins are expressed in eukaryotes, divided into muscle and nonmuscle myosins [112]. Known as "conventional myosin," nonmuscle myosin-2 (NM2) is present in almost every cell type, existing in three variants [113]. Therefore, it is not surprising that NM2 has been shown to play a role in the life cycle of numerous viruses [112].

3.5. Intermediate Filaments

The intermediate filaments (IFs) are the third component of the eukaryotic cytoskeletal network and are less studied than MTs and actin [114]. One reason is that the IFs are polymers of two, three, or more different proteins. These proteins include, among others, keratins, vimentin, lamins, and nestin, which form six subtypes of filaments [114]. Vimentin and nestin play a role in cell migration, but other proteins have diverse functions depending on the cell context [115]. Interestingly, IFs are formed in the cytoplasm and the nucleus [116]. So far, no motor proteins have been identified moving along IFs.

4. Viroplasm Interaction with the Host Cytoskeleton

Aside from the viral components, viroplasms interact with many cellular components, including lipid droplets, proteins, and host nucleic acids. In this context, viroplasms are found to recruit components of lipid droplets (LDs) during the replication cycle [117]. LDs are spherical organelles that play a significant role in lipid homeostasis and contain mostly perilipins [118]. Associations with LDs appear to be required to form viroplasms and infectious virus progeny by serving as a scaffold for viroplasm assembly and allowing the association between viroplasms and ER membranes [45,117,119].

However, viroplasms are also found to interact with many elements of the host-cell cytoskeleton. All three primary cytoskeletal components (actin, MTs, and IFs) are restructured during RV infection. The formation of viroplasms relies on several of these reorganizations [120–125].

The reorganization of the MT cytoskeleton has been shown to directly influence the coalescence and localization of viroplasms [52], which seems a trait common among many viruses inducing the formation of membraneless replication compartments, such as birnaviruses, reoviruses, or African swine fever viruses [98,126–129]. In this sense, MT depolymerization drugs harm both perinuclear condensation and coalescence of the viroplasms. On the other hand, MT stabilizing drugs, such as taxol, showed no effect. In fact (Figure 2a,b), RV infection increases stabilized MTs, as denoted by the rise of acetylated tubulin in viroplasms [52]. Collectively, RV can subvert the cytoskeleton to assemble and maintain viroplasms.

Indeed, RV NSP2 and NSP5 have been implicated in directly interacting with tubulin in coimmunoprecipitation assays followed by Western blot or mass spectrometry. However, while the interaction of NSP2 with tubulin appears very stable, the interaction between NSP5 and tubulin is shown to be weak [52,80,130,131]. NSP5 has been pulled down with tubulin as a contaminant in RV-infected cells due to its ability to bind to NSP2 [50,60]. It seems that NSP5 and tubulin compete for binding to the same positively charged grooves on the NSP2 octamer [130]. Interestingly, despite significant MT reorganization induced by NSP2 transfection, the study does not observe considerable colocalization of NSP2 and tubulin in NSP2-transfected cells [130]. Furthermore, NSP2 exhibits a robust binding to nonacetylated tubulin compared to acetylated tubulin [80]. Still, acetylated tubulin seems to accumulate in mature viroplasms [52].

A newly identified variant of NSP2 displays varying interactions with NSP5 and acetylated tubulin, depending on the phosphorylation status of NSP2 [80]. These two NSP2 conformations have been distinguished using two different monoclonal antibodies targeting different regions of NSP2. One conformation corresponds to viroplasmic NSP2 (vNSP2), which localizes in viroplasms. The second conformation is a cytosolic dispersed pool of NSP2 (dNSP2), which is phosphorylated at its C-terminus, specifically in S313. Additionally, dNSP2 is weakly colocalizing with NSP5 and vNSP2 in viroplasms. Interestingly, dNSP2 resulted in the unphosphorylated precursor of vNSP2, where dNSP2 is phosphorylated by CK1 alpha to generate vNSP2. Once NSP2 is phosphorylated (vNSP2), it can bind to acetylated tubulin and NSP5. On the other hand, vNSP2 interacts with phosphorylated NSP5 and only weakly with tubulin [80]. This outcome suggests a mechanism of viroplasm formation and assembly coordinated by the phosphorylation of NSP5 and NSP2, with VP2 and tubulin acetylation. In this model, dNSP2, phosphorylated by CK1 alpha, and VP2 can bind nonphosphorylated NSP5, triggering NSP5 phosphorylation at Ser67, also by CK1 alpha, leading to the initial nucleation steps required for viroplasm formation. Both NSP2 and VP2 associate with unphosphorylated NSP5 [68,84]. These events concomitantly initiate the reorganization of the MT network to induce favorable conditions. Following this model, the destabilization of MTs during the early stages of infection hinders the coalescence of viroplasms [52]. Additionally, the globular morphology of the viroplasms seems dependent on the phosphorylation of NSP5, since unphosphorylated NSP5 leads to aberrant viroplasms [68]. However, VP2 also plays a role in the morphology of viroplasms, in which the inhibition of TRiC chaperonin leads to defective VLS composed of NSP5 and VP2 without affecting VLS composed of NSP5 and VP2, suggesting that the proper folding of VP2 is required for viroplasm structure [48]. Similarly, the expression of VP2 harboring L124 mutated to alanine leads to defective viroplasm formation [84].

Interestingly, the interaction with the MT network is not only based on NSP2–tubulin associations. In experiments using VLSs induced by coexpression of NSP5 with either NSP2 or VP2 and treated with an MT-destabilizing drug, it was shown that NSP2 confers the coalescence properties while VP2 mediates the perinuclear condensation properties. Additional research provides evidence that transfected NSP4 also binds and reorganizes the MT network [132–134]. Overall, the interaction of NSP5 and NSP2 with tubulin and their phosphorylation-dependent effects on viroplasm formation remain to be fully discovered.

Some studies point to the involvement of dynein-mediated transport in the coalescence of viroplasms [135]. NSP2 can interact with the dynein intermediate chain (DIC), mediating the ability of the viroplasm to coalesce. These findings resemble measles virus replication compartments, whose liquid–liquid phase-separated replication organelles depend on dynein-mediated transport to form large inclusion and viral replication [136]—suggesting a conserved reliance on dynein-mediated transport among diverse viruses to organize replication structures. In addition, it has been shown that viroplasms can no longer coalesce or move to the perinuclear region when the molecular motor Eg5 of the kinesin-5 family is inhibited [52]. So far, however, no direct interaction partner has been identified, as VLS properties seem to be independent of the Eg5 function, regardless of VLS induction by NSP2 or VP2 [52]. Moreover, RV infection halts the host cell cycle in the S/G2 phase [137], a stage that correlates with a stabilized MT network [138]. The RV-induced cell cycle arrest relies on the kinesin motor Eg5 and the actin and MT networks. This connection underscores the significance of a stabilized MT network for viroplasm formation, linking it with the cell cycle arrest and, consequently, RV replication [137].

The actin cytoskeleton plays an additional important role in viroplasm dynamics and formation. In this context, actin has mainly been found to interact with VP4, but NSP4 has likewise been shown to induce actin remodeling [139–142]. VP4 is predominantly known as a structural spike protein but is also expressed as a soluble protein in the cytosol [139]. The interaction of VP4 and actin is well known [42,139,141]. It has been found that VP4 can induce actin remodeling when expressed in the absence of other virus proteins [141]. Thus, VP4 has an actin-binding domain (ABD, amino acid region 713 to 773) at its C-terminus and a coiled-coil domain, allowing association to actin filaments. The VP4 ABD is buried in the assembled particle, pointing to the importance of soluble VP4 in the cytoplasm [139]. The use of a recombinant RV harboring a BAP tag in the VP8 region of VP4 (rRV/VP4-BAP) (Figure 2c,d) demonstrated that cytosolic VP4 plays a critical role, either directly or indirectly, in interacting with actin filaments to facilitate viroplasm formation [142]. Similarly, as observed for Negri bodies in rabies virus (RABV)-infected cells [143], the treatment with cytochalasin B, an inhibitor of actin filament dynamics, leads to a reduced number of viroplasms in RV-infected cells.

Additional studies proved that VP4 associates with MTs, potentially in an early step of virus release [144]. Studies have also shown that VP4 colocalizes with β-tubulin in both RV-infected and VP4-transfected cells, an interaction susceptible to disruption through MT depolymerization [144]. Moreover, it has been hypothesized that VP4 is transported to the plasma membrane via MT molecular motors [144]. It is plausible that the VP4 intracellular transport is differentially regulated, depending on the specific component of the cytoskeleton. It is well known that various viruses, such as flaviviruses [145] or influenza viruses [146], shift from actin-mediated transport to MT-associated transport at different steps of their life cycle. Transportation along the actin cytoskeleton may direct VP4 towards viroplasms to facilitate viroplasm formation. This process might involve the regulation of actin filaments and stress-fiber formation by VP4, which are necessary for initiating viroplasm assembly. In contrast, the MT network may transport VP4 away from viroplasms for incorporation in the plasma membrane in an alternative TLP assembly pathway [144]. The potential of VP4 to undergo differential transport opens new questions regarding the regulation of host-cell factors.

Only sparse research is available on the interplay of RV infection and the intermediate filaments. Infection with RV induces substantial restructuring of vimentin in adherent kidney cells, whereas such reorganization is not observed in differentiated human intestinal epithelial cells. Conversely, differentiated human intestinal epithelial cells display rearrangement of other cytoskeletal elements, a phenomenon not observed in undifferentiated human intestinal epithelial cells. [122,147]. Further research on the role of intermediate filaments is needed, as this is a relatively unexplored area.

Despite significant progress in the research on the assembly and maintenance of viroplasms, there are still gaps in our understanding of the precise molecular mechanisms involved in their formation and organization, as well as the interplay between different cytoskeletal components and their regulatory mechanisms.

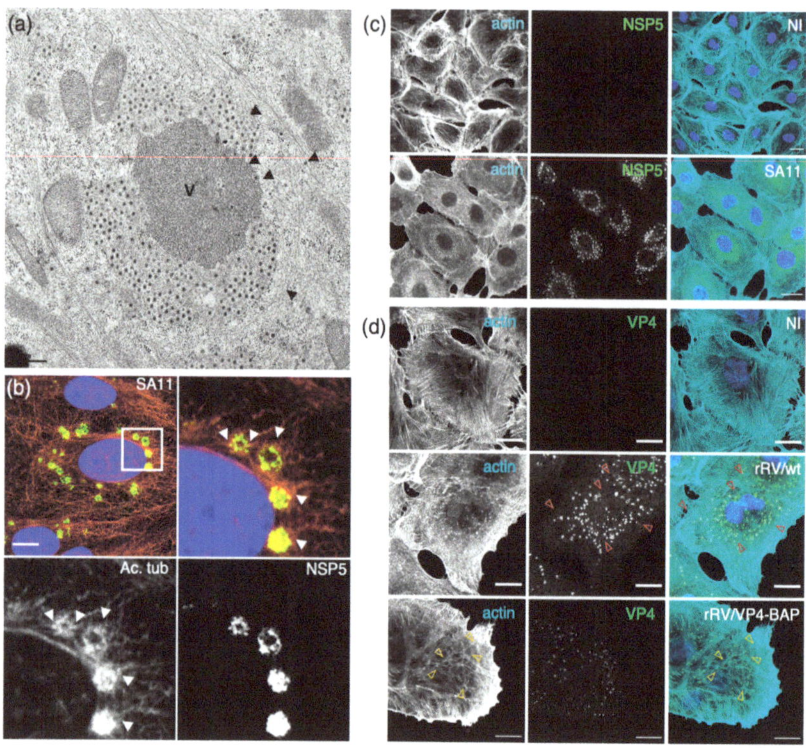

Figure 2. Association of viroplasms with microtubules and actin. (**a**) Electron microscopy of SA11-infected MA104 cells at 8 hpi, showing viroplasm. Black arrowheads indicate the MT bundles; viroplasms (V). Scale bar is 0.5 µm. (**b**) Immunofluorescence of SA11-infected MA104 cells at 6 hpi showing viroplasms (anti-NSP5, green), acetylated tubulin (mAb antiacetylated tubulin, red) and nucleus (DAPI, blue), upper left image. The white-boxed area shows an enlarged photomicrograph indicating the localization of the hyperacetylated MTs (white arrowheads) in the viroplasm region. Scale bar is 15 µm. From Eichwald et al., 2012 [52]. (**c**) Immunostaining of noninfected and SA11-infected MA104 cells. At 6 hpi, cells were fixed with methanol and immunostained to detect viroplasms (anti-NSP5, green) and actin cytoskeleton (antiactin, cyan). Nuclei were stained with DAPI (blue). The scale bar is 20 µm. Immunostaining of noninfected and rRV/wt- or rRV/VP4-BAP-infected MA104 cells. At 6 hpi, cells were fixed with methanol and immunostained for detection of (**d**) VP4 (anti-VP4, green) and actin cytoskeleton (antiactin, cyan). Nuclei were stained with DAPI (blue). The scale bar is 20 µm. Open yellow and red arrowheads point to stress fibers in the actin cytoskeleton and VP4 fibers, respectively. From Vetter et al., 2022 [142].

5. Interaction of Viral Factories with Host Components in Other dsRNA Viruses

Studies on other dsRNA viruses apart from RV, such as reoviruses or bluetongue virus, have revealed similar interactions between their viral factories and the host cell's cytoskeleton. Notably, research on mammalian reovirus (MRV) viral factories indicates their reliance on the MT network for their perinuclear condensation, movement, and structural assembly [98], properties observable in RV viroplasm formation as well [52]. It was found that both filamentous and globular MRV viral factories need an intact MT network for proper function with dynein localizing in both viral factories [98]. Moreover, the MT network is essential for forming large globular perinuclear inclusions via MRV non-structural protein µNS, as nocodazole treatment, a tubulin depolymerizing agent, was shown to disperse the filamentous viral factories into smaller inclusions [148]. These results appear consistent with studies on RV showing inhibition of perinuclear condensation and coalescence upon treatment with nocodazole, suggesting similarities in the interaction of viroplasms with the host cytoskeleton in other dsRNA viruses [52].

Further, the interaction between reovirus core protein μ2 and MTs stabilized by bundling and hyperacetylation of α-tubulin determined the filamentous shape of reovirus inclusion bodies, highlighting the dependency of a stabilized MT network for the distribution of MRV viral factories in cells [127,149]. The direct association between MRV and spindle tubules observed in L2 cells could explain the aggregates of the virus in extensive perinuclear inclusions, although this association is not necessary for viral replication [150]. Furthermore, studies on both RV and MRV have shown that their infection disrupts and reorganizes vimentin filaments without affecting MTs or microfilament bundles [122,151]. Likewise, bluetongue virus associates with the cytoskeleton. Linear arrays of virus particles around viral inclusion bodies were found to be formed upon treatment with a chemical compound (colchicine), leading to aggregation of the vimentin filament network in the perinuclear region, suggesting an association of the viral inclusion bodies with the intermediate filaments [152].

6. Concluding Remarks

This review describes the crucial interactions between RV proteins and the cellular cytoskeleton. It has become clear that phosphorylation, particularly the sequence of phosphorylation events, and other PTMs play a critical role in regulating the interaction between RV proteins and the cytoskeleton, particularly between NSP5, NSP2, and tubulin. Additionally, the new role of VP4 in regulating viroplasm formation, through its interaction with actin filaments as previously described, underscores the multifunctionality of RV proteins. This highlights the significance of host-cell factors on the dynamics of viroplasms and virus replication.

In conclusion, this review underlines promising research areas and aims to enrich the ongoing discussion surrounding viroplasm assembly and maintenance. Addressing these unresolved questions and conducting further studies in these areas will deepen our comprehension of the complex interplay between RV and the host-cell cytoskeleton, potentially leading to the development of novel therapeutic strategies for combating RV infections.

Author Contributions: Conceptualization, J.V., M.L. and C.E.; methodology, J.V. and M.L.; software, J.V., M.L. and C.E.; resources, J.V., M.L. and C.E.; data curation, J.V., M.L. and C.E.; writing—original draft preparation, J.V., M.L. and C.E.; writing—review and editing, J.V., M.L. and C.E.; visualization, J.V., M.L. and C.E.; supervision, C.E.; project administration, C.E.; funding acquisition, C.E. All authors have read and agreed to the published version of the manuscript.

Funding: This research was funded by the University of Zurich.

Institutional Review Board Statement: Not applicable.

Informed Consent Statement: Not applicable.

Data Availability Statement: Not applicable.

Acknowledgments: To permit open access, the authors have applied for a CC BY public copyright license for any version of this manuscript when accepted for publication.

Conflicts of Interest: The authors declare no conflicts of interest. The funders had no role in the design of this study; in the collection, analyses, or interpretation of data; in the writing of the manuscript; or in the decision to publish the results.

References

1. Adams, W.R.; Kraft, L.M. Epizootic Diarrhea of Infant Mice: Identification of the Etiologic Agent. *Science* **1963**, *141*, 359–360. [CrossRef] [PubMed]
2. Bishop, R.F.; Davidson, G.P.; Holmes, I.H.; Ruck, B.J. Virus particles in epithelial cells of duodenal mucosa from children with acute non-bacterial gastroenteritis. *Lancet* **1973**, *302*, 1281–1283. [CrossRef] [PubMed]

3. Crawford, S.E.; Ramani, S.; Tate, J.E.; Parashar, U.D.; Svensson, L.; Hagbom, M.; Franco, M.A.; Greenberg, H.B.; O'Ryan, M.; Kang, G.; et al. Rotavirus infection. *Nat. Rev. Dis. Primers* **2017**, *3*, 17083. [CrossRef] [PubMed]

4. Tate, J.E.; Burton, A.H.; Boschi-Pinto, C.; Steele, A.D.; Duque, J.; Parashar, U.D. 2008 estimate of worldwide rotavirus-associated mortality in children younger than 5 years before the introduction of universal rotavirus vaccination programmes: A systematic review and meta-analysis. *Lancet Infect. Dis.* **2012**, *12*, 136–141. [CrossRef] [PubMed]

5. Vesikari, T.; Karvonen, A.; Prymula, R.; Schuster, V.; Tejedor, J.C.; Cohen, R.; Meurice, F.; Han, H.H.; Damaso, S.; Bouckenooghe, A. Efficacy of human rotavirus vaccine against rotavirus gastroenteritis during the first 2 years of life in European infants: Randomised, double-blind controlled study. *Lancet* **2007**, *370*, 1757–1763. [CrossRef] [PubMed]

6. Hallowell, B.D.; Tate, J.; Parashar, U. An overview of rotavirus vaccination programs in developing countries. *Expert Rev. Vaccines* **2020**, *19*, 529–537. [CrossRef] [PubMed]

7. Fauquet, C.M. Taxonomy, Classification and Nomenclature of Viruses. In *Encyclopedia of Virology*; Academic Press: Cambridge, MA, USA, 2008.

8. Fauquet, C.M.; Mayo, M.A.; Maniloff, J.; Desselberger, U.; Ball, L.A. *Virus Taxonomy: VIIIth Report of the International Committee on Taxonomy of Viruses*; Academic Press: Cambridge, MA, USA, 2005.

9. Lefkowitz, E.J.; Dempsey, D.M.; Hendrickson, R.C.; Orton, R.J.; Siddell, S.G.; Smith, D.B. Virus taxonomy: The database of the International Committee on Taxonomy of Viruses (ICTV). *Nucleic Acids Res.* **2018**, *46*, D708–D717. [CrossRef]

10. Matthijnssens, J.; Theuns, S. Minutes of the 7th Rotavirus Classification Working Group (RCWG) Meeting. In Proceedings of the 12th International Double Stranded RNA Virus Symposium, Goa Marriott Beach Resort & Spa, Goa, India, 9 October 2015.

11. Johne, R.; Tausch, S.H.; Ulrich, R.G.; Schilling-Loeffler, K. Genome analysis of the novel putative rotavirus species K. *Virus Res.* **2023**, *334*, 199171. [CrossRef] [PubMed]

12. Johne, R.; Schilling-Loeffler, K.; Ulrich, R.G.; Tausch, S.H. Whole Genome Sequence Analysis of a Prototype Strain of the Novel Putative Rotavirus Species, L. *Viruses* **2022**, *14*, 462. [CrossRef]

13. Asensio-Cob, D.; Rodríguez, J.M.; Luque, D. Rotavirus Particle Disassembly and Assembly In Vivo and In Vitro. *Viruses* **2023**, *15*, 1750. [CrossRef]

14. Prasad, B.V.V.; Wang, G.J.; Clerx, J.P.M.; Chiu, W. Three-dimensional structure of rotavirus. *J. Mol. Biol.* **1988**, *199*, 269–275. [CrossRef] [PubMed]

15. Pesavento, J.B.; Crawford, S.E.; Estes, M.K.; Venkataram Prasad, B.V. Rotavirus Proteins: Structure and Assembly. In *Reoviruses: Entry, Assembly and Morphogenesis*; Roy, P., Ed.; Springer: Berlin/Heidelberg, Germany, 2006; pp. 189–219.

16. Arias, C.F.; Romero, P.; Alvarez, V.; López, S. Trypsin activation pathway of rotavirus infectivity. *J. Virol.* **1996**, *70*, 5832–5839. [CrossRef] [PubMed]

17. Gilbert, J.M.; Greenberg, H.B. Cleavage of rhesus rotavirus VP4 after arginine 247 is essential for rotavirus-like particle-induced fusion from without. *J. Virol.* **1998**, *72*, 5323–5327. [CrossRef] [PubMed]

18. Dormitzer, P.R.; Sun, Z.Y.; Wagner, G.; Harrison, S.C. The rhesus rotavirus VP4 sialic acid binding domain has a galectin fold with a novel carbohydrate binding site. *EMBO J.* **2002**, *21*, 885–897. [CrossRef] [PubMed]

19. Isa, P.; López, S.; Segovia, L.; Arias, C.F. Functional and structural analysis of the sialic acid-binding domain of rotaviruses. *J. Virol.* **1997**, *71*, 6749–6756. [CrossRef] [PubMed]

20. Isa, P.; Arias, C.F.; López, S. Role of sialic acids in rotavirus infection. *Glycoconj. J.* **2006**, *23*, 27–37. [CrossRef] [PubMed]

21. Huang, P.; Xia, M.; Tan, M.; Zhong, W.; Wei, C.; Wang, L.; Morrow, A.; Jiang, X. Spike Protein VP8* of Human Rotavirus Recognizes Histo-Blood Group Antigens in a Type-Specific Manner. *J. Virol.* **2012**, *86*, 4833–4843. [CrossRef] [PubMed]

22. Gutiérrez, M.; Isa, P.; Sánchez-San Martin, C.; Pérez-Vargas, J.; Espinosa, R.; Arias, C.F.; López, S. Different Rotavirus Strains Enter MA104 Cells through Different Endocytic Pathways: The Role of Clathrin-Mediated Endocytosis. *J. Virol.* **2010**, *84*, 9161–9169. [CrossRef] [PubMed]

23. Li, B.; Ding, S.; Feng, N.; Mooney, N.; Ooi, Y.S.; Ren, L.; Diep, J.; Kelly, M.R.; Yasukawa, L.L.; Patton, J.T.; et al. Drebrin restricts rotavirus entry by inhibiting dynamin-mediated endocytosis. *Proc. Natl. Acad. Sci. USA* **2017**, *114*, E3642–E3651. [CrossRef]

24. Arias, C.F.; Silva-Ayala, D.; López, S. Rotavirus Entry: A Deep Journey into the Cell with Several Exits. *J. Virol.* **2015**, *89*, 890–893. [CrossRef]

25. Díaz-Salinas Marco, A.; Silva-Ayala, D.; López, S.; Arias Carlos, F. Rotaviruses Reach Late Endosomes and Require the Cation-Dependent Mannose-6-Phosphate Receptor and the Activity of Cathepsin Proteases to Enter the Cell. *J. Virol.* **2014**, *88*, 4389–4402. [CrossRef] [PubMed]

26. Golantsova Nina, E.; Gorbunova Elena, E.; Mackow Erich, R. Discrete Domains within the Rotavirus VP5* Direct Peripheral Membrane Association and Membrane Permeability. *J. Virol.* **2004**, *78*, 2037–2044. [CrossRef] [PubMed]

27. Herrmann, T.; Torres, R.; Salgado, E.N.; Berciu, C.; Stoddard, D.; Nicastro, D.; Jenni, S.; Harrison, S.C. Functional refolding of the penetration protein on a non-enveloped virus. *Nature* **2021**, *590*, 666–670. [CrossRef] [PubMed]

28. Patton, J.T.; Silvestri, L.S.; Tortorici, M.A.; Carpio, V.-D.; Taraporewala, Z.F. Rotavirus genome replication and morphogenesis: Role of the viroplasm. In *Reoviruses: Entry, Assembly and Morphogenesis*; Springer: Berlin/Heidelberg, Germany, 2006; pp. 169–187.

29. Lawton, J.A.; Estes, M.K.; Prasad, B.V.V. Three-dimensional visualization of mRNA release from actively transcribing rotavirus particles. *Nat. Struct. Biol.* **1997**, *4*, 118–121. [CrossRef] [PubMed]

30. Petrie, B.L.; Graham, D.Y.; Hanssen, H.; Estes, M.K. Localization of rotavirus antigens in infected cells by ultrastructural immunocytochemistry. *J. Gen. Virol.* **1982**, *63*, 457–467. [CrossRef] [PubMed]

31. Arnold Michelle, M. The Rotavirus Interferon Antagonist NSP1, Many Targets, Many Questions. *J. Virol.* **2016**, *90*, 5212–5215. [CrossRef]
32. Padilla-Noriega, L.; Paniagua, O.; Guzmán-León, S. Rotavirus protein NSP3 shuts off host cell protein synthesis. *Virology* **2002**, *298*, 1477. [CrossRef] [PubMed]
33. Morelli, M.; Ogden, K.M.; Patton, J.T. Silencing the alarms: Innate immune antagonism by rotavirus NSP1 and VP3. *Virology* **2015**, *479–480*, 75–84. [CrossRef] [PubMed]
34. Gratia, M.; Sarot, E.; Vende, P.; Charpilienne, A.; Baron, C.H.; Duarte, M.; Pyronnet, S.; Poncet, D. Rotavirus NSP3 Is a Translational Surrogate of the Poly(A) Binding Protein-Poly(A) Complex. *J. Virol.* **2015**, *89*, 8773–8782. [CrossRef]
35. Taylor, J.A.; O'Brien, J.A.; Lord, V.J.; Meyer, J.C.; Bellamy, A.R. The RER-Localized Rotavirus Intracellular Receptor: A Truncated Purified Soluble Form Is Multivalent and Binds Virus Particles. *Virology* **1993**, *194*, 807–814. [CrossRef]
36. López, T.; Camacho, M.; Zayas, M.; Nájera, R.; Sánchez, R.; Arias Carlos, F.; López, S. Silencing the Morphogenesis of Rotavirus. *J. Virol.* **2005**, *79*, 184–192. [CrossRef] [PubMed]
37. Trask, S.D.; McDonald, S.M.; Patton, J.T. Structural insights into the coupling of virion assembly and rotavirus replication. *Nat. Rev. Microbiol.* **2012**, *10*, 165–177. [CrossRef] [PubMed]
38. Shah, P.N.M.; Gilchrist, J.B.; Forsberg, B.O.; Burt, A.; Howe, A.; Mosalaganti, S.; Wan, W.; Radecke, J.; Chaban, Y.; Sutton, G.; et al. Characterization of the rotavirus assembly pathway in situ using cryoelectron tomography. *Cell Host Microbe* **2023**, *31*, 604–615.e604. [CrossRef] [PubMed]
39. Musalem, C.; Espejo, R.T. Release of Progeny Virus from Cells Infected with Simian Rotavirus SA11. *J. Gen. Virol.* **1985**, *66*, 2715–2724. [CrossRef] [PubMed]
40. Cevallos Porta, D.; López, S.; Arias, C.F.; Isa, P. Polarized rotavirus entry and release from differentiated small intestinal cells. *Virology* **2016**, *499*, 65–71. [CrossRef] [PubMed]
41. Gardet, A.; Breton, M.; Trugnan, G.; Chwetzoff, S. Role for actin in the polarized release of rotavirus. *J. Virol.* **2007**, *81*, 4892–4894. [CrossRef] [PubMed]
42. Trejo-Cerro, Ó.; Eichwald, C.; Schraner, E.M.; Silva-Ayala, D.; López, S.; Arias, C.F. Actin-Dependent Nonlytic Rotavirus Exit and Infectious Virus Morphogenetic Pathway in Nonpolarized Cells. *J. Virol.* **2018**, *92*, e02076-17. [CrossRef] [PubMed]
43. Altenburg, B.C.; Graham, D.Y.; Kolb Estes, M. Ultrastructural study of rotavirus replication in cultured cells. *J. Gen. Virol.* **1980**, *46*, 75–85. [CrossRef] [PubMed]
44. Contin, R.; Arnoldi, F.; Campagna, M.; Burrone, O.R. Rotavirus NSP5 orchestrates recruitment of viroplasmic proteins. *J. Gen. Virol.* **2010**, *91*, 1782–1793. [CrossRef]
45. Cheung, W.; Gill, M.; Esposito, A.; Kaminski Clemens, F.; Courousse, N.; Chwetzoff, S.; Trugnan, G.; Keshavan, N.; Lever, A.; Desselberger, U. Rotaviruses Associate with Cellular Lipid Droplet Components To Replicate in Viroplasms, and Compounds Disrupting or Blocking Lipid Droplets Inhibit Viroplasm Formation and Viral Replication. *J. Virol.* **2010**, *84*, 6782–6798. [CrossRef]
46. Contin, R.; Arnoldi, F.; Mano, M.; Burrone, O.R. Rotavirus replication requires a functional proteasome for effective assembly of viroplasms. *J. Virol.* **2011**, *85*, 2781–2792. [CrossRef] [PubMed]
47. Lopez, T.; Silva-Ayala, D.; Lopez, S.; Arias, C.F. Replication of the Rotavirus Genome Requires an Active Ubiquitin-Proteasome System. *J. Virol.* **2011**, *85*, 11964–11971. [CrossRef] [PubMed]
48. Vetter, J.; Papa, G.; Tobler, K.; Rodriguez Javier, M.; Kley, M.; Myers, M.; Wiesendanger, M.; Schraner Elisabeth, M.; Luque, D.; Burrone Oscar, R.; et al. The recruitment of TRiC chaperonin in rotavirus viroplasms correlates with virus replication. *mBio* **2024**, *15*, e0049924. [CrossRef]
49. Gonzalez, R.A.; Espinosa, R.; Romero, P.; Lopez, S.; Arias, C.F. Relative localization of viroplasmic and endoplasmic reticulum-resident rotavirus proteins in infected cells. *Arch. Virol.* **2000**, *145*, 1963–1973. [CrossRef] [PubMed]
50. Eichwald, C.; Rodriguez, J.F.; Burrone, O.R. Characterization of rotavirus NSP2/NSP5 interactions and the dynamics of viroplasm formation. *J. Gen. Virol.* **2004**, *85*, 625–634. [CrossRef]
51. Carreño-Torres, J.J.; Gutiérrez, M.; Arias, C.F.; López, S.; Isa, P. Characterization of viroplasm formation during the early stages of rotavirus infection. *Virol. J.* **2010**, *7*, 350. [CrossRef]
52. Eichwald, C.; Arnoldi, F.; Laimbacher, A.S.; Schraner, E.M.; Fraefel, C.; Wild, P.; Burrone, O.R.; Ackermann, M. Rotavirus viroplasm fusion and perinuclear localization are dynamic processes requiring stabilized microtubules. *PLoS ONE* **2012**, *7*, e47947. [CrossRef]
53. Geiger, F.; Acker, J.; Papa, G.; Wang, X.; Arter, W.E.; Saar, K.L.; Erkamp, N.A.; Qi, R.; Bravo, J.P.K.; Strauss, S.; et al. Liquid–liquid phase separation underpins the formation of replication factories in rotaviruses. *EMBO J.* **2021**, *40*, e107711. [CrossRef] [PubMed]
54. Fabbretti, E.; Afrikanova, I.; Vascotto, F.; Burrone, O.R. Two non-structural rotavirus proteins, NSP2 and NSP5, form viroplasm-like structures in vivo. *J. Gen. Virol.* **1999**, *80 Pt 2*, 333–339. [CrossRef]
55. Borodavka, A.; Dykeman, E.C.; Schrimpf, W.; Lamb, D.C. Protein-mediated RNA folding governs sequence-specific interactions between rotavirus genome segments. *eLife* **2017**, *6*, e27453. [CrossRef]
56. Patton, J.T.; Jones, M.T.; Kalbach, A.N.; He, Y.W.; Xiaobo, J. Rotavirus RNA polymerase requires the core shell protein to synthesize the double-stranded RNA genome. *J. Virol.* **1997**, *71*, 9618–9626. [CrossRef] [PubMed]
57. Periz, J.; Celma, C.; Jing, B.; Pinkney, J.N.; Roy, P.; Kapanidis, A.N. Rotavirus mRNAS are released by transcript-specific channels in the double-layered viral capsid. *Proc. Natl. Acad. Sci. USA* **2013**, *110*, 12042–12047. [CrossRef] [PubMed]

58. Ruiz, M.C.; Leon, T.; Diaz, Y.; Michelangeli, F. Molecular Biology of Rotavirus Entry and Replication. *Sci. World J.* **2009**, *9*, 879851. [CrossRef] [PubMed]
59. Campagna, M.; Eichwald, C.; Vascotto, F.; Burrone, O.R. RNA interference of rotavirus segment 11 mRNA reveals the essential role of NSP5 in the virus replicative cycle. *J. Gen. Virol.* **2005**, *86*, 1481–1487. [CrossRef] [PubMed]
60. Martin, D.; Ouldali, M.; Ménétrey, J.; Poncet, D. Structural organisation of the rotavirus nonstructural protein NSP5. *J. Mol. Biol.* **2011**, *413*, 209–221. [CrossRef] [PubMed]
61. Martin, D.; Charpilienne, A.; Parent, A.; Boussac, A.; D'Autreaux, B.; Poupon, J.; Poncet, D. The rotavirus nonstructural protein NSP5 coordinates a [2Fe-2S] iron-sulfur cluster that modulates interaction to RNA. *FASEB J.* **2013**, *27*, 1074–1083. [CrossRef] [PubMed]
62. Eichwald, C.; Jacob, G.; Muszynski, B.; Allende, J.E.; Burrone, O.R. Uncoupling substrate and activation functions of rotavirus NSP5, phosphorylation of Ser-67 by casein kinase 1 is essential for hyperphosphorylation. *Proc. Natl. Acad. Sci. USA* **2004**, *101*, 16304–16309. [CrossRef] [PubMed]
63. Eichwald, C.; Vascotto, F.; Fabbretti, E.; Burrone, O.R. Rotavirus NSP5, mapping phosphorylation sites and kinase activation and viroplasm localization domains. *J. Virol.* **2002**, *76*, 3461–3470. [CrossRef] [PubMed]
64. Afrikanova, I.; Fabbretti, E.; Miozzo, M.C.; Burrone, O.R. Rotavirus NSP5 phosphorylation is up-regulated by interaction with NSP2. *J. Gen. Virol.* **1998**, *79 Pt 11*, 2679–2686. [CrossRef]
65. Campagna, M.; Budini, M.; Arnoldi, F.; Desselberger, U.; Allende, J.E.; Burrone, O.R. Impaired hyperphosphorylation of rotavirus NSP5 in cells depleted of casein kinase 1alpha is associated with the formation of viroplasms with altered morphology and a moderate decrease in virus replication. *J. Gen. Virol.* **2007**, *88*, 2800–2810. [CrossRef]
66. Torres-Vega, M.A.; González, R.A.; Duarte, M.; Poncet, D.; López, S.; Arias, C.F. The C-terminal domain of rotavirus NSP5 is essential for its multimerization, hyperphosphorylation and interaction with NSP6. *J. Gen. Virol.* **2000**, *81*, 821–830. [CrossRef] [PubMed]
67. Arnoldi, F.; Campagna, M.; Eichwald, C.; Desselberger, U.; Burrone, O.R. Interaction of rotavirus polymerase VP1 with nonstructural protein NSP5 is stronger than that with NSP2. *J. Virol.* **2007**, *81*, 2128–2137. [CrossRef] [PubMed]
68. Papa, G.; Venditti, L.; Arnoldi, F.; Schraner, E.M.; Potgieter, C.; Borodavka, A.; Eichwald, C.; Burrone, O.R. Recombinant rotaviruses rescued by reverse genetics reveal the role of NSP5 hyperphosphorylation in the assembly of viral factories. *J. Virol.* **2020**, *94*, 1–23. [CrossRef] [PubMed]
69. Bar-Magen, T.; Spencer, E.; Patton, J.T. An ATPase activity associated with the rotavirus phosphoprotein NSP5. *Virology* **2007**, *369*, 389–399. [CrossRef] [PubMed]
70. Silvestri, L.S.; Taraporewala, Z.F.; Patton, J.T. Rotavirus replication: Plus-sense templates for double-stranded RNA synthesis are made in viroplasms. *J. Virol.* **2004**, *78*, 7763–7774. [CrossRef] [PubMed]
71. Taraporewala, Z.; Chen, D.; Patton, J.T. Multimers formed by the rotavirus nonstructural protein NSP2 bind to RNA and have nucleoside triphosphatase activity. *J. Virol.* **1999**, *73*, 9934–9943. [CrossRef] [PubMed]
72. Jayaram, H.; Taraporewala, Z.; Patton, J.T.; Prasad, B.V.V. Rotavirus protein involved in genome replication and packaging exhibits a HIT-like fold. *Nature* **2002**, *417*, 311–315. [CrossRef] [PubMed]
73. Taraporewala Zenobia, F.; Jiang, X.; Vasquez-Del Carpio, R.; Jayaram, H.; Prasad, B.V.V.; Patton John, T. Structure-Function Analysis of Rotavirus NSP2 Octamer by Using a Novel Complementation System. *J. Virol.* **2006**, *80*, 7984–7994. [CrossRef] [PubMed]
74. Chamera, S.; Wycisk, K.; Czarnocki-Cieciura, M.; Nowotny, M. Cryo-EM structure of rotavirus B NSP2 reveals its unique tertiary architecture. *J. Virol.* **2024**, *98*, e0166023. [CrossRef]
75. Kattoura, M.D.; Clapp, L.L.; Patton, J.T. The rotavirus nonstructural protein, NS35, possesses RNA-binding activity in vitro and in vivo. *Virology* **1992**, *191*, 698–708. [CrossRef]
76. Kattoura, M.D.; Chen, X.; Patton, J.T. The Rotavirus RNA-Binding Protein NS35 (NSP2) Forms 10S Multimers and Interacts with the Viral RNA Polymerase. *Virology* **1994**, *202*, 803–813. [CrossRef] [PubMed]
77. Bravo, J.P.K.; Bartnik, K.; Venditti, L.; Acker, J.; Gail, E.H.; Colyer, A.; Davidovich, C.; Lamb, D.C.; Tuma, R.; Calabrese, A.N.; et al. Structural basis of rotavirus RNA chaperone displacement and RNA annealing. *Proc. Natl. Acad. Sci. USA* **2021**, *118*, e2100198118. [CrossRef] [PubMed]
78. Kumar, M.; Jayaram, H.; Vasquez-Del Carpio, R.; Jiang, X.; Taraporewala Zenobia, F.; Jacobson Raymond, H.; Patton John, T.; Prasad, B.V.V. Crystallographic and Biochemical Analysis of Rotavirus NSP2 with Nucleotides Reveals a Nucleoside Diphosphate Kinase-Like Activity. *J. Virol.* **2007**, *81*, 12272–12284. [CrossRef] [PubMed]
79. Criglar, J.M.; Anish, R.; Hu, L.; Crawford, S.E.; Sankaran, B.; Prasad, B.V.V.; Estes, M.K. Phosphorylation cascade regulates the formation and maturation of rotaviral replication factories. *Proc. Natl. Acad. Sci. USA* **2018**, *115*, E12015–E12023. [CrossRef] [PubMed]
80. Criglar, J.M.; Hu, L.; Crawford, S.E.; Hyser, J.M.; Broughman, J.R.; Prasad, B.V.; Estes, M.K. A novel form of rotavirus NSP2 and phosphorylation-dependent NSP2-NSP5 interactions are associated with viroplasm assembly. *J. Virol.* **2014**, *88*, 786–798. [CrossRef] [PubMed]
81. Criglar Jeanette, M.; Crawford Sue, E.; Zhao, B.; Smith Hunter, G.; Stossi, F.; Estes Mary, K. A Genetically Engineered Rotavirus NSP2 Phosphorylation Mutant Impaired in Viroplasm Formation and Replication Shows an Early Interaction between vNSP2 and Cellular Lipid Droplets. *J. Virol.* **2020**, *94*, e00972-20. [CrossRef]

82. Nichols, S.L.; Nilsson, E.M.; Brown-Harding, H.; LaConte, L.E.W.; Acker, J.; Borodavka, A.; McDonald Esstman, S. Flexibility of the Rotavirus NSP2 C-Terminal Region Supports Factory Formation via Liquid-Liquid Phase Separation. *J. Virol.* **2023**, *97*, e0003923. [CrossRef]

83. Montero, H.; Rojas, M.; Arias Carlos, F.; López, S. Rotavirus Infection Induces the Phosphorylation of eIF2α but Prevents the Formation of Stress Granules. *J. Virol.* **2008**, *82*, 1496–1504. [CrossRef]

84. Buttafuoco, A.; Michaelsen, K.; Tobler, K.; Ackermann, M.; Fraefel, C.; Eichwald, C. Conserved Rotavirus NSP5 and VP2 Domains Interact and Affect Viroplasm. *J. Virol.* **2020**, *94*, e01965-19. [CrossRef]

85. Akhmanova, A.; Steinmetz, M.O. Microtubule minus-end regulation at a glance. *J. Cell Sci.* **2019**, *132*, jcs227850. [CrossRef]

86. Mitchison, T.J. Localization of an Exchangeable GTP BZinding Site at the Plus End of Microtubules. *Science* **1993**, *261*, 1044–1047. [CrossRef] [PubMed]

87. Gittes, F.; Mickey, B.; Nettleton, J.; Howard, J. Flexural rigidity of microtubules and actin filaments measured from thermal fluctuations in shape. *J. Cell Biol.* **1993**, *120*, 923–934. [CrossRef] [PubMed]

88. Janke, C.; Bulinski, J.C. Post-translational regulation of the microtubule cytoskeleton: Mechanisms and functions. *Nat. Rev. Mol. Cell Biol.* **2011**, *12*, 773–786. [CrossRef] [PubMed]

89. Westermann, S.; Weber, K. Post-translational modifications regulate microtubule function. *Nat. Rev. Mol. Cell Biol.* **2003**, *4*, 938–948. [CrossRef] [PubMed]

90. Reed, N.A.; Cai, D.; Blasius, T.L.; Jih, G.T.; Meyhofer, E.; Gaertig, J.; Verhey, K.J. Microtubule Acetylation Promotes Kinesin-1 Binding and Transport. *Curr. Biol.* **2006**, *16*, 2166–2172. [CrossRef] [PubMed]

91. Alper Joshua, D.; Decker, F.; Agana, B.; Howard, J. The Motility of Axonemal Dynein Is Regulated by the Tubulin Code. *Biophys. J.* **2014**, *107*, 2872–2880. [CrossRef] [PubMed]

92. Goodson, H.V.; Jonasson, E.M. Microtubules and Microtubule-Associated Proteins. *Cold Spring Harb. Perspect. Biol.* **2018**, *10*, a022608. [CrossRef] [PubMed]

93. Bodakuntla, S.; Jijumon, A.S.; Villablanca, C.; Gonzalez-Billault, C.; Janke, C. Microtubule-Associated Proteins: Structuring the Cytoskeleton. *Trends Cell Biol.* **2019**, *29*, 804–819. [CrossRef] [PubMed]

94. Brandt, R.; Lee, G. Orientation, assembly, and stability of microtubule bundles induced by a fragment of tau protein. *Cell Motil.* **1994**, *28*, 143–154. [CrossRef]

95. Walczak, C.E.; Shaw, S.L. A MAP for Bundling Microtubules. *Cell* **2010**, *142*, 364–367. [CrossRef]

96. Greber, U.F.; Way, M. A Superhighway to Virus Infection. *Cell* **2006**, *124*, 741–754. [CrossRef] [PubMed]

97. Walsh, D.; Naghavi, M.H. Exploitation of Cytoskeletal Networks during Early Viral Infection. *Trends Microbiol.* **2019**, *27*, 39–50. [CrossRef]

98. Eichwald, C.; Ackermann, M.; Nibert, M.L. The dynamics of both filamentous and globular mammalian reovirus viral factories rely on the microtubule network. *Virology* **2018**, *518*, 77–86. [CrossRef] [PubMed]

99. Iwamoto, M.; Cai, D.; Sugiyama, M.; Suzuki, R.; Aizaki, H.; Ryo, A.; Ohtani, N.; Tanaka, Y.; Mizokami, M.; Wakita, T.; et al. Functional association of cellular microtubules with viral capsid assembly supports efficient hepatitis B virus replication. *Sci. Rep.* **2017**, *7*, 10620. [CrossRef] [PubMed]

100. Jouvenet, N.; Monaghan, P.; Way, M.; Wileman, T. Transport of African Swine Fever Virus from Assembly Sites to the Plasma Membrane Is Dependent on Microtubules and Conventional Kinesin. *J. Virol.* **2004**, *78*, 7990–8001. [CrossRef] [PubMed]

101. Dodding, M.P.; Way, M. Coupling viruses to dynein and kinesin-1. *EMBO J.* **2011**, *30*, 3527–3539. [CrossRef] [PubMed]

102. Roberts, A.J. Emerging mechanisms of dynein transport in the cytoplasm versus the cilium. *Biochem. Soc. Trans.* **2018**, *46*, 967–982. [CrossRef] [PubMed]

103. Roberts, A.J.; Kon, T.; Knight, P.J.; Sutoh, K.; Burgess, S.A. Functions and mechanics of dynein motor proteins. *Nat. Rev. Mol. Cell Biol.* **2013**, *14*, 713. [CrossRef]

104. King, S.M. The dynein microtubule motor. *Biochim. Biophys. Acta-Mol. Cell Res.* **2000**, *1496*, 60–75. [CrossRef]

105. Taylor, M.P.; Koyuncu, O.O.; Enquist, L.W. Subversion of the actin cytoskeleton during viral infection. *Nat. Rev. Microbiol.* **2011**, *9*, 427–439. [CrossRef]

106. Dominguez, R.; Holmes, K.C. Actin Structure and Function. *Annu. Rev. Biophys.* **2011**, *40*, 169–186. [CrossRef] [PubMed]

107. Tojkander, S.; Gateva, G.; Lappalainen, P. Actin stress fibers—Assembly, dynamics and biological roles. *J. Cell Sci.* **2012**, *125*, 1855–1864. [CrossRef] [PubMed]

108. Rajan, S.; Kudryashov, D.S.; Reisler, E. Actin Bundles Dynamics and Architecture. *Biomolecules* **2023**, *13*, 450. [CrossRef] [PubMed]

109. Meenderink, L.M.; Gaeta, I.M.; Postema, M.M.; Cencer, C.S.; Chinowsky, C.R.; Krystofiak, E.S.; Millis, B.A.; Tyska, M.J. Actin dynamics drive microvillar motility and clustering during brush border assembly. *Dev. Cell* **2019**, *50*, 545–556.e544. [CrossRef] [PubMed]

110. Lehtimäki, J.I.; Rajakylä, E.K.; Tojkander, S.; Lappalainen, P. Generation of stress fibers through myosin-driven reorganization of the actin cortex. *eLife* **2021**, *10*, e60710. [CrossRef] [PubMed]

111. Geiger, B.; Spatz, J.P.; Bershadsky, A.D. Environmental sensing through focal adhesions. *Nat. Rev. Mol. Cell Biol.* **2009**, *10*, 21–33. [CrossRef] [PubMed]

112. Matozo, T.; Kogachi, L.; de Alencar, B.C. Myosin motors on the pathway of viral infections. *Cytoskeleton* **2022**, *79*, 41–63. [CrossRef] [PubMed]

113. Sellers, J.R.; Heissler, S.M. Nonmuscle myosin-2 isoforms. *Curr. Biol.* **2019**, *29*, R275–R278. [CrossRef]

114. Leduc, C.; Etienne-Manneville, S. Intermediate filaments in cell migration and invasion: The unusual suspects. *Curr. Opin. Cell Biol.* **2015**, *32*, 102–112. [CrossRef]

115. Eriksson, J.E.; Dechat, T.; Grin, B.; Helfand, B.; Mendez, M.; Pallari, H.-M.; Goldman, R.D. Introducing intermediate filaments: From discovery to disease. *J. Clin. Investig.* **2009**, *119*, 1763–1771. [CrossRef]

116. Herrmann, H.; Bär, H.; Kreplak, L.; Strelkov, S.V.; Aebi, U. Intermediate filaments: From cell architecture to nanomechanics. *Nat. Rev. Mol. Cell Biol.* **2007**, *8*, 562–573. [CrossRef] [PubMed]

117. Desselberger, U. The significance of lipid droplets for the replication of rotaviruses and other RNA viruses. *J. Biol. Todays World* **2020**, *9*, 001–003.

118. Murphy, S.; Martin, S.; Parton, R.G. Lipid droplet-organelle interactions; sharing the fats. *Biochim. Biophys. Acta (BBA)-Mol. Cell Biol. Lipids* **2009**, *1791*, 441–447. [CrossRef] [PubMed]

119. Martínez, J.L.; Eichwald, C.; Schraner, E.M.; López, S.; Arias, C.F. Lipid metabolism is involved in the association of rotavirus viroplasms with endoplasmic reticulum membranes. *Virology* **2022**, *569*, 29–36. [CrossRef]

120. Mattion, N.M.; Cohen, J.; Aponte, C.; Estes, M.K. Characterization of an oligomerization domain and RNA-binding properties on rotavirus nonstructural protein NS34. *Virology* **1992**, *190*, 68–83. [CrossRef] [PubMed]

121. Hua, J.; Patton, J.T. The Carboxyl-Half of the Rotavirus Nonstructural Protein NS53 (NSP1) Is Not Required for Virus Replication. *Virology* **1994**, *198*, 567–576. [CrossRef] [PubMed]

122. Weclewicz, K.; Kristensson, K.; Svensson, L. Rotavirus causes selective vimentin reorganization in monkey kidney CV-1 cells. *J. Gen. Virol.* **1994**, *75*, 3267–3271. [CrossRef] [PubMed]

123. Brunet, J.-P.; Cotte-Laffitte, J.; Linxe, C.; Quero, A.-M.; Géniteau-Legendre, M.; Servin, A. Rotavirus infection induces an increase in intracellular calcium concentration in human intestinal epithelial cells: Role in microvillar actin alteration. *J. Virol.* **2000**, *74*, 2323–2332. [CrossRef] [PubMed]

124. Jourdan, N.; Brunet Jean, P.; Sapin, C.; Blais, A.; Cotte-Laffitte, J.; Forestier, F.; Quero, A.-M.; Trugnan, G.; Servin Alain, L. Rotavirus Infection Reduces Sucrase-Isomaltase Expression in Human Intestinal Epithelial Cells by Perturbing Protein Targeting and Organization of Microvillar Cytoskeleton. *J. Virol.* **1998**, *72*, 7228–7236. [CrossRef]

125. Weclewicz, K.; Svensson, L.; Billger, M.; Holmberg, K.; Wallin, M.; Kristensson, K. Microtubule-associated protein 2 appears in axons of cultured dorsal root ganglia and spinal cord neurons after rotavirus infection. *J. Neurosci. Res.* **1993**, *36*, 173–182. [CrossRef]

126. Campbell Elle, A.; Reddy Vishwanatha, R.A.P.; Gray Alice, G.; Wells, J.; Simpson, J.; Skinner Michael, A.; Hawes Philippa, C.; Broadbent Andrew, J. Discrete Virus Factories Form in the Cytoplasm of Cells Coinfected with Two Replication-Competent Tagged Reporter Birnaviruses That Subsequently Coalesce over Time. *J. Virol.* **2020**, *94*, e02107-19. [CrossRef] [PubMed]

127. Parker John, S.L.; Broering Teresa, J.; Kim, J.; Higgins Darren, E.; Nibert Max, L. Reovirus Core Protein μ2 Determines the Filamentous Morphology of Viral Inclusion Bodies by Interacting with and Stabilizing Microtubules. *J. Virol.* **2002**, *76*, 4483–4496. [CrossRef] [PubMed]

128. Heath, C.M.; Windsor, M.; Wileman, T. Aggresomes Resemble Sites Specialized for Virus Assembly. *J. Cell Biol.* **2001**, *153*, 449–456. [CrossRef] [PubMed]

129. Lahaye, X.; Vidy, A.; Pomier, C.; Obiang, L.; Harper, F.; Gaudin, Y.; Blondel, D. Functional Characterization of Negri Bodies (NBs) in Rabies Virus-Infected Cells: Evidence that NBs Are Sites of Viral Transcription and Replication. *J. Virol.* **2009**, *83*, 7948–7958. [CrossRef] [PubMed]

130. Martin, D.; Duarte, M.; Lepault, J.; Poncet, D. Sequestration of free tubulin molecules by the viral protein NSP2 induces microtubule depolymerization during rotavirus infection. *J. Virol.* **2010**, *84*, 2522–2532. [CrossRef] [PubMed]

131. Dhillon, P.; Tandra Varsha, N.; Chorghade Sandip, G.; Namsa Nima, D.; Sahoo, L.; Rao, C.D. Cytoplasmic Relocalization and Colocalization with Viroplasms of Host Cell Proteins, and Their Role in Rotavirus Infection. *J. Virol.* **2018**, *92*, e00612-18. [CrossRef]

132. Zambrano, J.L.; Sorondo, O.; Alcala, A.; Vizzi, E.; Diaz, Y.; Ruiz, M.C.; Michelangeli, F.; Liprandi, F.; Ludert, J.E. Rotavirus infection of cells in culture induces activation of RhoA and changes in the actin and tubulin cytoskeleton. *PLoS ONE* **2012**, *7*, e47612. [CrossRef]

133. Zhang, M.; Zeng Carl, Q.Y.; Morris Andrew, P.; Estes Mary, K. A Functional NSP4 Enterotoxin Peptide Secreted from Rotavirus-Infected Cells. *J. Virol.* **2000**, *74*, 11663–11670. [CrossRef] [PubMed]

134. Xu, A.; Bellamy, A.R.; Taylor, J.A. Immobilization of the early secretory pathway by a virus glycoprotein that binds to microtubules. *EMBO J.* **2000**, *19*, 6465–6474. [CrossRef]

135. Jing, Z.; Shi, H.; Chen, J.; Shi, D.; Liu, J.; Guo, L.; Tian, J.; Wu, Y.; Dong, H.; Ji, Z.; et al. Rotavirus Viroplasm Biogenesis Involves Microtubule-Based Dynein Transport Mediated by an Interaction between NSP2 and Dynein Intermediate Chain. *J. Virol.* **2021**, *95*, e0124621. [CrossRef]

136. Zhou, Y.; Su Justin, M.; Samuel Charles, E.; Ma, D. Measles Virus Forms Inclusion Bodies with Properties of Liquid Organelles. *J. Virol.* **2019**, *93*, e00948-19. [CrossRef] [PubMed]

137. Glück, S.; Buttafuoco, A.; Meier, A.F.; Arnoldi, F.; Vogt, B.; Schraner, E.M.; Ackermann, M.; Eichwald, C. Rotavirus replication is correlated with S/G2 interphase arrest of the host cell cycle. *PLoS ONE* **2017**, *12*, e0179607. [CrossRef] [PubMed]

138. Akhmanova, A.; Steinmetz, M.O. Control of microtubule organization and dynamics: Two ends in the limelight. *Nat. Rev. Mol. Cell Biol.* **2015**, *16*, 711–726. [CrossRef] [PubMed]

139. Condemine, W.; Eguether, T.; Couroussé, N.; Etchebest, C.; Gardet, A.; Trugnan, G.; Chwetzoff, S. The C Terminus of Rotavirus VP4 Protein Contains an Actin Binding Domain Which Requires Cooperation with the Coiled-Coil Domain for Actin Remodeling. *J. Virol.* **2018**, *93*, e01598-18. [CrossRef] [PubMed]
140. Berkova, Z.; Crawford, S.E.; Blutt, S.E.; Morris, A.P.; Estes, M.K. Expression of rotavirus NSP4 alters the actin network organization through the actin remodeling protein cofilin. *J. Virol.* **2007**, *81*, 3545–3553. [CrossRef] [PubMed]
141. Gardet, A.; Breton, M.; Fontanges, P.; Trugnan, G.; Chwetzoff, S. Rotavirus spike protein VP4 binds to and remodels actin bundles of the epithelial brush border into actin bodies. *J. Virol.* **2006**, *80*, 3947–3956. [CrossRef] [PubMed]
142. Vetter, J.; Papa, G.; Seyffert, M.; Gunasekera, K.; De Lorenzo, G.; Wiesendanger, M.; Reymond, J.L.; Fraefel, C.; Burrone, O.R.; Eichwald, C. Rotavirus Spike Protein VP4 Mediates Viroplasm Assembly by Association to Actin Filaments. *J. Virol.* **2022**, *96*, e0107422. [CrossRef] [PubMed]
143. Nikolic, J.; Le Bars, R.; Lama, Z.; Scrima, N.; Lagaudrière-Gesbert, C.; Gaudin, Y.; Blondel, D. Negri bodies are viral factories with properties of liquid organelles. *Nat. Commun.* **2017**, *8*, 58. [CrossRef] [PubMed]
144. Nejmeddine, M.; Trugnan, G.; Sapin, C.; Kohli, E.; Svensson, L.; Lopez, S.; Cohen, J. Rotavirus spike protein VP4 is present at the plasma membrane and is associated with microtubules in infected cells. *J. Virol.* **2000**, *74*, 3313–3320. [CrossRef]
145. Zhang, Y.; Gao, W.; Li, J.; Wu, W.; Jiu, Y. The Role of Host Cytoskeleton in Flavivirus Infection. *Virol. Sin.* **2019**, *34*, 30–41. [CrossRef]
146. Zhang, L.-J.; Xia, L.; Liu, S.-L.; Sun, E.-Z.; Wu, Q.-M.; Wen, L.; Zhang, Z.-L.; Pang, D.-W. A "Driver Switchover" Mechanism of Influenza Virus Transport from Microfilaments to Microtubules. *ACS Nano* **2018**, *12*, 474–484. [CrossRef] [PubMed]
147. Brunet, J.-P.; Jourdan, N.; Cotte-Laffitte, J.; Linxe, C.; Géniteau-Legendre, M.; Servin, A.; Quéro, A.-M. Rotavirus Infection Induces Cytoskeleton Disorganization in Human Intestinal Epithelial Cells: Implication of an Increase in Intracellular Calcium Concentration. *J. Virol.* **2000**, *74*, 10801–10806. [CrossRef] [PubMed]
148. Broering, T.J.; Parker, J.S.L.; Joyce, P.L.; Kim, J.; Nibert, M.L. Mammalian Reovirus Nonstructural Protein μNS Forms Large Inclusions and Colocalizes with Reovirus Microtubule-Associated Protein μ2 in Transfected Cells. *J. Virol.* **2002**, *76*, 8285–8297. [CrossRef] [PubMed]
149. Bussiere, L.D.; Choudhury, P.; Bellaire, B.; Miller, C.L. Characterization of a Replicating Mammalian Orthoreovirus with Tetracysteine-Tagged μNS for Live-Cell Visualization of Viral Factories. *J. Virol.* **2017**, *91*, e01371-17. [CrossRef] [PubMed]
150. Dales, S. Association Between The Spindle Apparatus and Reovirus. *Proc. Natl. Acad. Sci. USA* **1963**, *50*, 268–275. [CrossRef]
151. Sharpe, A.H.; Chen, L.B.; Fields, B.N. The interaction of mammalian reoviruses with the cytoskeleton of monkey kidney CV-1 cells. *Virology* **1982**, *120*, 399–411. [CrossRef]
152. Eaton, B.T.; Hyatt, A.D. Association of Bluetongue Virus with the Cytoskeleton. In *Virally Infected Cells*; Springer: Boston, MA, USA, 1989; Volume 15, pp. 233–273.

Review

Rotavirus NSP2: A Master Orchestrator of Early Viral Particle Assembly

Sarah L. Nichols [1], Cyril Haller [2], Alexander Borodavka [2,*] and Sarah M. Esstman [1,*]

[1] Department of Biology, Wake Forest University, Wake Downtown, 455 Vine Street, Winston-Salem, NC 27106, USA; nichsl20@wfu.edu

[2] Department of Chemical Engineering and Biotechnology, Cambridge University, Philippa Fawcett Drive, Cambridge CB3 0AS, UK; cjh253@cam.ac.uk

* Correspondence: ab2677@cam.ac.uk (A.B.); mcdonasm@wfu.edu (S.M.E.)

Abstract: Rotaviruses (RVs) are 11-segmented, double-stranded (ds) RNA viruses and important causes of acute gastroenteritis in humans and other animal species. Early RV particle assembly is a multi-step process that includes the assortment, packaging and replication of the 11 genome segments in close connection with capsid morphogenesis. This process occurs inside virally induced, cytosolic, membrane-less organelles called viroplasms. While many viral and cellular proteins play roles during early RV assembly, the octameric nonstructural protein 2 (NSP2) has emerged as a master orchestrator of this key stage of the viral replication cycle. NSP2 is critical for viroplasm biogenesis as well as for the selective RNA–RNA interactions that underpin the assortment of 11 viral genome segments. Moreover, NSP2's associated enzymatic activities might serve to maintain nucleotide pools for use during viral genome replication, a process that is concurrent with early particle assembly. The goal of this review article is to summarize the available data about the structures, functions and interactions of RV NSP2 while also drawing attention to important unanswered questions in the field.

Keywords: rotavirus; nonstructural protein 2; viroplasm; particle assembly; genome segment assortment; genome packaging; replication

Citation: Nichols, S.L.; Haller, C.; Borodavka, A.; Esstman, S.M. Rotavirus NSP2: A Master Orchestrator of Early Viral Particle Assembly. *Viruses* 2024, 16, 814. https://doi.org/10.3390/v16060814

Academic Editor: Feng Li

Received: 18 April 2024
Revised: 6 May 2024
Accepted: 16 May 2024
Published: 21 May 2024

1. Introduction

Rotaviruses (RVs) are 11-segmented, double-stranded (ds) RNA viruses belonging to the *Sedoreoviridae* family within the *Reovirales* order [1]. Nine genetically divergent species of RV (groups A-D and F-J) have been discovered to date; however, human infections are typically caused by group A strains [2,3]. Notably, group A RVs induce severe, dehydrating diarrhea and vomiting in infants and young children, which can be life-threatening in the absence of medical care [4]. Despite the availability of several live-attenuated vaccines against human group A RVs, it is estimated that infections still lead to ~128,000 child deaths each year in developing world regions [4]. Continued research on RVs is warranted because the outcomes of such work may inform new treatment measures to prevent childhood diarrhea. In addition to infecting humans, group A RVs also infect a broad range of avian and mammalian hosts in nature [5]. Because human strains are poorly cultivatable, the field has predominantly relied on animal strains as models (e.g., simian strains SA11 and RRV, bovine strains UK and RF, and porcine strain OSU) [6]. Structural and functional studies of these model strains and their individual viral proteins have shed light on many aspects of RV biology. In particular, the RV nonstructural protein 2 (NSP2) has been investigated by a cadre of laboratories over the past several decades. Results of this work have revealed multi-faceted roles for this protein during viral replication, particularly during the early stages of particle assembly. This review article seeks to summarize the growing body of published literature on RV NSP2 as well as emphasize the key gaps in knowledge for future research studies.

2. Virion, Genome and Replication Cycle

The RV virion is a ~100 nm non-enveloped, triple-layered particle that exhibits icosahedral symmetry (Figure 1A) [7,8]. The outermost layer of the virion is comprised of the VP7 glycoprotein, and it is embedded with numerous copies of the VP4 spike protein [7,8]. Proteolysis of VP4 by trypsin-like enzymes is required for RV attachment to cells and results in the formation of VP8* and VP5* fragments [9]. The intermediate layer of the virion is made up of VP6, and the innermost core shell is comprised of VP2 [7,8]. Within the VP2 core shell reside several copies of the VP1 RNA-dependent RNA polymerase and the VP3 RNA capping enzyme [8,10,11]. While the position of VP3 inside of the particle has not yet been validated, VP1 is bound beneath the VP2 core shell, just off-center from each icosahedral fivefold axis [8,11]. Also located within the VP2 core shell are the 11 dsRNA genome segments, which range in size from ~0.5 kb to ~3.3 kb, together comprising a ~18–23 kb viral genome [12] (Figure 1B). The extreme 5′ and 3′ termini of the 11 dsRNA genome segments are conserved, and they abut more variable non-coding regions (NCRs) [12]. For each segment, the NCRs surround a central open-reading frame (ORF) that typically encodes a single viral protein [12]. As such, most group A RV strains express 11 proteins: (i) 6 structural proteins (VP1–VP4, VP6 and VP7) that make up the virion particle and (ii) 5 nonstructural proteins (NSP1-NSP5) that play various roles during the viral replication cycle but are not incorporated into particles [12] (Figure 1B). Some group A RV strains express an additional protein (NSP6) from an alternative ORF in the NSP5-coding gene [12].

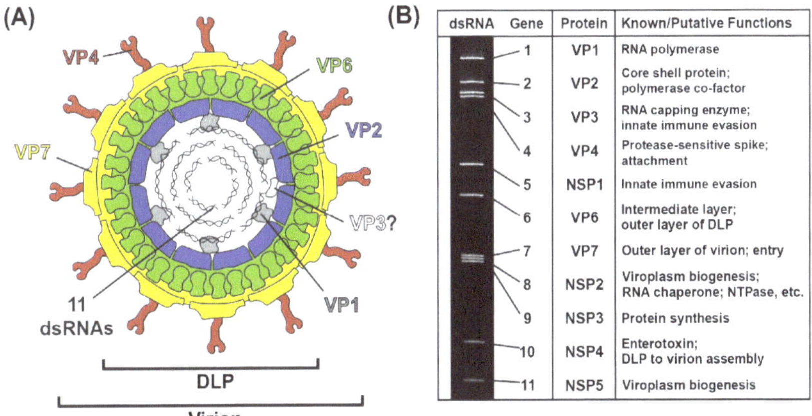

dsRNA Gene	Protein	Known/Putative Functions
1	VP1	RNA polymerase
2	VP2	Core shell protein; polymerase co-factor
3	VP3	RNA capping enzyme; innate immune evasion
4	VP4	Protease-sensitive spike; attachment
5	NSP1	Innate immune evasion
6	VP6	Intermediate layer; outer layer of DLP
7	VP7	Outer layer of virion; entry
8	NSP2	Viroplasm biogenesis; RNA chaperone; NTPase, etc.
9	NSP3	Protein synthesis
10	NSP4	Enterotoxin; DLP to virion assembly
11	NSP5	Viroplasm biogenesis

Figure 1. Rotavirus Virion, Genome and Proteins. (**A**) Cartoon image of the infectious rotavirus virion, which is made up of six viral proteins (VP1–VP4, VP6 and VP7). The location of VP3 is unknown. The 11 dsRNA genome segments are encased within the particle. (**B**) Rotavirus strain SA11 dsRNA genome segments are shown separated in a polyacrylamide gel. Each gene is numbered, and the encoded protein and putative/known functions are listed to the right.

Like all members of the *Reovirales* order, the RV replication cycle is entirely cytoplasmic [12] (Figure 2). In humans and lab animals, RVs primarily infect the epithelial cells of small intestinal villi [13,14]. However, most studies of RV replication are performed using transformed monkey kidney epithelial cell lines (e.g., MA104 and COS-7) [6,14]. RV infection is initiated by the binding of proteolytically activated virions (i.e., with cleaved VP4 spikes) to protein and/or carbohydrate receptors on the host cell surface [15]. The virus typically enters cells via receptor-mediated endocytosis [15–17]. In the low Ca^{2+} environment of the endosome, virions undergo shedding of their outer VP4/VP7 layer concurrent with penetration of the endosomal membrane, resulting in the deposition of double-layered particles (DLPs) into the cytosol [18–20]. As quickly as 15 min after penetration, the VP1 polymerases within the interior of the DLPs initiate viral RNA transcription [21]. More specifically, transcription is the synthesis of 11 different positive-sense, single-stranded RNA (+ssRNA) using the minus-strands (−ssRNAs) of the 11 dsRNA genome segments as templates [21]. The nascent +ssRNAs receive a 5′ 7-methylguanosine cap via the activities of VP3 prior to their egress from channels located at the fivefold axes of the capsid [21–24]. Serving as messenger RNAs (mRNAs), the +ssRNAs are translated into the 11 RV proteins by the host cell ribosomal machinery [25–27]. The exact sites of RV protein synthesis are unknown but likely include cytosolic polysomes and the rough endoplasmic reticulum (ER) [27].

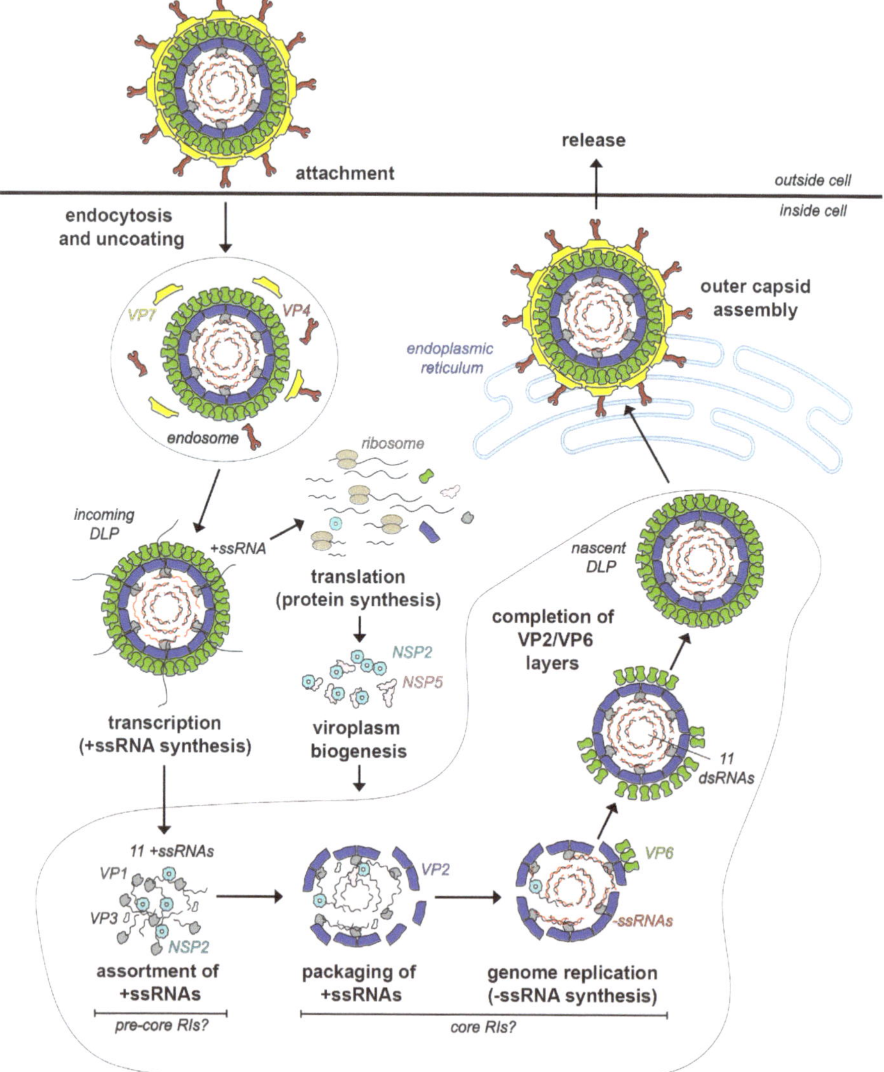

Figure 2. Rotavirus Replication Cycle. The rotavirus virion attaches to the host cell and enters via endocytosis. The outer VP7-VP4 layer is removed during the entry process, resulting in a transcriptionally active double-layered particle (DLP). Transcripts (+ssRNAs) first serve as mRNAs for protein synthesis. Once made, nonstructural proteins NSP2 and NSP5 interact to form viroplasms, where the early stages of particle assembly occur. The hypothetical steps and putative assembly intermediates involved in early assembly inside viroplasms are labeled. Specifically, the +ssRNAs

recruited to the viroplasm are thought to be assorted in the context of pre-core RIs, and then they are packaged into VP2-containing assembly intermediates called core RIs. In this core RI context, the +ssRNAs are used for −ssRNA (red) synthesis by VP1 to recreate the dsRNA genome segments inside a particle that morphs into a DLP. Nascent DLPs made in viroplasms acquire their outer VP7-VP4 layer in the endoplasmic reticulum prior to exiting from the cell.

Following protein synthesis, nonstructural proteins NSP2 and NSP5 nucleate the formation of cytoplasmic inclusions, called viroplasms, which are the sites of early RV particle assembly [28–30]. More simply, viroplasms can be thought of as "factories" wherein VP1, VP2, VP3, VP6 and the 11 viral RNA segments come together in a highly coordinated manner to form new DLPs (Figure 2). While the mechanistic details are incompletely understood, the DLP assembly process can be described as having 4 interconnected, synergistic steps: (i) assortment of the 11 distinct +ssRNAs, (ii) packaging of the 11 +ssRNAs along with the VP1/VP3 into a morphing VP2-containing particle, (iii) VP1-mediated −ssRNA synthesis (i.e., genome replication), converting the 11 +ssRNAs into the dsRNA genome segments and (iv) completed assembly of VP2 and VP6 capsid layers to create the intact DLP [18] (Figure 2). As we will detail in this review article, NSP2 plays several critical roles during early RV particle assembly—it nucleates viroplasms, mediates the assortment/packaging of the 11 +ssRNAs into the assembling capsid and may even help to maintain nucleotide substrate pools for genome replication. The final stages of RV particle assembly occur in the ER, where the VP4/VP7 capsid layer is added to newly formed DLPs, and the resulting virions exit from the cell via non-lytic or lytic pathways, depending on cell type [31,32].

3. Discovery and Early Characterization of NSP2

RV NSP2 was first described in the early 1980s when polypeptides from SA11-infected MA104 cells were analyzed via immunoprecipitation and partial proteolytic peptide mapping [33]. These experiments revealed the presence of a 35 kDa protein (originally called NS35 and later renamed NSP2) that failed to immunoprecipitate using antisera raised against DLPs, suggesting it to be a nonstructural protein [33]. Sequence analysis was used to deduce that the NSP2 protein was a conserved, highly basic protein and was coded for by strain SA11 segment 8; this assignment was experimentally confirmed by in vitro translation of viral mRNAs and RNA–RNA hybridization assays [34,35]. Following its discovery, multiple studies sought to better understand the possible function(s) of NSP2 during RV replication by investigating its intracellular localization and interactions. Immunocytochemistry and colloidal gold labeling of SA11-infected MA104 cells showed that NSP2 localized to viroplasms, which were judged by electron microscopy to be the sites of DLP assembly [36]. Moreover, NSP2 was reported to be a component of RV replication-assembly intermediates (RIs) that were purified from SA11-infected MA104 cells using either native gel agarose electrophoresis or immunoprecipitation [37–40]. Specifically, Patton and Gallegos reported that NSP2 was found in complexes called pre-core RIs, which lacked the VP2 core shell protein but contained VP1, VP3 and NSP5 [37]. NSP2 was also found in core RIs that contained VP2 along with the components of the pre-core RI [38]. Upon incubation with Mg^{2+} and NTPs, isolated core RIs (but not pre-core RIs) were found to be active for −ssRNA synthesis, creating the 11 dsRNA genome segments in vitro [38–40]. These results suggested that the assortment of the 11 genome segments (i.e., the 11 +ssRNAs), likely in the context of the pre-core RI, preceded core RI-mediated genome replication (Figure 2) [41]. Furthermore, these results indicated that the VP2 core shell protein was required for −ssRNA synthesis by the VP1 polymerase, a finding that has been since validated [42–46]. The observation that NSP2 was found in both pre-core RIs and core RIs suggested that it may function during the earliest DLP assembly steps, namely +ssRNA assortment/packaging and possibly genome replication. Supporting this notion were studies of a temperature-sensitive SA11 mutant virus that has a lesion mapping to segment 8 (*ts*E) [47]. *ts*E replicates well when grown in cell culture at the permissive temperature of 31 °C, but its replication is severely diminished at the non-permissive temperature of 39 °C [48,49]. Ramig and Petrie showed an increase in the formation of empty RV particles and a decrease in the number of viroplasms in *ts*E-infected MA104 cells at 39 °C versus 31 °C [49]. Moreover, *ts*E was reported to be phenotypically negative for −ssRNA synthesis at 39 °C, though it remains unclear whether this phenotype was an indirect consequence of failed +ssRNA packaging, which is pre-requisite to −ssRNA synthesis [50].

Early mechanistic insights into NSP2 functions were revealed by a series of biochemical studies. Kattoura et al. employed UV-crosslinking to demonstrate that the protein (i) assembles into higher-ordered oligomers in the cell, (ii) binds to +ssRNA (and to a lesser extent dsRNA) in a sequence-independent manner and (iii) interacts directly with the VP1 polymerase [51,52]. Both the oligomeric nature and RNA binding properties of NSP2 were confirmed by Taraporewala et al. using recombinant protein [53]. Specifically, by employing gel shift assays the authors were also able to show that recombinant, multimeric NSP2 bound to +ssRNA in discrete, cooperative steps [53]. Interestingly, the same authors also showed that the octameric form of NSP2 exhibited a Mg^{2+}-dependent nucleotide triphosphatase (NTPase) activity, meaning that it could hydrolyze the γ-phosphate from NTP to create an NDP in vitro [53,54]. Based on this result, the NTPase activity of NSP2 was hypothesized to serve as a source of energy for incorporating the 11 +ssRNAs into morphing RV particles [53,54]. Schuck et al. built upon the work of Taraporewala et al. and used a variety of biophysical techniques to confirm the octameric status of NSP2, and to show that it undergoes conformational changes upon binding to RNA, Mg^{2+} and NTP/NDPs [55]. In addition, NSP2 was reported to have RNA helix-destabilizing activity, as it disrupted RNA–RNA duplexes in vitro [56]. This function was predicted to be important for unwinding stem-loop structures in viral +ssRNAs during assortment and packaging [56]. Finally, NSP2 was reported to bind directly to NSP5, and co-expression of these two proteins alone induced the formation of viroplasm-like structures in the absence of infection [57,58]. Thus, in addition to roles during the DLP assembly pathway itself, NSP2 also seemed to help build the factories within which such assembly occurred. Altogether, these early studies created a strong foundation for the following decades of structural and functional work, revealing deeper insights into NSP2's multifaceted roles during RV replication.

4. Structural and Enzymatic Studies of NSP2

Several structures of recombinant, octameric group A NSP2 have been solved by either X-ray crystallography or by using single particle cryo-EM reconstruction techniques [59–64]. These studies show that each NSP2 monomer within the octameric unit has an N-terminal domain (residues ~1–140) and a C-terminal domain (residues ~156–313) that are connected by a short loop (residues ~141–155) (Figure 3A,B). Residues ~313–317 are unstructured (Figure 3A). The N-terminal domain is described as having two sub-domains: (i) the first is composed of two pairs of β-strands separated by two α-helices, and (ii) the second consists of four α-helices (Figure 3B). The two N-terminal subdomains are connected by a loop that is largely basic (Figure 3B). This loop and an α-helix from the N-terminal domain form a large portion of one side of a 25 Å-deep electropositive cleft on the monomer (Figure 3B). The C-terminal domain has a prominent twisted anti-parallel β-sheet followed by α-helices (Figure 3B). The other side of the electropositive cleft consists of a C-terminal α-helix and a loop that exists between residues ~248–265 (Figure 3B). The base of the cleft is composed of C-terminal anti-parallel β-strands (residues ~186–191; 226–230), along with a loop structure formed of residues ~221–226 (Figure 3B). The extreme C-terminal region (CTR; residues ~291–313) of NSP2 consists of a flexible linker region and a terminal α-helix (Figure 3A,B). The functional NSP2 octamer is formed via tail-to-tail stacking of two tetramers in a manner that creates a 35-Å wide central hole (Figure 3C,D).

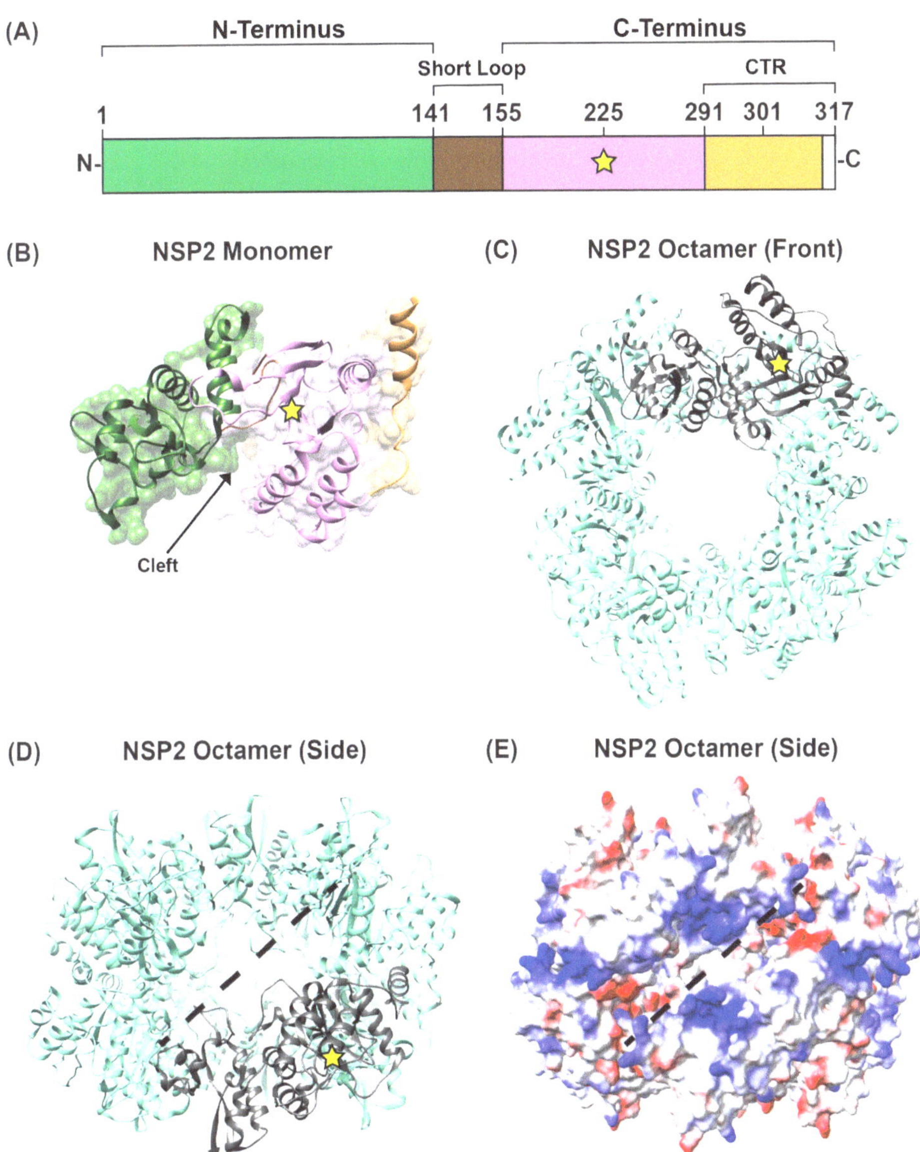

Figure 3. NSP2 Structure. (**A**) Linear schematic of strain SA11 NSP2 (317 amino acids in length). The protein is comprised of two domains: an N-terminal (green) and a C-terminal domain (pink), separated by a short loop (brown). The extreme C-terminal region (CTR; residues 291–317) is represented in orange; however, CTR residues 314 to 317 are unstructured (white). A yellow star represents the catalytic site H225. (**B**) SA11 NSP2 monomer (PDB no. 1L9V) is colored as in panel (**A**) and is shown in both surface and ribbon representation. An arrow indicates the electropositive cleft, and a yellow star represents the catalytic site H225. (**C**) SA11 NSP2 octamer structure (PDB no. 1L9V) is shown in ribbon representation (cyan), which a single monomer highlighted in gray. (**D**) SA11 NSP2 octamer from panel (**C**) is flipped "forward" 90 degrees to show the side view along the two-fold axis. The electropositive groove that comprises RNA/NSP5 binding site is shown as a dashed line. (**E**) SA11 NSP2 octameric structure from panel (**D**) is shown in electrostatic surface representation. Red indicates negative charge, while blue indicates positive charge.

In the functional octamer, the electropositive clefts of NSP2 monomers come together in the context of two highly basic, 25 Å deep and 30 Å wide grooves that run diagonally across each tetramer–tetramer interface [59] (Figure 3D,E). The histidine triad (HIT)-like motif of NSP2, which includes the catalytic residue H225 that mediates the hydrolysis of the γ-phosphate from an NTP to create NDP, is accessed within this region of the protein [59,65] (Figures 3D and 4A). Using structural and biochemical approaches, Kumar et al. showed that the hydrolyzed γ-phosphate is transferred to H225 of NSP2 [61]. The formation of this phosphohistidine intermediate was found to be part of an in vitro NDP kinase activity, whereby NSP2 converts NDP to NTP via transfer of the bound γ-phosphate [61] (Figure 4B). The observation that NSP2 has both NTPase and NDP kinase activities suggested a possible role for NSP2 in maintaining pools of nucleotides in the viroplasm for use during genome replication. It was further shown that recombinant NSP2 exhibits an in vitro RNA triphosphatase (RTPase) activity, whereby it hydrolyzes the γ-phosphate from an RNA molecule via a mechanism requiring H225 [62,65] (Figure 4A). Hu et al. investigated this RTPase mechanism by determining the X-ray crystal structure of recombinant NSP2 in complex with the 5′consensus sequence of RV −ssRNA (i.e., 5′GG) [62]. Consistent with previous studies, they found that the oligoribonucleotide interacted extensively with highly conserved residues in the enzymatic cleft of the monomers, which are accessed in the context of the grooves [62]. Indeed, cryo-EM reconstructions of NSP2 in complex with RNA support the notion that the grooves are also major RNA-binding sites [60]. Moreover, cryo-EM analysis of NSP2 in complex with a short fragment of NSP5 (residues 66–188) indicated that NSP5 may interact with NSP2 via the grooves, competing with RNA [60] (Figure 3D,E). The grooves have also been reported to bind tubulin, although it remains unclear whether NSP2 is involved in microtubule depolymerization [66]. It is possible that the highly charged nature of the groove and its geometry may accommodate the binding of multiple charged ligands, including small RNAs and negatively charged proteins.

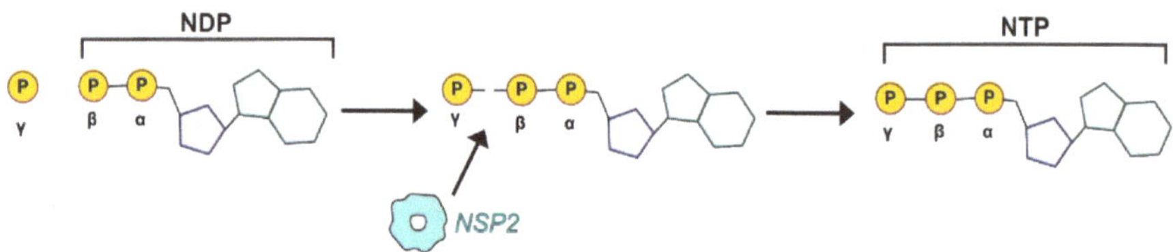

Figure 4. Enzymatic Activities of NSP2. (**A**) Nucleoside triphosphatase (NTPase) and RNA-triphosphatase (RTPase) activities of NSP2 are cartoon. In this case, the terminal γ phosphate of either a cellular NTP or a viral RNA is removed by NSP2, yielding inorganic phosphate and NDP. (**B**) Nucleoside diphosphate kinase (NDPase) activity of NSP2 is cartooned, whereby the terminal γ phosphate is added to a cellular NDP to form an NTP.

A separate, novel observation by Hu et al. was that the CTR of one NSP2 monomer in the octamer unit exhibited an "open" conformation that was flipped outward relative to the rest of the protein [62] (Figure 5A). The "open" conformation of the CTR was underpinned by flexible linker residues 293–295, allowing for a domain-swapping interaction that caused adjacent NSP2 octamers to interact within the crystal lattice [62] (Figure 5B). However, octamer chains were not found in the cryo-EM structures of NSP2, and it remains unknown whether such interactions occur in solution or in infected cells [64]. Still, as we discuss in detail in the following sections, the flexible CTR of NSP2 has emerged as an important functional domain of the protein for both viroplasm formation and for +ssRNA segment assortment.

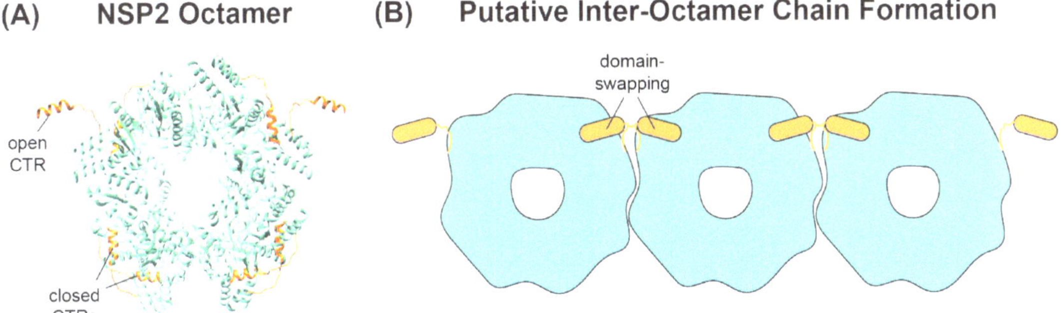

Figure 5. NSP2 CTR Conformations and Putative Inter-Octamer Chains. (**A**) SA11 NSP2 octamer (PDB no. 4G0A) from Hu et al. is shown in ribbon form (cyan), with the flexible CTRs highlighted in orange. Representative open and closed CTRs are labeled. (**B**) Cartoon image of putative NSP2 inter-octamer chains. Under some crystallography conditions, the open CTR of one octamer can interact with the body of a neighboring octamer (and vice-versa) via a domain-swapping interaction.

5. Role of NSP2 in Viroplasm Formation

As mentioned previously, viroplasms are cytosolic inclusions that serve as sites for the early stages of RV particle assembly, including the steps of +ssRNA assortment/packaging and −ssRNA synthesis (i.e., genome replication). Viroplasms can be seen microscopically as early as 2–4 h p.i., and they appear as small cytoplasmic puncta (~0.1–1 μm in diameter) [67–70] (Figure 6A). Over the course of infection, small viroplasms fuse together to become larger, with some reaching >5 μm in diameter [67–70]) (Figure 6A). Notably, Geiger et al. have shown that, at early stages, viroplasms are more dynamic and can be readily and reversibly dissolved by the low-concentration aliphatic diol treatments (e.g., 1,6-hexanediol or propylene glycol), suggesting that these structures are held by relatively weak, multivalent interactions between NSP2 and NSP5 (Figure 6B) [71]. Additionally, viroplasms contain all eleven types of +ssRNAs required for the assembly of nascent DLPs, as shown by multiplexed single-molecule fluorescence in situ hybridization (smFISH) analyses (Figure 6C) [72]. While viroplasms have also been shown to contain several viral and cellular proteins and lipids, NSP2 and NSP5 are sufficient for their nucleation [28–30,58,73]. Silencing of either NSP2 or NSP5 expression in RV-infected cells using RNA interference (RNAi) was shown to prevent viroplasm formation, further demonstrating their importance [74,75]. Interestingly, alanine mutation of the NSP2 catalytic histidine (H225A) did not affect the capacity of the protein to form viroplasms or viroplasm-like structures in cells [76,77]. Thus, the NTPase/RTPase/NDP kinase activities of NSP2 appear to be dispensable for viroplasm assembly.

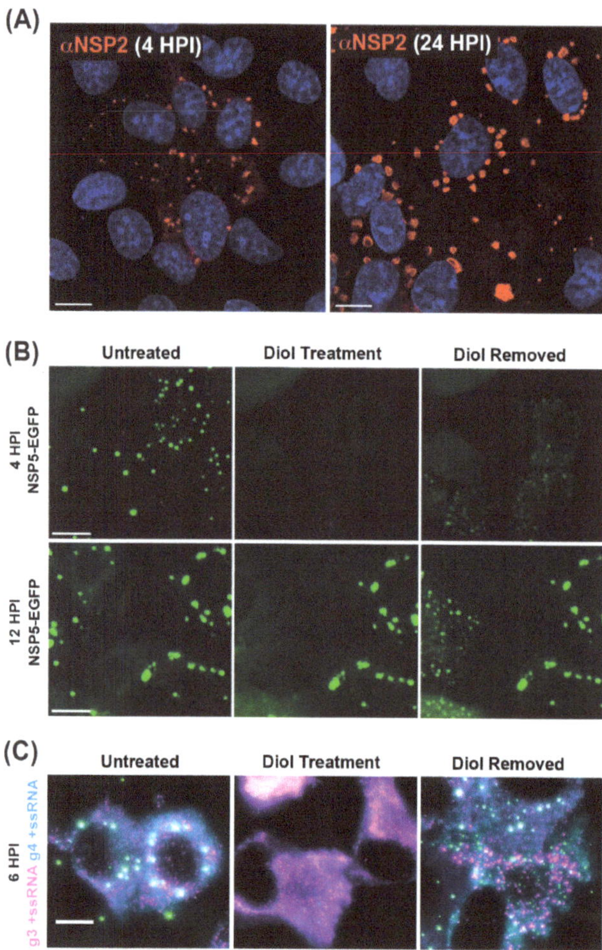

Figure 6. Viroplasms are Cytoplasmic Biomolecular Condensates Formed via LLPS. (**A**) Fixed immunofluorescence images of SA11-infected MA104 cells at 4 and 24 h post-infection (HPI). Viroplasms were stained with a polyclonal antibody against NSP2 (αNSP2; red), and cell nuclei were stained with Hoechst (blue). Scale bar = 10 μm. Images adapted from reference [70] with permission. (**B**) Viroplasms formed in MA104 cells stably expressing NSP5-EGFP and infected with strain SA11. Numerous small viroplasms can be dissolved when low concentrations of aliphatic diols (4.7% propylene glycol or 4% 1.6-hexane diol) are applied directly to the cell culture medium at 4 h post-infection (HPI). Removal of diols from the medium results in reassembly of multiple smaller granules dispersed in the cytosol (right panel). At 12 HPI, viroplasms become larger and less regular in shape. These larger viroplasms become resistant to the application of aliphatic diols. Scale bar = 50 μm. Images adapted from [71] with permission. Copyright Creative Commons Attribution License (https://creativecommons.org/licenses/by/4.0/). (**C**) RNA FISH imaging of gene segment 3 +ssRNA (magenta, g3 +ssRNA) and gene segment 4 +ssRNAs (cyan, g4 +ssRNA) in SA11-infected NSP5-EGFP-expressing MA104 cells fixed at 6 HPI. Viroplasms were treated with 4.7% (*v*/*v*) propylene glycol (middle) at 4 HPI, releasing +ssRNAs into the cytoplasm. These granules reformed after replacing the propylene glycol-containing cell culture medium, resulting in the rapid re-localization of g3 +ssRNA (magenta) and g4 +ssRNA (cyan) transcripts are detected via smFISH, and colocalizing RNAs (white). Scale bar = 10 μm. Images adapted from [72] with permission. Copyright Creative Commons Attribution License (https://creativecommons.org/licenses/by/4.0/).

How NSP2 nucleates viroplasms alongside NSP5 has been an active area of research in recent years. Using conformation-specific monoclonal antibodies (mAbs), Criglar et al. showed that two forms of NSP2 are present in RV-infected cells [78]. Specifically, a diffuse form of NSP2 (dNSP2) predominantly localizes in the cytosol, while another form of NSP2 (vNSP2) accumulates in viroplasms [78] (Figure 7A,B). The authors show evidence to support the notion that dNSP2 interacts primarily with a hypo-phosphorylated NSP5, while vNSP2 interacts with a hyper-phosphorylated form of NSP5 [78] (Figure 7C). Phosphorylation of serine 313 (S313) in the flexible C-terminus of NSP2 itself was found to be enriched in the vNSP2 preparation, suggesting that it might contribute to differential mAb recognition [78] (Figure 7C). Further structural characterization of the used mAbs and their modes of recognizing distinct NSP2 conformations would substantiate the proposed model. A separate study performed by the same group provided experimental evidence that S313 phosphorylation of NSP2 was mediated by cellular casein kinase 1 (CK1α) [79]. Interestingly, silencing of CK1α resulted in vNSP2 displaying a diffuse phenotype similar to that of dNSP2 [79]. This result led to the hypothesis that S313 phosphorylation might underpin the switch from dNSP2 to vNSP2. It should be noted that silencing of CK1α is also known to inhibit NSP5 phosphorylation, also resulting in similar viroplasm dispersal and morphology alteration [80]. However, a mutant SA11 RV bearing a phosphomimetic change at NSP2 position 313 (S313D) still produced dNSP2, indicating that the conformation change between these two forms of the protein may be more complicated than this single post-translational modification [63]. Nevertheless, S313 appears to be important for proper viroplasm formation, as the S313D mutant produced phenotypically "disorganized" viroplasms early times p.i., and it was delayed in its overall replication [63]. Several studies have shown that lipid droplet markers co-localize with viroplasms, and Criglar et al. revealed that phosphomimetic S313D NSP2 physically interacts with lipid droplet proteins [63,81–83]. Thus, post-translational modifications of NSP2, such as phosphorylation (and possibly other modifications), appear to regulate NSP2 interactions with NSP5 and with cellular proteins during viroplasm biogenesis [28,63,84].

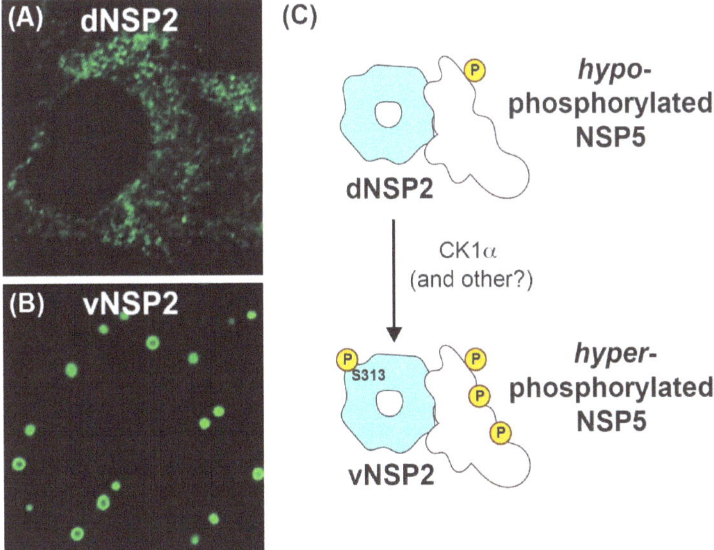

Figure 7. Intracellular Localizations and Interactions of vNSP2 and dNSP2. Fluorescence confocal micrographs of SA11-infected cells stained with monoclonal antibody against dNSP2 (**A**) and vNSP2 (**B**). Images taken from reference [78] with permission. (**C**) Cartoon model of vNSP2 and dNSP2 phosphorylation status and interactions with NSP5.

The NSP2 CTR is clearly an important determinant for viroplasm formation. Deletion of nearly the entire CTR (residues 293–301) abrogates the efficient formation of viroplasm-like structures in NSP2/NSP5 co-expressing cells [70,78]. As mentioned previously, the NSP2 CTR has the capacity to adopt an "open" conformation and participate in a domain-swapping interaction that links several octamers together, at least under crystallography conditions [62]. While not recapitulating the "open" NSP2 CTR, the crystal structure of S313D NSP2 showed enhanced lattice formation due to a hydrogen bond between D313 and R287 of a neighboring octamer [63]. Moreover, an SA11 RV bearing a lysine-to-glutamic acid change at the C-terminal position 294 (K294E) in NSP2 was found to exhibit viroplasm defects, including smaller and more numerous viroplasms under fixed- and live-cell conditions, as well as a delay in viroplasm fusion [70]. Residue K294 is located in a linker region that mediates the "open" vs. "closed" conformations of the CTR [62,70]. Molecular dynamics simulations of the K294E NSP2 monomer structure suggested that the mutation may have altered CTR flexibility, which in turn could have impacted inter-octamer associations [70]. Still, the proposed role of inter-octamer interactions observed in the crystal structure of NSP2 in the formation of viroplasms remains unclear and will require further investigation.

While the importance of NSP2 in the formation of viroplasms is undeniable, it is NSP5 that constitutes the primary component of viroplasms [85–87]. Using quantitative Western blotting of SA11 RV-infected cells, Geiger et al. estimated that the intracellular concentration of NSP5 exceeds that of NSP2, reaching >10 μM within 6 h p.i. [71]. NSP5 is a 22 kDa, serine/threonine-rich, intrinsically disordered and relatively acidic protein that can assemble into several higher-order oligomers [88–90]. During infection, this protein is O-glycosylated and differentially phosphorylated, causing it to migrate in sodium dodecyl sulfate (SDS)-polyacrylamide gels as phosphoisoforms ranging in size from 26 to 35 kDa [88–90]. Phosphorylation of NSP5 is required for proper viroplasm morphology and for viral replication [80,91,92]. Previous studies have shown that recombinant NSP5 can be phosphorylated by CK1α at serine 67 (S67), and only the hypo-phosphorylated isoform of NSP5 is observed in CK1α silenced cells [63,80]. However, further experiments revealed that CK1α is not sufficient for NSP5 hyper-phosphorylation [80]. Interestingly, only when NSP2 is co-expressed with NSP5 do the hyper-phosphorylated NSP5 isoforms appear, and in NSP2-silenced cells, only hypo-phosphorylated NSP5 is observed [57,63,93]. While it was proposed that NSP5 may undergo low levels of auto-phosphorylation through its autokinase activity, its phosphorylation was increased almost 5–10-fold upon incubation with recombinantly expressed NSP2 [76]. Interestingly, NSP2 mutants lacking NTPase activity were still capable of promoting NSP5 phosphorylation [76]. This result suggests that the mechanism of NSP5 phosphorylation may be linked to the conformational changes in NSP5 upon binding of NSP2, independent of its enzymatic activities. While an NSP2/NSP5 phosphorylation cascade is critical for viroplasm formation in infected cells, it is dispensable for viroplasm-like condensate formation in cells and in vitro [71,78]. Thus, phosphorylation may represent a regulatory mechanism controlling the timing of viroplasm formation, but it may not necessarily be a biophysical requirement for structure formation (see also below).

A recent model of viroplasm formation proposes that these replicative factories are formed via the process of liquid–liquid phase separation (LLPS), primarily driven by interactions between NSP2 and NSP5 [71]. Notably, Geiger et al. showed that recombinant NSP2 and NSP5 spontaneously form droplets in vitro with biophysical properties of LLPS condensates (Figure 8). Further evidence for RV viroplasms being biomolecular condensates formed via LLPS includes (i) the ability of these organelles to fuse and relax into larger droplets, (ii) the fast recovery after photobleaching of fluorescently labeled NSP5 and NSP2 and (iii) rapid and reversible dissolution of droplets upon treatments of RV-infected cells with low concentrations of aliphatic diols, including 1,6-hexanediol and 1,2-propanediol [71] (Figure 6B). Interestingly, the ability of aliphatic diols to dissolve viroplasms instantly and reversibly in cells decreased over the course of infection, with larger, less round viroplasms being resistant to these solvents [71]. The loss in sensitivity towards the compounds that interfere with weaker hydrophobic and Van der Waals interactions indicates changes in the nature of biomolecular interactions, potentially reflecting changes in the apparent affinities between NSP2 and NSP5 as viroplasms mature [71]. These findings suggest that viroplasms are RNA-rich biomolecular condensates that are nucleated by NSP5 (containing mostly intrinsically disordered regions) and the RNA-binding protein NSP2.

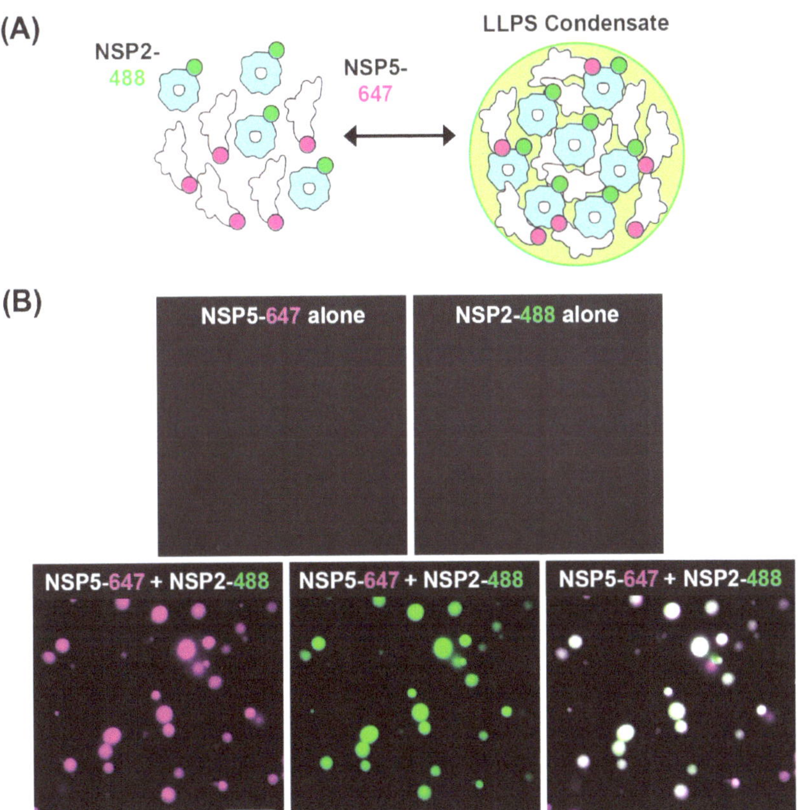

Figure 8. In Vitro LLPS Biomolecular Condensate Assay. (**A**) Recombinantly expressed, Atto 647-dye-labelled NSP5 (NSP5-647; magenta) and Atto488-dye-labelled NSP2 (NSP2-488; green) are mixed together in vitro to form droplets with LLPS characteristics. (**B**) Images showing either no condensate formation with individual proteins (top images) but efficient condensates when NSP5-647 and NSP2-488 are mixed (bottom images). Images modified from reference [71] with permission. Copyright Creative Commons Attribution License (https://creativecommons.org/licenses/by/4.0/).

Consistent with their liquid-like behavior, viroplasms initially coalesce over the course of infection and increase in size, with some reports suggesting that they migrate toward the perinuclear space [69–71]. Through experiments with microtubule destabilizing drugs, the movement and, ultimately, the fusion and condensation of viroplasms to the perinuclear space were inhibited, suggesting a role of microtubules in the movement and fusion of these structures [69,94]. Several studies corroborate this idea through immunofluorescence and electron microscopy, whereby viroplasms were shown to co-localize with microtubules [66,69,94]. This notion was further supported by co-immunoprecipitation experiments showing that NSP2 interacted with tubulin and dynein intermediate chain [95]. Thus, it is likely that the movement and coalescence of viroplasms over the course of infection is mediated by NSP2 directly interacting with dynein and kinesin motors; however, the functional significance of these interactions for RV replication remains unclear.

6. Role of NSP2 during +ssRNA Assortment

Within the viroplasm itself, NSP2 likely plays several key roles, particularly at the step of +ssRNA (i.e., genome segment) assortment/packaging in pre-core and core RIs. In particular, the assortment of distinct +ssRNAs is thought to be mediated by the RNA chaperoning activity of NSP2, including RNA helix destabilizing and RNA annealing activities [56,64,96]. NSP2 can bind both folded RNA stem-loops and less structured RNAs, exhibiting a helix-destabilizing activity that is independent of cofactors and energy requirements (i.e., Mg^{2+} or ATP) [56,64]. Originally, NSP2 was proposed to remove secondary structures in +ssRNAs that would impede their packaging into core RIs as well as their replication into dsRNA [56]. However, recent investigations of NSP2 uncovered its RNA strand annealing activity in the context of long +ssRNAs in vitro, which may be conducive to the formation of an RNA assortment complex containing all 11 distinct +ssRNAs [96] (Figure 9). Specifically, Borodavka et al. showed that when incubated together, +ssRNAs did not interact unless NSP2 was added in molar excess over +ssRNAs [96]. To confirm that the +ssRNAs were interacting due to the inter-molecular base-pairing, the authors removed NSP2 after the addition to the +ssRNA samples via proteinase K digestion [96]. They found that the formation of inter-segment RNA–RNA complexes was indeed mediated by NSP2 that can be removed afterwards, consistent with its role as an RNA chaperone [96]. Importantly, the RNA chaperone activity of NSP2 was unaffected by the addition of ATP to reactions, which suggested that it was independent of NTPase activity [96]. Collectively, this data provides evidence that sequence-specific inter-molecular base-pairing, mediated by NSP2 binding to viral +ssRNAs, governs inter-segment RNA interactions and suggests that NSP2 acts as a viral RNA chaperone (Figure 9).

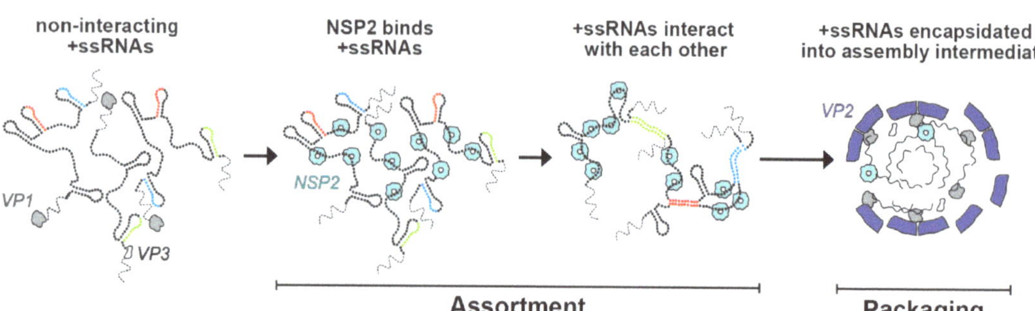

Figure 9. Model of NSP2-dependent +ssRNA Assortment. RV +ssRNAs (here only 3 RNAs are shown schematically) do not form stable RNA–RNA contacts with each other. NSP2 binding to +ssRNAs results in their structural rearrangements concomitant with the exposure of otherwise sequestered complementary sequences (interspersed sequences shown in red, blue and green) capable of inter-segment base-pairing. The exposed complementary sequences form stable sequence-specific inter-segment contacts (RNA helices shown in red, green and blue) during the +ssRNA assortment process. The resulting multi-RNA ribonucleoprotein complex (containing all 11 +ssRNAs) would then be encapsidated by a VP2 core assembly intermediate for genome packaging. Image modified from reference [96] with permission. Copyright Creative Commons Attribution License (https://creativecommons.org/licenses/by/4.0/).

While the determinants within NSP2 required for its RNA chaperone activity remain to be fully elucidated, Bravo et al. discovered that such activity also required the flexible CTR (residues 295–317) [64]. Using cryo-EM and structural proteomics studies, including hydrogen-deuterium exchange coupled with mass-spectrometry, the authors determined that the NSP2 CTR was not directly involved in RNA binding, a finding that had been previously reported [62,64]. Unexpectedly, the authors discovered that despite its reduced RNA chaperone activity, a mutant lacking the CTR (NSP2-ΔC) had enhanced capacity to unwind RNA stem-loops, and it was thus more efficient at destabilizing the RNA structure [64]. This result suggests that RNA helix unwinding by NSP2 must be fine-tuned in order for the protein to function as an efficient RNA chaperone. Indeed, the CTR harbors highly conserved negatively charged residues, which accelerate RNA dissociation from NSP2 [64]. Interestingly, these negatively charged residues appear to cluster next to the S313 residue involved in viroplasm formation [79]. In this manner, the CTR would acquire an extra negative charge due to phosphorylation, thus further promoting RNA dissociation via the charge repulsion mechanism. Taken together, the recent biophysical and structural data suggest that NSP2 may be a key RNA chaperone that relaxes intra-molecular RNA structure, increasing its propensity for inter-segment RNA base-pairing (Figure 9). This model is also in agreement with the results of in vitro RNA structure probing experiments that have recently confirmed the increased RNA backbone flexibility in the presence of NSP2 [97].

7. Future Directions

Taken together, the existing structural and functional studies of RV NSP2 have provided detailed mechanistic insights into the multifaceted roles that this protein plays during the viral replication cycle, particularly during early particle assembly. Notably, the current body of work strongly supports a critical role for octameric NSP2 in the biogenesis of viroplasms, which are cytoplasmic biomolecular condensates that serve as the sites of early particle assembly. However, there are several gaps in knowledge regarding how NSP2 nucleates viroplasms alongside NSP5 and other factors. For example, the conformational transition between the predominantly cytoplasmic dNSP2 and the viroplasm-localized vNSP2 remains unknown. Future studies could employ EM affinity grids and specific mAbs to capture dNSP2 vs. vNSP2 prior to cryo-EM single-particle reconstructions, thus revealing their different conformations [98]. Moreover, the role that NSP2 inter-octamer interactions play during viroplasm formation, if any, is not understood. Moving forward, it will be important to test whether NSP2 can mediate inter-octamer interactions in biologically relevant conditions. Approaches such as density gradient centrifugation, native gel electrophoresis and size-exclusion chromatography with multi-angle light scattering (SEC-MALS) could be used to detect the presence of NSP2 inter-octamer chains from lysates of infected cells and/or with recombinant protein. Mutagenesis could also be used to ablate the capacity of NSP2 to form inter-octamer chains in a manner that maintains other NSP2 interactions (e.g., with NSP5). Such experiments would be needed to unveil any effects of higher-ordered NSP2 multimerization on viroplasm biogenesis. Finally, studies seeking to identify other components that contribute to viroplasm formation (e.g., RNAs, cellular components and other viral proteins like VP2) will be important to pursue in future work. Quantitative in vitro reconstitution assays with individual interacting partners like that shown by Geiger et al. will likely prove valuable, as they would allow for direct probing of the sequence of events leading to the formation of these condensates [71].

In addition to its well-established role in viroplasm formation, the current literature also supports the notion that NSP2 acts as a viral RNA chaperone whereby it (i) binds viral +ssRNA, (ii) relaxes secondary structures via its helix-destabilizing activity and (iii) promotes +ssRNA assortment that may underpin the selective incorporation of the 11 RV genome segments. Still, the assortment/packaging signals within the +ssRNAs that allow for their selective enrichment into viroplasms and incorporation into particles are not known, and it remains a mystery how NSP2 binding to +ssRNA would reveal such signals [98]. Secondary structures within the +ssRNAs are beginning to be identified using biochemical approaches like SHAPE-Map [97]. Such experiments could be combined with the in vitro RNA–RNA interaction assay described by Borodavka et al. in order to validate which structures underpin selective assortment [96]. Of course, the functional significance of +ssRNA elements would need to be tested in the context of infected cells, perhaps by using small complimentary oligos that block specific RNA–RNA interactions.

In contrast to the well-established roles of NSP2 in viroplasm formation and +ssRNA assortment, the functional significance of NSP2 as an enzyme remains unclear. The NTPase/RTPase/NDP kinase enzymatic activities are dispensable for viroplasm formation as well as RNA chaperone activity, but they are clearly critical for viral replication.

Understanding how these activities support RV replication has been difficult to tackle experimentally, but it is interesting to speculate about several possibilities. One simple explanation that has been put forth is that the NTPase/NDP kinase activities of NSP2 are critical for maintaining pools of nucleotides for use during viral RNA synthesis, which occurs inside viroplasms and in connection with early particle assembly. Specifically, −ssRNA strand synthesis (i.e., genome replication) occurs in the context of a core RI assembly intermediate containing NSP2. In this case, it is possible that the NTPase/NDP kinase activities of NSP2 ensure that the polymerase is being "fed" sufficient NTPs for −ssRNA strand synthesis. The observation that NSP2 has an RTPase activity might be an in vitro artifact of its NTPase functionality. However, it is alternatively possible that NSP2 plays a "moonlighting" role in helping RVs evade the host immune response. More specifically, RV −ssRNAs are reported to lack a $5'$ γ-phosphate, which is a potent activator of RIG-I and, thus, the interferon pathway of the host cell [99,100]. Recombinant NSP2 was reported to bind VP1 near the dsRNA exit tunnel, at least in vitro, which could theoretically bring the RTPase in proximity with the $5'$ end of −ssRNA for γ-phosphate removal [101]. Nevertheless, enzymatic NSP2 could have a more direct function in the regulation of −ssRNA synthesis. NSP2 has been reported to have an inhibitory effect on VP1/VP2-mediated dsRNA synthesis in vitro, and it would be of interest to test whether the catalytic activity of NSP2 is required for this inhibition [102]. Future studies of NSP2 activities will also benefit from the fully plasmid reverse genetics system, but perhaps in combination with *trans*-complementation approaches, as NSP2 mutant viruses would likely have severe replication defects [103]. Altogether, such future work will enrich the already deep knowledge base of this critical and multifaceted NSP2 protein during the RV replication cycle. Such work may also serve to broadly inform an understanding of viral factory formation, viral enzymology and viral RNA chaperone activities, as well as provide a foundation for antiviral drug design.

Author Contributions: All authors contributed to all stages of manuscript preparation. All authors have read and agreed to the published version of the manuscript.

Funding: S.M.E. was supported by NIH R01-AI116815 and by the Wake Forest Robert and Debra Lee Faculty Fellowship. S.L.N. was supported by the NIH T32-AI146059. A.B. was supported by the Wellcome Trust (103068/Z/13/Z and 213437/Z/18/Z to A.B.), and C.H. acknowledges funding from the MRC-funded Doctoral Training Programme, University of Cambridge.

Institutional Review Board Statement: Not applicable.

Data Availability Statement: Not applicable.

Acknowledgments: The authors would like to thank members of the Esstman and Borodavka labs for their scientific and editorial suggestions.

Conflicts of Interest: The authors declare no conflict of interest. The funders had no role in the design of the study; in the collection, analyses or interpretation of data; in the writing of the manuscript, or in the decision to publish the results.

References

1. Matthijnssens, J.; Attoui, H.; Bányai, K.; Brussaard, C.P.; Danthi, P.; Del Vas, M.; Dermody, T.S.; Duncan, R.; Fang, Q.; Johne, R.; et al. ICTV Virus Taxonomy Profile: Sedoreoviridae 2022. *J. Gen. Virol.* **2022**, *103*. [CrossRef]
2. Matthijnssens, J.; Otto, P.H.; Ciarlet, M.; Desselberger, U.; Van Ranst, M.; Johne, R. VP6-sequence-based cutoff values as a criterion for rotavirus species demarcation. *Arch. Virol.* **2012**, *157*, 1177–1182. [CrossRef] [PubMed]
3. Mihalov-Kovács, E.; Gellért, Á.; Marton, S.; Farkas, S.L.; Fehér, E.; Oldal, M.; Jakab, F.; Martella, V.; Bányai, K. Candidate new rotavirus species in sheltered dogs, hungary. *Emerg. Infect. Dis.* **2015**, *21*, 660–663. [CrossRef] [PubMed]
4. Troeger, C.; Khalil, I.A.; Rao, P.C.; Cao, S.; Blacker, B.F.; Ahmed, T.; Armah, G.; Bines, J.E.; Brewer, T.G.; Colombara, D.V.; et al. Rotavirus Vaccination and the Global Burden of Rotavirus Diarrhea Among Children Younger Than 5 Years. *JAMA Pediatr.* **2018**, *172*, 958–965. [CrossRef]
5. Díaz Alarcón, R.G.; Liotta, D.J.; Miño, S. Zoonotic RVA: State of the Art and Distribution in the Animal World. *Viruses* **2022**, *14*, 2554. [CrossRef]

6. Arnold, M.; Patton, J.T.; McDonald, S.M. Culturing, storage, and quantification of rotaviruses. *Curr. Protoc. Microbiol.* **2009**, *15*. [CrossRef]
7. Settembre, E.C.; Chen, J.Z.; Dormitzer, P.R.; Grigorieff, N.; Harrison, S.C. Atomic model of an infectious rotavirus particle. *EMBO J.* **2011**, *30*, 408–416. [CrossRef] [PubMed]
8. Jenni, S.; Salgado, E.N.; Herrmann, T.; Li, Z.; Grant, T.; Grigorieff, N.; Trapani, S.; Estrozi, L.F.; Harrison, S.C. In situ Structure of Rotavirus VP1 RNA-Dependent RNA Polymerase. *J. Mol. Biol.* **2019**, *431*, 3124–3138. [CrossRef]
9. Dormitzer, P.R.; Greenberg, H.B.; Harrison, S.C. Proteolysis of monomeric recombinant rotavirus VP4 yields an oligomeric VP5* core. *J. Virol.* **2001**, *75*, 7339–7350. [CrossRef]
10. Kumar, D.; Yu, X.; Crawford, S.E.; Moreno, R.; Jakana, J.; Sankaran, B.; Anish, R.; Kaundal, S.; Hu, L.; Estes, M.K.; et al. 2.7 A cryo-EM structure of rotavirus core protein VP3, a unique capping machine with a helicase activity. *Sci. Adv.* **2020**, *6*, eaay6410. [CrossRef]
11. Ding, K.; Celma, C.C.; Zhang, X.; Chang, T.; Shen, W.; Atanasov, I.; Roy, P.; Zhou, Z.H. In situ structures of rotavirus polymerase in action and mechanism of mRNA transcription and release. *Nat. Commun.* **2019**, *10*, 2216. [CrossRef] [PubMed]
12. Estes, M.; Kapikian, A.Z. Rotaviruses and Their Replication. In *Fields Virology*, 5th ed.; Knipe, D.M., Howley, P.M., Eds.; Lippincott Williams and Wilkens: Philadelphia, PA, USA, 2007; pp. 1917–1974.
13. Davidson, G.P.; Goller, I.; Bishop, R.F.; Townley, R.R.; Holmes, I.H.; Ruck, B.J. Immunofluorescence in duodenal mucosa of children with acute enteritis due to a new virus. *J. Clin. Pathol.* **1975**, *28*, 263–266. [CrossRef] [PubMed]
14. Starkey, W.G.; Collins, J.; Wallis, T.S.; Clarke, G.J.; Spencer, A.J.; Haddon, S.J.; Osborne, M.P.; Candy, D.C.A.; Stephen, J. Kinetics, tissue specificity and pathological changes in murine rotavirus infection of mice. *J. Gen. Virol.* **1986**, *67 Pt 12*, 2625–2634. [CrossRef] [PubMed]
15. Arias, C.F.; López, S. Rotavirus cell entry: Not so simple after all. *Curr. Opin. Virol.* **2021**, *48*, 42–48. [CrossRef] [PubMed]
16. Díaz-Salinas, M.A.; Romero, P.; Espinosa, R.; Hoshino, Y.; López, S.; Arias, C.F. The Spike Protein VP4 Defines the Endocytic Pathway Used by Rotavirus To Enter MA104 Cells. *J. Virol.* **2013**, *87*, 1658–1663. [CrossRef] [PubMed]
17. Gutiérrez, M.; Isa, P.; Martin, C.S.-S.; Pérez-Vargas, J.; Espinosa, R.; Arias, C.F.; López, S. Different Rotavirus Strains Enter MA104 Cells through Different Endocytic Pathways: The Role of Clathrin-Mediated Endocytosis. *J. Virol.* **2010**, *84*, 9161–9169. [CrossRef] [PubMed]
18. Trask, S.D.; McDonald, S.M.; Patton, J.T. Structural insights into the coupling of virion assembly and rotavirus replication. *Nat. Rev. Microbiol.* **2012**, *10*, 165–177. [CrossRef] [PubMed]
19. Salgado, E.N.; Rodriguez, B.G.; Narayanaswamy, N.; Krishnan, Y.; Harrison, S.C. Visualization of Calcium Ion Loss from Rotavirus during Cell Entry. *J. Virol.* **2018**, *92*, e01327-18. [CrossRef] [PubMed]
20. Ludert, J.; Michelangeli, F.; Gil, F.; Liprandi, F.; Esparza, J. Penetration and uncoating of rotaviruses in cultured cells. *Intervirology* **1987**, *27*, 95–101. [CrossRef] [PubMed]
21. Lawton, J.A.; Estes, M.K.; Prasad, B.V. Mechanism of genome transcription in segmented dsRNA viruses. *Adv. Virus Res.* **2000**, *55*, 185–229.
22. Patton, J.T.; Chen, D. RNA-binding and capping activities of proteins in rotavirus open cores. *J. Virol.* **1999**, *73*, 1382–1391. [CrossRef]
23. Liu, M.; Mattion, N.M.; Estes, M.K. Rotavirus VP3 expressed in insect cells possesses guanylyltrans-ferase activity. *Virology* **1992**, *188*, 77–84. [CrossRef]
24. Pizarro, J.L.; Sandino, A.M.; Pizarro, J.M.; Fernandez, J.; Spencer, E. Characterization of rotavirus guan-ylyltransferase activity associated with polypeptide VP3. *J. Gen. Virol.* **1991**, *72 Pt 2*, 325–332. [CrossRef]
25. Vende, P.; Piron, M.; Castagne, N.; Poncet, D. Efficient translation of rotavirus mRNA requires simul-taneous interaction of NSP3 with the eukaryotic translation initiation factor eIF4G and the mRNA 3′ end. *J. Virol.* **2000**, *74*, 7064–7071. [CrossRef]
26. López, S.; Oceguera, A.; Sandoval-Jaime, C. Stress Response and Translation Control in Rotavirus Infection. *Viruses* **2016**, *8*, 162. [CrossRef] [PubMed]
27. Mitzel, D.N.; Weisend, C.M.; White, M.W.; Hardy, M.E. Translational regulation of rotavirus gene ex-pression. *J. Gen. Virol.* **2003**, *84*, 383–391. [CrossRef] [PubMed]
28. Papa, G.; Borodavka, A.; Desselberger, U. Viroplasms: Assembly and Functions of Rotavirus Replication Factories. *Viruses* **2021**, *13*, 1349. [CrossRef] [PubMed]
29. Patton, J.T.; Silvestri, L.S.; Tortorici, M.A.; Vasquez-Del Carpio, R.; Taraporewala, Z.F. Rotavirus genome replication and morphogenesis: Role of the viroplasm. *Curr. Top. Microbiol. Immunol.* **2006**, *309*, 169–187. [PubMed]
30. Taraporewala, Z.F.; Patton, J.T. Nonstructural proteins involved in genome packaging and replication of rotaviruses and other members of the Reoviridae. *Virus Res.* **2004**, *101*, 57–66. [CrossRef] [PubMed]
31. McNulty, M.S.; Curran, W.L.; McFerran, J.B. The morphogenesis of a cytopathic bovine rotavirus in madin-darby bovine kidney cells. *J. Gen. Virol.* **1976**, *33*, 503–508. [CrossRef]
32. Gardet, A.; Breton, M.; Fontanges, P.; Trugnan, G.; Chwetzoff, S. Rotavirus spike protein vp4 binds to and remodels actin bundles of the epithelial brush border into actin bodies. *J. Virol.* **2006**, *80*, 3947–3956. [CrossRef]
33. Ericson, B.L.; Graham, D.Y.; Mason, B.B.; Estes, M.K. Identification, synthesis, and modifications of simian rotavirus SA11 polypeptides in infected cells. *J. Virol.* **1982**, *42*, 825–839. [CrossRef] [PubMed]

34. Mason, B.B.; Graham, D.Y.; Estes, M.K. Biochemical mapping of the simian rotavirus SA11 genome. *J. Virol.* **1983**, *46*, 413–423. [CrossRef] [PubMed]
35. Both, G.W.; Bellamy, A.R.; Street, J.E.; Siegman, L.J. A general strategy for cloning double-stranded RNA: Nucleotide sequence of the Simian-11 rotavirus gene 8. *Nucleic Acids Res.* **1982**, *10*, 7075–7088. [CrossRef] [PubMed]
36. Petrie, B.L.; Greenberg, H.B.; Graham, D.Y.; Estes, M.K. Ultrastructural localization of rotavirus antigens using colloidal gold. *Virus Res.* **1984**, *1*, 133–152. [CrossRef] [PubMed]
37. Patton, J.T.; Gallegos, C.O. Structure and protein composition of the rotavirus replicase particle. *Virology* **1988**, *166*, 358–365. [CrossRef] [PubMed]
38. Gallegos, C.O.; Patton, J.T. Characterization of rotavirus replication intermediates: A model for the assembly of single-shelled particles. *Virology* **1989**, *172*, 616–627. [CrossRef] [PubMed]
39. Aponte, C.; Poncet, D.; Cohen, J. Recovery and characterization of a replicase complex in rota-virus-infected cells by using a monoclonal antibody against NSP2. *J. Virol.* **1996**, *70*, 985–991. [CrossRef]
40. Helmberger-Jones, M.; Patton, J.T. Characterization of subviral particles in cells infected with simian rotavirus SA11. *Virology* **1986**, *155*, 655–665. [CrossRef]
41. McDonald, S.M.; Patton, J.T. Assortment and packaging of the segmented rotavirus genome. *Trends Microbiol.* **2011**, *19*, 136–144. [CrossRef]
42. Mansell, E.A.; Patton, J.T. Rotavirus RNA replication: VP2, but not VP6, is necessary for viral replicase activity. *J. Virol.* **1990**, *64*, 4988–4996. [CrossRef] [PubMed]
43. Patton, J.T. Rotavirus VP1 alone specifically binds to the 3' end of viral mRNA, but the interaction is not sufficient to initiate minus-strand synthesis. *J. Virol.* **1996**, *70*, 7940–7947. [CrossRef] [PubMed]
44. Patton, J.T.; Jones, M.T.; Kalbach, A.N.; He, Y.W.; Xiaobo, J. Rotavirus RNA polymerase requires the core shell protein to synthesize the double-stranded RNA genome. *J. Virol.* **1997**, *71*, 9618–9626. [CrossRef] [PubMed]
45. McDonald, S.M.; Patton, J.T. Rotavirus VP2 core shell regions critical for viral polymerase activation. *J. Virol.* **2011**, *85*, 3095–3105. [CrossRef] [PubMed]
46. Long, C.P.; McDonald, S.M. Rotavirus genome replication: Some assembly required. *PLoS Pathog.* **2017**, *13*, e1006242. [CrossRef]
47. Gombold, J.L.; Estes, M.K.; Ramig, R.F. Assignment of simian rotavirus SA11 temperature-sensitive mutant groups B and E to genome segments. *Virology* **1985**, *143*, 309–320. [CrossRef] [PubMed]
48. Ramig, R.F. Isolation and genetic characterization of temperature-sensitive mutants that define five additional recombination groups in simian rotavirus SA11. *Virology* **1983**, *130*, 464–473. [CrossRef] [PubMed]
49. Ramig, R.F.; Petrie, B.L. Characterization of temperature-sensitive mutants of simian rotavirus SA11: Protein synthesis and morphogenesis. *J. Virol.* **1984**, *49*, 665–673. [CrossRef] [PubMed]
50. Chen, D.; Gombold, J.L.; Ramig, R.F. Intracellular RNA synthesis directed by temperature-sensitive mutants of simian rotavirus SA11. *Virology* **1990**, *178*, 143–151. [CrossRef]
51. Kattoura, M.D.; Clapp, L.L.; Patton, J.T. The rotavirus nonstructural protein, NS35, possesses RNA-binding activity in vitro and in vivo. *Virology* **1992**, *191*, 698–708. [CrossRef]
52. Kattoura, M.D.; Chen, X.; Patton, J.T. The rotavirus RNA-binding protein NS35 (NSP2) forms 10S multimers and interacts with the viral rna polymerase. *Virology* **1994**, *202*, 803–813. [CrossRef] [PubMed]
53. Taraporewala, Z.; Chen, D.; Patton, J.T. Multimers formed by the rotavirus nonstructural protein NSP2 bind to RNA and have nucleoside triphosphatase activity. *J. Virol.* **1999**, *73*, 9934–9943. [CrossRef]
54. Taraporewala, Z.F.; Schuck, P.; Ramig, R.F.; Silvestri, L.; Patton, J.T. Analysis of a Temperature-sensitive mutant rotavirus indicates that NSP2 octamers are the functional form of the protein. *J. Virol.* **2002**, *76*, 7082–7093. [CrossRef] [PubMed]
55. Schuck, P.; Taraporewala, Z.; McPhie, P.; Patton, J.T. Rotavirus nonstructural protein NSP2 self-assembles into octamers that undergo ligand-induced conformational changes. *J. Biol. Chem.* **2001**, *276*, 9679–9687. [CrossRef]
56. Taraporewala, Z.F.; Patton, J.T. Identification and characterization of the Helix-destabilizing activity of rotavirus nonstructural protein NSP2. *J. Virol.* **2001**, *75*, 4519–4527. [CrossRef]
57. Afrikanova, I.; Fabbretti, E.; Burrone, O.R.; Miozzo, M.C. Rotavirus NSP5 phosphorylation is up-regulated by interaction with NSP2. *J. Gen. Virol.* **1998**, *79*, 2679–2686. [CrossRef] [PubMed]
58. Fabbretti, E.; Afrikanova, I.; Vascotto, F.; Burrone, O.R. Two non-structural rotavirus proteins, NSP2 and NSP5, form viroplasm-like structures in vivo. *J. Gen. Virol.* **1999**, *80*, 333–339. [CrossRef] [PubMed]
59. Jayaram, H.; Taraporewala, Z.; Patton, J.T.; Prasad, B.V.V. Rotavirus protein involved in genome replication and packaging exhibits a HIT-like fold. *Nature* **2002**, *417*, 311–315. [CrossRef]
60. Jiang, X.; Jayaram, H.; Kumar, M.; Ludtke, S.J.; Estes, M.K.; Prasad, B.V.V. Cryoelectron microscopy structures of rotavirus NSP2-NSP5 and NSP2-RNA complexes: Implications for genome replication. *J. Virol.* **2006**, *80*, 10829–10835. [CrossRef]
61. Kumar, M.; Jayaram, H.; Vasquez-Del Carpio, R.; Jiang, X.; Taraporewala, Z.F.; Jacobson, R.H.; Patton, J.T.; Prasad, B.V. Crystallographic and biochemical analysis of rotavirus NSP2 with nucleotides reveals a nucleoside diphosphate kinase-like activity. *J. Virol.* **2007**, *81*, 12272–12284. [CrossRef]
62. Hu, L.; Chow, D.-C.; Patton, J.T.; Palzkill, T.; Estes, M.K.; Prasad, B.V.V. Crystallographic Analysis of Rotavirus NSP2-RNA Complex Reveals Specific Recognition of 5' GG Sequence for RTPase Activity. *J. Virol.* **2012**, *86*, 10547–10557. [CrossRef] [PubMed]

63. Criglar, J.M.; Anish, R.; Hu, L.; Crawford, S.E.; Sankaran, B.; Prasad, B.V.V.; Estes, M.K. Phosphorylation cascade regulates the formation and maturation of rotaviral replication factories. *Proc. Natl. Acad. Sci. USA* **2018**, *115*, E12015–E12023. [CrossRef] [PubMed]
64. Bravo, J.P.K.; Bartnik, K.; Venditti, L.; Acker, J.; Gail, E.H.; Colyer, A.; Davidovich, C.; Lamb, D.C.; Tuma, R.; Calabrese, A.N.; et al. Structural basis of rotavirus RNA chaperone displacement and RNA annealing. *Proc. Natl. Acad. Sci. USA* **2021**, *118*, e2100198118. [CrossRef] [PubMed]
65. Carpio, R.V.-D.; Gonzalez-Nilo, F.D.; Riadi, G.; Taraporewala, Z.F.; Patton, J.T. Histidine triad-like motif of the rotavirus NSP2 Octamer mediates both RTPase and NTPase activities. *J. Mol. Biol.* **2006**, *362*, 539–554. [CrossRef] [PubMed]
66. Martin, D.; Duarte, M.; Lepault, J.; Poncet, D. Sequestration of free tubulin molecules by the viral protein NSP2 induces microtubule depolymerization during rotavirus infection. *J. Virol.* **2010**, *84*, 2522–2532. [CrossRef] [PubMed]
67. Eichwald, C.; Rodriguez, J.F.; Burrone, O.R. Characterization of rotavirus NSP2/NSP5 interactions and the dynamics of viroplasm formation. *J. Gen. Virol.* **2004**, *85*, 625–634. [CrossRef] [PubMed]
68. Carreño-Torres, J.J.; Gutiérrez, M.; Arias, C.F.; López, S.; Isa, P. Characterization of viroplasm formation during the early stages of rotavirus infection. *Virol. J.* **2010**, *7*, 350. [CrossRef]
69. Eichwald, C.; Arnoldi, F.; Laimbacher, A.S.; Schraner, E.M.; Fraefel, C.; Wild, P.; Burrone, O.R.; Ackermann, M. Rotavirus viroplasm fusion and perinuclear localization are dynamic processes requiring stabilized microtubules. *PLoS ONE* **2012**, *7*, e47947. [CrossRef] [PubMed]
70. Nichols, S.L.; Nilsson, E.M.; Brown-Harding, H.; LaConte, L.E.W.; Acker, J.; Borodavka, A.; Esstman, S.M. Flexibility of the Rotavirus NSP2 C-Terminal Region Supports Factory Formation via Liquid-Liquid Phase Separation. *J. Virol.* **2023**, *97*, e0003923. [CrossRef]
71. Geiger, F.; Acker, J.; Papa, G.; Wang, X.; Arter, W.E.; Saar, K.L.; Erkamp, N.A.; Qi, R.; Bravo, J.P.; Strauss, S.; et al. Liquid-liquid phase separation underpins the formation of replication factories in rotaviruses. *EMBO J.* **2021**, *40*, e107711. [CrossRef]
72. Strauss, S.; Acker, J.; Papa, G.; Desiro, D.; Schueder, F.; Borodavka, A.; Jungmann, R. Principles of RNA recruitment to viral ribonucleoprotein condensates in a segmented dsRNA virus. *eLife* **2023**, *12*, e68670. [CrossRef] [PubMed]
73. Dhillon, P.; Tandra, V.N.; Chorghade, S.G.; Namsa, N.D.; Sahoo, L.; Rao, C.D. Cytoplasmic Relocalization and Colocalization with Viroplasms of Host Cell Proteins, and Their Role in Rotavirus Infection. *J. Virol.* **2018**, *92*, e00612-18. [CrossRef] [PubMed]
74. López, T.; Rojas, M.; Ayala-Bretón, C.; López, S.; Arias, C.F. Reduced expression of the rotavirus NSP5 gene has a pleiotropic effect on virus replication. *J. Gen. Virol.* **2005**, *86*, 1609–1617. [CrossRef] [PubMed]
75. Silvestri, L.S.; Taraporewala, Z.F.; Patton, J.T. Rotavirus replication: Plus-sense templates for double-stranded RNA synthesis are made in viroplasms. *J. Virol.* **2004**, *78*, 7763–7774. [CrossRef]
76. Carpio, R.V.-D.; González-Nilo, F.D.; Jayaram, H.; Spencer, E.; Prasad, B.V.V.; Patton, J.T.; Taraporewala, Z.F. Role of the histidine triad-like motif in nucleotide hydrolysis by the rotavirus RNA-packaging protein NSP2. *J. Biol. Chem.* **2004**, *279*, 10624–10633. [CrossRef] [PubMed]
77. Taraporewala, Z.F.; Jiang, X.; Vasquez-Del Carpio, R.; Jayaram, H.; Prasad, B.V.; Patton, J.T. Structure-function analysis of rotavirus NSP2 octamer by using a novel complementation system. *J. Virol.* **2006**, *80*, 7984–7994. [CrossRef]
78. Criglar, J.M.; Hu, L.; Crawford, S.E.; Hyser, J.M.; Broughman, J.R.; Prasad, B.V.V.; Estes, M.K. A novel form of rotavirus NSP2 and phosphorylation-dependent NSP2-NSP5 interactions are associated with viroplasm assembly. *J. Virol.* **2014**, *88*, 786–798. [CrossRef] [PubMed]
79. Criglar, J.M.; Crawford, S.E.; Zhao, B.; Smith, H.G.; Stossi, F.; Estes, M.K. A Genetically Engineered Rotavirus NSP2 Phosphorylation Mutant Impaired in Viroplasm Formation and Replication Shows an Early Interaction between vNSP2 and Cellular Lipid Droplets. *J. Virol.* **2020**, *94*, e00972-20. [CrossRef]
80. Eichwald, C.; Jacob, G.; Muszynski, B.; Allende, J.E.; Burrone, O.R. Uncoupling substrate and activation functions of rotavirus NSP5: Phosphorylation of Ser-67 by casein kinase 1 is essential for hyperphos-phorylation. *Proc. Natl. Acad. Sci. USA* **2004**, *101*, 16304–16309. [CrossRef]
81. Criglar, J.M.; Estes, M.K.; Crawford, S.E. Rotavirus-Induced Lipid Droplet Biogenesis Is Critical for Virus Replication. *Front. Physiol.* **2022**, *13*, 836870. [CrossRef]
82. Crawford, S.E.; Desselberger, U. Lipid droplets form complexes with viroplasms and are crucial for rotavirus replication. *Curr. Opin. Virol.* **2016**, *19*, 11–15. [CrossRef] [PubMed]
83. Cheung, W.; Gill, M.; Esposito, A.; Kaminski, C.F.; Courousse, N.; Chwetzoff, S.; Trugnan, G.; Keshavan, N.; Lever, A.; Desselberger, U. Rotaviruses associate with cellular lipid droplet components to replicate in viroplasms, and compounds disrupting or blocking lipid droplets inhibit viroplasm formation and viral replication. *J. Virol.* **2010**, *84*, 6782–6798. [CrossRef] [PubMed]
84. Campagna, M.; Marcos-Villar, L.; Arnoldi, F.; de la Cruz-Herrera, C.F.; Gallego, P.; González-Santamaría, J.; González, D.; Lopitz-Otsoa, F.; Rodriguez, M.S.; Burrone, O.R.; et al. Rotavirus viroplasm proteins interact with the cellular sumoylation system: Implications for viroplasm-like structure formation. *J. Virol.* **2013**, *87*, 807–817. [CrossRef] [PubMed]
85. Contin, R.; Arnoldi, F.; Campagna, M.; Burrone, O.R. Rotavirus NSP5 orchestrates recruitment of viroplasmic proteins. *J. Gen. Virol.* **2010**, *91*, 1782–1793. [CrossRef] [PubMed]
86. Mohan, K.V.K.; Muller, J.; Atreya, C.D.; Campbell, T.B.; Schneider, K.; Wrin, T.; Petropoulos, C.J.; Connick, E. The N- and C-terminal regions of rotavirus NSP5 are the critical determinants for the formation of viroplasm-like structures independent of NSP2. *J. Virol.* **2003**, *77*, 12105–12112. [CrossRef] [PubMed]

87. Buttafuoco, A.; Michaelsen, K.; Tobler, K.; Ackermann, M.; Fraefel, C.; Eichwald, C. Conserved Rotavirus NSP5 and VP2 Domains Interact and Affect Viroplasm. *J. Virol.* **2020**, *94*, e01965-19. [CrossRef] [PubMed]

88. Martin, D.; Ouldali, M.; Ménétrey, J.; Poncet, D. Structural organisation of the rotavirus nonstructural protein NSP5. *J. Mol. Biol.* **2011**, *413*, 209–221. [CrossRef] [PubMed]

89. Afrikanova, I.; Miozzo, M.C.; Giambiagi, S.; Burrone, O. Phosphorylation generates different forms of rotavirus NSP5. *J. Gen. Virol.* **1996**, *77 Pt 9*, 2059–2065. [CrossRef] [PubMed]

90. Eichwald, C.; Vascotto, F.; Fabbretti, E.; Burrone, O.R. Rotavirus NSP5: Mapping phosphorylation sites and kinase activation and viroplasm localization domains. *J. Virol.* **2002**, *76*, 3461–3470. [CrossRef]

91. Poncet, D.; Lindenbaum, P.; L'Haridon, R.; Cohen, J. In vivo and in vitro phosphorylation of rotavirus NSP5 correlates with its localization in viroplasms. *J. Virol.* **1997**, *71*, 34–41. [CrossRef]

92. Papa, G.; Venditti, L.; Arnoldi, F.; Schraner, E.M.; Potgieter, C.; Borodavka, A.; Eichwald, C.; Burrone, O.R. Recombinant Rotaviruses Rescued by Reverse Genetics Reveal the Role of NSP5 Hyperphosphorylation in the Assembly of Viral Factories. *J. Virol.* **2019**, *94*, e01110-19. [CrossRef] [PubMed]

93. Campagna, M.; Budini, M.; Arnoldi, F.; Desselberger, U.; Allende, J.E.; Burrone, O.R. Impaired hyper-phosphorylation of rotavirus NSP5 in cells depleted of casein kinase 1alpha is associated with the for-mation of viroplasms with altered morphology and a moderate decrease in virus replication. *J. Gen. Virol.* **2007**, *88*, 2800–2810. [CrossRef] [PubMed]

94. Cabral-Romero, C.; Padilla-Noriega, L. Association of rotavirus viroplasms with microtubules through NSP2 and NSP5. *Memórias Inst. Oswaldo Cruz* **2006**, *101*, 603–611. [CrossRef] [PubMed]

95. Jing, Z.; Shi, H.; Chen, J.; Shi, D.; Liu, J.; Guo, L.; Tian, J.; Wu, Y.; Dong, H.; Zhang, J.; et al. Rotavirus Viroplasm Biogenesis Involves Microtubule-Based Dynein Transport Mediated by an Interaction between NSP2 and Dynein Intermediate Chain. *J. Virol.* **2021**, *95*, e0124621. [CrossRef] [PubMed]

96. Borodavka, A.; Dykeman, E.C.; Schrimpf, W.; Lamb, D.C. Protein-mediated RNA folding governs sequence-specific interactions between rotavirus genome segments. *eLife* **2017**, *6*, e27453. [CrossRef]

97. Coria, A.; Wienecke, A.; Knight, M.L.; Desiro, D.; Laederach, A.; Borodavka, A. Rotavirus RNA chap-erone mediates global transcriptome-wide increase in RNA backbone flexibility. *Nucleic Acids Res.* **2022**, *50*, 10078–10092. [CrossRef]

98. Boudreaux, C.E.; Kelly, D.F.; McDonald, S.M. Electron microscopic analysis of rotavirus assembly-replication intermediates. *Virology* **2015**, *477*, 32–41. [CrossRef] [PubMed]

99. Imai, M.; Akatani, K.; Ikegami, N.; Furuichi, Y. Capped and conserved terminal structures in human rotavirus genome double-stranded RNA segments. *J. Virol.* **1983**, *47*, 125–136. [CrossRef]

100. McCrae, M.; McCorquodale, J. Molecular biology of rotaviruses V. terminal structure of viral RNA species. *Virology* **1983**, *126*, 204–212. [CrossRef]

101. Viskovska, M.; Anish, R.; Hu, L.; Chow, D.-C.; Hurwitz, A.M.; Brown, N.G.; Palzkill, T.; Estes, M.K.; Prasad, B.V.V. Probing the sites of interactions of rotaviral proteins involved in replication. *J. Virol.* **2014**, *88*, 12866–12881. [CrossRef]

102. Vende, P.; Tortorici, M.; Taraporewala, Z.F.; Patton, J.T. Rotavirus NSP2 interferes with the core lattice protein VP2 in initiation of minus-strand synthesis. *Virology* **2003**, *313*, 261–273. [CrossRef] [PubMed]

103. Kanai, Y.; Komoto, S.; Kawagishi, T.; Nouda, R.; Nagasawa, N.; Onishi, M.; Matsuura, Y.; Taniguchi, K.; Kobayashi, T. Entirely plasmid-based reverse genetics system for rotaviruses. *Proc. Natl. Acad. Sci. USA* **2017**, *114*, 2349–2354. [CrossRef] [PubMed]

Article

ML241 Antagonizes ERK 1/2 Activation and Inhibits Rotavirus Proliferation

Jinlan Wang [†], Xiaoqing Hu [†], Jinyuan Wu, Xiaochen Lin, Rong Chen, Chenxing Lu, Xiaopeng Song, Qingmei Leng, Yan Li, Xiangjing Kuang, Jinmei Li, Lida Yao, Xianqiong Tang, Jun Ye, Guangming Zhang, Maosheng Sun, Yan Zhou * and Hongjun Li *

Institute of Medical Biology, Chinese Academy of Medical Science & Peking Union Medical College, Yunnan Key Laboratory of Vaccine Research and Development on Severe Infectious Disease, Kunming 650118, China; lanlingyu@student.pumc.edu.cn (J.W.); huxiaoqing@imbcams.com.cn (X.H.); wujinyuan@imbcams.com.cn (J.W.); linxiaochen@imbcams.com.cn (X.L.); chenrong@imbcams.com.cn (R.C.); luchenxing@student.pumc.edu.cn (C.L.); igtheshy131@gmail.com (X.S.); lqm212855240@163.com (Q.L.); yjlz2314@163.com (Y.L.); kuangxiangjun@imbcams.com.cn (X.K.); lijinmei917@163.com (J.L.); adayao0926@163.com (L.Y.); tangxq8859@163.com (X.T.); yejun@imbcams.com.cn (J.Y.); zhangguangming@imbcams.com.cn (G.Z.); sunmaosheng@imbcams.com.cn (M.S.)

* Correspondence: zhouxiaobao_850@163.com (Y.Z.); lihj6912@163.com (H.L.); Tel.: +86-13888340684 (Y.Z.); +86-13888918945 (H.L.)

[†] These authors contributed equally to this work.

Abstract: Rotavirus (RV) is the main pathogen that causes severe diarrhea in infants and children under 5 years of age. No specific antiviral therapies or licensed anti-rotavirus drugs are available. It is crucial to develop effective and low-toxicity anti-rotavirus small-molecule drugs that act on novel host targets. In this study, a new anti-rotavirus compound was selected by ELISA, and cell activity was detected from 453 small-molecule compounds. The anti-RV effects and underlying mechanisms of the screened compounds were explored. In vitro experimental results showed that the small-molecule compound ML241 has a good effect on inhibiting rotavirus proliferation and has low cytotoxicity during the virus adsorption, cell entry, and replication stages. In addition to its in vitro effects, ML241 also exerted anti-RV effects in a suckling mouse model. Transcriptome sequencing was performed after adding ML241 to cells infected with RV. The results showed that ML241 inhibited the phosphorylation of ERK1/2 in the MAPK signaling pathway, thereby inhibiting IκBα, activating the NF-κB signaling pathway, and playing an anti-RV role. These results provide an experimental basis for specific anti-RV small-molecule compounds or compound combinations, which is beneficial for the development of anti-RV drugs.

Keywords: rotavirus; ML241 (hydrochloride); MAPK signaling pathway; ERK1/2; NF-κB

Citation: Wang, J.; Hu, X.; Wu, J.; Lin, X.; Chen, R.; Lu, C.; Song, X.; Leng, Q.; Li, Y.; Kuang, X.; et al. ML241 Antagonizes ERK 1/2 Activation and Inhibits Rotavirus Proliferation. *Viruses* **2024**, *16*, 623. https://doi.org/ 10.3390/v16040623

Academic Editors: Ulrich Desselberger and John T. Patton

Received: 4 February 2024
Revised: 27 March 2024
Accepted: 3 April 2024
Published: 17 April 2024

1. Introduction

Rotavirus (RV) is the main pathogen that causes severe diarrhea in infants and children under 5 years of age, with infection causing approximately 130,000 deaths annually [1]. Although licensed rotavirus vaccines provide more than 50% protection against rotavirus infection [2], currently, there are no specific antiviral treatments. The available treatments for the etiology of rotavirus-induced gastroenteritis are mainly symptomatic treatments and the correction of water and electrolyte imbalances using oral solutions to prevent or treat dehydration to reduce the duration and severity of diarrheal episodes [3,4]. Therefore, the control of rotavirus-induced gastroenteritis is of great importance for targeted interventions, such as the development of new small-molecule compound drugs to prevent and treat rotavirus-induced gastroenteritis.

Research on anti-RV drugs has shown that 2′-C-methylnucleoside [2CMC], 2′-C-methyladenosine [2CMA], 2′-C-methylguanosine [2CMG], and 7-deaza-2′-C-methyladenosine

[7DMA] can inhibit rotavirus, sapoviruses, and norovirus by inhibiting viral genome transcription [5]. Genipin, isolated from jasmine flowers, inhibits human rotavirus Wa strain and simian rotavirus SA-11 strain in vitro by inhibiting two different stages of the viral replication cycle: attachment and penetration (early stage) in pre-treatment and assembly and release (late stage) in post-treatment [6]. Deoxyshikonin can inhibit rotavirus replication by inducing low SIRT1, ac-Foxo1, Rab7, and VP6 protein levels, low RV titers, low autophagy, and oxidative stress [7]. The antiviral effect of Portulaca oleracea L. polysaccharide (POL-P), an active component of Portulaca oleracea L(POL), inhibits rotavirus replication by upregulating the expression of IFN-α [8]. Inhibitors of dihydroorotate dehydrogenase (the rate-limiting enzyme for de novo pyrimidine synthesis) (BQR) can resist rotavirus infection by inhibiting pyrimidine biosynthesis in cells and intestinal organoids [9]. The small-molecule compound ML-60218 is an RNA polymerase III inhibitor that inhibits viral replication by destroying the viral cytoplasmic structure (viroplasm) [10]. The organ transplant immunosuppressive drug, 6-thioguanine (6-TG), inhibits rotavirus replication in Caco-2 cells and HIEs by interacting with the cellular drug target Rac1. Thiazolactones inhibit viral proliferation by inhibiting the formation of viral cytoplasmic structures (viroplasms) [11]. Metformin hydrochloride significantly inhibited the expression of rotavirus mRNA and protein in Caco-2 cells, small intestinal organoids, and lactational mouse models [12]. Dyngo-4a can inhibit rotavirus infection in vivo and in vitro by affecting the formation of dynamin-2 oligomers [13]. These studies screened compounds from animal sources or laboratory rotavirus strains, explored the compounds' mechanisms of action on RV in vivo and in vitro, and provided treatment strategies for clinical symptoms caused by RV infection.

To find effective and low-toxicity anti-rotavirus small-molecule drugs, a wild human rotavirus ZTR-68 strain was used for drug screening from 453 small-molecule compounds. The anti-rotavirus activity was tested using an Enzyme-Linked Immunosorbent Assay (ELISA). The role of selected compounds in the adsorption, cell entry, and replication stages of the virus was studied using NSP3 real-time quantitative PCR (RT-qPCR) for rotaviral NSP3 and Western blot for rotaviral VP7 and NSP3. The antiviral mechanism of the compound was analyzed through transcriptome sequencing and WB, and the signaling pathway through which the compound exerted its inhibitory effect on rotavirus replication was determined. Suckling mice were used as a model to study the in vivo anti-RV effects of the compounds. In summary, this study discovered a small-molecule compound that effectively inhibits rotavirus replication and the mechanism underlying this.

2. Materials and Methods

2.1. Cell Culture

African green monkey embryonic kidney cells (MA104) were provided by the Molecular Biology Department of the Institute of Medical Biology, Chinese Academy of Medical Sciences, and Peking Union Medical College. The cells were cultured at 37 °C in a 5% carbon dioxide atmosphere in Dulbecco's Modified Eagle Medium (DMEM) containing 10% fetal bovine serum (FBS) and 1% double antibiotics (100 U/mL penicillin and 100 µg/mL streptomycin).

2.2. Rotavirus Amplification and Titer Determination

The genotype of rotavirus ZTR-68 is G1P [8], and the genotype of the SA11 strain is G3P [2]. They were isolated and preserved at the Molecular Biology Laboratory of the Institute of Medical Biology, Chinese Academy of Medical Sciences, and Peking Union Medical College. The virus titer was determined using the Kaerbar method with the following formula: $LgCCID50 = Xm - 1/2d + d \cdot \sum pi/100$. The cutoff value was 0.105. The titer of the ZTR-68 strain was found to be 7.5 $LgCCID_{50}$/mL, and that of the SA11 strain was 7.0 $LgCCID_{50}$/mL. The titer of the Wa strain was 6.5 $LgCCID_{50}$/mL and that of the Gottfried strain was 7.9 $LgCCID_{50}$/mL. To analyze whether ML241 affects the entry of RV into its host cells, the viruses were treated with ultraviolet irradiation at 220 nm (UV dose

22.5 mJ/cm^2). Irradiation with 220 nm of UV destroyed the nucleic acids in the viruses; therefore, RNA replication and protein translation could not be performed. However, this process did not affect virus entry into host cells or RNA release [14].

2.3. Enzyme-Linked Immunosorbent Assay

MA104 cells were transferred to a 96-well culture plate. When the cells grew to form a dense monolayer, the RV was activated with 20 μg/mL acetylase and 600 μg/mL CaCl$_2$. The multiplicity of virus infection was MOI = 0.1, and different concentrations were immediately added. After complete CPE was observed in the virus control group, the culture was frozen and thawed three times. While conducting the large-scale screening of the anti-RV small-molecule compounds and after exploring the optimal viral load of the reference virus to determine the optimal MOI through multiple preliminary experiments, we chose to use ELISA quantitative detection methods to screen the compounds [15–18]. A total of 453 small-molecule compounds were screened for rotavirus proliferation using an ELISA. The original solution of the inactivated rotavirus vaccine was used as the standard. The antigen content was 1236 EU (ELISA unit, EU)/mL. A standard curve was constructed using a 2-fold dilution of 12 standard gradients. A 50-fold-diluted standard was used as the internal reference. The OD$_{450-650}$ value was read using a microplate reader. GraphPad Prism 9.3.1 software was used to run the sigmoidal 4PL; the antigen content of the virus that proliferated was used as the virus control, and the virus that did not proliferate was used as the blank control to calculate the inhibition rate of the virus by the small-molecule compound. The calculation formula was as follows: inhibition rate (%) = ($A_{compound\ group}$ − $A_{virus\ group}$)/($A_{blank\ group}$ − $A_{virus\ group}$) × 100%. In this experiment, a purified goat anti-rotavirus G1P [8] antibody (batch number: RVAB2020101), preserved by the Institute of Medical Biology, Chinese Academy of Medical Sciences, was used as the primary antibody in the ELISA experiment, and the secondary antibody was an HRP-labeled goat anti-mouse purified antibody (batch number: RVAB2020101H).

2.4. Cell Viability Determination

A Cell Counting Kit-8 (CCK8) kit (CA1210, Solarbio, Beijing, China) was used to measure the toxic effects of small-molecule compounds on cell proliferation. After the small-molecule compounds were used for treatment with different concentration gradients for 48 h, 10% CCK8 solution was added, and the absorbance at 450 nm was measured using a microplate reader. The cell group without small-molecule compounds was used as a control, and the group without cultured cells was used as a blank control. Cell viability was calculated using the following formula: cell viability (%) = ($A_{compound\ group}$ − $A_{blank\ group}$)/($A_{cell\ group}$ − $A_{blank\ group}$) × 100%. Then, the toxicity of small-molecule compounds to cell proliferation was determined.

2.5. Immunofluorescence

We first transferred MA104 cells to a 12-well culture plate. When the cells grew to a dense monolayer, the RV was activated with 20 μg/mL acetylase and 600 μg/mL CaCl$_2$. The multiplicity of virus infection was MOI = 0.1, and a combined ELISA experiment was performed. Compounds with the optimal concentration measured in the CCK8 experiment were incubated for 16 h at 37 °C and 5% CO$_2$ and then taken out for immunofluorescence experiments. In this experiment, 4% paraformaldehyde containing 0.2% Triton (batch number RVAB2019101) was used. In this experiment, a purified goat anti-rotavirus antibody (batch number: RVAB2020101), preserved by the Institute of Medical Biology, Chinese Academy of Medical Sciences, was used as the primary antibody in the immunofluorescence experiment. The secondary antibody used was a fluorescein isothiocyanate (FITC)-labeled rabbit anti-goat antibody (Cat. No. 305-095-003, Jackson Immune Research, United States). 4′,6-diamidino-2-Phenylindole (DAPI) (Cat. No. C1005, Beyotime, Zhengzhou, China) was used to stain the cell nuclei, and then we observed and collected images using a fluorescence inverted microscope.

2.6. Real-Time Fluorescence Quantitative PCR

The viral genomic dsRNA was detected using RT-qPCR [19]. After extracting the viral genomic RNA, we used a HiScript® II One Step qRT-PCR SYBR Green Kit (Q222, Novozant, Nanjing, China) to detect the Ct value of the genomic dsRNA, which was also determined using RT-qPCR. In addition, to obtain a standard curve, the genomic dsRNA, which was used as the standard, was diluted in a gradient and the copy number of genomic dsRNA was detected using RT-qPCR. Finally, the number of virus copies was calculated based on the standard curve. We designed specific primers and probes targeting the highly conserved region of the NSP3 gene (Table 1).

The differentially expressed genes of the MA104 cells were also detected using RT-qPCR. We added the RVs (MOI = 0.1) and ML241 (20μM) to MA104 cells. After 20 h of infection, the cells were washed twice with PBS and then the RNA of the MA104 cells was extracted using trizol. After extracting the RNA from the MA104 cells, we used a HiScript® II One Step qRT-PCR SYBR Green Kit (Q221, Novozant, Nanjing, China) to detect the differentially expressed genes of the MA104 cells using RT-qPCR. We measured the relative expression level of the target gene using the gene of β-actin as the internal reference gene.

Table 1. NSP3 primer and probe sequence.

	Name	Sequence
	Forward primer	ACCATCTACACATGACCCTC
ZTR-68	Reverse primsr	GGTCACATAACGCCCC
	TaqMan probe	FAM-ATGAGCACAATAGTTAAAAGCTAACACTGTCAA-TAMRA
	Forward primer	GTTGTCATCTATGCATAACCCTC
SA11	Reverse primsr	ACATAACGCCCCTATAGCCA
	TaqMan probe	FAM-ATGAGCACAATAGTTAAAAGCTAACACTGTCAA-TAMRA

2.7. Western Blotting

Approximately 20 μM ML241 and RV were added to MA104 cells grown in a dense monolayer in sequence, with MOI = 0.1. After 20 h of incubation, the cell surface was gently washed twice with PBS, and a high-efficiency RIPA cell lysis buffer (R0010, Solarbio) was used to extract the total cell proteins. The bicinchoninic acid (BCA) protein concentration determination kit (P0012, Beyotime, Zhengzhou, China) was used to determine the protein concentration, and then Western blotting was performed.

2.8. Animal Experiments

The experimental protocol was approved (DWLL202208006) by the Experimental Animal Welfare Ethics Committee of the Institute of Medical Biology within the Chinese Academy of Medical Sciences (Beijing, China). The SA11 strain was used to establish a suckling mouse model to evaluate the in vivo anti-RV effects of ML241. The groups are listed in Table 2. The diarrhea score was based on the scoring rules for diarrhea in suckling rats proposed by BOSHUTZENJA et al. [20]. Diarrhea in suckling rats was scored from 0 to 4 based on the color, hardness, and quantity of feces. The score for no feces discharged is 0 points; the score for brown formed stool is 1 point; the score for brown soft stool is 2 points; the score for yellow soft stool is 3 points; the score for yellow watery stool is 4 points; and the score for perianal fecal contamination is 4 points. A score greater than 2 points was considered an indication of diarrhea.

Table 2. Grouping of suckling mice by gavage.

Group	Quantity	Virus (SA11) Dose	The Medicine Dose (mg/kg)	Frequency of Administration	Route of Administration
RV−	11	PBS (100 μL)	−	−	gavage
RV+	11	10^5 pfu	−	−	gavage
ML241 (1 h) + RV	11	10^5 pfu	20	QD	gavage
ML241 + RV	11	10^5 pfu	20	QD	gavage
RV (24 h) + ML241	11	10^5 pfu	20	QD	gavage

2.9. HE Staining Experiment for Small Intestinal Tissue

The small intestinal tissue of neonatal mice was dissected and immediately placed in a tissue fixative (Cat. No.: G1101, Servicebio, Wuhan, China), fixed for 24 h, dehydrated, soaked in wax, embedded in paraffin, and then cooled on a −20 °C freezing table. Paraffin sections were 4 μm thick. The paraffin sections were then dewaxed, covered with water, stained with hematoxylin and eosin in sequence, dehydrated, and mounted for microscopic observation to collect images.

2.10. Transmission Electron Microscopy Experiment of Small Intestinal Tissue

The small intestinal tissue of neonatal mice was dissected to a size of 1 mm^3 and stored in an electron microscope fixative (Cat. No.: G1102; Servicebio, Wuhan, China) at 4 °C. Then, 1% osmic acid was prepared in 0.1 M phosphate buffer PB (PH 7.4) to protect the samples from light and fixed for 2 h. After that, they were rinsed with 0.1 M phosphate buffer (PB) (pH 7.4) and dehydrated at 24 °C. After permeation, embedding, polymerization, and staining, transmission electron microscopy was used to observe the small intestinal tissue and to collect images.

2.11. Statistical Analyses

GraphPad Prism 9.3.1 (GraphPad, La Jolla, CA, USA) was used for data analyses and mapping. Experimental results are expressed as the geometric mean ± standard error. Between-group differences were analyzed using the two-tailed Student's *t*-test or Prapey multiple comparison test. $p < 0.05$ was considered significant.

3. Results

3.1. Screening of Anti-RV Small-Molecule Compounds

It is critical to determine the viral infection dose for screening compounds. Through pre-experimental screening, the small-molecule compounds that resisted the proliferation of the rotavirus ZTR-68 strain were screened from a library of 453 small-molecule compounds, and the optimal viral infection dose MOI = 0.1 was found. After determining the amount of infectious virus, five concentration gradients of 10 μM, 1 μM, 100 nM, 10 nM, and 1 nM were set according to the recommended concentrations of the compound library. The ELISA method was used to determine the effect of small-molecule compounds on inhibiting rotavirus proliferation. The 126 compounds that could significantly inhibit the proliferation of the rotavirus ZTR-68 strain at a concentration of approximately 10 μM were further tested through cell toxicity testing. Compounds with high toxicity to MA104 cells were removed, and the remaining five compounds with relatively low toxicity were subjected to a second round of screening.

Furthermore, five small-molecule compounds, namely 4-D10, 3-F8, 5-E9, 4-C4, and ML241, were used in experiments on the inhibitory effect on rotavirus proliferation and cell proliferation (survival). Toxicity testing was undertaken with eight concentration gradients of 70, 60, 50, 40, 30, 20, 10, and 1 μM. The results showed that at a concentration of 20 μM, compared with the other four compounds, ML241 had the best inhibitory effect on rotavirus (Figure 1A) and was less toxic to MA104 cells, the host cells of rotavirus (Figure 1B). Based on these results, ML241 was screened out. The molecular structural formula of ML241 is

C23H25CIN4O (Figure 1C). It was calculated and measured that the half-toxic concentration of the ML241 drug was CC50 = 45.42 ± 1.03μM, the half inhibitory concentration of the drug IC50 = 24.38 ± 4.33 μM, and SI (CC50/IC50) = 1.93 ± 0.36 (Figure 1D).

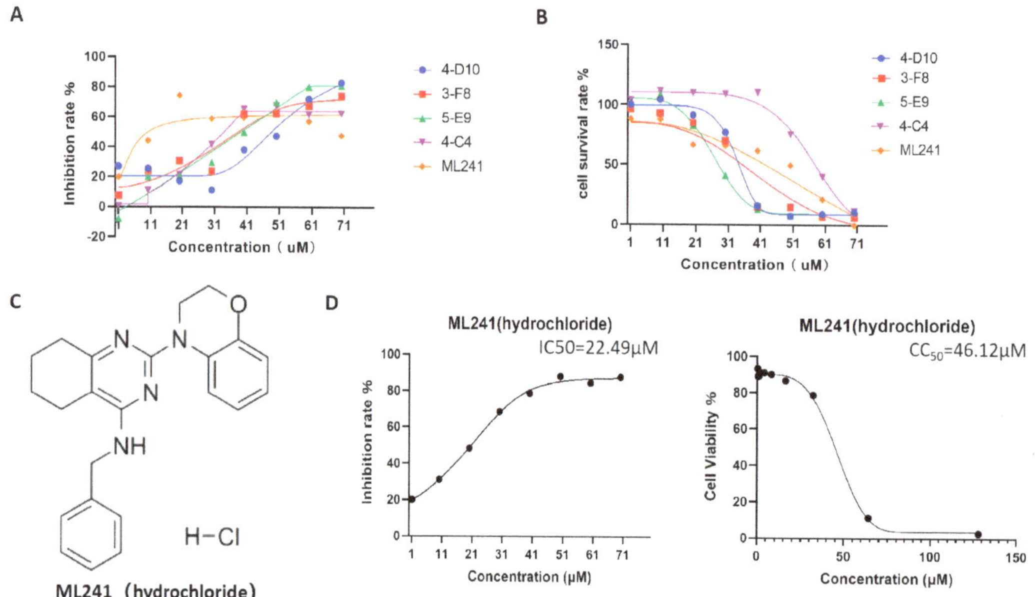

Figure 1. Screening of anti-RV small-molecule compounds. (**A**) ELISA was used to detect the inhibitory rate of five compounds against rotavirus. (**B**) CCK8 was used to measure the toxic effects of the five compounds on the cells. (**C**) The structural formula of ML241 (hydrochloride). (**D**) Half of the inhibitory rate of ML241 against rotavirus and half of its toxic effect on MA104 cells.

3.2. In Vitro Effects of ML241 on Rotavirus

To analyze the effect of the small-molecule compound ML241 on RV, immunofluorescence, R-qPCR, and WB were used to detect viral protein expression and viral replication 20 h after the addition of the drug and virus. The results showed that, compared with the control group, the addition of ML241 inhibited the expression of viral proteins and viral replication (Figure 2A–C).

To analyze whether ML241 affected the process by which RV entered a cell, the virus copy number and NSP3 protein expression 2 h after viral infection were detected using RT-qPCR and WB. The results showed that after adding ML241 for 2 h, the virus copy number decreased (Figure 3A), and the expression of the NSP3 protein decreased (Figure 3B). To analyze whether ML241 affected the process of RV entry into its host cells, the viruses were treated with ultraviolet irradiation at 220 nm (UV dose 22.5 mJ/cm^2) to disrupt their nucleic acids and prevent them from RNA replication and protein translation. Compared with the RV group without UV irradiation, the NSP3 copy number was significantly reduced after UV irradiation (Figure 3C). After adding ML241 to the UV-irradiated RV group, NSP3 also decreased compared with the UV-irradiated RV group. This decrease (Figure 3D) suggests that ML241 affected the process of RV entry into its host cells. RT-qPCR was used to detect the NSP3 copy number at different time points, and it was found that ML241 had a significant inhibitory effect at the early stage of RV infection (Figure 3E), and its inhibition of rotavirus proliferation was still statistically significant until 48 h (Figure 2D).

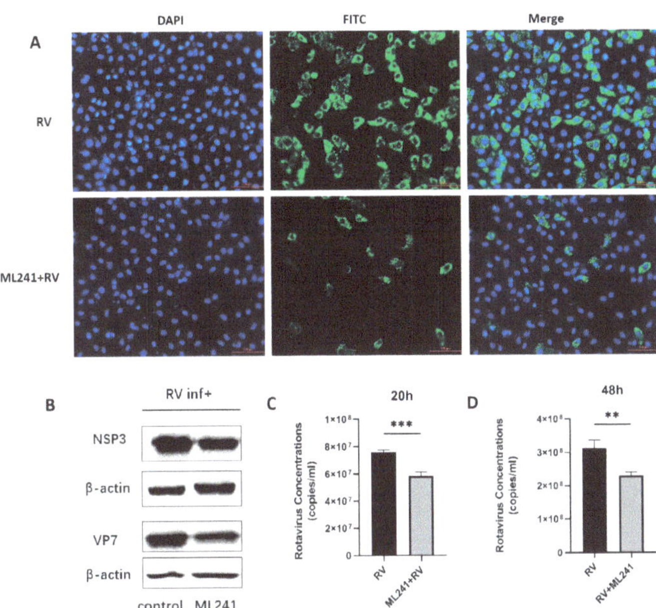

Figure 2. In vitro effects of ML241 on rotavirus. (**A**) Immunofluorescence experiments verified the inhibitory effect of ML241 on RV. (**B**) Western blotting was used to detect the expression of NSP3 and VP7 after adding ML241 for 20 h. (**C**) The RV copy number was measured by RT-qPCR using ML241 after 20 h of infection. (**D**) The RV copy number was measured by RT-qPCR using ML241 after 48 h of infection. Data are presented as mean $\pm$ SD. Significant differences were determined by an unpaired *t* test (** $p < 0.01$, *** $p < 0.001$).

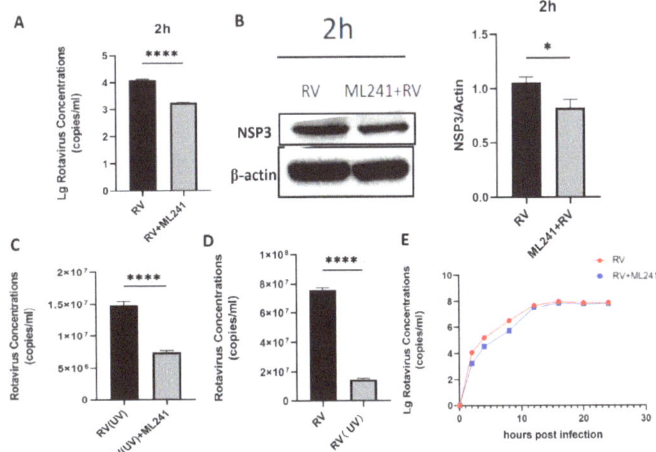

Figure 3. In vitro inhibitory effects of ML241 on rotavirus. (**A**) The RV copy number was measured by RT-qPCR using ML241 after 2 h of infection. (**B**) The expression of NSP3 was detected after adding ML241 for 2 h by a Western blotting experiment and the value of NSP3/β-actin was 1.30 ± 0.18. (**C**) RT-qPCR detection, with ML241, increased the rotavirus (RV) copy number following 20 h of UV irradiation. (**D**) RT-qPCR detects the copy number of RV and UV-irradiated RV at 20 h. (**E**) RT-qPCR is used to detect the copy number of RV at different times after the addition of ML241. Data are presented as mean $\pm$ SD. Significant differences were determined by an unpaired *t* test (* $p < 0.05$, **** $p < 0.0001$).

To verify the inhibitory effect of ML241 on other rotavirus strains, we measured the amount of antigen in different rotavirus strains after the administration of ML241 by ELISA. The results showed that ML241 had inhibitory effects on the RV of SA11, UK, Wa, and Gottfried strains (Figure 4).

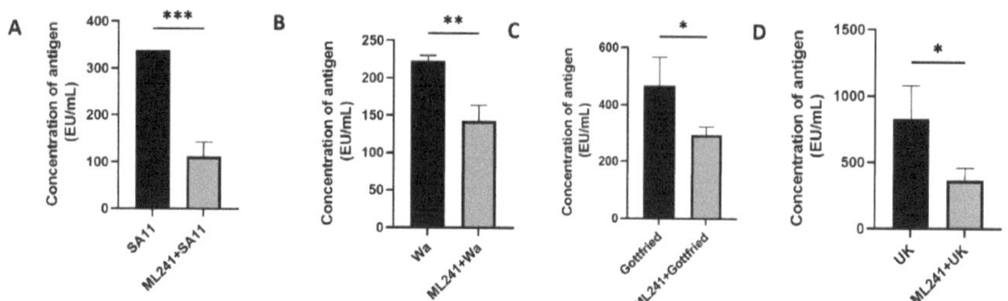

Figure 4. In vitro inhibitory effects of ML241 on rotavirus SA11, Wa, Gottfried, and UK strains. The anti-rotavirus activity was tested by Enzyme-Linked Immunosorbent Assay (ELISA). (**A**) In vitro inhibitory effects of ML241 on rotavirus SA11 strains. (**B**) In vitro inhibitory effects of ML241 on rotavirus Wa strains. (**C**) In vitro inhibitory effects of ML241 on rotavirus Gottfried strains. (**D**) In vitro inhibitory effects of ML241 on rotavirus UK strains. Data are presented as mean $\pm$ SD. Significant differences were determined by an unpaired t test (* $p < 0.05$, ** $p < 0.01$, *** $p < 0.001$).

3.3. In Vivo Effects of ML241 on Rotavirus

A 5-day-old BALB/c suckling mouse diarrhea model was established to test the inhibitory effect of ML241 on rotavirus in vivo. The grouping information is presented in Table 1. The body weight of the suckling mice was measured before and 24 h, 48 h, 72 h, 96 h, and 120 h after challenge with the SA11 strain of RV, and their diarrhea scores were calculated. The results showed that, compared with the normal control group, the weight gain of mice in the SA11 challenge group (model group) was slower, whereas the weight gain of the ML241-treated group was significantly higher than that of the model group (Figure 5A). Before the challenge, there was no statistical difference in diarrhea scores between the groups (Figure 5B). Twenty-four hours after the challenge, the diarrhea score of the RV model group was significantly higher than that of the normal control group, indicating that a suckling mouse diarrhea model of RV infection was successfully created (Figure 5C). After 48 h, compared to the model group, the scores of compound groups decreased, and there were significant differences in all scores (Figure 5D), with the most obvious being observed at 72 h (Figure 5E). There was no difference between the groups at 96 h and 120 h (Figure 5F,G). The results showed that prevention or treatment with ML241 can reduce the degree of diarrhea in suckling mice.

Two suckling mice were randomly dissected at 24 h, 48 h, 72 h, 96 h, and 120 h after the challenge, and their hearts, livers, spleens, lungs, kidneys, stomachs, and intestines were collected. Electron microscopy results at 72 h showed (Figure 6A) that the microvilli in the small intestine of the unchallenged group (normal control group) of suckling mice were densely arranged and neatly structured. The small intestinal microvilli of the challenge group (model group) were shortened, loosely arranged, and disordered; the basal layer was loose; the small intestinal villi were severely vacuolated; and in some places, they even fell off and caused gaps. The microvilli in the small intestine of suckling mice in the ML241 intervention and challenge groups were slightly shortened and loosely arranged; however, the situation was significantly better than that in the non-intervention challenge group.

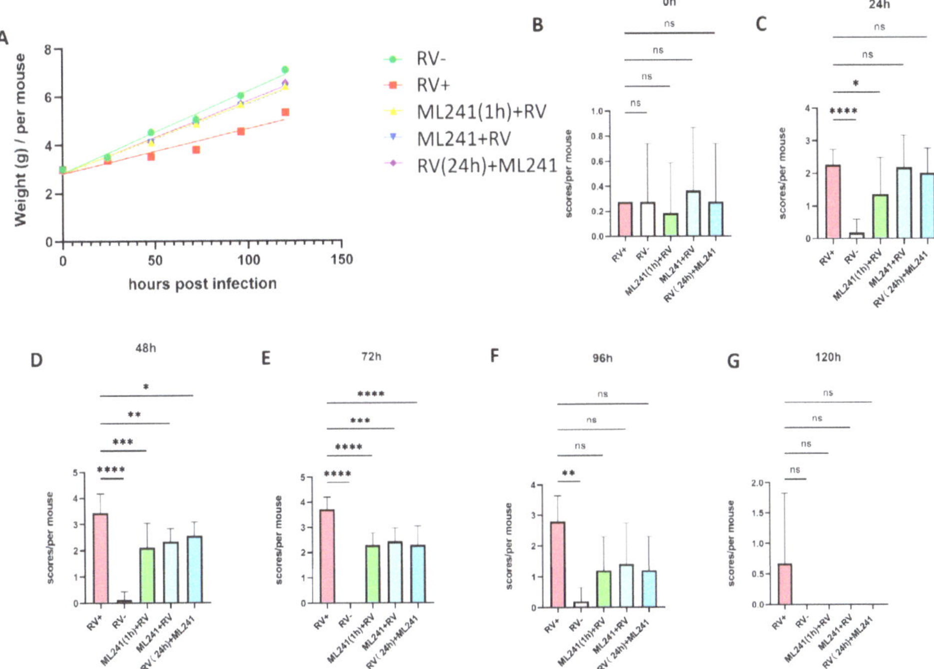

Figure 5. In vivo effects of ML241 on rotavirus. (**A**) Body weights of the suckling mice in each group. (**B**) Diarrhea scores of suckling mice in each group before challenge. (**C**) Diarrhea scores of suckling mice in each group 24 h after challenge. (**D**) Diarrhea scores of suckling mice in each group 48 h after challenge. (**E**) Diarrhea scores of suckling mice in each group 72 h after challenge. (**F**) Diarrhea scores of suckling mice in each group 96 h after challenge. (**G**) Diarrhea scores of suckling mice in each group 120 h after challenge. Data are presented as mean $\pm$ SD. Significant differences were determined by an unpaired *t* test (ns $p > 0.05$, * $p < 0.05$, ** $p < 0.01$, *** $p < 0.001$, **** $p < 0.001$).

HE staining of the small intestinal tissue of suckling mice (Figure 6B) showed that the small intestinal tissue of the unchallenged mice (normal control group) had a normal length of intestinal villi (yellow arrow) and abundant intestinal glands in the lamina propria, which were densely arranged and of a short tubular shape. The structure of the muscle layer was clear and the muscle cells were regularly arranged. In the challenge group (model group), the intestinal villous epithelium was occasionally lost in the small intestinal tissue of the suckling mice (yellow arrow), a small amount of intestinal villous epithelium was separated from the lamina propria (black arrow), the gap was widened, and the intestinal glands in the lamina propria were numerous and densely arranged. A short tubular shape was observed, with occasional scattered granulocytic infiltration (green arrow). The small intestinal tissue of the ML241 intervention group showed long intestinal villi, abundant intestinal villi, and an intact intestinal villus epithelium. Occasionally, the top of the intestinal villous epithelium separated from the lamina propria (black arrow), and the gap widened. There was a high number of intestinal glands in the lamina propria, which was large; it was in the shape of a short tube, with a small amount of vascular congestion (green arrow). Occasionally, a small focal accumulation of lymphocytes (gray arrow) was observed, along with a clear muscle layer structure and a regular arrangement of muscle cells. This shows that ML241 can significantly improve lesions in the small intestines of suckling mice, reduce diarrhea symptoms, and play a protective role in suckling mice.

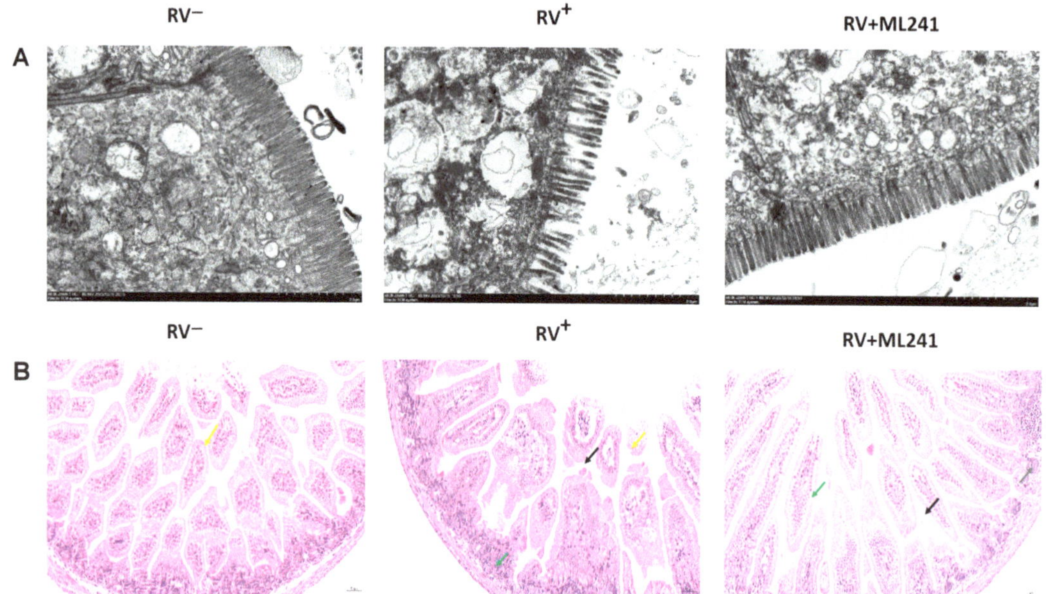

Figure 6. In vivo effects of ML241 on rotavirus. (**A**) Electron microscopic observation of small intestinal lesions in the different treatment groups. (**B**) HE staining was used to observe small intestinal lesions in the different treatment groups.

3.4. ML241 Antagonizes ERK 1/2 Activation of the MAPK Signaling Pathway by RV and Inhibits Rotavirus Replication

To analyze the mechanism by which ML241 inhibits rotavirus replication, we performed transcriptome sequencing (RNA-seq) in three groups: cell, RV, and ML241 + RV. When using FC $\geq$ 2.0, compared with the RV group, there were 195 genes with upregulated expression and 201 genes with downregulated expression in the group of ML241+RV (Figure 7A). We performed RT-qPCR verification analysis on the top 15 genes with upregulated and downregulated expression (FC $\geq$ 2.0) in each group of sequencing results, and the results showed that they were consistent with the RV group; the addition of ML241 caused an increase in the expression of interferon- and interleukin-related transcription factors, such as GADD45G, IFNL1, IRF8, KLF4, RGS2, and RSADZ genes (Figure 7B). A gene ontology (GO) enrichment of differentially expressed genes was performed (Figure 8A). A set analysis showed that compared with the RV group, after adding ML241, the molecular function was mostly the activation of cytokines, the cellular composition was the activation of protein phosphatase type I complex, and the biological process was negative for transcription. The differential gene Encyclopedia of Genes and Genomes (KEGG) analysis showed (Figure 8B) that after the addition of ML241, the differential genes were mostly enriched in the MAPK signaling pathway. We speculated that the inhibitory effect of ML241 on RV proliferation may be mediated through the MAPK signaling pathway, which plays a role. After clarifying that the MAPK signaling pathway may be involved, we detected the key proteins in the MAPK signaling pathway through WB. The results showed that after adding RV, RV significantly activated the phosphorylation of extracellular signal-regulated kinase 1/2 (ERK1/2), and its downstream IκBα was significantly increased due to RV infection. When ML241 was added, ERK phosphorylation was weakened (Figure 9A), IκBα expression was reduced (Figure 9B), and NF-κB and pNF-κB were increased (Figure 9C).

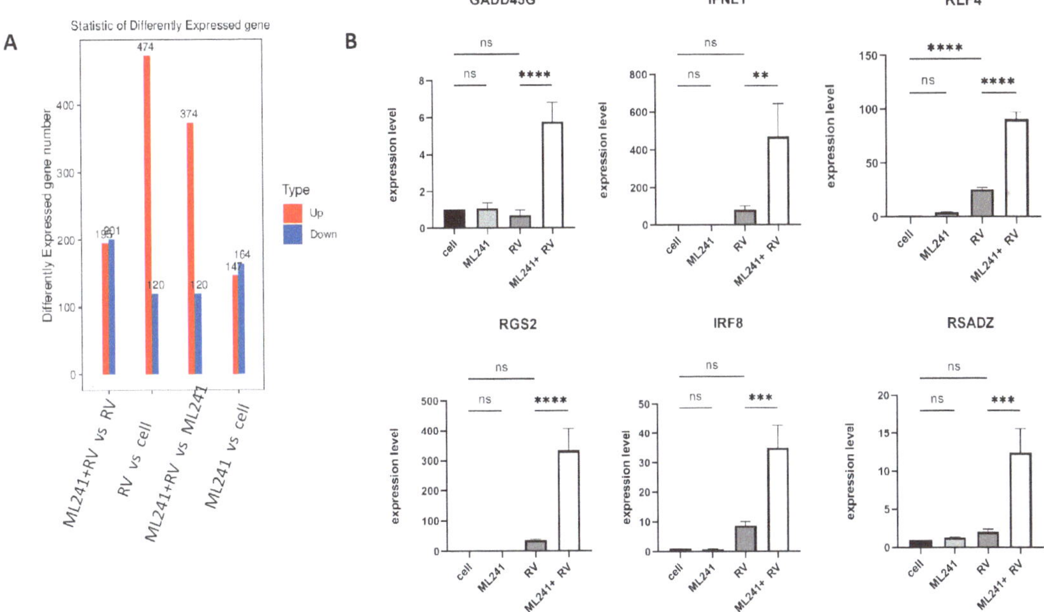

Figure 7. ML241 antagonizes ERK 1/2 activation of the MAPK signaling pathway via RV and inhibits rotavirus replication. (**A**) Number of differentially expressed genes in each group. (**B**) Relative expression of the differentially expressed genes in each group. (ns $p > 0.05$, ** $p < 0.01$, *** $p < 0.001$, **** $p < 0.001$).

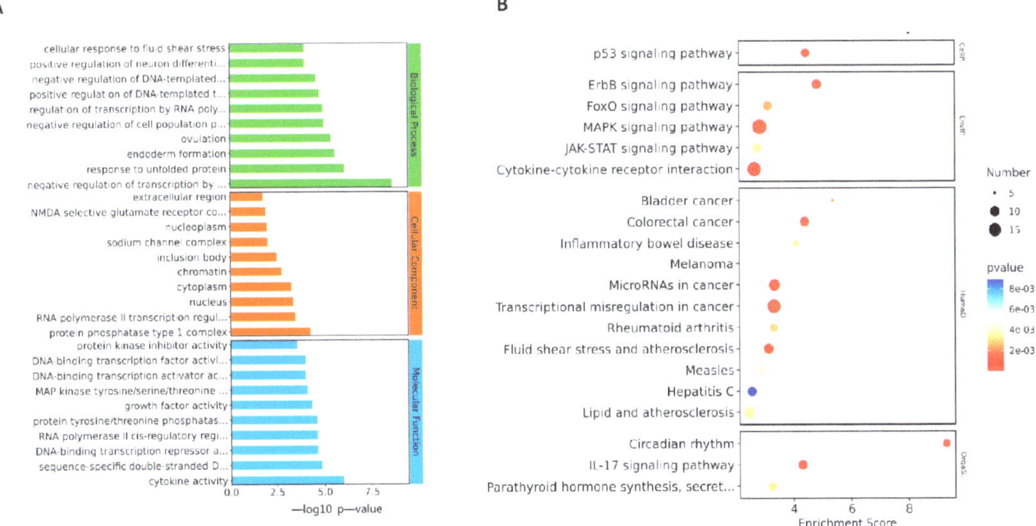

Figure 8. ML241 antagonizes ERK 1/2 activation of the MAPK signaling pathway via RV and inhibits rotavirus replication. (**A**) GO enrichment analysis of the top 30 genes with upregulated expression in ML241+RV vs. RV. (**B**) KEGG enrichment analysis of the top 20 genes with upregulated expression in ML241+RV vs. RV.

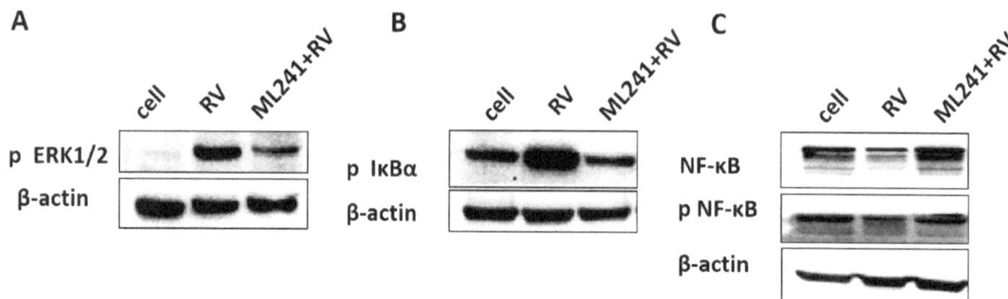

Figure 9. ML241 antagonizes ERK 1/2 activation of the MAPK signaling pathway via RV and inhibits rotavirus replication. (**A**) Western blotting was used to detect the expression of phosphorylated ERK1/2. (**B**) Western blotting was used to detect the expression of phosphorylated IκBα. (**C**) Western blotting was used to detect the expression of NF-κB and phosphorylated NF-κB.

4. Discussion

In this study, we screened 453 small-molecule compounds for anti-RV wild strain ZTR-68, which was isolated from humans using the ELISA assay. It was found that the small-molecule compound ML241 (hydrochloride) can inhibit the replication of the human rotavirus ZTR-68 strain, and the cytotoxicity test results showed that it has low toxicity to MA104 cells, which is the RV host cell. In vitro experiments showed that its inhibitory effect is particularly obvious in the early stages of RV infection, and it has inhibitory effects on the virus adsorption, cell entry, and replication stages. The antiviral mechanism of ML241 was analyzed through transcriptome sequencing and WB, and it was found that ML241 antagonizes ERK 1/2 activation and inhibits rotavirus proliferation. Using suckling mice as a model, we studied the in vivo anti-RV effect of ML241 and found that ML241 could reduce the severity of diarrhea in suckling mice and improve the degree of lesions in the small intestines of suckling mice. This study discovered a small-molecule compound that effectively inhibits rotavirus replication and studied its mechanism of action.

Small-molecule compounds are biologically active compounds with a molecular weight of less than 1000 Da (especially less than 500 Da). They can enter cells through the cell membrane, regulate targets in organelles, and carry out their corresponding biological functions. Compared with macromolecular compounds, small-molecule compounds have more advantages in terms of their targets (enzymes, ion channels, and receptors), their preparations, their costs, and patient compliance and have been widely used in virology, oncology, immunology, and neurology. Important research areas include biology, epigenetics, stem cells, organoids, apoptosis, ion channels, and signal transduction [21]. Antiviral small-molecule compounds mainly exert antiviral effects on virus adsorption, invasion, replication, assembly, and release by regulating host proteins or directly inhibiting viral proteins [22]. A variety of small-molecule drugs targeting SARS-CoV-2 have made breakthrough progress [23–25], and a variety of therapeutic drugs have entered Phase III clinical trials. To date, there are no specific antiviral therapies or marketed anti-rotavirus drugs against rotavirus. The development of anti-RV drug treatments can effectively prevent severe disease caused by viral infection, shorten the course of the disease, and alleviate symptoms.

ML241 (hydrochloride), CAS 2070015-13-1, chemical formula C23H25ClN4O, screened in this experiment, is an effective AAA ATPase p97 inhibitor [26]. The compound's half-inhibitory concentration value is 100 nM and it is widely used in anti-tumor and anti-inflammatory research. AAA ATPase p97 maintains eukaryotic cell proteostasis by promoting the degradation of ubiquitinated proteins via the proteasome and the maturation of autophagosomes [27]. In this study, we found that ML241 inhibited RV proliferation in vivo and in vitro, especially at the early stages of RV infection. The in vivo experimental results showed that there was little difference in the therapeutic effect when ML241 was administered before and after viral infection.

Compared to the viral infection group, the differentially expressed genes were mainly clustered in the MAPK signaling pathway. Further analysis revealed that they mainly clustered in the mitogen-activated protein extracellular signal-regulated kinase/extracellular-regulated kinase (MEK/ERK) signaling cascade. This signaling pathway mediates a variety of processes, including cell adhesion, cell cycle progression, cell migration, cell survival, differentiation, inflammation, metabolism, proliferation, and transcription [28]. Studies have shown that RV promotes replication by regulating the MEK/ERK signaling pathway [29]. The RV-induced apoptosis observed in the early stages of infection is inhibited by RV nonstructural protein 1 through the activation of the PI3K/Akt and NF-κB pro-survival pathways [30–32]. Many viruses, including DNA and RNA viruses, utilize the MEK/ERK pathway to promote different stages of their life cycles [28]. In this study, after adding ML241, the phosphorylation of ERK in the MAPK signaling pathway was downregulated compared with that in the RV group. ML241 antagonizes the activation of ERK phosphorylation induced by RV and inhibits viral proliferation.

NSP1 is an RNA-binding protein [33] that evades the innate immune response and delays early apoptosis by inhibiting interferon (IFN) induction and activating the PI3K/Akt pathway [34,35]. NSP1 interacts with TRAF2 to inhibit interferon-induced atypical NF-κB activation and antagonizes virus-induced cytokine responses to promote virus reproduction [36]. In this study, we found that the downstream protein IκBα of ERK was inhibited. IκBα is an inhibitory protein in the nuclear factor-κB (NF-κB) signaling pathway [37,38]. Cells respond to inflammatory stimuli via the NF-κB signaling pathway. When IκBα is inhibited, the NF-κB signaling pathway is activated, which is consistent with our detection of the expression of numerous inflammation-related genes.

5. Conclusions

In conclusion, in this study, a compound that effectively inhibited the proliferation of the human rotavirus ZTR-68 strain at multiple replication stages was selected. The results of the signaling pathways analysis showed that ML241 could inhibit viral proliferation by antagonizing the activation of ERK in the MAPK pathway. Further, by using suckling mice as an animal model, the in vivo effects of ML241 were studied, and it was found that ML241 also has a good effect on inhibiting the proliferation of rotavirus in vivo and has a good protective and therapeutic effect on suckling mice. This study helps us to further understand the pathogenesis of rotavirus and provides research ideas for the development of drugs to inhibit rotavirus, which is of significance for the development of clinical drugs for the treatment of rotavirus diarrhea.

Author Contributions: J.W. (Jinlan Wang): Conceptualization, Writing—Original draft preparation writing, Investigation, Data curation, Methodology, Formal analysis. X.H. and J.W. (Jinyuan Wu): Data curation, Formal analysis. X.L. and R.C.: Investigation, Formal analysis. C.L. and X.S.: Visualization. Q.L. and Y.L.: Data curation. X.K., J.L. and L.Y.: Investigation. X.T. and J.Y.: Software. G.Z.: Supervision. M.S.: Supervision, Validation. Y.Z.: Investigation, Conceptualization, Writing—Reviewing and Editing. H.L.: Conceptualization, Writing—Reviewing and Editing. All authors have read and agreed to the published version of the manuscript.

Funding: This work was supported by the [Major Science and Technology Special Project of Yunnan Province (Biomedicine) #1] under Grant [number 202202AA100006]; [Science and Technology Project of Yunnan Province—general program #2] under Grant [number 202201AT070236]; [CAMS Innovation Fund for Medical Sciences (CIFMS) #3] under Grant [number 2021-I2M-1-043]; [Yunnan Province Innovative Vaccine Technology and Industrial Transformation Platform #5] under Grant [number 202002AA100009].

Institutional Review Board Statement: The animal study protocol was approved by the Experimental Animal Welfare Ethics Committee of the Institute of Medical Biology within the Chinese Academy of Medical Sciences (DWLL202208006 on 24 August 2022).

Informed Consent Statement: Not applicable.

Data Availability Statement: Data are contained within the article.

Conflicts of Interest: The authors declare that they have no conflicts of interest.

References

1. Troeger, C.; Khalil, I.A.; Rao, P.C.; Cao, S.; Blacker, B.F.; Ahmed, T.; Armah, G.; Bines, J.E.; Brewer, T.G.; Colombara, D.V.; et al. Rotavirus Vaccination and the Global Burden of Rotavirus Diarrhea Among Children Younger Than 5 Years. *JAMA Pediatr.* **2018**, *172*, 958–965. [CrossRef]
2. Crawford, S.E.; Ramani, S.; Tate, J.E.; Parashar, U.D.; Svensson, L.; Hagbom, M.; Franco, M.A.; Greenberg, H.B.; O'Ryan, M.; Kang, G.; et al. Rotavirus infection. *Nat. Rev. Dis. Primers* **2017**, *3*, 17083. [CrossRef] [PubMed]
3. Gandhi, G.R.; Barreto, P.G.; Lima, B.d.S.; Quintans, J.d.S.S.; Araújo, A.A.d.S.; Narain, N.; Quintans-Júnior, L.J.; Gurgel, R.Q. Medicinal plants and natural molecules with in vitro and in vivo activity against rotavirus: A systematic review. *Phytomedicine* **2016**, *23*, 1830–1842. [CrossRef]
4. Michałek, D.; Kołodziej, M.; Konarska, Z.; Szajewska, H. Efficacy and safety of gelatine tannate for the treatment of acute gastroenteritis in children: Protocol of a randomised controlled trial: Table 1. *BMJ Open* **2016**, *6*, e010530. [CrossRef]
5. Van Dycke, J.; Arnoldi, F.; Papa, G.; Vandepoele, J.; Burrone, O.R.; Mastrangelo, E.; Tarantino, D.; Heylen, E.; Neyts, J.; Rocha-Pereira, J. A Single Nucleoside Viral Polymerase Inhibitor Against Norovirus, Rotavirus, and Sapovirus-Induced Diarrhea. *J. Infect. Dis.* **2018**, *218*, 1753–1758. [CrossRef]
6. Kim, J.-H.; Kim, K.; Kim, W. Genipin inhibits rotavirus-induced diarrhea by suppressing viral replication and regulating inflammatory responses. *Sci. Rep.* **2020**, *10*, 15836. [CrossRef] [PubMed]
7. Huang, H.; Liao, D.; He, B.; Pu, R.; Cui, Y.; Zhou, G. Deoxyshikonin inhibited rotavirus replication by regulating autophagy and oxidative stress through SIRT1/FoxO1/Rab7 axis. *Microb. Pathog.* **2023**, *178*, 106065. [CrossRef]
8. Zhou, X.; Li, Y.; Li, T.; Cao, J.; Guan, Z.; Xu, T.; Jia, G.; Ma, G.; Zhao, R. Portulaca oleracea L. Polysaccharide Inhibits Porcine Rotavirus In Vitro. *Animals* **2023**, *13*, 2306. [CrossRef]
9. Chen, S.; Ding, S.; Yin, Y.; Xu, L.; Li, P.; Peppelenbosch, M.P.; Pan, Q.; Wang, W. Suppression of pyrimidine biosynthesis by targeting DHODH enzyme robustly inhibits rotavirus replication. *Antivir. Res.* **2019**, *167*, 35–44. [CrossRef]
10. Eichwald, C.; De Lorenzo, G.; Schraner, E.M.; Papa, G.; Bollati, M.; Swuec, P.; de Rosa, M.; Milani, M.; Mastrangelo, E.; Ackermann, M. Identification of a Small Molecule That Compromises the Structural Integrity of Viroplasms and Rotavirus Double-Layered Particles. *J. Virol.* **2018**, *92*, e01943-17. [CrossRef]
11. La Frazia, S.; Ciucci, A.; Arnoldi, F.; Coira, M.; Gianferretti, P.; Angelini, M.; Belardo, G.; Burrone, O.R.; Rossignol, J.-F.; Santoro, M.G. Thiazolides, a New Class of Antiviral Agents Effective against Rotavirus Infection, Target Viral Morphogenesis, Inhibiting Viroplasm Formation. *J. Virol.* **2013**, *87*, 11096–11106. [CrossRef] [PubMed]
12. Zhang, R.; Feng, C.; Luo, D.; Zhao, R.; Kannan, P.R.; Yin, Y.; Iqbal, M.Z.; Hu, Y.; Kong, X. Metformin Hydrochloride Significantly Inhibits Rotavirus Infection in Caco2 Cell Line, Intestinal Organoids, and Mice. *Pharmaceuticals* **2023**, *16*, 1279. [CrossRef] [PubMed]
13. Zhang, Q.; Zhang, Q.; Xu, Z.; Tang, Q.; Liu, X.; Niu, D.; Gao, X.; Lan, K.; Wu, S. Dyngo-4a protects mice from rotavirus infection by affecting the formation of dynamin 2 oligomers. *Sci. Bull.* **2020**, *65*, 1796–1799. [CrossRef]
14. Araud, E.; Fuzawa, M.; Shisler, J.L.; Li, J.; Nguyen, T.H. UV Inactivation of Rotavirus and Tulane Virus Targets Different Components of the Virions. *Appl. Environ. Microbiol.* **2020**, *86*, e02436-19. [CrossRef]
15. RWard, R.L.; Kapikian, A.Z.; Goldberg, K.M.; Knowlton, D.R.; Watson, M.W.; Rappaport, R. Serum Rotavirus Neutralizing-Antibody Titers Compared by Plaque Reduction and Enzyme-Linked Immunosorbent Assay-Based Neutralization Assays. *J. Clin. Microbiol.* **1996**, *34*, 983–985. [CrossRef]
16. Fix, A.D.; Harro, C.; McNeal, M.; Dally, L.; Flores, J.; Robertson, G.; Boslego, J.W.; Cryz, S. Safety and immunogenicity of a parenterally administered rotavirus VP8 subunit vaccine in healthy adults. *Vaccine* **2015**, *33*, 3766–3772. [CrossRef] [PubMed]
17. Agarwal, S.; Hickey, J.M.; Sahni, N.; Toth, R.T.; Robertson, G.A.; Sitrin, R.; Cryz, S.; Joshi, S.B.; Volkin, D.B. Recombinant Subunit Rotavirus Trivalent Vaccine Candidate: Physicochemical Comparisons and Stability Evaluations of Three Protein Antigens. *J. Pharm. Sci.* **2020**, *109*, 380–393. [CrossRef] [PubMed]
18. Groome, M.J.; Fairlie, L.; Morrison, J.; Fix, A.; Koen, A.; Masenya, M.; Jose, L.; Madhi, S.A.; Page, N.; McNeal, M.; et al. Safety and immunogenicity of a parenteral trivalent P2-VP8 subunit rotavirus vaccine: A multisite, randomised, double-blind, placebo-controlled trial. *Lancet Infect. Dis.* **2020**, *20*, 851–863. [CrossRef] [PubMed]
19. Eric, M.K.; Mathew, D.E.; Rashi, G.; Michael, D.B. Development of a Real-Time Reverse Transcription-PCR Assay To Detect and Quantify Group A Rotavirus Equine-Like G3 Strains. *J. Clin. Microbiol.* **2021**, *59*, e02602-20.
20. Boshuizen, J.A.; Reimerink, J.H.J.; Korteland-van Male, A.M.; van Ham, V.J.J.; Koopmans, M.P.G.; Büller, H.A.; Dekker, J.; Einerhand, A.W.C. Changes in Small Intestinal Homeostasis, Morphology, and Gene Expression during Rotavirus Infection of InfantMice. *J. Virol.* **2003**, *77*, 13005–13016. [CrossRef]
21. Hayashi, M.A.F.; Ducancel, F.; Konno, K. Natural Peptides with Potential Applications in Drug Development, Diagnosis, and/or Biotechnology. *Int. J. Pept.* **2012**, *2012*, 757838. [CrossRef] [PubMed]
22. Islam, K.U.; Anwar, S.; Patel, A.A.; Mirdad, M.T.; Mirdad, M.T.; Azmi, M.I.; Ahmad, T.; Fatima, Z.; Iqbal, J. Global Lipidome Profiling Revealed Multifaceted Role of Lipid Species in Hepatitis C Virus Replication, Assembly, and Host Antiviral Response. *Viruses* **2023**, *15*, 464. [CrossRef] [PubMed]

23. Al Adem, K.; Shanti, A.; Stefanini, C.; Lee, S. Inhibition of SARS-CoV-2 Entry into Host Cells Using Small Molecules. *Pharmaceuticals* **2020**, *13*, 447. [CrossRef] [PubMed]
24. Zhao, L.; Li, S.; Zhong, W. Mechanism of Action of Small-Molecule Agents in Ongoing Clinical Trials for SARS-CoV-2: A Review. *Front. Pharmacol.* **2022**, *13*, 840639. [CrossRef] [PubMed]
25. Pandey, A.; Nikam, A.N.; Shreya, A.B.; Mutalik, S.P.; Gopalan, D.; Kulkarni, S.; Padya, B.S.; Fernandes, G.; Mutalik, S.; Prassl, R. Potential therapeutic targets for combating SARS-CoV-2: Drug repurposing, clinical trials and recent advancements. *Life Sci.* **2020**, *256*, 117883. [CrossRef] [PubMed]
26. Chou, T.F.; Li, K.; Frankowski, K.J.; Schoenen, F.J.; Deshaies, R.J. Structure–Activity Relationship Study Reveals ML240 and ML241 as Potent and Selective Inhibitors of p97 ATPase. *ChemMedChem* **2013**, *8*, 297–312. [CrossRef] [PubMed]
27. Tillotson, J.; Zerio, C.J.; Harder, B.; Ambrose, A.J.; Jung, K.S.; Kang, M.; Zhang, D.D.; Chapman, E. Arsenic Compromises Both p97 and Proteasome Functions. *Chem. Res. Toxicol.* **2017**, *30*, 1508–1514. [CrossRef]
28. Wortzel, I.; Seger, R. The ERK Cascade: Distinct Functions within Various Subcellular Organelles. *Genes Cancer* **2011**, *2*, 195–209. [CrossRef] [PubMed]
29. Rossen, J.W.A.; Bouma, J.; Raatgeep, R.H.C.; Büller, H.A.; Einerhand, A.W.C. Inhibition of Cyclooxygenase Activity Reduces Rotavirus Infection at a Postbinding Step. *J. Virol.* **2004**, *78*, 9721–9730. [CrossRef]
30. Bagchi, P.; Dutta, D.; Chattopadhyay, S.; Mukherjee, A.; Halder, U.C.; Sarkar, S.; Kobayashi, N.; Komoto, S.; Taniguchi, K.; Chawla-Sarkar, M. Rotavirus Nonstructural Protein 1 Suppresses Virus-Induced Cellular Apoptosis To Facilitate Viral Growth by Activating the Cell Survival Pathways during Early Stages of Infection. *J. Virol.* **2010**, *84*, 6834–6845. [CrossRef]
31. Bagchi, P.; Nandi, S.; Nayak, M.K.; Chawla-Sarkar, M. Molecular Mechanism behind Rotavirus NSP1-Mediated PI3 Kinase Activation: Interaction between NSP1 and the p85 Subunit of PI3 Kinase. *J. Virol.* **2013**, *87*, 2358–2362. [CrossRef]
32. Halasz, P.; Holloway, G.; Coulson, B.S. Death mechanisms in epithelial cells following rotavirus infection, exposure to inactivated rotavirus or genome transfection. *J. Gen. Virol.* **2010**, *91*, 2007–2018. [CrossRef]
33. Esona, M.D.; Gautam, R. Rotavirus. *Clin. Lab. Med.* **2015**, *35*, 363–391. [CrossRef]
34. Feng, N.; Sen, A.; Nguyen, H.; Vo, P.; Hoshino, Y.; Deal, E.M.; Greenberg, H.B. Variation in Antagonism of the Interferon Response to Rotavirus NSP1 Results in Differential Infectivity in Mouse Embryonic Fibroblasts. *J. Virol.* **2009**, *83*, 6987–6994. [CrossRef]
35. Barro, M.; Patton, J.T. Rotavirus NSP1 Inhibits Expression of Type I Interferon by Antagonizing the Function of Interferon Regulatory Factors IRF3, IRF5, and IRF7. *J. Virol.* **2007**, *81*, 4473–4481. [CrossRef]
36. Bagchi, P.; Bhowmick, R.; Nandi, S.; Kant Nayak, M.; Chawla-Sarkar, M. Rotavirus NSP1 inhibits interferon induced non-canonical NFκB activation by interacting with TNF receptor associated factor 2. *Virology* **2013**, *444*, 41–44. [CrossRef]
37. Chen, M.; Lin, X.; Zhang, L.; Hu, X. Effects of nuclear factor-κB signaling pathway on periodontal ligament stem cells under lipopolysaccharide-induced inflammation. *Bioengineered* **2022**, *13*, 7951–7961. [CrossRef]
38. Ye, J.; Ye, C.; Huang, Y.; Zhang, N.; Zhang, X.; Xiao, M. Ginkgo biloba sarcotesta polysaccharide inhibits inflammatory responses through suppressing both NF-κB and MAPK signaling pathway. *J. Sci. Food Agric.* **2018**, *99*, 2329–2339. [CrossRef]

Review

Rotavirus Sickness Symptoms: Manifestations of Defensive Responses from the Brain

Arash Hellysaz and Marie Hagbom *

Division of Molecular Medicine and Virology, Department of Biomedical and Clinical Sciences, Linköping University, 581 85 Linköping, Sweden; arash.hellysaz@liu.se
* Correspondence: marie.hagbom@liu.se

Abstract: Rotavirus is infamous for being extremely contagious and for causing diarrhea and vomiting in infants. However, the symptomology is far more complex than what could be expected from a pathogen restricted to the boundaries of the small intestines. Other rotavirus sickness symptoms like fever, fatigue, sleepiness, stress, and loss of appetite have been clinically established for decades but remain poorly studied. A growing body of evidence in recent years has strengthened the idea that the evolutionarily preserved defensive responses that cause rotavirus sickness symptoms are more than just passive consequences of illness and rather likely to be coordinated events from the central nervous system (CNS), with the aim of maximizing the survival of the individual as well as the collective group. In this review, we discuss both established and plausible mechanisms of different rotavirus sickness symptoms as a series of CNS responses coordinated from the brain. We also consider the protective and the harmful nature of these events and highlight the need for further and deeper studies on rotavirus etiology.

Keywords: rotavirus; gastroenteritis; CNS; sickness symptoms; behavioral responses; evolution

1. Introduction

Rotavirus infection is one of the leading causes of pediatric viral gastroenteritis. It is highly contagious and has been estimated to infect every newborn child in the world at least once before the age of five [1]. In healthy older children and adults, rotavirus infections can be mild or even asymptomatic. Therefore, its role as a pathogen in adults has been vastly underappreciated [2]. Nonetheless, it was recently found that in 1/3 of the families of children hospitalized with rotavirus, a caregiver also becomes ill and suffers from gastroenteritis [3]. Outbreaks among adults have also been reported from various parts of the world recently [4–6].

The hallmark symptoms of rotavirus infection are diarrhea and vomiting, which in children can lead to dehydration associated death if left untreated [7]. As such, most studies on rotavirus have focused on the intestinal mechanisms of diarrhea. Conversely, other relevant and common sickness symptoms like fever, nausea, malaise, headache, abdominal discomfort, myalgias, fatigue and loss of appetite [8] have received far less attention. Severe adult rotavirus gastroenteritis with multi-organ failure and critical management have in rare cases also been reported [9], further highlighting the complexity of rotavirus etiology.

Most of our knowledge of rotavirus pathophysiology comes from animal studies, which have established that the infection is commonly restricted to mature enterocytes of the small intestine [10,11]. Extraintestinal spread only occurs in rare cases [12–14]. Contrary to what could be expected from the symptomology, infection does not, however, induce a distinctive cytokine response [10,15–18]. Instead, a growing body of evidence points towards gut–brain crosstalk driving the defensive response against rotavirus infection [19].

The involvement of the brain in mediating rotavirus sickness symptoms has only recently started to be investigated. This includes vomiting through activation of the

Citation: Hellysaz, A.; Hagbom, M. Rotavirus Sickness Symptoms: Manifestations of Defensive Responses from the Brain. *Viruses* **2024**, *16*, 1086. https://doi.org/10.3390/v16071086

Academic Editors: Ulrich Desselberger and John T. Patton

Received: 20 May 2024
Revised: 1 July 2024
Accepted: 2 July 2024
Published: 6 July 2024

vomiting center in the medulla oblongata of the brainstem [20], but also increased intestinal motility by downregulation of the sympathetic nervous system, specifically to the ileum, but not the duodenum or jejunum [21]. These new findings together with previous knowledge about how sickness symptoms generally occur suggest the central nervous system (CNS) as a driving force of defensive responses and consequently sickness symptoms during rotavirus infection.

In this review, we focus on the clinically established rotavirus symptomology beyond diarrhea. Although the precise mechanisms of most rotavirus sickness symptoms have not been elucidated, we discuss them as results of a series of coordinated systemic events, that are likely synchronized by the CNS, against the pathogen, to increase the survival of the individual host as well as the collective group. We also consider evolutionary perspectives and the advantages and the trade-offs of these defensive strategies, and discuss the harmful and protective nature of rotavirus sickness symptoms.

2. Gut–Brain Crosstalk

The idea that the gut and the brain are connected has been a medically accepted subject for a long time. Already in the 10th century, the great Iranian physician Ibn Sina, also known as Avicenna (980-1037 AD), anatomically described the nervous connection (Figure 1) of the peritoneum to the spine and the brain [22]. In his extended medical encyclopedia Qānūn fī al-Tibb (the Canon of Medicine), which was written in 1025 AD and used as the main medical reference textbook in Europe until the 17th century [22–25], Ibn Sina explained how imbalance in the gut could, via specific direct and indirect pathways, be relayed to the brain and contribute to the pathogenesis of a number of diseases including headache, melancholia, nausea, bowel incontinence [i.e., diarrhea], and vomiting [26].

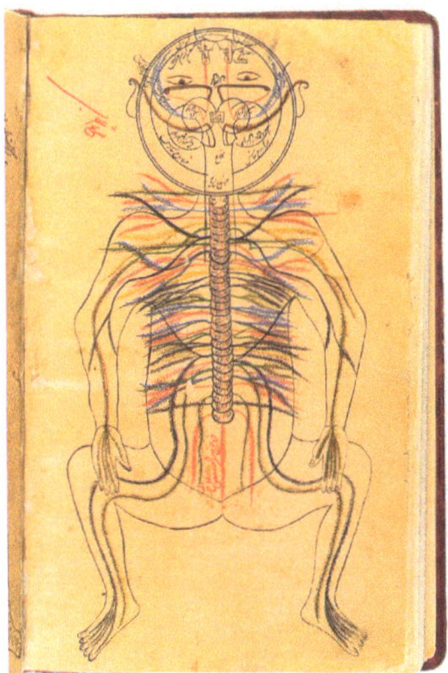

Figure 1. The nervous system innervates the peritoneum. An illustration of the nervous system of the human body by Ibn Sina (Avicenna) in Qānūn fī al-Tibb (the Canon of Medicine). Image available on the internet, open source, courtesy of the Wellcome Collection (Link: https://wellcomecollection.org/works/mx97zpqj [accessed on 4 June 2024]).

Many aspects of these early medieval descriptions persist in current anatomy [27]. Molecular biological and mechanistic knowledge about the interconnectivity of the gastrointestinal tract and the CNS through direct and indirect ascending and descending pathways (Figure 2) have been vastly extended in the last century, and the concept is well established in modern medicine [28,29]. These pathways can explain how viral gastroenteritis can cause various sickness symptoms. Surprisingly, gut–brain crosstalk has remained poorly studied in the context of rotavirus pathogenesis [19].

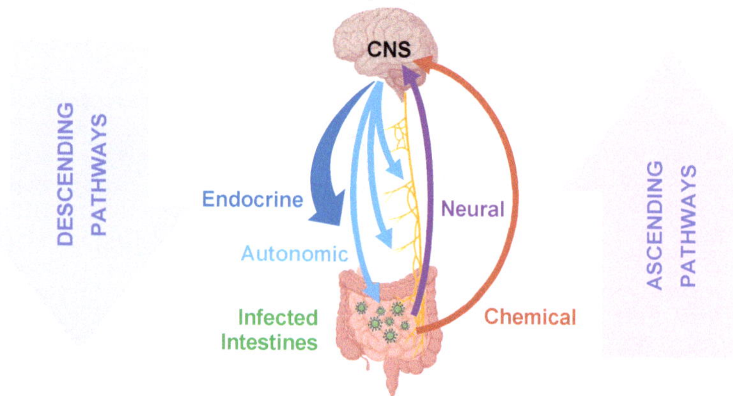

Figure 2. Bidirectional gut–brain crosstalk can occur via multiple independent chemical and electrical pathways. During rotavirus infection, the information of the pathogenic presence can reach the central nervous system (CNS) through ascending neural pathways, which include the vagus nerve and the spinal pathways. Released chemicals, including peptides, hormones and toxins, can also carry information and reach the brain through vascular or lymphatic systems. The brain processes these signals and coordinates the defense and output back to the periphery through descending pathways. While the autonomic nervous system, consisting of the sympathetic and the parasympathetic systems, provides direct, specific, and rapid access to multiple organs, the endocrine system, by releasing circulating hormones, performs the same role for broader systemic regulation at slower speed.

2.1. Ascending Pathways

Information to the CNS about the presence of a gastrointestinal pathogen can reach the brain through three major types of pathways, which consist of (a) electrical nervous signaling, (b) direct invasion by the pathogen into the brain, and (c) chemical signaling by means of various messenger molecules like cytokines, hormones, peptides, and toxins, which are transported to the brain in the vascular or lymphatic pathways [19]. Importantly, these pathways operate at different speeds and temporal resolutions. They can occur at different time points and last for shorter or longer periods of time over the course of the disease. While nervous signaling provides instant communication that could elicit a quick and short defensive response and consequent sickness symptoms, like vomiting, within a short timeframe after infection, circulating messenger molecules like toxins, cytokines, and hormones, which could affect multiple organs and for instance alter feeding behavior [30], operate at a slower rate but persist for longer [19].

Of course, multiple ascending and local pathways could also interact to create the complex rotavirus pathophysiology that is clinically observed. For instance, the rotavirus non-structural protein 4 (NSP4) has been found to stimulate enterochromaffin (EC) cells of the small intestine and induce local release of the neurotransmitter serotonin (5HT), which activates the vagus nerve to elicit vomiting [20,31]. Released serotonin is also involved in local regulation of intestinal motility by activating primary afferent nerves of the myenteric plexus, which stimulate the nerves of the submucosa plexus to release vasoactive intestinal peptide (VIP) from nerve endings adjacent to crypt cells [32,33].

2.1.1. Nervous Signaling from the Gastrointestinal Tract

The vagus nerve, extending from its origin in the brainstem, extensively innervates different parts of the gastro-intestinal tract and provides nervous feedback from the digestive system to the brain. This includes information from the enteric nervous system (ENS) as well as direct sensory information from mechano-, chemo-, and tension receptors [34].

The cell bodies of sensory afferents to the small intestine are located in the nodose ganglion and project directly into the nucleus of the solitary tract (NTS).

There are similar spinal ascending pathways that directly project into other brain areas, including the parabrachial area, the hypothalamus, and the amygdala [35]. The presence of an infectious pathogen in the intestines can therefore be detected through multiple nervous pathways and rapidly conveyed to the CNS for further processing. This also includes possible interactions between the pathogen and the gut microbiome, which could elicit nervous signaling to the CNS [36,37]. Locally produced cytokines can also activate primary afferent nerves, such as the vagal nerves, during abdominal and visceral infections [38].

2.1.2. Direct Invasion of the Brain

Both viremia and extraintestinal infections of rotavirus in the liver, lungs, and kidneys have been clinically reported [12]. In a prospective study of acute encephalitis in children from Sweden, rotavirus RNA could be identified via PCR in stool samples in 10% of the patients in the cohort [39]. Encephalopathy [40], acute cerebellitis [41], and other CNS infection-associated complications have also been found to be concurrent with rotavirus gastroenteritis. Based on these reports, direct invasion by rotavirus of the brain has been theorized [42]. Nonetheless, the underlying disease causalities in these reports remain unknown, and to the best of our knowledge, no substantiated evidence of rotavirus infection of the human brain has been provided yet.

Recent immunohistochemical data from mice clearly show that the epidemic diarrhea of infant mice (EDIM), rotavirus is restricted to the gastrointestinal tract up to 72 h post-infection [43] and does not extend to the brain even though altered brain activity is observed earlier [21]. Early studies performed by L. M. Kraft on EDIM before it was recognized as a rotavirus [44] did however identify infectious virus in the lungs, liver, spleen, kidneys, bladder, brain, and blood by 72 h post-infection [45–47]. Since these organs are highly vascular, it was assumed that the presence of virus reflected the presence of blood rather than actual infection of these organs. Nonetheless, extraintestinal rotavirus infection in humans remains a controversy and is also likely to be attributed to premorbid characteristics like immunodeficiency [45] or hereditary factors [39].

2.1.3. Chemical Signaling

Cytokines, hormones, and peptides function as signaling molecules and can be transported in the vascular and lymphatic systems from peripheral organs to the CNS. The CNS constantly monitors for new or altered signaling molecules. Any induced alteration caused by pathogenic infection can potentially elicit a CNS response. Cytokines are a response to infection, and the main cytokines involved in sickness responses are the pro-inflammatory cytokines IL-1 and TNF-α [38]. Moreover, anti-inflammatory cytokines regulate the intensity and duration of sickness behavior.

Infection can affect the ENS and potentially alter regulating gut and brain peptides and hormones like peptide YY (PYY), glucagon-like peptide-1 (GLP-1), cholecystokinin (CCK), leptin, or ghrelin [30]. Cytokines have, through both direct and indirect actions, a profound effect on systemic metabolism and the development of sickness behaviors [48].

2.2. Descending Pathways

The brain can coordinate defensive responses and output to the periphery through either neural or humoral pathways [30]. The sympathetic and parasympathetic branches of the autonomic nervous system are the main neural outputs that provide the brain direct regulatory access to peripheral organs, including the gastrointestinal tract [49]. These two antagonistic systems operate like "gas and brakes" for a multitude of peripheral organs and systems.

Sympathetic activation increases, for instance, intestinal motility, blood sugar levels, heart rate, blood pressure, and breathing rate, which can make one ready to combat or fly from immediate danger. Activation of the parasympathetic nervous system, on the other hand, has opposing effects and is therefore known to drive the rest and digest conditions [50]. Importantly, stable conditions are maintained through proper balance between the two systems. For instance, increased intestinal motility can be achieved by either increasing the sympathetic or reducing the parasympathetic signaling to the intestines. The autonomic nervous system is an important part of an organism's survival in managing external danger, but also in combating internal pathogens [51].

The neuroendocrine system constitutes the main humoral output of the brain. It is mainly regulated by the neurons of the hypothalamus, which, either directly or indirectly through specialized endocrine cells of the pituitary, release hormones into circulating blood where they can reach the entire body [52].

The neuronal and humoral pathways provide the brain with great versatility to coordinate defense mechanisms against infections. While neuronal pathways provide a near-instant signaling that can be directed to specific organs, the humoral pathways provide the means to, albeit at a slower speed, direct broad systemic signaling to the entire body. It is also well accepted that the CNS is, through both neural and humoral pathways, involved in complex bidirectional communication with the immune system and can control peripheral immunity [53].

3. Central Coordination of Defense

Brain areas that receive primary ascending afferents to drive the peripheral inputs to other brain regions are known as first-order nuclei [54]. The NTS is therefore often referred to as a first-order nuclei for the vagus nerve [30]. Conversely, the area postrema (AP), which contains a chemoreceptor trigger zone, is the first-order nuclei to detect circulating emetic agents [55]. From these entry points, the signal can be relayed to a multitude of other brain areas, including the amygdala, the thalamus, and the hypothalamus [49], which are involved in regulating various outputs including mood [56], behavior, and homeostatic balance.

Modulation of various brain structures following rotavirus infection in mice has recently been reported. This includes increased cFos expression in the NTS and AP [20] as well as reduced phosphorylated signal transducer and activator of transcription 5 (pSTAT5) expression in the bed nucleus of the stria terminalis (BNST) [21]. The common denominator for these structures is their role as first-order nuclei.

The NTS in the dorsal medulla is a major input for vagal afferents from various organs including the gastrointestinal tract. It is part of the dorsal vagal complex and responsible for triggering vomiting [49]. Furthermore, the NTS also projects into a large number of other regions, including the hypothalamus, from which several other rotavirus sickness symptoms like fever, fatigue, sleepiness, stress, and loss of appetite, could arise. Similarly, the BNST receives direct vagal afferents from the periphery [35]. It is a center of integration for limbic information and highly associated with reward, stress, and anxiety. Furthermore, the BNST acts as a relay to the hypothalamic–pituitary–adrenal axis, which regulates acute stress response.

The AP is a circumventricular organ located caudal to the floor of the fourth ventricle in the medulla [55]. It is highly vascular, lacks a blood–brain barrier, and can sense circulating chemicals in the blood and the cerebrospinal fluid. It projects into various other brain regions including the NTS.

How the signal that arises from rotavirus infection propagates from these nuclei has not been elucidated yet, and systemic investigation of the brain during rotavirus infection is lacking. Nonetheless, these three nuclei together can potentially relay the signal from a rotavirus infection to both neural and hormonal pathways. This does, however, not exclude the possibility of the involvement of other first-order nuclei in rotavirus pathogenesis.

4. Sickness Symptoms

The feeling of sickness is not easily measurable and sometimes vaguely defined as the perception of not feeling well. Sickness behaviors follow sickness feelings and include lethargy, social withdrawal, depression, and reduced exploration, but also loss of appetite, sleepiness, and hyperalgesia [57]. It is now abundantly clear that sickness is an active process and not merely a passive consequence of systemic infections [58]. Inducing sickness feelings and monitoring behavioral responses in infants and children are not ethically defensible; however, outbreaks and volunteer studies in adults have shown that sickness symptoms and behaviors occur during rotavirus infection [2].

In a study of 18 rotavirus infected adult volunteers, Kapikan et al. [59] found that 22% develop sickness feelings and 28% shed rotavirus. Observed symptoms among the subjects included diarrhea (22%), vomiting (11%), headache (22%), anorexia (22%), malaise (17%), abdominal cramping (11%), and elevated body temperature (17%). In another study of 83 college students from a rotavirus outbreak [60], the vast majority of the subjects suffered from diarrhea (93%), abdominal pain or discomfort (90%), loss of appetite (83%), and nausea (81%). Furthermore, more than 50% showed the following symptoms: fatigue, vomiting, headache, myalgia, chills, and low-grade fever.

4.1. Diarrhea

Diarrhea is one of the hallmark symptoms of rotavirus infection [17]. It is an evolutionarily preserved defense mechanism [61] for effectively driving pathogenic clearance from the lower gastrointestinal tract [62], but simultaneously also very costly to the host, as it can rapidly cause malnutrition and fatal dehydration if sustained over a long period of time [63]. Diarrhea is very effective when it comes to flushing substances from the intestinal lumen, but less so when it comes to getting rid of intracellular pathogens like viral agents, which replicate inside the cells. Nonetheless, reduction of free infectious virus as well as viral toxins from the lumen should some extent be beneficial to the host. In infants and young children, who are less resilient to loss of electrolytes and fluid imbalance, diarrhea is, however, more often harmful.

The onset of rotavirus diarrhea is between 24–48 h post-infection and has been found to last on average for 6 days among Swedish children [64]. From a cross-sectional study including five hospital-based studies, diarrhea average from the time of presentation ranged from 2.3 to 7.4 days [65].

Rotavirus diarrhea was initially regarded to be osmotically driven and emerging by excessive fluid and electrolyte loss due to malabsorption [66,67]. This model did not, however, provide a full explanation for rotavirus etiology. In 1999, the NSP4 enterotoxin was shown to participate in the diarrhea response to rotavirus [68], and in 2000 Lundgren et al. showed for the first time that rotavirus infection induce activation of the enteric nervous system [18]. Today it is well accepted that the underlying mechanisms of rotavirus diarrhea are multifactorial and involve activation of the enteric nervous system (ENS) [18,69,70]. Although the ENS can work independently and regulate many of the intestinal functions such as motility and secretion, there is an autonomic control by the brain to ensure regulation of intestinal homeostasis [71]. Recently, it has been shown that following infection, descending sympathetic nerves from the CNS participate in rotavirus diarrhea by increasing ileal motility [21]. This increase was observed in mice 16 h post-infection, which is before the onset of diarrhea. The central pathways that induce this autonomic response during rotavirus infection have, however, not yet been studied.

4.2. Vomiting

Evolutionarily, our physiological capabilities developed at a time when one had to race across the savannah to hunt or catch the next meal [72]. It is only during the last century in modern times that humans have been surrounded by a plethora of food which is highly nutritious and largely free from pathogens and toxins. During evolution, vomiting was probably the most important defense mechanism against food poisoning and uptake of noxious chemicals. Vomiting rapidly and effectively empties the stomach and eliminates the risk of gastric uptake of harmful particles [72]. The cost of falsely inducing vomiting, when no harmful agent is present, is only a few calories [73]. However, the penalty for a single miss, when harmful toxins or pathogens are present in the food, is extremely high and might even lead to death.

Vomiting is commonly the first symptom of rotavirus infection and occurs simultaneously with fever [74]. In a study of rotavirus-infected children, vomiting was the onset symptom in 55% of the patients, preceding diarrhea by 24 h [64]. Interestingly, vomiting usually lasts for only 2 days [64,75], even though virus replication continues for longer and shedding persists for up to 10 days post-infection [76]. It is thus less likely for circulating emetic agents to drive rotavirus vomiting through activation of chemoreceptors in AP. The symptomology instead suggests direct nervous signaling to drive rotavirus vomiting.

Indeed, investigations in mice have identified activation of the NTS following rotavirus infection, and it is now well accepted that rotavirus vomiting occurs through serotonergic activation of the vagus nerve, which relays the signal to the NTS [20]. The serotonergic surge in the intestines is caused by the release of rotavirus NSP4 enterotoxin, which induces calcium increase in the enterochromaffin cells of the intestines, and therefore the release of serotonin [20]. How the body quickly moderates this NSP4-5-HT vagal pathway to alleviate rotavirus vomiting remains elusive. The anti-emetic 5-HT$_3$ receptor antagonist ondansetron, which blocks 5-HT binding to vagal afferent nerves, is clinically used to viral gastroenteritis, although not an established treatment alternative for gastroenteritis. There have been clinical trials of ondansetron use against viral gastroenteritis [77,78], but more studies are needed to confirm positive or negative effects from inhibiting the host vomiting response.

4.3. Fever

Fever is a common response to infection that has been conserved in vertebrates for over 600 million years [79]. Fever even occurs in cold-blooded vertebrates like reptiles and fish, which raise their core temperature during infection by altering their behavior and seeking warmer environments, despite the risk of predation.

Fever is very common during rotavirus infection [64,75,80]. Rotavirus illness usually begins with acute onset of fever and vomiting, and about 30–40% of children may experience temperatures above 39 °C [74]. In a prospective study of acute gastroenteritis in Swedish children, about 84% of rotavirus-infected patients showed signs of fever, which on average lasted for 2.2 days. Interestingly, fever was found to be much more frequent in rotavirus-infected patients compared to patients with enteric adenovirus (44%) or bacterial gastroenteritis (69%). In another Scandinavian study comprising 118 rotavirus-infected children [75], elevated body temperature was measured in 86–90% of the patients (depending on subtype), with a mean duration of 2.1–2.3 days. Of these children, 47% were in the range of 37.7–38.9 °C and 43% had temperatures above 39 °C.

An American study showed that children suffering from acute rotavirus gastroenteritis exhibit higher levels of IL-6 in serum when also experiencing fever, and children with both fever and more episodes of diarrhea also exhibit higher levels of TNF-α [81]. The study also found that the levels of IL-6, IL-10, and IFN-γ were significantly higher in children with acute rotavirus infection compared to healthy children without diarrhea [81].

Another study, from Turkey, comparing bacterial and rotavirus gastroenteritis found that elevated IL-6 and TNF-α was lower in rotavirus-infected children compared to children with bacterial gastroenteritis [80]. Rotavirus-infected children also had slightly elevated fever compared to bacterially infected children and did not have any C-reactive protein (CRP), leucocytes, or blood in their feces. It should be noted that these differences in cytokine responses during rotavirus infection are detectable when compared to bacterial infections. Standalone, cytokine and inflammatory elevations during rotavirus infection is mostly modest [10,15–18]. Little is known about the role of cytokines in the pathogenesis of rotavirus disease and needs to be further investigated.

The temperature homeostasis in mammals is regulated by neurons located in the preoptic area of the hypothalamus, which function as a thermostat that keeps the body temperature within an optimal preset range, despite varying ambient temperature [82,83].

These neurons can respond to nervous input, cytokines, or circulating hormones like prostaglandin, and slightly raise this setpoint to increase the efficacy of the immune response during a pathogenic invasion [84,85]. Although acutely effective, sustained high temperature can rapidly become harmful and even fatal [86].

Depression, loss of appetite, and anorexia, which reduce the overall metabolic rate, are also common during fever [58,87]. This is rather counter-intuitive, as a systemic increase of body temperature is extremely costly and requires a 10–12.5% increase of metabolic rate for a single degree Celsius increase of body temperature [79]. However, the fact that fever has been retained throughout vertebrate evolution strongly suggests that febrile temperatures confer a survival advantage.

The induction and maintenance of fever during infection involves a tightly coordinated interplay between the innate immune system and neuronal circuitries within the central and peripheral nervous systems. Acetylcholine contributes to fever by stimulating muscle myocytes to induce shivering, and norepinephrine elevates body temperature by increasing thermogenesis in brown adipose tissue as well as inducing vasoconstriction to prevent passive heat loss [79]. Release of prostaglandin E2 (PGE2) as well as pyrogenic cytokines like IL-1, IL-6, and TNF-α, from innate immune cells, can also induce fever [88]. Where and how these systems are perturbated during rotavirus infection remain elusive.

4.4. Fatigue and Sleepiness

Fatigue appears central in sickness behavior because it is a strong signal to rest and focus the energy to combat infection or inflammation and regain health. Fatigue is a result of the modification of metabolic pathways through an endocrine loop, initiated in the hypothalamus and amplified by the pituitary gland. The release of corticotrophin-releasing hormones, cortisol, and adreno-corticotropic hormones affect nutrient homeostasis in tissues such as the liver, muscles, and adipose tissue. A meta-analysis of human sickness behavior showed that fatigue was one of the most commonly reported symptoms and is associated with IL-6 and IL-1 [89].

Fatigue should not be confused with sleepiness, i.e., sleep propensity, which is the normal signal for sleep, and although sleepiness and fatigue are often used interchangeably, they are two distinct responses [90]. Sleepiness is a driving force related to the need for sleep, while fatigue is a more general signal to avoid activity [90].

Sleep is altered during infection, but it is unknown why [91]. However, it is known that infection-induced alterations in sleep is a CNS response to IL-1 and TNF-α. Neurons immunoreactive for IL-1 and TNF and involved in the regulation of sleep and wake behavior are notably located in the hypothalamus, but also in the hippocampus and the brainstem [91]. It is known that the regulation of body temperature is coupled to sleep and the changes in sleep architecture that occur during infection, suppressed REM (rapid eye moment) sleep and increased but more fragmented non-REM sleep, might have fever-promoting benefits [91]. Sleep is proposed to be an acute-phase response and a sickness behavior that promotes recovery and increases survival. Information about fatigue is not easily accessed in children but in a rotavirus outbreak study among college students, 50% of the 83 students with rotavirus infection declare that they experienced fatigue [2].

4.5. Stress

The sympathetic system works as fight or flight response and parasympathetic activation induces the rest response. Rotavirus infection, like other viral pathogens, cause infection-induced stress response, likely by activation of the sympathetic pathway to the hypothalamus and the release of cortisol. Activation of the hypothalamic–pituitary–adrenal (HPA) axis is well known to subserve the body's response to a stressor, and viral infections, in general, are physiologically stressful [88].

Catecholamines, the end-product of sympathetic nervous system activation, and glucocorticoids orchestrate the "fight or flight" response, with rapid mobilization of energy to critical muscles and the brain, concomitant with increased heart rate, blood pressure, and breathing rate to facilitate rapid transport of nutrients and oxygen to relevant tissues. At the same time the HPA axis also assists in shunting metabolic resources from growth, digestion, reproduction, and certain aspects of immunity to the more immediate and acute functions.

The physiological function of stress–induced increases in glucocorticoid levels is to protect against the normal defense reactions, e.g., immune response/inflammation that are activated by stress; glucocorticoids accomplish this function by turning off those defense reactions, thus preventing them from overshooting. This can be seen as the glucocorticoid actions may help mediate the recovery from the stress response, rather than mediate the stress response itself. In a systematic review of 15 studies of acute illness [92], it was found that cortisol levels were more than 3-fold higher in the group with severe gastroenteritis than in the control group [92]. Activation of the HPA axis is well known to subserve the body's response to a stressor, and viral infections, in general, are physiologically stressful, as indicated by the concomitant activation of the HPA axis [88,93].

4.6. Loss of Appetite

Loss or lack of appetite, also referred to as anorexia, are common during infection conditions, and have been reported as high as 83% in an rotavirus outbreak among college students [2]. The brain continuously receives information from the periphery regarding energy stores and energy needs and processes this information in order to regulate feeding behavior [94]. The brain also senses and responds to peripheral infection and inflammatory processes, which in turn, affect feeding and metabolism, indicating that during inflammation and or infection the immune response largely depends on the energy status of the host [94]. Understanding the mechanisms through which the brain regulates appetite and feeding behavior will provide insights into the metabolic adaptation for therapeutic intervention [94].

Reducing the level of food intake is a good strategy to rest and enable evacuation of existing pathogens from the gastric system. It also reduces the risk of introducing new pathogens. This might be perceived as counter-intuitive, as access to nutrients is crucial for successfully combating an infectious pathogen.

The hypothalamic melanocortin system is heavily involved in the regulation of appetite, and has been considered as a promising target to control appetite during disease conditions [94]. Another key region controlling appetite is the dorsal vagal complex (DVC), a brain structure that comprises the area postrema (AP), the nucleus of the solitary tract (NTS) and the dorsal motor nucleus of the vagus (DMV) [94]. Several peripheral mechanisms, including motor functions of the stomach as well as released peptides and hormones, provide feedback to the hypothalamic circuitry and vagal complexes of the brain to regulate appetite and thereby balance consumed and expended body energy [95]. There are cytokines, and prostaglandin E, involved in appetite regulation and these signals can reach the brain through vagal sensory nerve receptor or by the circulation [94].

Although several microorganisms are known for their ability to manipulate host defenses to their own advantage, the role of anorexia induced by pathogens is not clear yet. It may also be that access to nutrition can have different effects depending on the pathogen itself. While fasting appear protective during bacterial infection, it seems to have detrimental effects during viral infection, highlighting divergent metabolic requirements [96]. Low nutrition status has been associated with risk for more severe rotavirus infection, however not yet proved and studies have shown opposing results [97,98]. However, in children with metabolic disorders, rotavirus infection was associated with increased morbidity and mortality [99]. Weather if the host response of reduced appetite with a fasting state promotes rotavirus recovery, or if nutrition supplement is beneficial, needs to be further investigated.

5. Evolution of Defense Strategies

Characteristics that benefit our survival have been preserved during evolution. Humans and animals evolved with viruses, and up to 2/3 of the human genome is derived from viruses and transposable elements [100]. Early in human evolution, the means to properly store food was lacking, and food quality was uncontrolled. The ability to handle ingestion of toxic and pathogenic agents was literally a matter of life and death.

Successful host defense strategies against viral pathogens were, and to large extent still are, strongly favorable for the survival of the species and thus evolutionary conserved [58]. These facts have led to the evolution of complex physiological and behavioral defense strategies against ingulfed harmful substances, like rotavirus.

Defensive responses like fever, sleepiness and loss of appetite are not weaknesses due to the infection per se, but rather an acquired adaptive strategy for survival [58]. Counter-intuitively, viral infections, which normally enter cells, are considered more likely to induce vomiting, than bacterial infections, which normally do not enter cells [64]. It should, however, be noted that the mechanism underlying rotavirus vomiting is at least partly driven by the NSP4 enterotoxin, and in an essence similar to the mechanism of toxin producing bacteria that also induce vomiting [101], indicating that the neuronal pathways have evolved against toxins rather than specific pathogenic agents. Interestingly, virus is the Latin word for poison.

Gastrointestinal bacterial infections usually cause prolonged bloody diarrhea [64]. Acute viral gastroenteritis, on the other hand, is characterized by watery diarrhea and a low inflammatory response with mild elevation of serum inflammatory markers [102]. These characteristics are also recapitulated in the symptomology as viral gastroenteritis are more acute, but resolve faster than bacterial infections [102].

6. Conclusions

The question of whether sickness symptoms are harmful or protective during rotavirus gastroenteritis has no straightforward answer. It is case-specific and depends on many variables. From an evolutionary point of view, sickness symptoms are indeed manifestations of survival strategies and adaptive host defensive responses against different kinds of danger. This should be considered, particularly when prescribing therapies and treatments against sickness symptoms, rather than the pathogen and its toxic components. While it is possible in modern medicine to identify specific pathogens like rotavirus and provide targeted therapies, adequate knowledge about the body defense mechanisms is in many aspects largely lacking.

Blocking sickness symptoms, without adequate knowledge about the underlying mechanisms, could have negative impact, introduce adverse effects, and lead to prolonged recovery. In cases when a therapy does not have an obvious and direct benefit, it might thus be better to let the host carry on its evolved adaptive responses, despite any temporary discomfort that might arise. At the same time, excessive defensive host measures could also become harmful. Such is the case with prolonged rotavirus diarrhea and vomiting, which arguably provide little to no benefit to the host during the later stages of the disease but pose an imminent risk of fatal outcome. In such a case, blocking the sickness symptom is directly beneficial and should be considered.

Funding: This research was funded by the Swedish Research Council [2023-02720].

Conflicts of Interest: The authors declare no conflicts of interest.

References

1. Grimwood, K.; Lambert, S.B. Rotavirus vaccines: Opportunities and challenges. *Hum. Vaccines* **2009**, *5*, 57–69. [CrossRef] [PubMed]
2. Anderson, E.J.; Weber, S.G. Rotavirus infection in adults. *Lancet Infect. Dis.* **2004**, *4*, 91–99. [CrossRef] [PubMed]
3. Tran, A.N.; Husberg, M.; Bennet, R.; Brytting, M.; Carlsson, P.; Eriksson, M.; Storsaeter, J.; Osterlin, B.; Johansen, K. Impact on affected families and society of severe rotavirus infections in Swedish children assessed in a prospective cohort study. *Infect. Dis.* **2018**, *50*, 361–371. [CrossRef] [PubMed]
4. Chia, G.; Ho, H.J.; Ng, C.G.; Neo, F.J.; Win, M.K.; Cui, L.; Leo, Y.S.; Chow, A. An unusual outbreak of rotavirus G8P[8] gastroenteritis in adults in an urban community, Singapore, 2016. *J. Clin. Virol.* **2018**, *105*, 57–63. [CrossRef]
5. Niendorf, S.; Ebner, W.; Marques, A.M.; Bierbaum, S.; Babikir, R.; Huzly, D.; Maassen, S.; Grundmann, H.; Panning, M. Rotavirus outbreak among adults in a university hospital in Germany. *J. Clin. Virol.* **2020**, *129*, 104532. [CrossRef] [PubMed]

6. Pacilli, M.; Cortese, M.M.; Smith, S.; Siston, A.; Samala, U.; Bowen, M.D.; Parada, J.P.; Tam, K.I.; Rungsrisuriyachai, K.; Roy, S.; et al. Outbreak of Gastroenteritis in Adults Due to Rotavirus Genotype G12P[8]. *Clin. Infect. Dis.* **2015**, *61*, e20–e25. [CrossRef]
7. Tate, J.E.; Burton, A.H.; Boschi-Pinto, C.; Steele, A.D.; Duque, J.; Parashar, U.D.; MBBS the WHO-Coordinated Global Rotavirus Surveillance Network. 2008 estimate of worldwide rotavirus-associated mortality in children younger than 5 years before the introduction of universal rotavirus vaccination programmes: A systematic review and meta-analysis. *Lancet Infect. Dis.* **2012**, *12*, 136–141. [CrossRef]
8. Hellysaz, A.; Neijd, M.; Vesikari, T.; Svensson, L.; Hagbom, M. Viral gastroenteritis: Sickness symptoms and behavioral responses. *mBio* **2023**, *14*, e0356722. [CrossRef]
9. Edwards, N.; Abasszade, J.H.; Nan, K.; Abrahams, T.; La, P.B.D.; Tinson, A.J. Severe adult rotavirus gastroenteritis: A rare case with multi-organ failure and critical management. *Am. J. Case Rep.* **2023**, *24*, e940967. [CrossRef]
10. Ramig, R.F. Pathogenesis of intestinal and systemic rotavirus infection. *J. Virol.* **2004**, *78*, 10213–10220. [CrossRef]
11. Estes, M.K.; Kang, G.; Zeng, C.Q.; Crawford, S.E.; Ciarlet, M. Pathogenesis of rotavirus gastroenteritis. *Novartis Found Symp.* **2001**, *238*, 82–96; discussion 96–100.
12. Li, N.; Wang, Z.Y. Viremia and extraintestinal infections in infants with rotavirus diarrhea. *Di Yi Jun Yi Da Xue Xue Bao* **2003**, *23*, 643–648. [PubMed]
13. Gilger, M.A.; Matson, D.O.; Conner, M.E.; Rosenblatt, H.M.; Finegold, M.J.; Estes, M.K. Extraintestinal rotavirus infections in children with immunodeficiency. *J. Pediatr.* **1992**, *120*, 912–917. [CrossRef]
14. Paul, S.P.; Candy, D.C. Extra-intestinal manifestation of rotavirus infection....Beyond the gut. *Indian J. Pediatr.* **2014**, *81*, 111–113. [CrossRef] [PubMed]
15. Lundgren, O.; Svensson, L. Pathogenesis of rotavirus diarrhea. *Microbes Infect.* **2001**, *3*, 1145–1156. [CrossRef]
16. Morris, A.P.; Estes, M.K. Microbes and microbial toxins: Paradigms for microbial-mucosal interactions VIII. Pathological consequences of rotavirus infection and its enterotoxin. *Am. J. Physiol. Gastrointest. Liver Physiol.* **2001**, *281*, G303–G310. [PubMed]
17. Greenberg, H.B.; Estes, M.K. Rotaviruses: From pathogenesis to vaccination. *Gastroenterology* **2009**, *136*, 1939–1951. [CrossRef]
18. Lundgren, O.; Peregrin, A.T.; Persson, K.; Kordasti, S.; Uhnoo, I.; Svensson, L. Role of the enteric nervous system in the fluid and electrolyte secretion of rotavirus diarrhea. *Science* **2000**, *287*, 491–495. [CrossRef]
19. Hellysaz, A.; Hagbom, M. Understanding the central nervous system symptoms of rotavirus: A qualitative review. *Viruses* **2021**, *13*, 658. [CrossRef] [PubMed]
20. Hagbom, M.; Istrate, C.; Engblom, D.; Karlsson, T.; Rodriguez-Diaz, J.; Buesa, J.; Taylor, J.A.; Loitto, V.M.; Magnusson, K.E.; Ahlman, H.; et al. Rotavirus stimulates release of serotonin (5-HT) from human enterochromaffin cells and activates brain structures involved in nausea and vomiting. *PLoS Pathog.* **2011**, *7*, e1002115. [CrossRef] [PubMed]
21. Hellysaz, A.; Svensson, L.; Hagbom, M. Rotavirus downregulates tyrosine hydroxylase in the noradrenergic sympathetic nervous system in ileum, early in infection and simultaneously with increased intestinal transit and altered brain activities. *mBio* **2022**, *13*, e0138722. [CrossRef]
22. Dadmehr, M.; Amini-Behbahani, F.; Eftekhar, B.; Minaei, B.; Bahrami, M. Peritoneum as an origin of epilepsy from the viewpoint of Avicenna. *Neurol. Sci.* **2018**, *39*, 1121–1124. [CrossRef]
23. Bahrami, M.; Shokri, S.; Mastery Farahani, R.; Dadmehr, M. A brief historical overview of the anatomy of fascia in medieval Persian medicine. *J. Med. Ethics Hist. Med.* **2020**, *13*, 7. [CrossRef] [PubMed]
24. Gorji, A.; Khaleghi Ghadiri, M. History of headache in medieval Persian medicine. *Lancet Neurol.* **2002**, *1*, 510–515. [CrossRef]
25. Shoja, M.M.; Tubbs, R.S. The history of anatomy in Persia. *J. Anat.* **2007**, *210*, 359–378. [CrossRef]
26. Dadmehr, M.; Seif, F.; Bahrami, M.; Amini-Behbahni, F.; Minaii Zangi, B.; Tavakol, C. A Historical overview of the neurological disorders associated with gastrointestinal ailments from the viewpoint of Avicenna. *Acta Med. Hist. Adriat.* **2024**, *21*, 307–319. [PubMed]
27. Mazengenya, P.; Bhikha, R. Revisiting Avicenna's (980–1037 AD) anatomy of the abdominal viscera from the Canon of Medicine. *Morphologie* **2018**, *102*, 225–230. [CrossRef] [PubMed]
28. Mayer, E.A. Gut feelings: The emerging biology of gut-brain communication. *Nat. Rev. Neurosci* **2011**, *12*, 453–466. [CrossRef] [PubMed]
29. Mayer, E.A.; Nance, K.; Chen, S. The Gut-Brain Axis. *Annu. Rev. Med.* **2022**, *73*, 439–453. [CrossRef]
30. Li, S.; Liu, M.; Cao, S.; Liu, B.; Li, D.; Wang, Z.; Sun, H.; Cui, Y.; Shi, Y. The mechanism of the gut-brain axis in regulating food intake. *Nutrients* **2023**, *15*, 3728. [CrossRef]
31. Bialowas, S.; Hagbom, M.; Nordgren, J.; Karlsson, T.; Sharma, S.; Magnusson, K.E.; Svensson, L. Rotavirus and serotonin cross-talk in Diarrhoea. *PLoS ONE* **2016**, *11*, e0159660. [CrossRef] [PubMed]
32. Crawford, S.E.; Ramani, S.; Tate, J.E.; Parashar, U.D.; Svensson, L.; Hagbom, M.; Franco, M.A.; Greenberg, H.B.; O'Ryan, M.; Kang, G.; et al. Rotavirus infection. *Nat. Rev. Dis. Primers* **2017**, *3*, 17083. [CrossRef] [PubMed]
33. Kordasti, S.; Sjovall, H.; Lundgren, O.; Svensson, L. Serotonin and vasoactive intestinal peptide antagonists attenuate rotavirus diarrhoea. *Gut* **2004**, *53*, 952–957. [CrossRef] [PubMed]
34. Berthoud, H.R.; Neuhuber, W.L. Functional and chemical anatomy of the afferent vagal system. *Auton. Neurosci.* **2000**, *85*, 1–17. [CrossRef] [PubMed]
35. Menetrey, D.; De Pommery, J. Origins of spinal ascending pathways that reach central areas involved in visceroception and visceronociception in the rat. *Eur. J. Neurosci.* **1991**, *3*, 249–259. [CrossRef] [PubMed]

36. Aburto, M.R.; Cryan, J.F. Gastrointestinal and brain barriers: Unlocking gates of communication across the microbiota-gut-brain axis. *Nat. Rev. Gastroenterol. Hepatol.* **2024**, *21*, 222–247. [CrossRef]
37. Ahmed, H.; Leyrolle, Q.; Koistinen, V.; Karkkainen, O.; Laye, S.; Delzenne, N.; Hanhineva, K. Microbiota-derived metabolites as drivers of gut-brain communication. *Gut Microbes.* **2022**, *14*, 2102878. [CrossRef] [PubMed]
38. Dantzer, R.; O'Connor, J.C.; Freund, G.G.; Johnson, R.W.; Kelley, K.W. From inflammation to sickness and depression: When the immune system subjugates the brain. *Nat. Rev. Neurosci.* **2008**, *9*, 46–56. [CrossRef]
39. Fowler, A.; Ygberg, S.; Svensson, E.; Bergman, K.; Cooray, G.; Wickstrom, R. Prospective evaluation of childhood encephalitis: Predisposing factors, prevention and outcome. *Pediatr. Infect. Dis. J.* **2020**, *39*, e417–e422. [CrossRef]
40. Nakagomi, T.; Nakagomi, O. Rotavirus antigenemia in children with encephalopathy accompanied by rotavirus gastroenteritis. *Arch. Virol.* **2005**, *150*, 1927–1931. [CrossRef]
41. Takanashi, J.; Miyamoto, T.; Ando, N.; Kubota, T.; Oka, M.; Kato, Z.; Hamano, S.; Hirabayashi, S.; Kikuchi, M.; Barkovich, A.J. Clinical and radiological features of rotavirus cerebellitis. *Am. J. Neuroradiol.* **2010**, *31*, 1591–1595. [CrossRef] [PubMed]
42. Lee, K.Y. Rotavirus infection-associated central nervous system complications: Clinicoradiological features and potential mechanisms. *Clin. Exp. Pediatr.* **2022**, *65*, 483–493. [CrossRef] [PubMed]
43. Hellysaz, A.; Nordgren, J.; Neijd, M.; Marti, M.; Svensson, L.; Hagbom, M. Microbiota do not restrict rotavirus infection of colon. *J. Virol.* **2023**, *97*, e0152623. [CrossRef] [PubMed]
44. Adams, W.R.; Kraft, L.M. Epizootic diarrhea of infant mice: Indentification of the etiologic agent. *Science* **1963**, *141*, 359–360. [CrossRef] [PubMed]
45. Kraft, L.M. Observations on the control and natural history of epidemic diarrhea of infant mice (EDIM). *Yale J. Biol. Med.* **1958**, *31*, 121–137. [PubMed]
46. Kraft, L.M. *The Problems of Laboratory Animal Disease*; Harris, R.J.C., Ed.; Academic Press: New York, NY, USA, 1962.
47. Kraft, L.M. *The Mouse in Biomedical Research*; Foster, H.L., Fox, J.G., Small, D.J., Eds.; Academic Press: New York, NY, USA, 1982; Volume 2, pp. 159–191.
48. Krapic, M.; Kavazovic, I.; Wensveen, F.M. Immunological mechanisms of sickness behavior in viral infection. *Viruses* **2021**, *13*, 2245. [CrossRef] [PubMed]
49. Browning, K.N.; Travagli, R.A. Central nervous system control of gastrointestinal motility and secretion and modulation of gastrointestinal functions. *Compr. Physiol.* **2014**, *4*, 1339–1368.
50. Gibbons, C.H. Basics of autonomic nervous system function. *Handb. Clin. Neurol.* **2019**, *160*, 407–418. [PubMed]
51. Cook, T.M.; Mansuy-Aubert, V. Communication between the gut microbiota and peripheral nervous system in health and chronic disease. *Gut Microbes* **2022**, *14*, 2068365. [CrossRef]
52. Flament-Durand, J. The hypothalamus: Anatomy and functions. *Acta Psychiatr. Belg.* **1980**, *80*, 364–375. [PubMed]
53. Peruzzotti-Jametti, L.; Donega, M.; Giusto, E.; Mallucci, G.; Marchetti, B.; Pluchino, S. The role of the immune system in central nervous system plasticity after acute injury. *Neuroscience* **2014**, *283*, 210–221. [CrossRef] [PubMed]
54. Sherman, S.M.; Guillery, R.W. Functional organization of thalamocortical relays. *J. Neurophysiol.* **1996**, *76*, 1367–1395. [CrossRef] [PubMed]
55. MacDougall, M.R.; Sharma, S. Physiology, Chemoreceptor Trigger Zone. In *Disclosure Sandeep Sharma Declares No Relevant Financial Relationships with Ineligible Companies*; StatPearls: Treasure Island, FL, USA, 2024.
56. Forsythe, P.; Sudo, N.; Dinan, T.; Taylor, V.H.; Bienenstock, J. Mood and gut feelings. *Brain Behav. Immun.* **2010**, *24*, 9–16. [CrossRef]
57. Prather, A.A. Sickness Behavior. In *Encyclopedia of Behavioral Medicine*; Gellman, M.D., Turner, J.R., Eds.; Springer: New York, NY, USA, 2013.
58. Hart, B.L. Biological basis of the behavior of sick animals. *Neurosci. Biobehav. Rev.* **1988**, *12*, 123–137. [CrossRef] [PubMed]
59. Kapikian, A.Z.; Wyatt, R.G.; Levine, M.M.; Black, R.E.; Greenberg, H.B.; Flores, J.; Kalica, A.R.; Hoshino, Y.; Chanock, R.M. Studies in volunteers with human rotaviruses. *Dev. Biol. Stand.* **1983**, *53*, 209–218. [PubMed]
60. Centers for Disease Control and Prevention. Foodborne outbreak of group a rotavirus gastroenteritis among college students—District of Columbia, March–April 2000. *MMWR Morb. Mortal Wkly. Rep.* **2000**, *49*, 1131–1133.
61. Williams, G.N.R. *Why We Get Sick: The New Science of Darwinian Medicine*; Vintage Books: New York, NY, USA, 1996.
62. Tsai, P.Y.; Zhang, B.; He, W.Q.; Zha, J.M.; Odenwald, M.A.; Singh, G.; Tamura, A.; Shen, L.; Sailer, A.; Yeruva, S.; et al. IL-22 upregulates epithelial Claudin-2 to drive diarrhea and enteric pathogen clearance. *Cell Host Microbe* **2017**, *21*, 671–681.e4. [CrossRef] [PubMed]
63. Siciliano, V.; Nista, E.C.; Rosa, T.; Brigida, M.; Franceschi, F. Clinical management of infectious Diarrhea. *Rev. Recent Clin. Trials* **2020**, *15*, 298–308. [CrossRef]
64. Uhnoo, I.; Olding-Stenkvist, E.; Kreuger, A. Clinical features of acute gastroenteritis associated with rotavirus, enteric adenoviruses, and bacteria. *Arch. Dis. Child.* **1986**, *61*, 732–738. [CrossRef]
65. Diarrhoea and Vomiting Caused by Gastroenteritis: Diagnosis, Assessment and Management in Children Younger than 5 Years. In *National Collaborating Centre for Women's and Children's Health (UK)*; RCOG Press at the Royal College of Obstetricians and Gynaecologists: London, UK, 2009.
66. Graham, D.Y.; Estes, M.K. Pathogenesis and treatment of rotavirus diarrhea. *Gastroenterology* **1991**, *101*, 1140–1141. [CrossRef]

67. Graham, D.Y.; Sackman, J.W.; Estes, M.K. Pathogenesis of rotavirus-induced diarrhea. Preliminary studies in miniature swine piglet. *Dig. Dis. Sci.* **1984**, *29*, 1028–1035. [CrossRef] [PubMed]
68. Morris, A.P.; Scott, J.K.; Ball, J.M.; Zeng, C.Q.; O'Neal, W.K.; Estes, M.K. NSP4 elicits age-dependent diarrhea and Ca(2+)mediated I(-) influx into intestinal crypts of CF mice. *Am. J. Physiol.* **1999**, *277*, G431–G444.
69. Lundgren, O.; Svensson, L. The enteric nervous system and infectious diarrhea. In *Viral Gastroenteritis*; Desselberger, U., Gray, J., Eds.; Elsevier: Amsterdam, The Netherlands, 2003; Volume 9, pp. 51–67.
70. Istrate, C.; Hagbom, M.; Vikstrom, E.; Magnusson, K.E.; Svensson, L. Rotavirus infection increases intestinal motility but not permeability at the onset of diarrhea. *J. Virol.* **2014**, *88*, 3161–3169. [CrossRef]
71. Furness, J.B.; Callaghan, B.P.; Rivera, L.R.; Cho, H.J. The enteric nervous system and gastrointestinal innervation: Integrated local and central control. *Adv. Exp. Med. Biol.* **2014**, *817*, 39–71. [PubMed]
72. Horn, C.C. Why is the neurobiology of nausea and vomiting so important? *Appetite* **2008**, *50*, 430–434. [CrossRef] [PubMed]
73. Nesse, R.M.; Williams, G.C. Evolution and the origins of disease. *Sci. Am.* **1998**, *279*, 86–93. [CrossRef] [PubMed]
74. Parashar, U.D.; Nelson, E.A.; Kang, G. Diagnosis, management, and prevention of rotavirus gastroenteritis in children. *BMJ* **2013**, *347*, f7204. [CrossRef] [PubMed]
75. Uhnoo, I.; Svensson, L. Clinical and epidemiological features of acute infantile gastroenteritis associated with human rotavirus subgroups 1 and 2. *J. Clin. Microbiol.* **1986**, *23*, 551–555. [CrossRef] [PubMed]
76. Kapikian, A.Z.; Shope, R.E. Rotaviruses, Reoviruses, Coltiviruses, and Orbiviruses. In *Medical Microbiology*, 4th ed.; Baron, S., Ed.; University of Texas Medical Branch at Galveston: Galveston, TX, USA, 1996.
77. Hagbom, M.; Novak, D.; Ekstrom, M.; Khalid, Y.; Andersson, M.; Lindh, M.; Nordgren, J.; Svensson, L. Ondansetron treatment reduces rotavirus symptoms-A randomized double-blinded placebo-controlled trial. *PLoS ONE* **2017**, *12*, e0186824. [CrossRef]
78. Bonvanie, I.J.; Weghorst, A.A.; Holtman, G.A.; Russchen, H.A.; Fickweiler, F.; Verkade, H.J.; Kollen, B.J.; Berger, M.Y. Oral ondansetron for paediatric gastroenteritis in primary care: A randomised controlled trial. *Br. J. Gen. Pract.* **2021**, *71*, e728–e735. [CrossRef]
79. Evans, S.S.; Repasky, E.A.; Fisher, D.T. Fever and the thermal regulation of immunity: The immune system feels the heat. *Nat. Rev. Immunol.* **2015**, *15*, 335–349. [CrossRef]
80. Kutukculer, N.; Caglayan, S. Tumor necrosis factor-alpha and interleukin-6 in stools of children with bacterial and viral gastroenteritis. *J. Pediatr. Gastroenterol. Nutr.* **1997**, *25*, 556–557.
81. Jiang, B.; Snipes-Magaldi, L.; Dennehy, P.; Keyserling, H.; Holman, R.C.; Bresee, J.; Gentsch, J.; Glass, R.I. Cytokines as mediators for or effectors against rotavirus disease in children. *Clin. Diagn. Lab. Immunol.* **2003**, *10*, 995–1001. [CrossRef] [PubMed]
82. Biddle, C. The neurobiology of the human febrile response. *AANA J.* **2006**, *74*, 145–150. [PubMed]
83. *The Pituitary*, 3rd ed.; Academic Press: London, UK, 2011.
84. Morrison, S.F. Central control of body temperature. *F1000Research* **2016**, *5*, 880. [CrossRef] [PubMed]
85. *Clinical Manual of Fever in Children*; Springer: Berlin/Heidelberg, Germany, 2009.
86. Harden, L.M.; Kent, S.; Pittman, Q.J.; Roth, J. Fever and sickness behavior: Friend or foe? *Brain Behav. Immun.* **2015**, *50*, 322–333. [CrossRef]
87. Szelenyi, Z.; Szekely, M. Sickness behavior in fever and hypothermia. *Front. Biosci.* **2004**, *9*, 2447–2456. [CrossRef] [PubMed]
88. Silverman, M.N.; Pearce, B.D.; Biron, C.A.; Miller, A.H. Immune modulation of the hypothalamic-pituitary-adrenal (HPA) axis during viral infection. *Viral Immunol* **2005**, *18*, 41–78. [CrossRef] [PubMed]
89. Shattuck, E.C.; Muehlenbein, M.P. Towards an integrative picture of human sickness behavior. *Brain Behav. Immun.* **2016**, *57*, 255–262. [CrossRef]
90. Hossain, J.L.; Ahmad, P.; Reinish, L.W.; Kayumov, L.; Hossain, N.K.; Shapiro, C.M. Subjective fatigue and subjective sleepiness: Two independent consequences of sleep disorders? *J. Sleep Res.* **2005**, *14*, 245–253. [CrossRef] [PubMed]
91. Imeri, L.; Opp, M.R. How (and why) the immune system makes us sleep. *Nat. Rev. Neurosci.* **2009**, *10*, 199–210. [CrossRef]
92. Rezai, M.; Fullwood, C.; Hird, B.; Chawla, M.; Tetlow, L.; Banerjee, I.; Patel, L. Cortisol levels during acute illnesses in children and adolescents: A systematic review. *JAMA Netw. Open* **2022**, *5*, e2217812. [CrossRef] [PubMed]
93. Dunn, A.J.; Powell, M.L.; Meitin, C.; Small, P.A., Jr. Virus infection as a stressor: Influenza virus elevates plasma concentrations of corticosterone, and brain concentrations of MHPG and tryptophan. *Physiol. Behav.* **1989**, *45*, 591–594. [CrossRef] [PubMed]
94. Aviello, G.; Cristiano, C.; Luckman, S.M.; D'Agostino, G. Brain control of appetite during sickness. *Br. J. Pharmacol.* **2021**, *178*, 2096–2110. [CrossRef] [PubMed]
95. Camilleri, M. Peripheral mechanisms in appetite regulation. *Gastroenterology* **2015**, *148*, 1219–1233. [CrossRef] [PubMed]
96. Wang, A.; Huen, S.C.; Luan, H.H.; Yu, S.; Zhang, C.; Gallezot, J.D.; Booth, C.J.; Medzhitov, R. Opposing effects of fasting metabolism on tissue tolerance in bacterial and viral inflammation. *Cell* **2016**, *166*, 1512–1525.e12. [CrossRef] [PubMed]
97. Verkerke, H.; Sobuz, S.; Ma, J.Z.; Petri, S.E.; Reichman, D.; Qadri, F.; Rahman, M.; Haque, R.; Petri, W.A., Jr. Malnutrition is associated with protection from rotavirus diarrhea: Evidence from a longitudinal birth cohort study in Bangladesh. *J. Clin. Microbiol.* **2016**, *54*, 2568–2574. [CrossRef]
98. Burnett, E.; Parashar, U.D.; Tate, J.E. Rotavirus Infection, Illness, and vaccine performance in malnourished children: A review of the literature. *Pediatr. Infect Dis. J.* **2021**, *40*, 930–936. [CrossRef] [PubMed]
99. Smith, A.; Mannion, M.; O'Reilly, P.; Crushell, E.; Hughes, J.; Knerr, I.; Gavin, P.; Monavari, A. Rotavirus gastroenteritis is associated with increased morbidity and mortality in children with inherited metabolic disorders. *Ir. Med. J.* **2017**, *110*, 546.

100. Broecker, F.; Moelling, K. What viruses tell us about evolution and immunity: Beyond Darwin? *Ann. N. Y. Acad. Sci.* **2019**, *1447*, 53–68. [CrossRef]
101. Hu, D.L.; Nakane, A. Mechanisms of staphylococcal enterotoxin-induced emesis. *Eur. J. Pharmacol.* **2014**, *722*, 95–107. [CrossRef]
102. Stuempfig, N.D.; Seroy, J. Viral Gastroenteritis. In *Disclosure: Justin Seroy Declares No Relevant Financial Relationships with Ineligible Companies*; StatPearls: Treasure Island, FL, USA, 2024.

Article

Isolation and Pathogenicity Analysis of a G5P[23] Porcine Rotavirus Strain

Liguo Gao [1], Hanqin Shen [2,3], Sucan Zhao [1], Sheng Chen [1], Puduo Zhu [1], Wencheng Lin [1] and Feng Chen [1,*]

[1] College of Animal Science, South China Agricultural University, Guangzhou 510642, China; gaoliguo@stu.scau.edu.cn (L.G.); zhaosc@stu.scau.edu.cn (S.Z.); chens@stu.scau.edu.cn (S.C.); zpd@stu.scau.edu.cn (P.Z.); wenchenglin@scau.edu.cn (W.L.)
[2] Wen's Food Group, Yunfu 527300, China; hqshen@wens.com.cn
[3] Guangdong Jingjie Inspection and Testing Co., Ltd., Yunfu 527300, China
* Correspondence: fengch@scau.edu.cn

Abstract: (1) Background: Group A rotaviruses (RVAs) are the primary cause of severe intestinal diseases in piglets. Porcine rotaviruses (PoRVs) are widely prevalent in Chinese farms, resulting in significant economic losses to the livestock industry. However, isolation of PoRVs is challenging, and their pathogenicity in piglets is not well understood. (2) Methods: We conducted clinical testing on a farm in Jiangsu Province, China, and isolated PoRV by continuously passaging on MA104 cells. Subsequently, the pathogenicity of the isolated strain in piglets was investigated. The piglets of the PoRV-infection group were orally inoculated with 1 mL of 1.0×10^6 TCID50 PoRV, whereas those of the mock-infection group were fed with an equivalent amount of DMEM. (3) Results: A G5P[23] genotype PoRV strain was successfully isolated from one of the positive samples and named RVA/Pig/China/JS/2023/G5P[23](JS). The genomic constellation of this strain was G5-P[23]-I5-R1-C1-M1-A8-N1-T1-E1-H1. Sequence analysis revealed that the genes *VP3, VP7, NSP2*, and *NSP4* of the JS strain were closely related to human RVAs, whereas the remaining gene segments were closely related to porcine RVAs, indicating a reassortment between porcine and human strains. Furthermore, infection of 15-day-old piglets with the JS strain resulted in a diarrheal rate of 100% (8 of 8) and a mortality rate of 37.5% (3 of 8). (4) Conclusions: The isolated G5P[23] genotype rotavirus strain, which exhibited strong pathogenicity in piglets, may have resulted from recombination between porcine and human strains. It may serve as a potential candidate strain for developing vaccines, and its immunogenicity can be tested in future studies.

Keywords: porcine rotavirus; G5; virus isolation; pathogenicity

Citation: Gao, L.; Shen, H.; Zhao, S.; Chen, S.; Zhu, P.; Lin, W.; Chen, F. Isolation and Pathogenicity Analysis of a G5P[23] Porcine Rotavirus Strain. *Viruses* **2024**, *16*, 21. https://doi.org/10.3390/v16010021

Academic Editors: Ulrich Desselberger and John T. Patton

Received: 2 November 2023
Revised: 12 December 2023
Accepted: 17 December 2023
Published: 22 December 2023

1. Introduction

Rotaviruses (RVs) are the most common cause of viral gastroenteritis in infants and animals. RVs are primarily transmitted through the fecal–oral route and infected individuals typically present with symptoms such as diarrhea, vomiting, and decreased appetite. The infection can spread between humans and animals, posing a serious threat to both human health and animal husbandry [1–4]. In 1969, bovine RV was first isolated and cultured in cells and was confirmed to be the primary cause of diarrhea in cattle [5]. The first human RV was discovered in 1973 [6]. RVs were subsequently identified in pigs and poultries [7,8]. Cultivating RVs in cell cultures was quite challenging until 1984. Subsequently, RVs were successfully cultured in African Green Monkey kidney cells (MA104) after treatment with trypsin, which accelerated the processes of RV isolation and cultivation [9]. Furthermore, subsequent studies have reported that RVs could replicate in cells such as Vero cells [10,11].

Porcine rotavirus infection mainly causes diarrhea in suckling piglets, manifesting as watery diarrhea, weight loss, and dehydration [12]. Porcine rotavirus infection occurs in piglets aged 1 to 4 weeks, and PoRV is co-infected with other viral pathogens such as Porcine Epidemic Diarrhea Virus (PEDV) and Transmissible Gastroenteritis Virus (TGEV), as well as bacterial pathogens such as *Escherichia coli*, causing significant losses to the pig farming industry [13,14]. Porcine rotaviruses (PoRVs) infection can damage the small intestines in pigs, leading to villous atrophy and reduced digestive absorption capacity [15]. The most important mechanism for RVA to cause secretory diarrhea is that NSP4 serves as an enterotoxin that increases chloride secretion through a calcium-dependent mechanism, and it also activates the enteric nervous system and blocks the intestinal sodium/glucose cotransporter [16,17].

The genus rotavirus belongs to the family Reoviridae and order Reovirales. Their genome consists of 11 double-stranded RNA segments, encoding 11 proteins, including 6 structural proteins (virion protein [*VP*]*1–VP4*, *VP6*, and *VP7*) and 5 nonstructural proteins (*NSP1–NSP5/6*) [15]. RVs are classified into 10 subgroups (A–J) based on the *VP6* antigen [18]. The outer shell proteins *VP7* and *VP4* induce neutralizing antibodies and form the basis for the G and P dual classification system [19]. The most common groups infecting Porcine are Group A RV (RVA), Group B RV (RVB), and Group C RV (RVC), with RVA having the highest prevalence and causing the most significant harm [15]. For RVA strains with high genetic diversity, the dual (G/P) typing system was expanded in 2008 to a complete genome sequence-based classification system. A nomenclature for the comparison of complete rotavirus genomes was considered in which the notations Gx-P[x]-Ix-Rx-Cx-Mx-Ax-Nx-Tx-Ex-Hx are used for the *VP7-VP4-VP6-VP1-VP2-VP3-NSP1-NSP2-NSP3-NSP4-NSP5/6* encoding genes, respectively. Subsequently, the Rotavirus Classification Working Group (RCWG) was established to develop classification guidelines and maintain the proposed classification system of RVs [20]. To date, 42 "G" and 58 "P" genotypes have been reported in the RCWG database (https://rega. kuleuven.be/cev/viralmetagenomics/virus-classification/rcwg, accessed on 1 October 2023). RVA strains with G3, G5, G9, and G11 genotypes in combination with P[9], P[6], P[13], P[23], P[27], and P[28] genotypes are considered the most common worldwide [21].

In this study, a G5P[23] genotype PoRV strain was isolated from the diarrheal samples of piglets on a farm in Jiangsu Province, China, and named RVA/Pig/China/JS/2023/G5P[2](JS). The whole genome of the JS strain was sequenced, and its genetic evolution was determined. The JS strain was also used to infect 15-day-old piglets to investigate its pathogenicity.

2. Materials and Methods

2.1. Clinical Samples, Cells, and Antibodies

On 6 January 2023, 28 diarrheal samples were collected from piglets in a pig farm infected by PoRV in Jiangsu Province, China. The virus isolated from these samples was identified as PoRV through Reverse Transcription-Polymerase Chain Reaction (RT–PCR) and Quantitative real-time polymerase chain reaction (qRT–PCR) using *VP6* (Supplementary Table S1) [22–24]. Samples of the intestinal contents were homogenized in serum-free Dulbecco's modified Eagle medium [DMEM] (Invitrogen, Carlsbad, CA, USA) containing 1% penicillin–streptomycin [10,000 units/mL penicillin and 10,000 µg/mL streptomycin] (Gibco™, New York, MT, USA) and 0.3% trypsin phosphate broth (Sigma-Aldrich, St. Louis, MO, USA). The samples were then centrifuged for 30 min at 3500 rpm and 4 °C. The supernatant was filtered through a 0.22 µm pore filter (Merck Millipore, Darmstadt, Germany) to remove the bacteria and was stored at −80 °C until use as an inoculum for virus isolation. MA104 cells were cultured in DMEM supplemented with 10% heat-inactivated fetal bovine serum [FBS] (Invitrogen, Carlsbad, CA, USA) and 1% antibiotics [10,000 units/mL penicillin, 10,000 µg/mL streptomycin, and 25 µg/mL Fungizone®] (Gibco™, New York, MT, USA). These cells were maintained at 37 °C in a humidified incubator under 5% CO_2. Mouse monoclonal antibodies (McAbs) against PoRV *VP6* protein (Zoonogen®, Beijing, China) were purchased and stored in the laboratory.

2.2. Virus Isolation Assay

PoRV isolation was performed using MA104 cells via a previously described method with modifications [25]. Specifically, prior to inoculation, cells were rinsed thrice with sterile phosphate-buffered saline (PBS; pH 7.2) to remove FBS completely. Simultaneously, the stored inoculum was briefly vortexed and used for cultivation. In total, 300 μL of the inoculum was added to 6-well plates. After incubation at 37 °C for 1 h, the inoculum was removed, and the maintenance medium containing trypsin at a final concentration of 4 μg/mL was added. The inoculated cells were maintained at 37 °C in a humidified incubator under 5% CO_2, and the CPEs were monitored daily. When CPEs were observed in 90% of the cells, the flask was subjected to three rounds of freezing and thawing. The cells and the supernatant were mixed with a pipette, aliquoted, and stored at −80 °C. The harvested cell culture was used as the seed stock for the next generation. For successive passages, the scale of the culture was gradually increased until the PoRV strains were properly propagated and continuously passaged using T-25 flasks. Subsequently, the virus was purified using the limited-dilution method.

2.3. Titration and Growth Curve Determination

The growth curve of the 10th generation of the virus was constructed. Initially, the virus was inoculated onto a 96-well plate at a multiplicity of infection (MOI) of 0.01. Cell culture supernatants were collected at 6, 12, 18, 24, 30, 36, 42, and 48 hpi. Subsequently, based on the Reed–Muench method [26], the viral titers of these samples were determined using the TCID50 assay. After washing thrice with PBS, 100 μL of the experimental supernatant was mixed with 900 μL of DMEM to obtain serial dilutions ranging from 10^{-1} to 10^{-10}. Each dilution was added to a monolayer of MA104 cells in 10 vertical wells of a 96-well plate. The plate was then placed in a cell culture incubator at 37 °C for 5 days, and the viral titers were determined by observing the CPEs under a microscope. Finally, the growth curves for each virus were constructed based on the viral titers at different time points post-infection.

2.4. Immunofluorescence Assay (IFA)

MA104 cells cultured in 6-well plates were either mock-infected or infected with PoRV at a MOI of 0.01. At 0, 12, and 24 hpi, the cells were fixed with 4% paraformaldehyde at 4 °C for 30 min, followed by permeabilization with 0.25% Triton X-100 (Solarbio, Beijing, China) for 10 min at room temperature (RT). Blocking was performed using 5% bovine serum albumin (Solarbio, Beijing, China) in PBS at RT for 1 h. McAbs against the PoRV *VP6* protein and Alexa Fluor® 488-conjugated goat anti-mouse IgG (Abcam, Cambridge, UK) were used as primary and secondary antibodies, respectively. Nuclei were stained with 4′,6-diamidino-2-phenylindole (DAPI) (VectorLabs, Newark, CA, USA) for 20 min at RT. After washing with PBS, the stained cells were observed under a fluorescence microscope (Olympus, Tokyo, Japan).

2.5. Transmission Electron Microscopy (TEM)

To visualize PoRV particles, MA104 cells were harvested when CPEs were observed in 90% of the cells. The cell culture was subjected to three freeze–thaw cycles, followed by centrifugation for 30 min at $4000 \times g$ and 4 °C. The supernatant was filtered through a 0.22 mm filter to remove cell debris and was then mixed overnight with polyethylene glycol 8000 [PEG-8000] (Solarbio, BeiJing, China) at a final concentration of 10%. The mixture was subsequently ultracentrifuged for 2 h at $118,000 \times g$ and 4 °C to obtain the PoRV particles. These particles were then resuspended in Tris-buffered saline and negatively stained with 2% phosphotungstic acid. The viruses within the infected MA104 cells were examined using a transmission electron microscope (TEM) (Leica, Wetzlarm, Germany), and further fixation and imaging were performed based on the methods described in previous studies.

2.6. Sequence Analysis

The complete genome of the RVA strain JS was sequenced using the MiSeq high-throughput sequencing platform (Illumina, San Diego, CA, USA). The total RNA from the cell culture supernatant was extracted using TRIzol reagent (Invitrogen, Carlsbad, CA, USA) following the manufacturer's instructions. rRNA was removed using RiBo-Zero Magnetic

Gold kit (Epicenter Biotechnologies, Madison, WI, USA), and the remaining RNA was sequenced using NEBNext® Ultra™ Directional RNA Library Prep Kit (New England Biolabs, Ipswich, MA, USA) and the Illumina MiSeq platform (GENEWIZ, Guangzhou, China).

The obtained raw reads were trimmed using Trimmomatic v0.39 software [16] and then aligned with the Sscrofa 11.1 reference genome using Bowtie2 v2.4.1 software [17] to eliminate the reads corresponding to the sequences of the pig genome. The remaining reads were reassembled into contigs using MEGAHIT v1.2.9 software [18]. Viral sequences were identified using BLASTn and BLASTx searches against a custom-built virus nucleotide reference database from GenBank or UniProt virus classification database.

The genotypes of the isolated strains were determined using the automatic genotyping tool provided by ViPR (https://www.viprbrc.org/, accessed on 6 October 2023), with a nucleotide truncation threshold of 80% [27]. To further explore the genetic origin of JS, phylogenetic trees were constructed using the neighbor-joining method using MEGA11 software with the default settings (1000 bootstrap replicates) based on the PoRV *VP4* and *VP7* gene sequences sequenced in this study and reference sequences collected in GenBank.

2.7. Recombination Analysis

The sequences involved in recombination were analyzed using RDP4 software version 4 [28]. DP4: Detection and analysis of recombination patterns in virus genomes. Virus Evol. 1, vev003 [28]. The RDP, GENECONV, Chimaera, MaxChi, Bootscan, SiScan, and 3Seq methods were employed with their default parameters.

2.8. Animal Experiments

Sixteen conventional piglets, 15 days old, were purchased from a commercial pig farm with no history of PoRV outbreaks or vaccination. Additionally, various virus tests, including qRT-PCR and enzyme-linked immunosorbent assay (ELISA), yielded negative results for these piglets. The tested viruses included PoRV, PEDV, TGEV, Porcine Reproductive and Respiratory Syndrome Virus (PRRSV), Porcine Circovirus (PCV), Classical Swine Fever Virus (CSFV), and Pseudorabies Virus (PRV). All animal experiments were approved by the Ethics Committee of the South China Agricultural University, Guangzhou, China (approval ID: SYXK-2019-0136). The 18 piglets were randomly divided into 2 groups, with 8 piglets in each group: PoRV-infection and mock-infection groups. The piglets of the PoRV-infection group were orally inoculated with 1 mL of 1.0×10^6 TCID50 PoRV, whereas those of the mock-infection group were fed with an equivalent amount of DMEM. The animals were manually fed with a milk replacer every 4 h during the experiment. In piglets showing signs of anorexia, 60–100 mL of milk replacer was administered by gavage every 4 h. Clinical symptoms such as diarrhea, vomiting, anorexia, and depression were recorded daily. Fecal consistency was assessed daily using a scoring system based on solid (0 points), paste-like (1 point), semiliquid (2 points, mild diarrhea), and liquid (3 points, severe diarrhea) standards. Three piglets from the virus-infected group and three piglets from the mock group were euthanized at 9 d post-infection (dpi).

Histopathological and immunohistochemical examinations were performed on the small intestines. Intestinal tissue samples from a randomly selected piglet were examined and collected 9 dpi. After fixation in 4% formaldehyde at RT for 48 h, the processed tissue samples were embedded in paraffin, sectioned using a microtome (Leica, Germany), deparaffinized with xylene, and washed with reducing concentrations of ethanol. Subsequently, conventional Hematoxylin and Eosin (BaSo, Zhuhai, China) [H&E] staining was performed for HE, and PoRV-VP6-specific monoclonal antibody was used IHC [29].

Rectal swabs were collected daily for the qRT–PCR detection. The animals were euthanized 9 dpi, and the tissues of the lungs, duodenum, jejunum, ileum, etc., were collected for HE, IHC, and qRT–PCR examination.

2.9. Statistical Analysis

All statistical analyses were performed using the GraphPad Prism software version 8.0 (graphpad.com). The statistical significance of the differences between the experimental groups was ascertained using the Student's *t*-test. The differences were considered statistically significant at $p < 0.05$.

3. Results

3.1. Virus Isolation

The collected piglet diarrhea samples were tested using RT-PCR and qRT-PCR, and all 28 diarrhea samples showed positive results for porcine rotavirus. Inoculate these 28 positive samples into MA104 cells for three blind passages to isolate the virus. Only one strain was successfully isolated from positive samples, while the remaining samples could not be continuously passaged on MA104 cells.

The PoRV-positive samples at a 10-fold dilution were inoculated into MA104 cells. Visible cytopathic effects (CPEs) were observed in the second generation, 24 h post-infection [hpi]. Complete CPE was observed at 48 hpi (Figure 1a). Compared with mock-inoculated cells, the inoculated cells were initially characterized by syncytium and vacuole formation, followed by elongation, detachment, and cessation of growth (Figure 1a). The isolate was named RVA/Pig/China/JS/2023/G5P[23](JS).

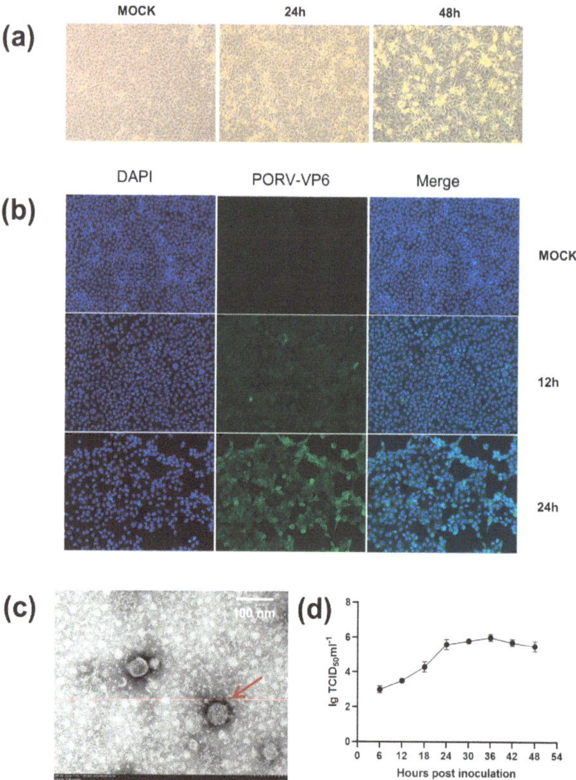

Figure 1. (**a**) Observation of CPEs in MA104 cells infected with the PoRV JS strain at 24 and 48 h post-infection (hpi). (**b**) Immunofluorescence staining (green) of PoRV *VP6* in MA104 cells infected with the JS strain at 12 and 24 hpi. (**c**) Electron micrograph of the PoRV particles detected in the culture medium of MA104 cells infected with the JS strain, as indicated by arrows. Scale bar = 100 nm. (**d**) Growth curve of the JS strain at the 12th passage of MA104 cells. Multiplicity of infection (MOI) = 0.01.

Furthermore, the isolated strain was identified via IFA using PoRV-VP6-specific monoclonal antibodies. From 12 hpi, pronounced green signals were observed in the infected MA104 cells but not in the uninfected cells. These signals tended to intensify markedly with time (0, 12, and 24 hpi) (Figure 1b).

TEM was also conducted to identify PoRV particles obtained from the MA104 cell culture medium and MA104 cells infected with the JS strain. The typical RV particles were observed in the cell culture medium (Figure 1c). The viral particles appeared wheel-shaped, with a diameter of 70–100 nm, exhibiting characteristic surface projections unique to RVs. TEM confirmed the successful reproduction of PoRV in MA104 cells. The growth curves of the JS strain at different time points post-infection were plotted based on the 50% tissue culture infective dose (TCID50). The results indicated that the viral replication titer peaked at ~1×10^6/mL TCID50, and the titer enhanced rapidly from 6 to 36 hpi (Figure 1d).

3.2. Sequence Analysis

According to the latest classification and naming system established by the RCWG, the nucleotide identity cutoff values for 11 gene segments of Rotavirus Group A are as follows: 80% (G), 80% (P), 85% (I), 83% (R), 84% (C), 81% (M), 79% (A), 85% (N), 85% (T), 85% (E), and 91% (H) [30]. The genotype of JS was identified as G5-[P23]-I5-R1-C1-M1-A8-N1-T1-E1-H1 (Supplementary Table S2). The sequences of all 11 gene segments were uploaded to GenBank with accession numbers OR644644 to OR644654.

Homology and phylogenetic analyses were conducted between the closely related reference RVA strains of the same genotype. Several gene segments were highly homologous with RV strains of porcine origin, including *VP1*, *VP2*, *VP4*, *VP6*, *NSP1*, *NSP3*, and *NSP5*. Additionally, some segments such as *VP3*, *VP7*, *NSP2*, and *NSP4* exhibited an even higher homology with RV strains of human origin, suggesting that the JS-G5P[23] strain was likely a recombinant human–porcine virus (Supplementary Table S2).

Sequence analysis of *VP7* revealed that it comprised 981 nucleotides (nts), encoding 326 amino acids (aa). A phylogenetic tree was constructed using the full-length *VP7* sequence (981 nts) and selected G genotype sequences obtained from GenBank. The *VP7* of the JS strain is grouped within the G5 genotype (glycolated), primarily consisting of strains of porcine origin. However, the segment from the JS strain formed a smaller clade within this genotype, containing sequences derived from both pig and human sources. The most closely related sequence to the JS strain fragment was identified to be derived from a human RVA strain, RVA/Pig-wt/THA/CMP-001-12/2012/G5P[13] (KT727252.1), which was isolated from a patient in Thailand in 2014 (Figure 2a) [31]. The nucleotide identity rates between the selected G5 genotypes in GenBank range from 86.6% to 91.4%, and the amino acid identity rates range from 92.0% to 95.6%. Furthermore, using ViPR automatic genotyping tool, we revealed that RVA/Pig/China/JS/2023/G5P[23] was closely related to multiple RVA G5 strains [27]; this sequence belongs to the G5 genotype (glycolated), with the most similar query sequence being RVA/Human-tc/BRA/IAL28/1992/G5P[8] (EF672588) (Supplementary Table S2) [32].

Sequence analysis of *VP4* revealed that it comprised 2331 nts, encoding 776 aa. A phylogenetic tree was constructed using the full-length *VP4* sequence from the JS strain and selected P genotype sequences obtained from GenBank. JS was most closely related to a porcine-origin RVA strain, RVA/Pig/CHN/SX-2021/P23 (OP650547.1), isolated from China (Figure 2b). The nucleotide identity rates between the selected [P13] genotypes in GenBank range from 80.0% to 89.6%, and the amino acid identity rates range from 72.0% to 83.1%. Additionally, the ViPR automatic genotyping tool revealed that RVA/Pig/China/JS/2023/G5P[23] was closely related to RVA P[23] strains [21]; this sequence belongs to the P[23] genotype, with the highest similarity to the PoRV strain P23-RVA/Pig-tc/VEN/A34/1985/G5P[23] (AY174094) (Supplementary Table S2) [27,33].

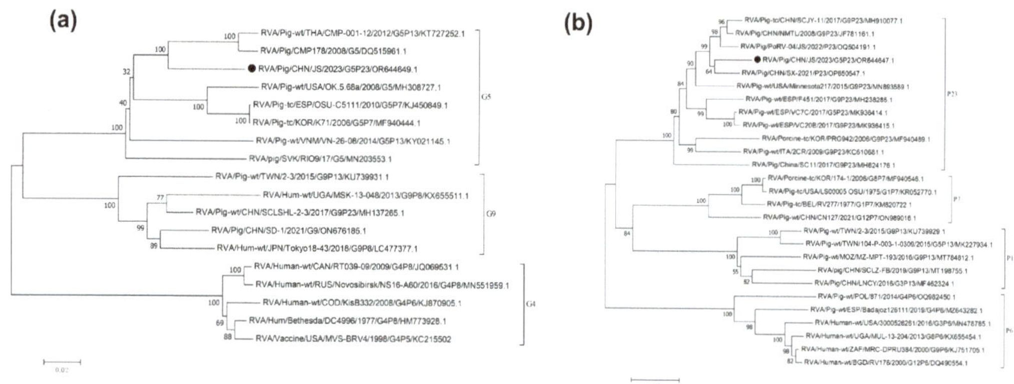

Figure 2. (**a**) Phylogenetic tree constructed using the segments of *VP7*; G = glycolated. (**b**) Phylogenetic tree constructed using the segments of *VP4*; P = protease-sensitive. ● Display JS strain sequence.

3.3. Recombination Analysis

To analyze the association between *NSP2* from JS and the existing isolates further, a genetic analysis was conducted between them using RDP4 software [28]. The sequence of *NSP2* from JS is indicated in Figure 3. Crossover points for a potential recombination zone were located at nucleotides 1–494 and 939–994 of *NSP2* (Figure 3), but no recombination was detected in other genes encoding viral proteins. The major parental strain for the recombination was RVA/Human-tc/KOR/CAU14-1-262/2014/G3P[9] (KR262156.1), and the minor strain was RVA/Dog-wt/GER/88977/2013/G8P1/(KJ940158.1), indicating that C1 of JS was the gene recombination product of C3 and C2 RVAs [34,35].

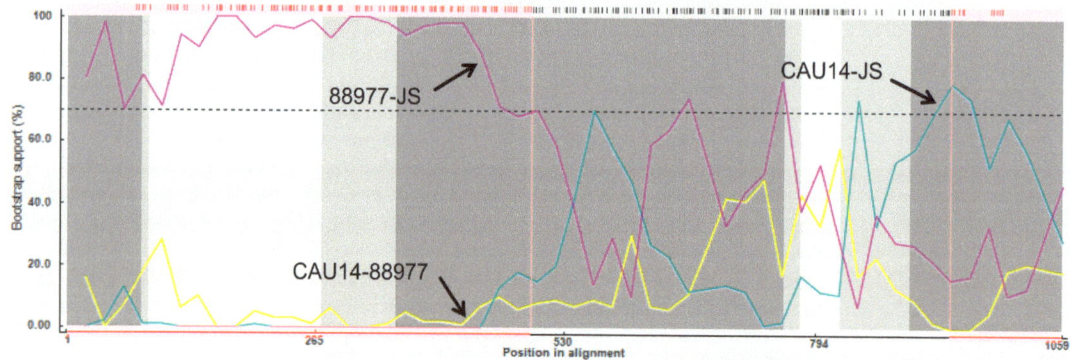

Figure 3. Analysis of the recombination of *NSP2* from RVA/Pig/China/JS/2023/G5P[23] and other PoRV strains such as RVA/Human-tc/KOR/CAU14-1-262/2014/G3P9/(KR262156.1) and RVA/Dog-wt/GER/88977/2013/G8P1/(KJ940158.1). JS represents RVA/Pig/China/JS/2022/G3P[7] (OR232953), CAU14 represents RVA/Human-tc/KOR/CAU14-1-262/2014/G3P[9] (KR262156.1), and 88977. represents RVA/Dog-wt/GER/88977/2013/G8P1/(KJ940158.1). Recombination crossover points are located at nucleotides 1–494 and 939–994.

3.4. Clinical Signs and Histological Changes

The piglets of the JS strain infection group showed diarrhea symptoms, manifested as dark green, watery feces (Figure 4a). One piglet succumbed on day 3 post-infection and two piglets on day 6. On autopsy, the lungs showed congestion and swelling, the intestinal wall became thin and transparent, and the intestines bulged and were filled with yellow water-like liquid (Figure 4a). On the contrary, in the simulated infection group, except for a few piglets whose feces briefly appeared pasty, none experienced diarrhea, and none succumbed at the end of the experimental cycle. No pathological changes were found in the lungs and intestinal tissues of these piglets during autopsy (Figure 4a).

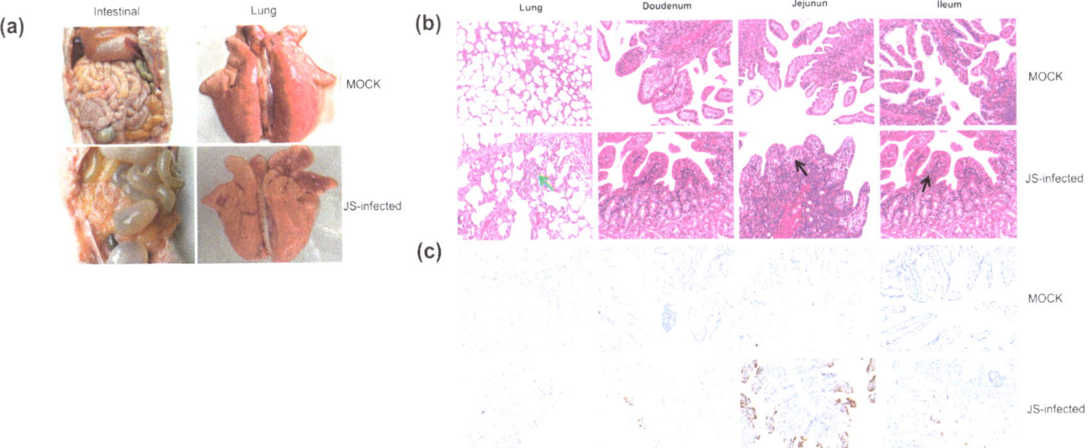

Figure 4. (**a**) Intestinal and lung sections of the piglets in the mock-infected and JS-infected groups. (**b**) Histopathological sections of the piglets in both groups. (Green indicates thickening of alveolar walls, while black indicates an increase in inflammatory cells) (**c**) Immunohistochemical staining of the tissues collected from the piglets in both groups.

The HE results indicate that the lung exhibited thickening of the alveolar walls, suggesting interstitial pneumonia (as shown by the green arrow in Figure 4b). The villi lamina propia shows an increased number of inflammatory cells, predominantly lymphocytes (as shown by the black arrow in Figure 4b). The intestinal tissue pathology of piglets in the control group was normal (Figure 4b). In addition, IHC examination revealed that in specific segments of the small intestine with villous atrophy, cytoplasmic staining of the PoRV antigen was dominant, with the highest antigen content in the jejunum (Figure 4c). No PoRV antigen was detected in the small intestines of piglets in the negative control group (Figure 4c).

3.5. Viral Load in Stool Samples and Tissues

One piglet developed diarrhea on day 2 post-oral inoculation with the JS strain (Figure 5a). By day 3, all eight piglets exhibited diarrhea; one of them succumbed to the infection on day 3 and two on day 6 (Figure 5b). From day 7 onward, most piglets began to eat, which continued until day 9, when the symptoms of diarrhea began to subside. The piglets infected with JS had a diarrheal rate of 100% (8 of 8; Figure 5a) and a mortality rate of 37.5% (3 of 8; Figure 5b). In contrast, the control group, apart from a brief period of loose stools in a few piglets, did not show any significant signs of diarrhea, and no deaths occurred (Figure 5a,b).

qRT–PCR results showed that the viral RNA was detectable in the feces of infected piglets as early as 24 h post-infection (hpi). Viral shedding in rectal swabs of piglets continued to increase from day 1 to day 5 post-infection, reaching its peak on day 5 (Figure 5c). From day 7, viral shedding began to decline and decreased until the end of the experiment. Additionally, the viral load in the lungs was significantly lower than in various parts of the small intestine ($p < 0.05$), with the highest viral load detected in the ileum (Figure 5d). No viral shedding was detected in the feces of the control group, and no viral load was identified in the intestinal tissues (Figure 5c,d).

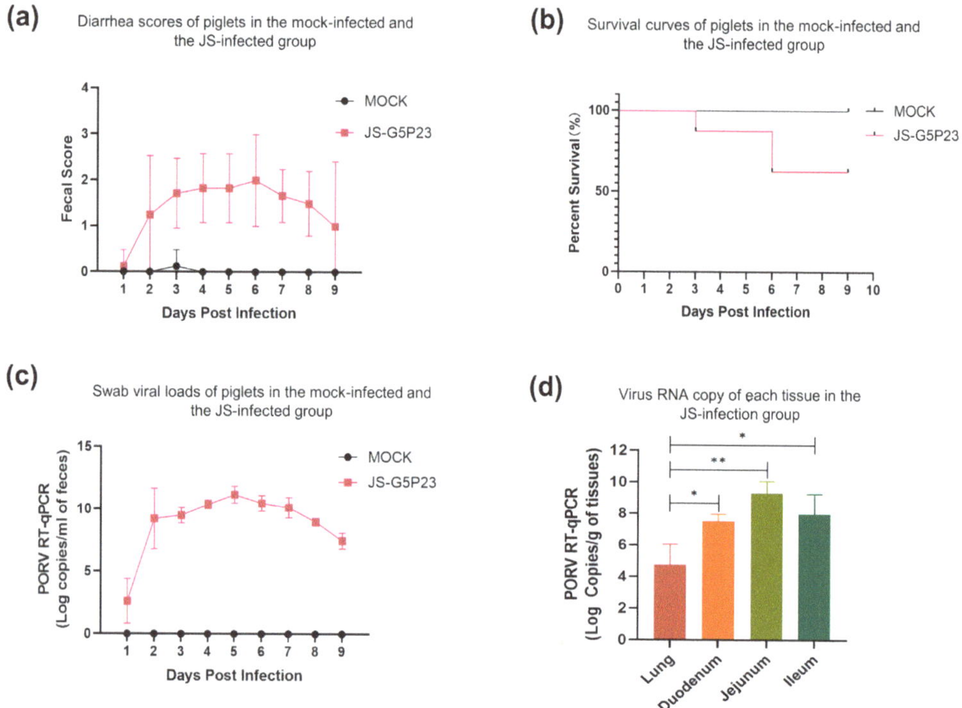

Figure 5. (**a**) Diarrhea scores of the piglets in the mock-infected and JS-infected groups. (**b**) Survival curves of the piglets in both groups. (**c**) Viral loads in the swabs of the piglets in both groups. (**d**) The viral RNA copy number per mg of each tissue collected from the piglets in both groups. Asterisk (*) indicates significant differences between different tissues. (* $p < 0.05$; ** $p < 0.01$).

4. Discussion

Approximately 50 years ago, RVs were considered the primary cause of diarrhea in infants and young animals [6]. According to the World Health Organization, RVs cause ~450,000 deaths annually, with 90% of them occurring in developing countries in Asia and Africa [4,36,37]. PoRVs are prevalent worldwide and result in significant economic losses to the swine industry [15]. Isolation and cell-based cultivation of PoRV can help understand their pathogenicity in piglets [38]. In this study, PoRV-positive samples from a pig farm, suffering from diarrhea in suckling piglets, were treated with 10 μg/mL trypsin for 1 h, and the optimal conditions for the isolation of PoRV were established by adding trypsin to a final concentration of 4 μg/mL in the maintenance medium. A PoRV strain was successfully isolated under these conditions and named JS. After three-blind-passaging in MA104 cells, stable CPEs, including cell elongation and detachment, were observed (Figure 1a), similar to previous studies [12,25]. IFA confirmed the replication of the 10th generation of the JS strain in MA104 cells (Figure 1b). TEM revealed typical RV particles of 70–100 nm in diameter in the cell culture medium (Figure 1c). The viral titer at the 10th passage was ~1×10^6/mL TCID50 (Figure 1d).

The complete genome sequence of JS was obtained using high-throughput sequencing. Analysis of genetic evolution revealed that the JS strain was a G5P[23] genotype RV. The genotype G5 PoRV is more common in pigs than in horses, cows, or other animals. It was identified in children in Brazil in the 1980s [39,40]. Although the G5 genotype RV has been reported in multiple countries, research on its pathogenicity is limited [19]. ViPR-based analysis revealed that the genotype of JS strain is G5-P[23]-I5-R1-C1-M1-A8-N1-T1-E1-H1. The structural proteins *VP3* and *VP7*, along with the nonstructural proteins *NSP2* and *NSP4* of the RVA/Pig/China/JS/2023/G5P[23], were closely related to human RVAs. However, the structural proteins *VP1*, *VP2*, *VP4*, and *VP6*, along with the nonstructural proteins *NSP1*, *NSP3*, and *NSP5*, were closely related to porcine RVAs. This suggests that the JS strain is a recombinant strain of human and porcine RVAs (Supplementary Table S2). The *NSP2* of the JS strain may be a product of recombination between human and dog strains (Figure 3).

A previous study reported severe watery diarrhea in 3-day-old piglets within 24 h of infection with the HN03 (G9P[23]) strain and recovery after 72 h [12]. Similarly, diarrhea was observed in 4-day-old piglets at 16–24 h after infection with the JS-01-20149 (G9P[7]) strain, followed by skin redness and death [41]. In a previous study, diarrhea was reported in piglets infected with PRG942 (G9P[23]) and PRG9121 (G9P[7]) strains at 1–8 dpi [42]. Furthermore, Miao et al. reported diarrhea in 1-day-old piglets infected with the CN127 strain after 6–24 h, which continued until euthanization at 48 h, but none of them died due to the infection [38].

In this study, the pathogenicity of the JS strain was investigated by infecting 15-day-old piglets. Diarrhea and dark green, foamy feces were observed on day 2 post-infection. On day 3, one piglet succumbed to the infection (Figure 4a). A postmortem examination of the piglets in the JS-infected group showed transparent, distended intestines filled with yellowish fluid and the accumulation of gastric and intestinal gases. RVA infections are not restricted to the intestines and further induce the formation of extraintestinal lesions in humans and animals [43]. After dissection, the lungs of infected piglets also showed significant lesions, manifested as local congestion (Figure 4a). The HE results indicate that the lung exhibited thickening of the alveolar walls, suggesting interstitial pneumonia. The villi lamina propia shows an increased number of inflammatory cells, predominantly lymphocytes (Figure 4b). The intestinal tissue pathology of piglets in the control group was normal (Figure 4b). In addition, IHC examination revealed that in specific segments of the small intestine with villous atrophy, cytoplasmic staining of the PoRV antigen was dominant, with the highest antigen content in the jejunum (Figure 4c). No PoRV antigen was detected in the small intestines of piglets in the negative control group (Figure 4c).

At the end of the experimental period, the diarrheal and mortality rates of infected piglets were 100% (8 of 8) and 38.5% (3 of 8), respectively. In contrast, piglets in the control group did not exhibit any apparent symptoms of diarrhea, and no deaths occurred (Figure 5a,b). TaqMan qRT–PCR detected viral RNA shedding in the feces of infected piglets as early as 24 hpi. Viral shedding in the rectal swabs of piglets continued to increase from day 1 to day 5 post-infection but decreased from day 6 onward (Figure 5c). The viral load in the lungs was significantly lower than in various parts of the small intestine, such as the duodenum, jejunum, and ileum ($p < 0.05$). The highest viral load was detected in the ileum, consistent with the IHC results (Figure 5d). Throughout the entire experiment, no viral shedding was detected in the feces, and no viral load was identified in the intestinal tissues of the control group animals (Figure 5c,d).

There are differences between using artificial milk feeding instead of breastfeeding during the experimental period and production in this study. The study is based on samples collected from a specific farm in Jiangsu Province, China. The findings may not be representative of PoRVs in other regions or countries. In clinical pathology, the infection of porcine rotavirus is more complex, and it occurs in different age groups of pigs [15]. In addition, PoRV is co-infected with other viral pathogens, such as PEDV and TGEV, as well as bacterial pathogens, such as *Escherichia coli* [13]. These phenomena have brought new challenges to this research study.

5. Conclusions

In summary, we successfully isolated and identified a strain of G5P[23] genotype PoRV. This strain exhibited excellent adaptability to MA104 cells and demonstrated high virulence in piglets. These findings hold significant importance for understanding the characteristics of PoRV in China and developing novel and effective PoRV vaccines.

Supplementary Materials: The following supporting information can be downloaded at https://www. mdpi.com/article/10.3390/v16010021/s1, Table S1. Sequences of primers and probes used in this study; Table S2. Comparison of nucleotide and amino acid characteristics of RVA/Pig/China/JS/2023/G5P[23] strain with the closest strains in the GenBank database.

Author Contributions: L.G.: Writing—review draft and validation. H.S.: Methodology and resources. S.Z.: Writing—review, and editing. S.C.: Methodology. P.Z.: Writing, review, and editing. W.L.: Writing, review, and editing. F.C.: Methodology and supervision. All authors have read and agreed to the published version of the manuscript.

Funding: This study was supported by the Science and Technology Program of Yunfu City, Guangdong Province, China (2022020202).

Institutional Review Board Statement: This study was approved by the Animal Care Committee of the South China Agricultural University (approval ID: SYXK-2019-0136). All study procedures and animal care activities were conducted per the recommendations of the Guide for the Care and Use of Laboratory Animals of the Ministry of Science and Technology of the People's Republic of China.

Informed Consent Statement: Not applicable.

Data Availability Statement: The data that support the findings of this study are available from the corresponding author upon reasonable request. GenBank accession numbers for the sample 14 RV sequences are OR644644–OR644654.

Acknowledgments: We thank Qingfeng Zhou for their assistance in the animal study.

Conflicts of Interest: The authors declare no conflict of interest. There is no conflict of interest between this research and Wen's Food Group/Guangdong Jingjie Inspection and Testing Co., Ltd.

Abbreviations

BLASTn: Basic local alignment search tool for nucleotides; CPE: cytopathic effect; NCBI: National Center for Biotechnology Information; RCWG: Rotavirus Classification Working Group; *NSP*: nonstructural protein; PCR: polymerase chain reaction; qRT–PCR: quantitative reverse transcription polymerase chain reaction; RNA: ribonucleic acid; RVA: Group A rotavirus; *VP*: virion protein.

References

1. Anderson, E.J.; Weber, S.G. Rotavirus infection in adults. *Lancet Infect. Dis.* **2004**, *4*, 91–99. [CrossRef] [PubMed]
2. Rheingans, R.D.; Antil, L.; Dreibelbis, R.; Podewils, L.J.; Bresee, J.S.; Parashar, U.D. Economic Costs of Rotavirus Gastroenteritis and Cost-Effectiveness of Vaccination in Developing Countries. *J. Infect. Dis.* **2009**, *200*, S16–S27. [CrossRef] [PubMed]
3. Cook, N.; Bridger, J.; Kendall, K.; Gomara, M.I.; El-Attar, L.; Gray, J. The zoonotic potential of rotavirus. *J. Infect.* **2004**, *48*, 289–302. [CrossRef] [PubMed]
4. Martella, V.; Bányai, K.; Matthijnssens, J.; Buonavoglia, C.; Ciarlet, M. Zoonotic aspects of rotaviruses. *Vet. Microbiol.* **2010**, *140*, 246–255. [CrossRef] [PubMed]
5. Mebus, C.A.; Nr, U.; Rhodes, M.B.; Twiehaus, M.J. *Calf Diarrhea (Scours): Reproduced with a Virus from a Field Outbreak*; University of Nebraska: Lincoln, NE, USA, 1969.
6. Bishop, R.F.; Davidson, G.P.; Holmes, I.H.; Ruck, B.J. Virus particles in epithelial cells of duodenal mucosa from children with acute non-bacterial gastroenteritis. *Lancet* **1973**, *2*, 1281–1283. [CrossRef] [PubMed]
7. Chasey, D.; Bridger, J.C.; McCrae, M.A. A new type of atypical rotavirus in pigs. *Arch. Virol.* **1986**, *89*, 235–243. [CrossRef]
8. McNulty, M.S.; Allan, G.M.; McFerran, J.B. Prevalence of antibody to conventional and atypical rotaviruses in chickens. *Vet. Rec.* **1984**, *114*, 219. [CrossRef]
9. Bohl, E.H.; Theil, K.W.; Saif, L.J. Isolation and serotyping of porcine rotaviruses and antigenic comparison with other rotaviruses. *J. Clin. Microbiol.* **1984**, *19*, 105–111. [CrossRef]
10. Welter, M.W.; Welter, C.J.; Chambers, D.M.; Svensson, L. Adaptation and serial passage of porcine group C rotavirus in ST-cells, an established diploid swine testicular cell line. *Arch. Virol.* **1991**, *120*, 297–304. [CrossRef]
11. Esona, M.D.; Foytich, K.; Wang, Y.H.; Shin, G.; Wei, G.; Gentsch, J.R.; Glass, R.I.; Jiang, B.M. Molecular characterization of human rotavirus vaccine strain CDC-9 during sequential passages in Vero cells. *Hum. Vaccines* **2010**, *6*, 247–253. [CrossRef]
12. Wang, Z.Y.; Lv, C.C.; Xu, X.; Li, X.D.; Yao, Y.L.; Gao, X.J.; Sun, Z.; Wang, Y.Z.; Sun, Y.J.; Xiao, Y.; et al. The dynamics of a Chinese porcine G9P 23 rotavirus production in MA-104 cells and intestines of 3-day-old piglets. *J. Vet. Med. Sci.* **2018**, *80*, 790–797. [CrossRef] [PubMed]

13. Mertens, N.; Theuss, T.; Köchling, M.; Dohmann, K.; Lillie-Jaschniski, K. Pathogens Detected in 205 German Farms with Porcine Neonatal Diarrhea in 2017. *Vet. Sci.* **2022**, *9*, 44. [CrossRef]
14. Jacobson, M. On the Infectious Causes of Neonatal Piglet Diarrhoea-A Review. *Vet. Sci.* **2022**, *9*, 422. [CrossRef] [PubMed]
15. Vlasova, A.N.; Amimo, J.O.; Saif, L.J. Porcine Rotaviruses: Epidemiology, Immune Responses and Control Strategies. *Viruses* **2017**, *9*, 48. [CrossRef] [PubMed]
16. Díaz, Y.; Chemello, M.E.; Peña, F.; Aristimuño, O.C.; Zambrano, J.L.; Rojas, H.; Bartoli, F.; Salazar, L.; Chwetzoff, S.; Sapin, C.; et al. Expression of Nonstructural Rotavirus Protein NSP4 Mimics Ca^{2+} Homeostasis Changes Induced by Rotavirus Infection in Cultured Cells. *J. Virol.* **2008**, *82*, 11331–11343. [CrossRef]
17. Lundgren, O.; Peregrin, A.T.; Persson, K.; Kordasti, S.; Uhnoo, I.; Svensson, L. Role of the enteric nervous system in the fluid and electrolyte secretion of rotavirus diarrhea. *Science* **2000**, *287*, 491–495. [CrossRef] [PubMed]
18. Matthijnssens, J.; Otto, P.H.; Ciarlet, M.; Desselberger, U.; Van Ranst, M.; Johne, R. VP6-sequence-based cutoff values as a criterion for rotavirus species demarcation. *Arch. Virol.* **2012**, *157*, 1177–1182. [CrossRef]
19. Papp, H.; László, B.; Jakab, F.; Ganesh, B.; De Grazia, S.; Matthijnssens, J.; Ciarlet, M.; Martella, V.; Bányai, K. Review of group A rotavirus strains reported in swine and cattle. *Vet. Microbiol.* **2013**, *165*, 190–199. [CrossRef]
20. Maes, P.; Matthijnssens, J.; Rahman, M.; Van Ranst, M. RotaC: A web-based tool for the complete genome classification of group A rotaviruses. *BMC Microbiol.* **2009**, *9*, 238. [CrossRef]
21. Ren, X.L.; Saleem, W.; Haes, R.; Xie, J.X.; Theuns, S.; Nauwynck, H.J. Milk lactose protects against porcine group A rotavirus infection. *Front. Microbiol.* **2022**, *13*, 989242. [CrossRef]
22. Zhu, J.H.; Rawal, G.; Aljets, E.; Yim-Im, W.; Yang, Y.L.; Huang, Y.W.; Krueger, K.; Gauger, P.; Main, R.; Zhang, J.Q. Development and Clinical Applications of a 5-Plex Real-Time RT-PCR for Swine Enteric Coronaviruses. *Viruses* **2022**, *14*, 1536. [CrossRef] [PubMed]
23. Huang, X.; Chen, J.N.; Yao, G.; Guo, Q.Y.; Wang, J.Q.; Liu, G.L. A TaqMan-probe-based multiplex real-time RT-qPCR for simultaneous detection of porcine enteric coronaviruses. *Appl. Microbiol. Biotechnol.* **2019**, *103*, 4943–4952. [CrossRef] [PubMed]
24. Marthaler, D.; Homwong, N.; Rossow, K.; Culhane, M.; Goyal, S.; Collins, J.; Matthijnssens, J.; Ciarlet, M. Rapid detection and high occurrence of porcine rotavirus A, B, and C by RT-qPCR in diagnostic samples. *J. Virol. Methods* **2014**, *209*, 30–34. [CrossRef] [PubMed]
25. Park, G.N.; Kim, D.I.; Choe, S.; Shin, J.; An, B.H.; Kim, K.S.; Hyun, B.H.; Lee, J.S.; An, D.J. Genetic Diversity of Porcine Group A Rotavirus Strains from Pigs in South Korea. *Viruses* **2022**, *14*, 2522. [CrossRef] [PubMed]
26. Reed, L.J.; Muench, H. A simple method of estimating fifty per cent endpoints. *Am. J. Epidemiol.* **1938**, *27*, 493–497. [CrossRef]
27. Pickett, B.E.; Sadat, E.L.; Zhang, Y.; Noronha, J.M.; Squires, R.B.; Hunt, V.; Liu, M.Y.; Kumar, S.; Zaremba, S.; Gu, Z.P.; et al. ViPR: An open bioinformatics database and analysis resource for virology research. *Nucleic Acids Res.* **2012**, *40*, D593–D598. [CrossRef]
28. Martin, D.P.; Murrell, B.; Golden, M.; Khoosal, A.; Muhire, B. RDP4: Detection and analysis of recombination patterns in virus genomes. *Virus Evol.* **2015**, *1*, vev003. [CrossRef]
29. Liu, X.S.; Zhang, Q.L.; Zhang, L.P.; Zhou, P.; Yang, J.; Fang, Y.Z.; Dong, Z.L.; Zhao, D.H.; Li, W.Y.; Feng, J.X.; et al. A newly isolated Chinese virulent genotype GIIb porcine epidemic diarrhea virus strain: Biological characteristics, pathogenicity and immune protective effects as an inactivated vaccine candidate. *Virus Res.* **2019**, *259*, 18–27. [CrossRef]
30. Matthijnssens, J.; Ciarlet, M.; Rahman, M.; Attoui, H.; Bányai, K.; Estes, M.K.; Gentsch, J.R.; Iturriza-Gómara, M.; Kirkwood, C.D.; Martella, V.; et al. Recommendations for the classification of group A rotaviruses using all 11 genomic RNA segments. *Arch. Virol.* **2008**, *153*, 1621–1629. [CrossRef]
31. Yodmeeklin, A.; Khamrin, P.; Chuchaona, W.; Saikruang, W.; Kongkaew, A.; Vachirachewin, R.; Kumthip, K.; Okitsu, S.; Ushijima, H.; Maneekarn, N. Great genetic diversity of rotaviruses detected in piglets with diarrhea in Thailand. *Arch. Virol.* **2016**, *161*, 2843–2849. [CrossRef]
32. Heiman, E.M.; McDonald, S.M.; Barro, M.; Taraporewala, Z.F.; Bar-Magen, T.; Patton, J.T. Group A Human Rotavirus Genomics: Evidence that Gene Constellations Are Influenced by Viral Protein Interactions. *J. Virol.* **2008**, *82*, 11106–11116. [CrossRef] [PubMed]
33. Liprandi, F.; Gerder, M.; Bastidas, Z.; López, J.A.; Pujol, F.H.; Ludert, J.E.; Joelsson, D.B.; Ciarlet, M. A novel type of VP4 carried by a porcine rotavirus strain. *Virology* **2003**, *315*, 373–380. [CrossRef] [PubMed]
34. Jeong, S.; Than, V.T.; Lim, I.; Kim, W. Whole-Genome Analysis of a Rare Human Korean G3P 9 Rotavirus Strain Suggests a Complex Evolutionary Origin Potentially Involving Reassortment Events between Feline and Bovine Rotaviruses. *PLoS ONE* **2014**, *9*, e97127. [CrossRef] [PubMed]
35. Sieg, M.; Rückner, A.; Köhler, C.; Burgener, I.; Vahlenkamp, T.W. A bovine G8P 1 group A rotavirus isolated from an asymptomatically infected dog. *J. Gen. Virol.* **2015**, *96*, 106–114. [CrossRef]
36. Tate, J.E.; Burton, A.H.; Boschi-Pinto, C.; Steele, A.D.; Duque, J.; Parashar, U.D.; S, W.H.O.C.G.R. 2008 estimate of worldwide rotavirus-associated mortality in children younger than 5 years before the introduction of universal rotavirus vaccination programmes: A systematic review and meta-analysis. *Lancet Infect. Dis.* **2012**, *12*, 136–141. [CrossRef] [PubMed]
37. Parashar, U.D.; Nelson, E.A.S.; Kang, G. Diagnosis, management, and prevention of rotavirus gastroenteritis in children. *BMJ-Br. Med. J.* **2013**, *347*, f7204. [CrossRef]

38. Miao, Q.; Pan, Y.D.; Gong, L.; Guo, L.J.; Wu, L.; Jing, Z.Y.; Zhang, G.H.; Tian, J.; Feng, L. Full genome characterization of a human-porcine reassortment G12P 7 rotavirus and its pathogenicity in piglets. *Transbound. Emerg. Dis.* **2022**, *69*, 3506–3517. [CrossRef]

39. da Silva, M.F.M.; Tort, L.F.L.; Goméz, M.M.; Assis, R.M.S.; Volotao, E.D.; de Mendonça, M.C.L.; Bello, G.; Leite, J.P.G. VP7 Gene of Human Rotavirus A Genotype G5: Phylogenetic Analysis Reveals the Existence of Three Different Lineages Worldwide. *J. Med. Virol.* **2011**, *83*, 357–366. [CrossRef]

40. Gouvea, V.; de Castro, L.; Timenetsky, M.C.; Greenberg, H.; Santos, N. Rotavirus serotype G5 associated with diarrhea in Brazilian children. *J. Clin. Microbiol.* **1994**, *32*, 1408–1409. [CrossRef]

41. Zhang, H.W.; Zhang, Z.; Wang, Y.F.; Wang, X.; Xia, M.Q.; Wu, H. Isolation, molecular characterization and evaluation of the pathogenicity of a porcine rotavirus isolated from Jiangsu Province, China. *Arch. Virol.* **2015**, *160*, 1333–1338. [CrossRef]

42. Kim, H.H.; Park, J.G.; Matthijnssens, J.; Kim, H.J.; Kwon, H.J.; Son, K.Y.; Ryu, E.H.; Kim, D.S.; Lee, W.S.; Kang, M.I.; et al. Pathogenicity of porcine G9P 23 and G9P 7 rotaviruses in piglets. *Vet. Microbiol.* **2013**, *166*, 123–137. [CrossRef] [PubMed]

43. Ciarlet, M.; Conner, M.E.; Finegold, M.J.; Estes, M.K. Group A rotavirus infection and age-dependent diarrheal disease in rats: A new animal model to study the pathophysiology of rotavirus infection. *J. Virol.* **2002**, *76*, 41–57. [CrossRef] [PubMed]

Article

Recent Molecular Characterization of Porcine Rotaviruses Detected in China and Their Phylogenetic Relationships with Human Rotaviruses

Mengli Qiao [1,2,†], Meizhen Li [1,†], Yang Li [2,†], Zewei Wang [3], Zhiqiang Hu [4], Jie Qing [2], Jiapei Huang [2], Junping Jiang [5], Yaqin Jiang [5], Jinyong Zhang [2], Chunliu Gao [2], Chen Yang [2], Xiaowen Li [1,2,5,*] and Bin Zhou [1,*]

[1] MOE Joint International Research Laboratory of Animal Health and Food Safety, College of Veterinary Medicine, Nanjing Agricultural University, Nanjing 210014, China; qiaoml2024@163.com (M.Q.); 15005658994@163.com (M.L.)

[2] Shandong Engineering Research Center of Pig and Poultry Health Breeding and Important Disease Purification, Shandong New Hope Liuhe Co., Ltd., Qingdao 266000, China; youglion@163.com (Y.L.); qingjie94@163.com (J.Q.); sicaujiapeihuang@163.com (J.H.); zhangjy3@newhope.cn (J.Z.); gaochunliu1@newhope.cn (C.G.); 18864808615@163.com (C.Y.)

[3] Beef Cattle Industry Development Center, Fangshan 033100, China; 18404982338@163.com

[4] College of Animal Science, Xichang University, Xichang 615012, China; zhiqianghu0624@163.com

[5] China Agriculture Research System-Yangling Comprehensive Test Station, Xianyang 712100, China; jiangjunping1@newhope.cn (J.J.); jiangyq@newhope.cn (Y.J.)

* Correspondence: lxw8272@163.com (X.L.); zhoubin@njau.edu.cn (B.Z.)

† These authors contributed equally to this work.

Citation: Qiao, M.; Li, M.; Li, Y.; Wang, Z.; Hu, Z.; Qing, J.; Huang, J.; Jiang, J.; Jiang, Y.; Zhang, J.; et al. Recent Molecular Characterization of Porcine Rotaviruses Detected in China and Their Phylogenetic Relationships with Human Rotaviruses. *Viruses* **2024**, *16*, 453. https://doi.org/10.3390/v16030453

Academic Editors: Ulrich Desselberger and John T. Patton

Received: 6 February 2024
Revised: 10 March 2024
Accepted: 12 March 2024
Published: 14 March 2024

Abstract: Porcine rotavirus A (PoRVA) is an enteric pathogen capable of causing severe diarrhea in suckling piglets. Investigating the prevalence and molecular characteristics of PoRVA in the world, including China, is of significance for disease prevention. In 2022, a total of 25,768 samples were collected from 230 farms across China, undergoing porcine RVA positivity testing. The results showed that 86.52% of the pig farms tested positive for porcine RVA, with an overall positive rate of 51.15%. Through the genetic evolution analysis of VP7, VP4 and VP6 genes, it was revealed that G9 is the predominant genotype within the VP7 segment, constituting 56.55%. VP4 genotypes were identified as P[13] (42.22%), P[23] (25.56%) and P[7] (22.22%). VP6 exhibited only two genotypes, namely I5 (88.81%) and I1 (11.19%). The prevailing genotype combination for RVA was determined as G9P[23]I5. Additionally, some RVA strains demonstrated significant homology between VP7, VP4 and VP6 genes and human RV strains, indicating the potential for human RV infection in pigs. Based on complete genome sequencing analysis, a special PoRVA strain, CHN/SD/LYXH2/2022/G4P[6]I1, had high homology with human RV strains, revealing genetic reassortment between human and porcine RV strains in vivo. Our data indicate the high prevalence, major genotypes, and cross-species transmission of porcine RVA in China. Therefore, the continuous monitoring of porcine RVA prevalence is essential, providing valuable insights for virus prevention and control, and supporting the development of candidate vaccines against porcine RVA.

Keywords: porcine rotavirus A (RVA); serotypes; molecular characteristics; prevalence; cross-species transmission

1. Introduction

Rotaviruses (RVs), which belong to the *Reoviridae* family, are a significant cause of diarrhea in children and young animals globally [1,2]. RVs were identified more than 60 years ago in rectal swabs from monkeys and intestinal biopsies from mice [3]. Soon after, in 1973, human RV was discovered in duodenal biopsies from nine children [4]. Rotavirus A (RVA) infects humans and animals and is the most common subtype that causes diarrhea in sucking piglets. It accounts for more than 90% of diarrhea cases caused by RV in commercial pig populations [5,6]. The World Health Organization (WHO) estimates that RV causes approximately 450,000 deaths each year, with more than 90% of these deaths occurring in developing countries in Asia and Africa [3]. Porcine RV infection is a common enteric infectious disease with watery diarrhea, vomiting, anorexia, and dehydration as the main clinical features [7]. Current studies have suggested that porcine RVs may be the pathogen of RV infection in humans, dogs and other species [8,9]. Patterns of RV genotypes have evolved through interspecies transmission and recombination events [10].

The mature RV particle encapsulates a genome of 11 segments of double-stranded (ds) RNA encoding six structural proteins (VP1 to VP4, VP6 and VP7) and five or six non-structural viral proteins (NSP1 to NSP5/6) [5]. RVs are classified into ten groups (A–J) based on the antigenic relationships of the VP6 protein, which can stimulate the body to produce IgA and determine the specificity of the RV serotype [11,12]. The structural proteins VP7 and VP4, which determine the G and P genotypes of RV, together form the outer capsid of RV, and are important antigens that neutralize and induce the production of neutralizing antibodies [13]. Porcine RVA (PoRVA) encompasses a diverse array of genotypes, including G, P and I genotypes. Twelve G genotypes (G1 to G6, G8 to G12 and G26), fifteen P genotypes (P[1] to P[8], P[11], P[13], P[19], P[23] and P[26] to P[28]) and four I genotypes (I1, I2, I5 and I14) of porcine RVA have been reported worldwide [14–16]. G genotypes (i.e., G3 to G5, G9 and G11) are often freely combined with P genotypes (i.e., P[5] to P[7], P[13] and P[28]). G5P[7], G4P[6] and G4P[7] were the most common genotype combinations worldwide. Few studies have focused on the genotype of VP6 in pigs, resulting in limited reports on the combination of dominant G/P/I genotypes [5].

The diversity of RV strains poses significant challenges for vaccine development [10]. The trivalent live attenuated vaccine, based on the pandemic G/P combination, was found to be safe and effective in protecting piglets against diarrhea caused by homologous virulent strains. However, bivalent vaccines (containing G5P[7] and G9P[7] strains) were not effective in preventing infections caused by G8P[1], G9P[23] and G8P[7] RVA strains [17]. Therefore, understanding genetic diversity is essential for the development, optimization and improvement of vaccines, as there is limited cross-protection against heterologous strains. Therefore, in this study, we reported the prevalence of RVA in fecal samples collected from 23 provinces in China in 2022, and analyzed the proportions of genotypes in different regions to understand the prevalence characteristics of RVA in China, the largest pig-farming country.

2. Materials and Methods

2.1. Sample Preparation

In 2022, a total of 25,768 fecal samples from sucking piglets with diarrhea were collected from 230 different industrial farms in 23 provinces across China. These samples were diluted with two volumes of cold phosphate-buffered saline (PBS) and purified by centrifugation at $5000 \times g$ for 1 min. A 300 µL aliquot of supernatant was extracted from each sample, and the total RNA was then extracted using the Virus DNA/RNA Extraction Kit from Bioer (Hangzhou, China), following the manufacturer's instructions.

2.2. Reverse Transcription-Quantitative PCR (RT-qPCR)

To determine the viral load in each sample, RT-qPCR was performed by using the TransScript Probe One-Step qRT-PCR Kit (TransGen, Beijing, China) on a Step One Plus instrument (ABI). Briefly, the procedure involved an initial reverse transcription at 45 °C for 5 min, followed by pre-denaturation at 94 °C for 30 s. The qPCR reactions were run for 40 cycles with denaturation at 94 °C for 5 s, annealing and extension at 60 °C for 30 s. The sample was considered positive for PoRVA if the CT value was less than 35 or between 35 and 40 in two repeated assays. The farm was considered positive for PoRVA if one or more diarrheal samples tested positive for PoRVA. The primers and probes utilized for the detection of PoRVA nucleic acid were specifically designed to target the conserved region of the NSP3 gene, as outlined in Supplementary Table S1.

2.3. Amplification of PoRVA Genes

Specific primers targeting the VP1 to VP4, VP6, VP7 and NSP1 to NSP5 genes of PoRVA were designed based on the conserved regions (Supplementary Table S1) and synthesized by Sangon Bioengineering Co., Ltd. (Shanghai, China). The reverse transcription and amplification of the selected RNAs were performed by using the HiScript II One-Step RT-PCR Kit (Dye Plus) following the manufacturer's instructions (Vazyme, Nanjing, China). The standard program was below: initial reverse transcription at 45 °C for 25 min, pre-denaturation at 94 °C for 5 min, 32 cycles of denaturation at 94 °C for 30 s, annealing at 55 °C for 30 s, extension at 72 °C for 1, 2, or 1.5 min (for VP7, VP4 and VP6 genes, respectively) and final extension at 72 °C for 8 min. All RT-PCR products were subjected to gel electrophoresis on a 1.5% agarose gel, and the target bands were finally verified using a UV transilluminator. The fragments of VP7, VP4 and VP6 genes were cloned into the pMD-18T vector (TaKaRa, Beijing, China) followed by successful transformation into *E. coli* DH5α competent cells. The plasmids were sequenced using the Sanger approach by Sangon Bioengineering Co., Ltd.

2.4. Genotyping and Phylogenetic Analysis

The genotypes of individual genes of the study strains were determined with the Virus Pathogen Resource (ViPR) automated genotyping tool (https://www.viprbrc.org/brc/rvaGenotyper.spg?method=ShowCleanInputPage&decorator=reo, accessed on 17 June 2023). The sequence similarities were analyzed by using BLAST (http://blast.ncbi.nlm.nih.gov/Blast.cgi, accessed on 12 April 2023). The complete sequences of the RVA VP7, VP6 and VP4 genes, along with the complete genome-wide sequences of the LYXH2, were utilized to construct a phylogenetic tree. Reference sequences downloaded from GenBank were also incorporated in the analysis. The neighbor-joining method with 1000 bootstrap replicates for each gene was applied using MEGA 11 software. The initial tree was drawn to scale, with branch lengths representing the number of substitutions per site. Visualization was conducted using iTOL v6 (Interactive Tree of Life, http://itol.embl.de/, accessed on 28 February 2024). To assess the genomic characteristics of the LYXH2, its nucleotide sequences were compared to the complete genome sequences of rotavirus A (RVA) present in the GenBank database.

2.5. Nucleotide Sequence Accession Numbers

The nucleotide sequence data obtained in this study have been uploaded to the GenBank database. The accession numbers for VP7, VP4 and VP6 gene sequences are OQ743847-OQ743991, OQ799656-OQ799745 and OQ799746-OQ799879, respectively (Supplementary Table S2). The accession numbers of the CHN/SD/LYXH2/2022/G4P6I1 genome (except for VP4, VP6 and VP7) are OQ799880-OQ799887.

3. Results

3.1. Prevalence of PoRVA in China, 2022

In 2022, a total of 25,768 diarrhea samples were collected from 230 pig farms in 23 provinces across China and tested for PoRVA by qRT-PCR. The results showed that the prevalence among provinces significantly varied with positive rates ranging from 29.94% to 87.10%. Additionally, the average frequency of positive samples was found to be 51.15%. More than half of the samples tested positive for PoRVA in 52.17% (12/23) of the provinces. Furthermore, the prevalence of positive farms ranged from 70% to 100% in all of the tested provinces. Moreover, in 82.61% (19/23) of the provinces, more than 80% of the farms tested positive for PoRVA. The average prevalence rate of PoRVA at the farm level in China in 2022 was 86.09% (Table 1).

Table 1. PoRVA positive samples and farms in China, 2022.

China Region	Province	No. of Tested Samples	No. of Positive Samples	Positive Rate at the Sample Level (%)	No. of Tested Farms	No. of Positive Farms	Positive Rate at the Farm Level (%)
Northern	Hebei	627	250	39.87	6	5	83.33
	Heilongjiang	155	135	87.10	2	2	100.00
	Liaoning	1490	891	59.80	21	19	90.48
	Inner Mongolia	915	558	60.98	11	10	90.91
	Tianjin	182	97	53.30	10	7	70.00
	Total	3369	1931	57.32	50	43	86.00
Central	Gansu	1585	819	51.67	14	13	92.86
	Henan	1440	843	58.54	11	10	90.91
	Shanxi	1411	1217	86.25	4	4	100.00
	Shaanxi	1894	567	29.94	9	8	88.89
	Hubei	1682	1351	80.32	11	10	90.91
	Total	8012	4797	59.87	49	45	91.84
Eastern	Anhui	894	475	53.13	5	4	80.00
	Shandong	1302	452	34.72	17	13	76.47
	Jiangsu	1212	538	44.39	10	9	90.00
	Zhejiang	1310	563	42.98	12	11	91.67
	Total	4718	2028	42.98	44	37	84.09
Southern	Fujian	329	208	63.22	4	3	75.00
	Guangdong	1333	735	55.14	7	6	85.71
	Guangxi	1645	816	49.60	16	13	81.25
	Hainan	264	97	36.74	5	4	80.00
	Hunan	1548	466	30.10	11	10	90.91
	Jiangxi	1597	728	45.59	12	11	91.67
	Total	6716	3050	45.41	55	47	85.45
Southwestern	Sichuan	1746	677	38.77	21	18	85.71
	Guizhou	1100	649	59.00	7	6	85.71
	Yunnan	107	48	44.86	4	3	75.00
	Total	2953	1374	46.53	32	27	84.38
Total		25,768	13,180	51.15	230	199	86.52

To analyze the patterns of PoRVA prevalence, the pig farms were categorized into five groups based on their geographical location: northern, central, eastern, southern and southwestern China (Figure 1). The positive rates of PoRVA were higher in central China, both in samples (59.87%) and farms (91.84%), compared to other regions. The prevalence of PoRVA in eastern China was relatively lower than in the other four regions, as indicated by the positive rate at the sample and farm levels (Table 1).

3.2. Genotypes and Distribution of PoRVA

To elucidate the genetic characteristics of PoRVA circulating in China, the VP7, VP4 and VP6 genes of 180 positive samples (CT value less than 30) from different farms were amplified using one-step RT-PCR. Full-length VP7 (n = 145), VP4 (n = 90) and VP6 (n = 134) genes were sequenced, followed by phylogenetic analysis presented in Figures 2a, 3a and 4a. The VP7 gene sequences were clustered into nine branches: G9 (56.55%), G5 (14.48%), G1 (8.97%), G26 (6.90%), G4 (6.21%), G3 (3.45%), G12 (2.07%), G11 (0.69%) and G2 (0.69%) (Figure 2a,c). The P genotypes were clustered into five branches, including P[13] (42.22%), P[23] (25.56%), P[7] (22.22%), P[6] (8.89%) and P[3] (1.11%) (Figure 3a,c). Only two I-genotypes (I5 and I1) were identified with proportions of 88.81% and 11.19% (Figure 4a,c).

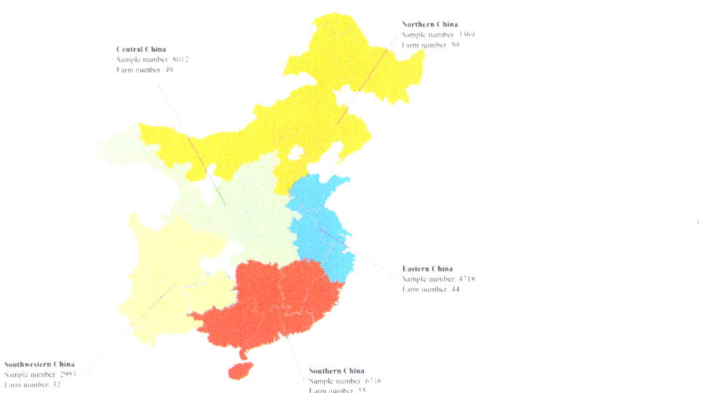

Figure 1. Numbers of collected samples and farms in each of the geographical divisions of China defined in this study. Northern China included Hebei, Heilongjiang, Liaoning, Inner Mongolia and Tianjin. Central China included Gansu, Henan, Shanxi, Shaanxi and Hubei. Eastern China including Anhui, Shandong, Jiangsu and Zhejiang. Southern China included Fujian, Guangdong, Guangxi, Hainan, Hunan and Jiangxi. Southwestern China included Guizhou, Sichuan and Yunnan.

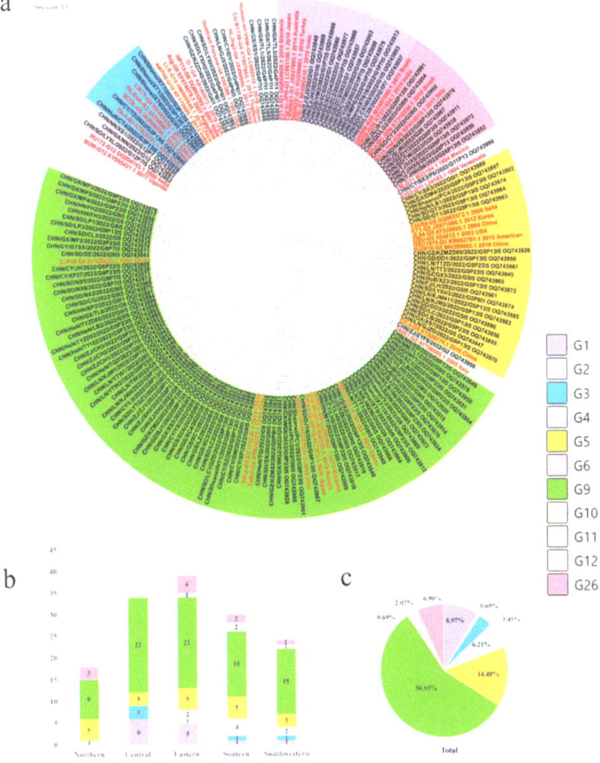

Figure 2. (**a**) The genotype distribution and genetic and phylogenetic analysis of the VP7 gene in different regions from diarrheal pigs. The trees of the VP7 gene were created via neighbor-joining analysis using MEGA 11 software with 1000 bootstrap replicates. Sequences of the various G genotypes of reference RVA strains were obtained from GenBank. The reference sequence was shown in red. (**b**) The genotype proportions of the VP7 gene at the farm level in different regions in 2022. (**c**) Proportional distribution of genotypes in the VP7 gene in 2022.

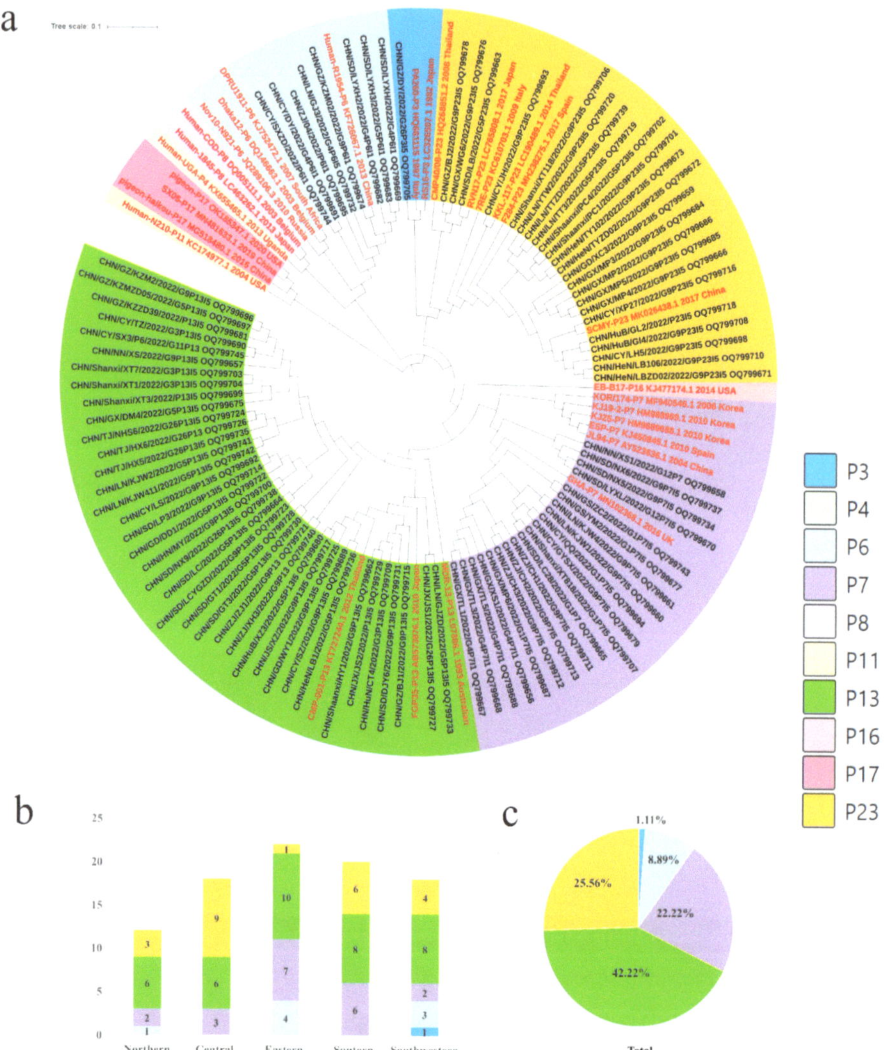

Figure 3. (**a**) Genotype distribution and genetic and phylogenetic analysis of the VP4 gene in different regions from diarrheal pigs. The trees of the VP4 gene were created via neighbor-joining analysis using MEGA 11 software with 1000 bootstrap replicates. Sequences of the various P genotypes of reference RVA strains were obtained from GenBank. The reference sequence was shown in red. (**b**) The genotype proportions of the VP4 gene at the farm level in different regions in 2022. (**c**) Proportional distribution of genotypes in the VP4 gene in 2022.

3.3. Distribution of PoRVA Genotypes in Different Regions in China

The G9 genotype was predominant in all five regions, with proportions ranging from 50.00% to 64.71%. The G5 genotype was also detected in five regions, but its prevalence was significantly lower than that of G9. The G1, G4 and G26 genotypes showed regional distribution in China. G1 was not detected in northern China, and G4 and G26 were not found in central China (Figure 2b). P[13] was the major genotype in all regions except central China, where P[23] was the dominant genotype. The proportion of P[13] and P[23], along with P[7], was 90%. These strains were circulating in all five regions of China. The other two genotypes of VP4, P[6] and P[3], were found in three regions and one region of China, respectively (Figure 3b). The I5 genotype was the dominant genotype in all of the regions, ranging from 80.65% to 100%. The I1 genotype of PoRVA was only circulating in southern, eastern, and southwestern China (Figure 4b).

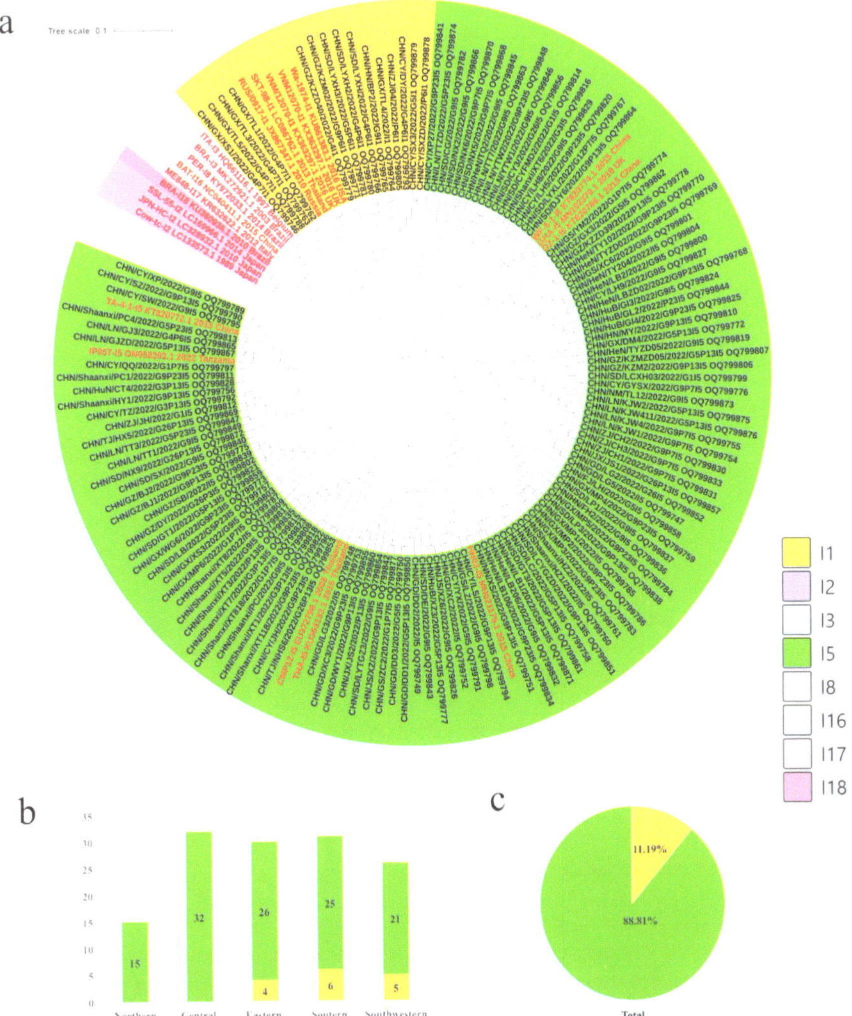

Figure 4. (**a**) Genotype distribution and genetic and phylogenetic analysis of the VP6 gene in different regions from diarrheal pigs. The trees of the VP6 gene were created via neighbor-joining analysis using MEGA 11 software with 1000 bootstrap replicates. Sequences of the various I genotypes of reference RVA strains were obtained from GenBank. The reference sequence was shown in red. (**b**) The genotype proportions of the VP6 gene at the farm level in different regions in 2022. (**c**) Proportional distribution of genotypes in the VP6 gene in 2022.

3.4. Main Genotype Combinations of VP7, VP4 and VP6 Genes

The combinations of the VP7, VP4 and VP6 genes of PoRVA were obtained, and 79 samples were detected for the G/P/I genotypes. G9P[23]I5 (22.78%) was the most common combination of PoRVA, followed by G9P[13]I5 (15.19%), G5P[13]I5 (13.92%) and G9P[7]I5 (10.13%) (Table 2). The results demonstrated that the combinations of VP7 and

VP4 genes exhibited significant diversity. Nonetheless, it was noticeable that the majority of these combinations were linked with the I5 genotype of VP6.

Table 2. PoRVA genotype combination analysis.

	P3	P6	P7	P[13]	P[23]	
G1						I1
			5			I5
G3						I1
				4		I5
G4		3	4			I1
		1				I5
G5		1				I1
				11	4	I5
G9		1				I1
			8	12	18	I5
G11				1		I1
						I5
G12						I1
			1			I5
G26						I1
	1			4		I5

3.5. Homology Analysis of PoRVA in China, 2022

The highest sequence similarity of the VP7, VP4 and VP6 genes obtained in this study was submitted to the standard nucleotide BLAST program on NCBI. As shown in Supplementary Table S2, approximately 73.79% of the VP7 sequences, 93.33% of the VP4 sequences and 78.36% of the VP6 sequences were closely related to the PoRVAs reported previously. Interestingly, it was found that 18.62% of VP7 sequences, 6.67% of VP4 sequences and 21.64% of VP6 sequences were similar to human RVA sequences. All three genes of two strains (G4P[6]I1) from Shandong and one strain (G9[P6]I1) from Guizhou showed high homology with human RVA strains instead of PoRVA strains. Among the analyzed PoRVA strains, eight strains exhibited the highest similarity to the VP7 gene of the strains isolated from giant pandas. Additionally, three RVA strains demonstrated the highest homology of VP7 genes with the strains isolated from dogs. These data indicate the possibility of interspecific transmission.

3.6. Complete Genomic Analysis of the Special PoRVA Strain

We attempted to sequence the complete genomes of three RVA strains that were suspected to have originated in humans. These strains displayed the greatest homology with human RVA strains, specifically in relation to the VP7, VP4 and VP6 genes. However, only the G4P[6]I1 strain from Shandong province was successfully sequenced. This strain was designated as CHN/SD/LYXH2/2022/G4P6I1 (LYXH2 for short), and its genotypes were identified as G4-P[6]-I1-R1-C1-M1-A8-N1-T1-E1-H1 (Figure 5).

Phylogenetic analyses were conducted between the closely related reference RVA strains of the same genotype. The sequence most closely related to the VP7 gene fragment of LYXH2 was identified from the human RVA strain RVA/human-WT/CHN/SZ18-2049/2018/G4P[6] (Figure 5a). Homology analysis revealed that the strain with the highest similarity to the VP7 fragment was also SZ18-2049 (Supplementary Table S3). The strain most closely associated with the genetic evolution of VP4 and NSP1 fragments was RVA/Human-wt/CHN/R1954/2013/G4P[6] (Figure 5b,g), which was documented as a porcine-like human strain in 2015 [18]. The VP6 gene fragment exhibited a high degree of similarity to the R946 strain (Figure 5c), showing 100% amino acid homology (Supplementary Table S3). The RVA/Human-wt/CHN/R946/2006/G3P[6] strain was identified in 2015 as a potential recombinant, incorporating a human RVA-like gene fragment into the genetic background of the porcine RVA strain [18]. Gene fragments VP1, VP2, NSP2 and NSP5 shared a close genetic relationship with RVA/Human-wt/CHN/LL3354/2000/G5P[6] (Figure 5d,e,h,k), an isolate obtained from the fecal samples of children under 5 years old suffering from diarrhea in China [19]. The remaining genes showed close relationship to porcine strains (Supplementary Table S3).

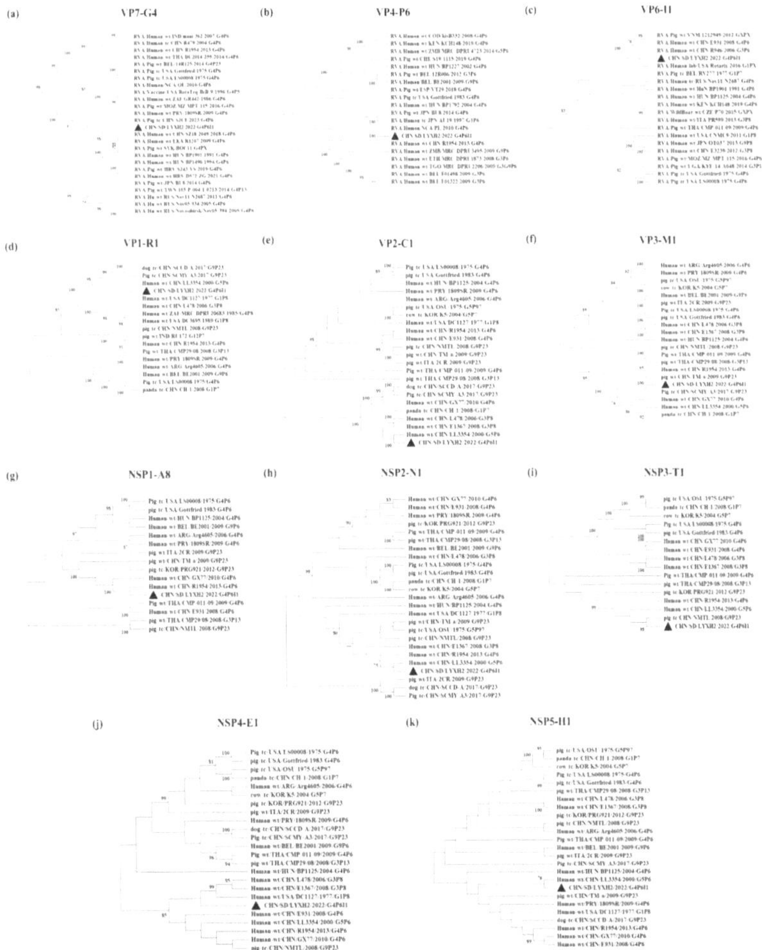

Figure 5. Phylogenetic dendrogram based on the nucleotide sequences of VP7 (**a**), VP4 (**b**), VP6 (**c**), VP1 (**d**), VP2 (**e**), VP3 (**f**), NSP1 (**g**), NSP2 (**h**), NSP3 (**i**), NSP4 (**j**) and NSP5 (**k**) from LYXH2 and reference strains. The trees were created via neighbor-joining analysis using MEGA 11 software with 1000 bootstrap replicates. ▲ The sequences in this study.

Remarkably, seven genes of the LYXH2 genome, including VP1, VP4, VP6, VP7, NSP2, NSP3 and NSP5, showed a high degree of homology with human RVA strains rather than other previously-reported PoRVA strains (Table 3). These data provide evidence for interspecies transmission and reassortment events between porcine and human RVAs.

Table 3. Comparison of CHN/SD/LYXH2/2022/G4P6I genes with those of human and animal rotaviruses.

Strain	Host	Genotypes of Viral Protein Genes and Nucleotide Sequence Identities (% *) to LYXH2																				
		VP7		VP4		VP6		VP1		VP2		VP3		NSP1		NSP2		NSP3		NSP4		NSP5	
LYXH2	Pig	G4	-#	P6	-	I1	-	R1	-	C1	-	M1	-	A8	-	N1	-	T1	-	E1	-	H1	-
ET8B/2015	Pig	G5	72.4	P[13]	63.7	I5	84.2	R1	85.8	C1	86.9	M1	86.5	A8	89.7	N1	89.6	T7	83.6	E1	90.8	H1	98.3
HeNNY	Pig	G4	85.2	P[23]	71.1	I5	82.8	R1	90.4	C1	86.9	M1	97.4	A8	81.2	N1	89.9	T1	93.6	E1	95.3	H1	96.1
LNCY	Pig	G3	73.6	P[13]	68.2	I5	83.5	R1	86.0	C1	87.5	M1	97.0	A8	81.4	N1	89.6	T1	94.2	E1	98.7	H1	97.6
CMP-011	Pig	G4	84.0	P6	93.9	I1	90.8	R1	86.8	C1	86.6	M1	87.8	A8	85.4	N1	89.5	T1	91.4	E1	89.2	H1	96.6
SD-1	Pig	G9	74.9	P[23]	71.4	I5	83.6	R1	85.7	C1	95.5	M1	95.3	A8	81.9	N1	90.3	T1	94.0	E1	94.5	H1	97.3
CN127	Pig	G12	72.1	P7	68.3	I1	94.6	R1	85.9	C1	86.3	M1	95.1	A8	88.7	N1	87.8	T1	88.0	E1	96.5	H1	97.0
SCLSHL	Pig	G9	74.6	P[23]	71.9	I1	&	R1	88.2	C1	86.5	M1	95.6	A8	95.8	N1	93.0	T1	94.1	E1	92.5	H1	97.8
SCMY	Pig	G9	74.4	P[23]	72.2	I5	83.5	R1	92.8	C1	86.6	M1	97.5	A8	82.8	N1	87.6	T1	92.3	E1	87.9	H1	97.0
LS00008	Pig	G4	83.6	P6	84.5	I1	90.8	R1	86.1	C1	86.5	M1	86.6	A8	87.2	N1	90.4	T1	87.6	E1	89.5	H1	97.0
GX54	Human	G4	84.7	P6	96.3	I1	94.6	R1	88.1	C1	86.9	M1	96.4	A8	95.3	N1	89.4	T1	88.1	E1	93.0	H1	98.0
SCLS-R3	Human	G3	74.3	P[13]	69.9	I5	81.3	R1	88.6	C1	86.6	M1	93.4	A8	95.7	N1	88.6	T7	86.6	E1	93.2	H1	98.0
E931	Human	G4	84.3	P6	96.5	I1	96.5	R1	88.6	C1	87.4	M1	85.0	A8	86.1	N1	89.9	T1	88.3	E1	93.7	H1	98.0
R946	Human	G3	72.9	P6	94.9	I1	96.7	R1	88.2	C1	86.9	M1	84.8	A1	77.1	N1	94.5	T1	94.4	E1	92.1	H1	97.0
R1954	Human	G4	84.8	P6	96.7	I1	96.6	R1	88.4	C1	86.5	M1	95.8	A8	95.8	N1	94.7	T1	94.6	E1	93.0	H1	97.5
R1207	Human	G4	96.0	P6	94.7	I1	96.6	R1	86.1	C1	87.2	M1	95.1	A1	77.0	N1	91.6	T1	87.6	E1	89.0	H1	96.8
LL3354	Human	G5	73.1	P6	95.5	I5	83.7	R1	93.8	C1	95.5	M1	95.8	A1	77.0	N1	96.0	T1	94.4	E1	93.7	H1	98.5
R479	Human	G4	84.4	P6	95.2	I5	82.9	R1	88.6	C1	86.9	M1	95.3	A1	77.5	N1	88.9	T7	85.5	E1	89.2	H1	98.5
KNA/08979	Simian	G5	72.1	PX	&	I5	84.3	R1	86.7	C1	86.7	M1	86.5	A8	89.7	N1	89.2	T7	83.6	E1	90.8	H1	98.3
SCCD-A	Dog	G9	75.2	P[23]	72.2	I5	83.3	R1	92.9	C1	86.5	M1	97.3	A8	80.4	N1	87.4	T7	86.8	E1	87.9	H1	97.8

* The nucleotide (nt) sequence identity value (nt/aa), expressed as a percentage, was calculated based on the complete ORF sequence of each genetic fragment. The distance matrix for nucleotide (nt) sequences was generated using the p-distance algorithm in Mega 7.0. # Self-homology is 100%, which is meaningless. & Gene sequences not obtained in the original research.

4. Discussion

RV is a zoonotic pathogen that causes severe diarrhea in young animals, weakening their immune system and making them susceptible to other pathogens, which results in substantial economic losses in the pig industry [20]. Epidemiological survey data have shown that PoRVA is endemic and widespread in large-scale pig farms worldwide [15,16]. Until now, there has been no nationwide epidemiological investigation specifically targeting multiple genes of PoRVA in China, except a few studies that reported the positive rate of pig samples and farms in individual provinces [21–23]. For instance, a study conducted from 2013 to 2019 reported an RVA positive rate of 7% in Zhejiang, Shandong, Jiangsu and Shanxi provinces [24]. Another study conducted from 2017 to 2019 found an RVA positive rate of 16.83% (100/394) in East China [23]. In 2022, a study conducted in Heilongjiang province reported an RVA positive rate of 4.3% (12/280) [25]. Additionally, a study conducted in Jiangsu province in 2022 reported an RVA positive rate of 12.5% (11/88) [26]. For a comprehensive understanding of the prevalence of PoRVA in China, a large number of porcine diarrheal samples were collected from the major provinces in pig production. Our study showed that the positive rates of the samples and farms in 2022 were 51.70% and 86.26%, suggesting that PoRVA infections are widespread in China (Table 1). The high positive rates observed further emphasize the significance of addressing the potential harm caused by rotavirus infection, highlighting the importance of further research and intervention. The significant increase in the prevalence of RVA in recent years can be attributed to various factors. The complexity of pig diseases, the constant turnover of pig herds resulting in varying levels of neutralizing antibodies in sows and the increasingly complex overall disease burden on pig farms have all contributed to the rise in morbidity and high viral load in the environment. This makes it challenging to completely eliminate RV viral nucleic acids with disinfectants, and even after disinfection, the virus persists, leading to piglet infection due to viral enrichment. Additionally, existing inactivated RVA vaccines have shown limited efficacy in boosting neutralizing antibody levels in sows. Cross-protection limitations also exist between different RVA genotypes, rendering vaccines targeting specific genotypes ineffective against others. This scenario could potentially drive the emergence of new genotypes. Furthermore, the increasing occurrence of rotavirus recombination in recent years can also contribute to higher infection rates [2,27,28].

RVs are always distinguished by different G, P and I genotypes based on VP7, VP4 and VP6 genes. Due to the segmented genome, different genotypes of RVs tend to form random combinations. G9 as an emerging genotype in pigs and humans worldwide is often associated with P[7], P[13], P[19] and P[23] [29]. G9P[23] was the dominant genotype combination in Germany, Japan and Korea [13,30,31]. Interestingly, we found that the most prevalent genotypes for VP7, VP4 and VP6 in China were G9, P[13] and I5. In addition, the dominant genotype combination was G9P[23]I5, followed by G9P[13]I5 and G5P[13]I5 (Table 2). The dominance of the G9 and I5 genotypes was evident in all five regions of China, while the P[13] genotype was prominent, but not the sole leader. It has been reported that the potential for cross-protection between different genotypes of RV strains is limited [32]. Given the constant evolution of rotavirus genotypes [5], it is imperative to develop vaccines based on the prevailing epidemic genotypes. Therefore, our results would contribute to understanding the prevalence of RVA in pigs in China and to develop novel vaccines against the dominant genotypes. In addition, only two I genotypes were detected in this study, suggesting that universal methods could be developed to target VP6 nucleic acids or antibodies for assessing RV infection in pigs. The success rates of sequencing targeting the VP7, VP4 and VP6 genes were 80.56%, 50.00% and 74.44%, respectively. More conservative and genotype-specific primers should be developed in future studies.

Previous studies have shown that certain PoRVA strains are highly similar to human RVAs and often experience interspecies recombination and reassortment events [8,33]. The existence of interspecific transmission and recombination between pigs and humans could be demonstrated by conducting whole genome sequencing of RVs [8,18,27]. In line with this evidence, the homology analysis of our data showed that one or more genes of certain strains exhibited a higher homology with human RVA strains rather than porcine RVA strains. Furthermore, based on the whole genome of the LYXH2 strain from a piglet with RVA-induced diarrhea, seven out of eleven genes showed high homology to human RVA strains, but not to PoRVA strains. In addition, genetic evolutionary analysis indicated that several gene segments of LYXH2 shared close relationships with human strains R1954, R946 and LL3354, all of which have been identified as recombinant strains originating from both humans and pigs [18,19]. This phenomenon suggests instances where pig strains reinfect pigs following recombination with human strains, underscoring the dynamic interplay between human and porcine rotaviruses. These data suggest that during the long evolutionary process, human and porcine RVAs have continuously crossed species barriers and frequently recombined with each other. Despite interspecific barriers and limitations in host range, interspecific transmission plays a significant role in the diversity and evolution of rotaviruses. Pigs have served as hosts for potential zoonotic disease transmission and the emergence of new genotypes [2,27,28]. Future research should focus on the virulence and transmissibility of recombinant strains between humans and pigs.

We conducted an epidemiological survey of probable RVA-positive samples collected from 23 provinces in 2022. The results showed that PoRVA was widespread in different regions of China, with a high prevalence in both samples and farms. The dominant genotypes were G9, P[13] and I5. The most prevalent genotype combination was G9P[23]I5. Furthermore, interspecies transmission of RVA and complex reassortment involving human and porcine RVAs were demonstrated. Monitoring the epidemiology of PoRVA can provide insights for the development of vaccines and other preventive measures.

Supplementary Materials: The following supporting information can be downloaded at: https://www.mdpi.com/article/10.3390/v16030453/s1, Table S1: Primers used for PoRVA detection and genotyping; Table S2: Homology analysis of PoRVA identified in this study; Table S3: Whole genomic analyses of CHN/SD/LYXH2/2022/G4P6I1 detected in diarrheal piglets.

Author Contributions: M.Q.: Investigation, Writing—original draft; M.L.: Investigation, Writing—review & editing; Y.L.: Investigation, Writing—review & editing; Z.W.: Investigation; Z.H.: Investigation; J.Q.: Sample collection; J.H.: Methodology; J.J.: Investigation; Y.J.: Investigation; J.Z.: Methodology; C.G.: Methodology; C.Y.: Methodology; X.L.: Investigation, conceptualization; B.Z.: Writing—review, conceptualization. All authors have read and agreed to the published version of the manuscript.

Funding: This work was supported by the Taishan Industry Leadership Talent Project of Shandong Province in China, China Agriculture Research System [CARS-35] and the Natural Science Foundation of Shandong Province [ZR2022MC158].

Institutional Review Board Statement: Ethical review and approval were waived for this study because fecal samples were collected as part of a routine laboratory diagnosis of enteric pathogens implicated in sucking piglets.

Informed Consent Statement: Not applicable.

Data Availability Statement: All data analyzed during this study are included in this published article. The raw data generated during the current study are available from the corresponding author on reasonable request. The nucleotide sequences were deposited in GenBank under the accession numbers OQ743847-OQ743991 (VP7), OQ799656-OQ799745 (VP4), OQ799746-OQ799879 (VP6), and OQ799880-OQ799887 (LYXH2 VP1-NSP5).

Acknowledgments: We thank the staff at the pig farms belonging to New Hope Liuhe Co., Ltd. for collecting clinical samples.

Conflicts of Interest: Authors M.Q.; Y.L.; J.Q.; J.H.; J.J.; Y.J.; J.Z.; C.G.; C.Y. and X.L. were employed by the company New Hope Liuhe Co., Ltd. The remaining authors declare that the research was conducted in the absence of any commercial or financial relationships that could be construed as a potential conflict of interest.

References

1. Tate, J.E.; Burton, A.H.; Boschi-Pinto, C.; Parashar, U.D.; World Health Organization—Coordinated Global Rotavirus Surveillance Network. Global, Regional, and National Estimates of Rotavirus Mortality in Children <5 Years of Age, 2000–2013. *Clin. Infect. Dis.* **2016**, *62* (Suppl. S2), S96–S105. [PubMed]
2. Wu, F.T.; Liu, L.T.; Jiang, B.; Kuo, T.Y.; Wu, C.Y.; Liao, M.H. Prevalence and diversity of rotavirus A in pigs: Evidence for a possible reservoir in human infection. *Infect. Genet. Evol.* **2022**, *98*, 105198. [CrossRef] [PubMed]
3. Sadiq, A.; Bostan, N.; Yinda, K.C.; Naseem, S.; Sattar, S. Rotavirus: Genetics, pathogenesis and vaccine advances. *Rev. Med. Virol.* **2018**, *28*, e2003. [CrossRef] [PubMed]
4. Bishop, R.F.; Davidson, G.P.; Holmes, I.H.; Ruck, B.J. Virus particles in epithelial cells of duodenal mucosa from children with acute non-bacterial gastroenteritis. *Lancet* **1973**, *2*, 1281–1283. [CrossRef] [PubMed]
5. Vlasova, A.N.; Amimo, J.O.; Saif, L.J. Porcine Rotaviruses: Epidemiology, Immune Responses and Control Strategies. *Viruses* **2017**, *9*, 48. [CrossRef] [PubMed]
6. Midgley, S.E.; Bányai, K.; Buesa, J.; Halaihel, N.; Hjulsager, C.K.; Jakab, F.; Kaplon, J.; Larsen, L.E.; Monini, M.; Poljšak-Prijatelj, M.; et al. Diversity and zoonotic potential of rotaviruses in swine and cattle across Europe. *Vet. Microbiol.* **2012**, *156*, 238–245. [CrossRef] [PubMed]
7. Patel, M.M.; Pitzer, V.E.; Alonso, W.J.; Vera, D.; Lopman, B.; Tate, J.; Viboud, C.; Parashar, U.D. Global seasonality of rotavirus disease. *Pediatr. Infect. Dis. J.* **2013**, *32*, e134–e147. [CrossRef]
8. Wandera, E.A.; Hatazawa, R.; Tsutsui, N.; Kurokawa, N.; Kathiiko, C.; Mumo, M.; Waithira, E.; Wachira, M.; Mwaura, B.; Nyangao, J.; et al. Genomic characterization of an African G4P6 human rotavirus strain identified in a diarrheic child in Kenya: Evidence for porcine-to-human interspecies transmission and reassortment. *Infect. Genet. Evol.* **2021**, *96*, 105133. [CrossRef]
9. Yan, N.; Tang, C.; Kan, R.; Feng, F.; Yue, H. Genome analysis of a G9P[23] group A rotavirus isolated from a dog with diarrhea in China. *Infect. Genet. Evol.* **2019**, *70*, 67–71. [CrossRef]
10. Sadiq, A.; Bostan, N.; Jadoon, K.; Aziz, A. Effect of rotavirus genetic diversity on vaccine impact. *Rev. Med. Virol.* **2022**, *32*, e2259. [CrossRef]
11. Banyai, K.; Kemenesi, G.; Budinski, I.; Foldes, F.; Zana, B.; Marton, S.; Varga-Kugler, R.; Oldal, M.; Kurucz, K.; Jakab, F. Candidate new rotavirus species in Schreiber's bats, Serbia. *Infect. Genet. Evol.* **2017**, *48*, 19–26. [CrossRef]
12. Caddy, S.L.; Vaysburd, M.; Wing, M.; Foss, S.; Andersen, J.T.; O'Connell, K.; Mayes, K.; Higginson, K.; Iturriza-Gomara, M.; Desselberger, U.; et al. Intracellular neutralisation of rotavirus by VP6-specific IgG. *PLoS Pathog.* **2020**, *16*, e1008732. [CrossRef] [PubMed]
13. Wenske, O.; Ruckner, A.; Piehler, D.; Schwarz, B.A.; Vahlenkamp, T.W. Epidemiological analysis of porcine rotavirus A genotypes in Germany. *Vet. Microbiol.* **2018**, *214*, 93–98. [CrossRef] [PubMed]
14. Amimo, J.O.; Junga, J.O.; Ogara, W.O.; Vlasova, A.N.; Njahira, M.N.; Maina, S.; Okoth, E.A.; Bishop, R.P.; Saif, L.J.; Djikeng, A. Detection and genetic characterization of porcine group A rotaviruses in asymptomatic pigs in smallholder farms in East Africa: Predominance of P8 genotype resembling human strains. *Vet. Microbiol.* **2015**, *175*, 195–210. [CrossRef]

15. Pham, H.A.; Carrique-Mas, J.J.; Nguyen, V.C.; Ngo, T.H.; Nguyet, L.A.; Do, T.D.; Vo, B.H.; Phan, V.T.; Rabaa, M.A.; Farrar, J.; et al. The prevalence and genetic diversity of group A rotaviruses on pig farms in the Mekong Delta region of Vietnam. *Vet. Microbiol.* **2014**, *170*, 258–265.

16. Theuns, S.; Vyt, P.; Desmarets, L.M.B.; Roukaerts, I.D.M.; Heylen, E.; Zeller, M.; Matthijnssens, J.; Nauwynck, H.J. Presence and characterization of pig group A and C rotaviruses in feces of Belgian diarrheic suckling piglets. *Virus Res.* **2016**, *213*, 172–183. [CrossRef] [PubMed]

17. Park, J.G.; Alfajaro, M.M.; Cho, E.H.; Kim, J.Y.; Soliman, M.; Baek, Y.B.; Park, C.H.; Lee, J.H.; Son, K.Y.; Cho, K.O.; et al. Development of a live attenuated trivalent porcine rotavirus A vaccine against disease caused by recent strains most prevalent in South Korea. *Vet. Res.* **2019**, *50*, 2. [CrossRef] [PubMed]

18. Zhou, X.; Wang, Y.-H.; Ghosh, S.; Tang, W.-F.; Pang, B.-B.; Liu, M.-Q.; Peng, J.-S.; Zhou, D.-J.; Kobayashi, N. Genomic characterization of G3P6, G4P6 and G4P8 human rotaviruses from Wuhan, China: Evidence for interspecies transmission and reassortment events. *Infect. Genet. Evol.* **2015**, *33*, 55–71. [CrossRef]

19. Li, D.D.; Duan, Z.J.; Zhang, Q.; Liu, N.; Xie, Z.P.; Jiang, B.; Steele, D.; Jiang, X.; Wang, Z.S.; Fang, Z.Y. Molecular characterization of unusual human G5P6 rotaviruses identified in China. *J. Clin. Virol.* **2008**, *42*, 141–148. [CrossRef]

20. Shui, I.M.; Baggs, J.; Patel, M.; Parashar, U.D.; Rett, M.; Belongia, E.A.; Hambidge, S.J.; Glanz, J.M.; Klein, N.P.; Weintraub, E. Risk of intussusception following administration of a pentavalent rotavirus vaccine in US infants. *JAMA* **2012**, *307*, 598–604. [CrossRef]

21. Xue, R.; Tian, Y.; Zhang, Y.; Zhang, M.; Li, Z.; Chen, S.; Liu, Q. Diversity of group Arotavirus of porcine rotavirus in Shandong province China. *Acta Virol.* **2018**, *62*, 229–234. [CrossRef] [PubMed]

22. Zhang, F.; Luo, S.; Gu, J.; Li, Z.; Li, K.; Yuan, W.; Ye, Y.; Li, H.; Ding, Z.; Song, D.; et al. Prevalence and phylogenetic analysis of porcine diarrhea associated viruses in southern China from 2012 to 2018. *BMC Vet. Res.* **2019**, *15*, 470. [CrossRef]

23. Tao, R.; Chang, X.; Zhou, J.; Zhu, X.; Yang, S.; Li, K.; Gu, L.; Zhang, X.; Li, B. Molecular epidemiological investigation of group A porcine rotavirus in East China. *Front. Vet. Sci.* **2023**, *10*, 1138419. [CrossRef]

24. Nan, P.; Wen, D.; Opriessnig, T.; Zhang, Q.; Yu, X.; Jiang, Y. Novel universal primer-pentaplex PCR assay based on chimeric primers for simultaneous detection of five common pig viruses associated with diarrhea. *Mol. Cell. Probes* **2021**, *58*, 101747. [CrossRef] [PubMed]

25. Werid, G.M.; Zhang, H.; Ibrahim, Y.M.; Pan, Y.; Zhang, L.; Xu, Y.; Zhang, W.; Wang, W.; Chen, H.; Fu, L.; et al. Development of a Multiplex RT-PCR Assay for Simultaneous Detection of Four Potential Zoonotic Swine RNA Viruses. *Vet. Sci.* **2022**, *9*, 176. [CrossRef]

26. Zhang, L.; Jiang, Z.; Zhou, Z.; Sun, J.; Yan, S.; Gao, W.; Shao, Y.; Bai, Y.; Wu, Y.; Yan, Z.; et al. A TaqMan Probe-Based Multiplex Real-Time PCR for Simultaneous Detection of Porcine Epidemic Diarrhea Virus Subtypes G1 and G2, and Porcine Rotavirus Groups A and C. *Viruses* **2022**, *14*, 1819. [CrossRef] [PubMed]

27. Yahiro, T.; Takaki, M.; Chandrasena, T.G.A.N.; Rajindrajith, S.; Iha, H.; Ahmed, K. Human-porcine reassortant rotavirus generated by multiple reassortment events in a Sri Lankan child with diarrhea. *Infect. Genet. Evol.* **2018**, *65*, 170–186. [CrossRef]

28. Flores, P.S.; Costa, F.B.; Amorim, A.R.; Mendes, G.S.; Rojas, M.; Santos, N. Rotavirus A, C, and H in Brazilian pigs: Potential for zoonotic transmission of RVA. *J. Vet. Diagn. Investig.* **2020**, *33*, 129–135. [CrossRef]

29. Wang, Z.; Lv, C.; Xu, X.; Li, X.; Yao, Y.; Gao, X.; Sun, Z.; Wang, Y.; Sun, Y.; Xiao, Y.; et al. The dynamics of a Chinese porcine G9P[23] rotavirus production in MA-104 cells and intestines of 3-day-old piglets. *J. Vet. Med. Sci.* **2018**, *80*, 790–797. [CrossRef]

30. Kim, Y.; Chang, K.O.; Straw, B.; Saif, L.J. Characterization of group C rotaviruses associated with diarrhea outbreaks in feeder pigs. *J. Clin. Microbiol.* **1999**, *37*, 1484–1488. [CrossRef]

31. Miyazaki, A.; Kuga, K.; Suzuki, T.; Kohmoto, M.; Katsuda, K.; Tsunemitsu, H. Genetic diversity of group A rotaviruses associated with repeated outbreaks of diarrhea in a farrow-to-finish farm: Identification of a porcine rotavirus strain bearing a novel VP7 genotype, G26. *Vet. Res.* **2011**, *42*, 112. [CrossRef] [PubMed]

32. Hoshino, Y.; Sereno, M.M.; Midthun, K.; Flores, J.; Kapikian, A.Z.; Chanock, R.M. Independent segregation of two antigenic specificities (VP3 and VP7) involved in neutralization of rotavirus infectivity. *Proc. Natl. Acad. Sci. USA* **1985**, *82*, 8701–8704. [CrossRef] [PubMed]

33. Theingi Win, M.; Hlaing Myat, T.; Ye Myint, K.; Khin Mar, A.; Mo Mo, W.; Htin, L.; Thin Thin, S.; Win, M.; Khin Khin, O.; Kyaw Zin, T. Sentinel surveillance for rotavirus in children <5 years of age admitted for Diarrheal illness to Yangon Children's Hospital, Myanmar, 2009–2014. *Vaccine* **2018**, *36*, 7832–7835.

Communication

Histo-Blood Group Antigen-Producing Bacterial Cocktail Reduces Rotavirus A, B, and C Infection and Disease in Gnotobiotic Piglets

Sergei A. Raev [1], Maryssa K. Kick [1], Maria Chellis [1], Joshua O. Amimo [2], Linda J. Saif [1] and Anastasia N. Vlasova [1,*]

[1] Center for Food Animal Health, Department of Animal Sciences, College of Food, Agricultural and Environmental Sciences, The Ohio State University, Wooster, OH 44691, USA; raev.1@osu.edu (S.A.R.); kick.28@osu.edu (M.K.K.); chellis.15@osu.edu (M.C.); saif.2@osu.edu (L.J.S.)

[2] GIVAX Inc., Philadelphia, PA 19106, USA; jamimo@givax.bio

* Correspondence: vlasova.1@osu.edu

Abstract: The suboptimal performance of rotavirus (RV) vaccines in developing countries and in animals necessitates further research on the development of novel therapeutics and control strategies. To initiate infection, RV interacts with cell-surface *O*-glycans, including histo-blood group antigens (HBGAs). We have previously demonstrated that certain non-pathogenic bacteria express HBGA⁻ like substances (HBGA⁺) capable of binding RV particles in vitro. We hypothesized that HBGA⁺ bacteria can bind RV particles in the gut lumen protecting against RV species A (RVA), B (RVB), and C (RVC) infection in vivo. In this study, germ-free piglets were colonized with HBGA⁺ or HBGA⁻ bacterial cocktail and infected with RVA/RVB/RVC of different genotypes. Diarrhea severity, virus shedding, immunoglobulin A (IgA) Ab titers, and cytokine levels were evaluated. Overall, colonization with HBGA⁺ bacteria resulted in reduced diarrhea severity and virus shedding compared to the HBGA⁻ bacteria. Consistent with our hypothesis, the reduced severity of RV disease and infection was not associated with significant alterations in immune responses. Additionally, colonization with HBGA⁺ bacteria conferred beneficial effects irrespective of the piglet HBGA phenotype. These findings are the first experimental evidence that probiotic performance in vivo can be improved by including HBGA⁺ bacteria, providing decoy epitopes for broader/more consistent protection against diverse RVs.

Keywords: probiotics; rotavirus infection; histo-blood group antigens; glycans; diarrhea; shedding

Citation: Raev, S.A.; Kick, M.K.; Chellis, M.; Amimo, J.O.; Saif, L.J.; Vlasova, A.N. Histo-Blood Group Antigen-Producing Bacterial Cocktail Reduces Rotavirus A, B, and C Infection and Disease in Gnotobiotic Piglets. *Viruses* **2024**, *16*, 660. https://doi.org/10.3390/v16050660

Academic Editors: Ulrich Desselberger and John T. Patton

Received: 1 April 2024
Revised: 17 April 2024
Accepted: 19 April 2024
Published: 24 April 2024

1. Introduction

Rotavirus (RV) is the major causative agent of acute gastroenteritis and is associated with an increased risk of secondary bacterial infections in children and young animals globally [1]. In children younger than 5 years of age, severe RV-induced diarrhea may lead to hospitalization and even death [2,3]. RV mainly targets the mature terminally differentiated intestinal epithelial cells (IECs), primarily of the ileum and jejunum [4,5]. Among a variety of RV receptors, cellular glycans have been shown to play a major role as attachment sites [5]. Specifically, histo-blood group antigens (HBGAs), including the antigens of the ABO blood group system, have been shown to play a critical role in determining RV species/genotype-specific binding and disease [5–8]. O and A but not B have been described for pigs (AO system) [9,10]. This is determined by the presence of only two alleles, *A* and *O*, in the porcine *ABO* gene [11,12], resulting in the existence of four phenotypes: A, A^weak, O, and "H⁻A⁻" [10]. Based on reactivity with "anti-A" and "anti-H" antibodies, pigs can be H⁻A+ (A phenotype), H⁺A⁺ (A^weak phenotype), H⁺A⁻ (O phenotype), and H⁻A⁻ (H⁻A⁻ phenotype) [12]. Our previous studies have demonstrated that the piglet HBGA phenotype affects RVA/RVC replication levels in vitro (in porcine ileal enteroids) [6,7]. This underscores the importance of considering the AO phenotype as an important factor influencing RV replication in vivo.

Before reaching its principal target, IECs, RV must penetrate the mucus layer, which protects IECs against enteric pathogens, including RV [13], and provides a niche for intestinal commensals [14]. There is growing evidence that several members of non-pathogenic bacteria produce glycans recognized by human HBGA-specific monoclonal antibodies [15–19]. Thus, while cellular HBGAs aid RV attachment, bacterial HBGAs might act as decoy epitopes, preventing RV attachment to the IECs. Our recent study has demonstrated the ability of some Gram-positive and Gram-negative non-pathogenic bacteria to express a variety of HBGAs and bind RV of different species (RVA/RVC) and genotypes in a genotype-specific manner [16,20], suggesting the potential role of these bacteria as decoy receptors for RV (Table S1). However, the impact of these HBGA-expressing bacteria on RV infection and disease in vivo remains unknown.

While probiotic supplementation is generally beneficial in terms of the overall performance of livestock animals [21], feed conversion efficiency, and in reducing post-weaning diarrhea in pigs [21,22], it does not always meet producer expectations mostly due to the inconsistent outcomes [23–26]. This is likely due to variable dosages and types of probiotics used, animal diet, and age. While most studies on the impact of probiotics demonstrate immune-mediated decreases in viral shedding and clinical disease severity [20,27–29], the data on the role of direct bacteria–RV interactions are limited. We hypothesized that colonization of germ-free (GF) piglets with HBGA expressing (HBGA$^+$) vs. non-expressing (HBGA$^-$) bacteria would lead to decreased replication of RV, resulting in reduced diarrhea severity and virus shedding after virus inoculation, and that these beneficial effects will be independent of the probiotic-induced immunomodulation. Thus, the goal of this study was to evaluate the protective effects of HBGA$^+$ vs. HBGA$^-$ bacteria against RVA, RVB, and RVC infection in vivo.

2. Materials and Methods

2.1. Commensal Bacteria

We used two commensal facultative anaerobic bacteria (*L. brevis*, *S. bovis*) and four obligate anaerobes (*B. adolescentis*, *B. longum*, *B. thetaiotaomicron*, and *C. clostridioforme*) previously isolated from the gut of healthy pigs (kindly provided by Dr. David Francis, South Dakota State University, Brookings, SD, USA). An additional facultative anaerobe, *E. coli* G58 (kindly provided by Dr. Carlton Gyles, University of Guelph, Guelph, ON, Canada) was also included in this study. All strains were cultured under aerobic (*E. coli* G58) and anaerobic conditions (*S. bovis*, *B. thetaiotaomicron*, *B. adolescentis*, *L. brevis*, *C. clostridioforme*, *B. longum*); the latter were generated using the GasPakTM EZ Anaerobe Container System Sachets (BD, Franklin Lakes, NJ, USA). All bacteria were enumerated as described previously [30]. Selected media and growth conditions for preparing bacterial cultures were reported previously [16].

2.2. Rotaviruses

Intestinal contents of GF piglets containing rotavirus A (RVA): Wa G1P[8] [31], RV0084 G9P[13] [32], Gottfried G4P[6] [33], OSU G5P[7]; rotavirus B (RVB): Ohio [34] (non-typed); and rotavirus C (RVC): Cowden G1P[1] [35]; RV0104 G3P[18]; RV0143 G6P[5] [32], were used to orally inoculate piglets at a dose of 1×10^6 fluorescent focus units (FFU).

2.3. Animal Experiments

All our animal experiments were approved by the Institutional Animal Care and Use Committee (IACUC) at Ohio State University (protocols #2009A0146, #2010A00000088). Near-term sows (Landrace × Yorkshire × Duroc crossbred) were purchased from the Ohio State University swine center facility/Shoup Brothers Farm LTD, Orville, OH, USA. GF piglets were derived by cesarean section and maintained as described previously [36]. On the 2nd day of life, rectal swabs were taken from all the piglets, and sterility was confirmed by culturing of rectal swabs in blood agar plates and thioglycolate broth culture. The presence of bacteria in the intestine vs. GF conditions has been shown to play a crucial role in nutrient absorption [37], immune system development [38], glycosylation profiles, and maintaining intestinal epithelial cell integrity [39], thus impacting immune responses to pathogens, including RV [40–42]. Therefore, to evaluate the anti-RV properties of the HBGA$^+$ bacterial cocktail, instead of using non-colonized piglets, we used the HBGA$^-$ bacterial cocktail as a control. Five- to seven-day-old GF piglets were supplemented for 5 consecutive days with one of the commensal bacteria cocktails (1×10^5 colony-forming units, CFU, of each strain per pig) containing (1) HBGA-expressing bacteria (HBGA$^+$): *Escherichia coli G-58*, *Bifidobacterium adolescentis*, *Bacteroides thetaiotaomicron*, *Streptococcus bovis*, and *Clostridium clostridioforme*; (2) HBGA-non-expressing bacteria (HBGA$^-$): *Lactobacillus brevis* and *Bifidobacterium longum*. The HBGA expression profiles of the bacterial strains used in this study were evaluated in our previous study [16]. On day 5 of supplementation, rectal swabs were collected for enumeration of fecal bacterial shedding [30] in the colonized pigs. Additionally, the presence of each bacterial strain was confirmed using PCR, as described previously [43] (primers [44–49] are listed in Table S2). On day 5 of supplementation, all piglets were inoculated with individual RVA/RVB/RVC strains at a dose of 1×10^6 FFU/piglet. After the RV challenge, rectal swabs were collected daily to assess RV shedding and diarrhea severity, as previously described [50]. Blood samples were collected on days 0, 3, 7, and 11 post-infection (dpi) to evaluate the IgA Ab titers and canonical innate and pro-inflammatory cytokine responses to RV infection. An innate immune response early-response cytokine IFN-α was evaluated at dpi 0 and dpi 3, while TNF-α, IL-10, and IL-22 were assessed at dpi 0 and dpi 11 to capture the late phase of the immune response [51–53]. All piglets were euthanized at dpi 11, and small and large intestinal contents (SIC and LIC) were collected, resuspended at a 1:1 ratio in MEM with a protease inhibitor cocktail containing 250 µg/mL of trypsin inhibitor and 50 µg/mL of leupeptin (Sigma, Saint Louis, MO, USA), and stored at $-70\,^\circ$C to evaluate the local IgA response. Ileum sections were collected to determine porcine HBGA phenotype.

2.4. Rotavirus Fecal Shedding

Cell culture immunofluorescence (CCIF) assay was used to quantify RVA as previously described [54]. The final RV titers were calculated and expressed as the reciprocal of the highest dilution showing positive fluorescing cells. To detect/quantify RVC and RVB, real-time RT-PCR was used as previously described [55,56] (primers are listed in Table S2).

2.5. Rotavirus A-Specific Antibody (Ab) ELISA Assay

Cell-culture-adapted RVA OSU G5P[7] and Wa G1P[8] strains were used to inoculate MA-104 cells, as described previously [57]. Infected cells were frozen/thawed 3 times, and after centrifugation, the supernatant was used as an antigen for IgA Ab ELISA (mock-infected MA-104 cells were used as a control). RV IgA ELISA was performed as described previously [58]. The RVA-specific IgA Ab titers were expressed as the reciprocal of the highest dilution that had a corrected optical density (OD)$_{450}$ value (sample OD$_{450}$ in the RVA antigen-coated well minus sample OD$_{450}$ in the mock antigen-coated well) greater than the cut-off value (the mean + three standard deviations of negative control samples).

2.6. RVC ELISA

Ninety-six-well plates (Nunc Maxisorp, Thermo Scientific Pierce, Rockford, IL, USA) were coated with lysates (normalized for total protein content) of High-Five cells (Mock) or High-Five cells infected with the recombinant baculovirus containing the VP6 gene of RVC G1P[1] diluted 1:50 in carbonate–bicarbonate (coating) buffer (pH 9.6). After overnight incubation at 4 $^\circ$C, the plates were rinsed twice with PBS containing 0.05% Tween-20 (PBST) and blocked with PBST containing 2% non-fat dry milk, and then incubated at 37 $^\circ$C for 1 h. After rinsing the plates 5 times with PBST, seven 3-fold dilutions of serum samples (starting at 1:5) were added to both Mock/VP6-coated plates and incubated at 37 $^\circ$C for

1 h. The plates were rinsed 5 times with PBST, and the secondary antibody, horseradish peroxidase-conjugated goat anti-porcine IgA antibody (Bio-Rad, Hercules, CA, USA), was added and incubated at 37 °C for 1 h. Then, the plates were rinsed 5 times with PBST, developed with TMB 2-Component Microwell Peroxidase Substrate Kit, and stopped with TMB Stop Solution (both from SeraCare Life Sciences Inc. Milford, MA, USA), and the OD values were read at 450 nm using SoftMax Pro 7.1 (Molecular Devices, LLC., San Jose, CA, USA). The antibody titers were determined as described previously [57].

2.7. Cytokine ELISA

Porcine TNF-α, INF-α, IL-10, and IL-22 ELISA kits were used as described in the manufacturer's recommendations (Thermo Scientific Pierce, Rockford, IL, USA).

2.8. HBGA Immunohistochemistry

To determine porcine HBGA phenotype, formalin-fixed ileal sections [59] were stained with HBGA-A- or -H-specific mouse monoclonal antibodies (mAb) (Biolegend, San Diego, CA, USA) as described previously [6].

2.9. Statistical Analysis

All statistical analyses were performed using GraphPad Prism version 8 (GraphPad Software, Inc., La Jolla, CA, USA). The mean duration of diarrhea and fecal RV shedding post-challenge were analyzed using an unpaired t-test. Log-transformed RV-specific IgA Ab antibodies were analyzed using two-way ANOVA followed by Duncan's multiple range test. The area under the curve (AUC) analysis was conducted to compare diarrhea severity and shedding among the groups [60]. A Kruskal–Wallis rank sum test was then performed to compare the total AUC values between the groups. Differences were considered significant at $p \leq 0.05$.

3. Results

3.1. HBGA-Expressing Bacteria (HBGA⁺) Cocktail Reduces Diarrhea Severity and Virus Shedding

The presence of individual strains of both HBGA⁺ and HBGA⁻ bacterial cocktails in rectal swabs was confirmed by using species-specific primers in PCR analysis. Total aerobic bacterial counts in piglets colonized with HBGA⁻ vs. HBGA⁺ bacteria did not differ, while the numbers of anaerobic bacteria were significantly higher in the piglets colonized with HBGA⁻ bacteria (Figure S1, $p < 0.05$). Diarrhea onset after infection with RVA (G5P[7], G4P[6]), RVC G3P[18], and RVB strains was significantly delayed in the HBGA⁺ vs. HBGA⁻ piglets (Table 1). Further, the RV-induced diarrhea lasted significantly longer in the piglets colonized with HBGA⁻ bacteria, in groups infected with RVA G4P[6] and RVB strains (Table 1, $p < 0.05$). In addition, piglets colonized with HBGA⁺ bacteria displayed a significant reduction in the mean cumulative fecal score (Table 1, $p < 0.05$), in the RVB-infected piglets. A significantly lower AUC value was noted in the piglets colonized with the HBGA⁺ bacterial cocktail after infection with RVA G4P[6], G9P[13], and RVB (Table 1, $p < 0.05$). Further, statistically significant decreases in diarrhea severity were noted at dpi 1 for the RVB-infected piglets (Figure 1E, $p < 0.05$); at dpi 6 for the RVA G9P[13]-infected piglets (Figure 1D, $p < 0.01$), and at dpi 7 for the RVA G4P[6]- and G9P[13]-infected piglets (Figure 1C,D, $p < 0.05$).

Table 1. Diarrhea in piglets orally inoculated with virulent RVs.

RV Strains	N		Mean Days to Diarrhea Onset [1]		Mean Diarrhea Duration (Days) [2]		Mean Cumulative Fecal Score [3]		AUC [4]	
	HBGA⁺	HBGA⁻	HBGA⁺	HBGA⁻	HBGA⁺	HBGA⁻	HBGA⁺	HBGA⁻	HBGA⁺	HBGA⁻
Wa G1P[8]	4	4	2.0	2.8	3.3	4.8	13.0	13.8	12.50	12.88
OSU G5P[7]	4	4	3.3	2.5	5.5	6.0	15.0	17.0	14.38	16.50
Gottfried G4P[6]	3	3	2.0	1.0	**3.3**	**5.7**	11.7	15.3	**11.17**	**14.83**
RV0084 G9P[13]	4	4	1.0	1.0	2.3	3.8	8.8	13.8	**8.625**	**13.75**
RVB Ohio	4	4	2.0	1.0	**5.3**	**7.3**	**13.5**	**19.0**	**13.50**	**19.00**
Cowden G1P[1]	4	4	2.7	2.8	2.7	4.5	10.3	13.5	9.833	12.75
RV0104 G3P[18]	4	4	4.5	3.8	3.8	5.3	14.3	16.3	13.75	15.83
RV0143 G6P[5]	4	8	2.3	2.1	6.0	5.3	16.3	16.0	15.75	15.50

[1] Diarrhea onset is defined as the number of days between the virus inoculation and the first manifestation of diarrhea (e.g., fecal consistency score of $\geq$2). [2] Duration of diarrhea is defined as the number of days that the fecal consistency score was $\geq$2. Fecal diarrhea was scored as follows: 0, normal; 1, pasty; 2, semiliquid; 3, liquid. [3] Mean cumulative fecal score [(sum of fecal consistency score for 11 days postinoculation)/N], where N is the number of pigs receiving the inoculation. Means in the same row were analyzed by unpaired *t*-test. [4] Area under the curve (AUC) was calculated using the area under the curve analysis function in the Prism software. A Kruskal–Wallis rank sum test was then performed to compare the total AUC between the groups. Significant differences (bold) are indicated as calculated by an unpaired *t*-test.

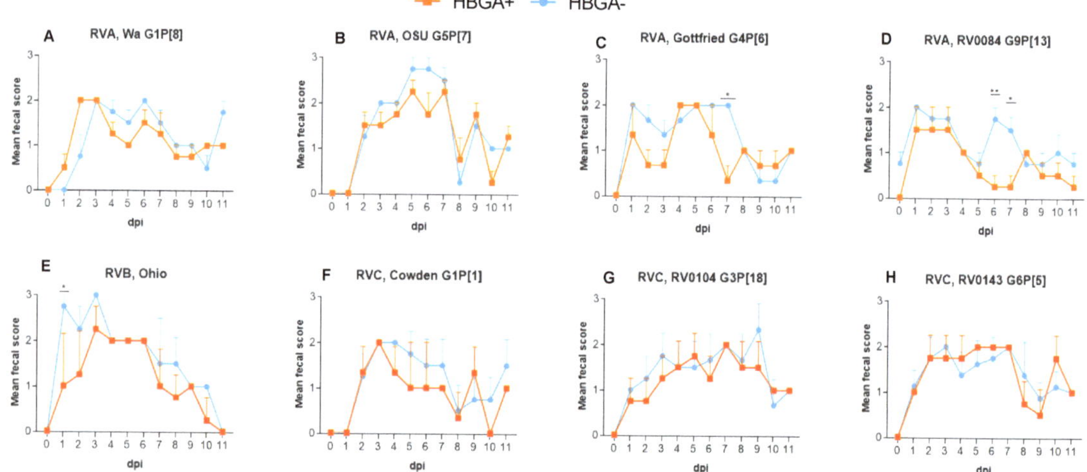

Figure 1. Diarrhea in piglets following RVA (**A–D**), RVB (**E**), and RVC (**F–H**) inoculation. Individual RV strains were used to inoculate (1×10^6 FFU) piglets after 5 consecutive days of supplementation with HBGA⁺ or HBGA⁻ bacteria. Fecal consistency was scored as follows: 0, normal; 1, pasty; 2, semiliquid; 3, liquid; and diarrhea was considered as a score of $\geq$2. The error bars represent the standard deviations; significant differences (* $p < 0.05$, ** $p < 0.01$) are indicated as calculated by two-way ANOVA followed by Duncan's multiple comparisons test.

Consistent with the clinical data, we observed a significantly delayed onset of virus shedding in the piglets colonized with HBGA⁺ bacteria after infection with RVC strains G6P[5] and G3P[18] (Table 2, $p < 0.05$) compared to the piglets colonized with HBGA⁻ bacteria. In addition, HBGA+-inoculated piglets had a significantly shortened duration of virus shedding in piglets infected with RVA G1P[8] and RVC G6P[5] (Table 2, $p < 0.05$). Significantly lower viral shedding titers in this group were observed on dpi 1 after infection with RVA G4P[6], G9P[13], RVB (Figure 2C–E, $p < 0.001$), and RVC G6P[5] (Figure 2H, $p < 0.05$); on dpi 2 after infection with RVA G4P[6] (Figure 2C, $p < 0.05$); and at dpi 6 after infection with RVA G5P[7] and RVC G1P[1] (Figure 2B,F, $p < 0.05$).

Table 2. Virus shedding in piglets orally inoculated with virulent RVs.

	N		Mean Days to Shedding Onset		Mean Shedding Duration (Days)		Avg Peak Titer (FFU/mL)		AUC [1]	
RV Strains	HBGA+	HBGA−	HBGA+	HBGA−	HBGA+	HBGA−	HBGA+	HBGA−	HBGA+	HBGA−
Wa G1P[8]	4	4	2.0	2.0	**6.3**	**8.5**	5.58×10^4	6.14×10^4	6.04×10^4	2.50×10^5
OSU G5P[7]	4	4	2.0	2.0	6.8	6.5	7.41×10^6	9.01×10^6	2.16×10^7	2.18×10^7
Gottfried G4P[6]	3	3	2.0	1.7	4.7	5.7	6.63×10^5	4.83×10^6	6.62×10^5	4.82×10^6
RV0084 G9P[13]	4	4	1.3	1.0	7.5	8.3	8.02×10^6	9.85×10^6	8.04×10^6	9.87×10^6
RVB Ohio	4	4	1.0	1.0	7.0	7.0	7.00×10^2	1.74×10^3	2.07×10^3	2.64×10^3
Cowden G1P[1]	4	4	2.0	2.0	10.0	10.0	3.76×10^5	4.12×10^5	9.60×10^5	8.29×10^5
RV0104 G3P[18]	4	4	**2.0**	**1.0**	9.8	10.3	5.07×10^4	3.94×10^4	1.32×10^5	9.23×10^4
RV0143 G6P[5]	4	8	**3.5**	**1.1**	**8.0**	**10.5**	1.60×10^5	1.83×10^6	3.97×10^5	3.53×10^6

Means in the same row were analyzed by unpaired *t*-test. [1] Area under the curve (AUC) was calculated using the area under the curve analysis function in the Prism software. A Kruskal–Wallis rank sum test was then performed to compare the total AUC between the groups. Significant differences (bold) are indicated as calculated by an unpaired *t*-test.

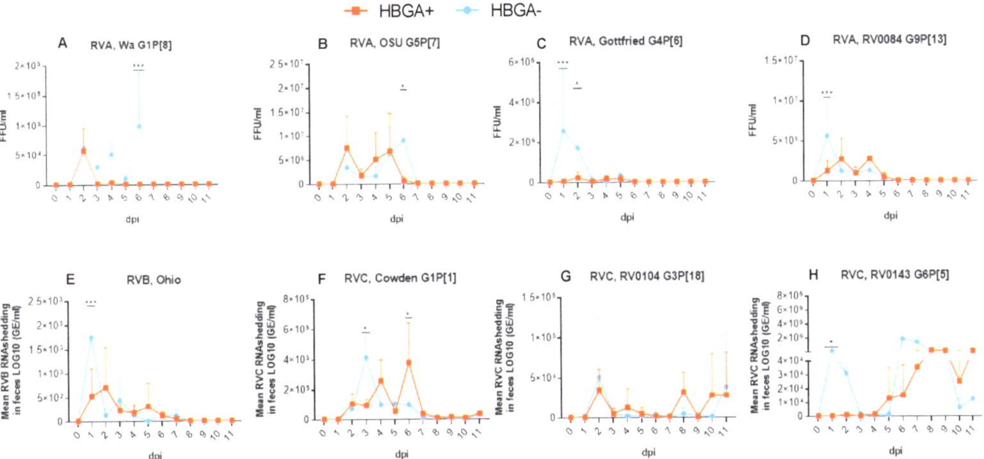

Figure 2. Mean virus shedding titers in piglets following RVA (**A–D**), RVB (**E**), and RVC (**F–H**) inoculation. Individual RV strains were used to colonize (1×10^6 FFU) piglets after 5 consecutive days of supplementation with HBGA+ or HBGA− bacteria. The error bars represent the standard deviations; significant differences (* $p < 0.05$, *** $p < 0.001$) are indicated as calculated by two-way ANOVA followed by Duncan's multiple comparisons test.

3.2. There Was No Evidence That the Protective Effect of HBGA+ Bacteria Was Immune-Mediated

To confirm that the observed protective effect of HBGA+ bacteria on RV infections was not associated with bacteria-mediated immunomodulation, we evaluated the local (intestinal content) and systemic (serum) RV-specific IgA Ab responses. Data on the mean RVA/RVC-specific IgA Ab titers in the blood (Figure 3) and intestinal contents (Figure 4) revealed that colonization with HBGA+ bacteria did not result in significantly higher IgA Ab titers (compared to piglets colonized with HBGA− bacteria) after infection with RVA/RVC strains. In addition, significantly lower IgA Ab titers in the blood of HBGA+

bacteria-colonized piglets were observed after infection with RVA G1P[8] (Figure 3A, $p < 0.01$), RVA G5P[7] (Figure 3B, $p < 0.001$), and RVC G1P[1] (Figure 3E, $p < 0.05$). This coincided with significantly lower IgA Ab titers in the intestinal contents of the piglets colonized with HBGA$^+$ bacteria after infection with RVA G1P[8] and G9P[13] (Figure 4A,D, $p < 0.01$).

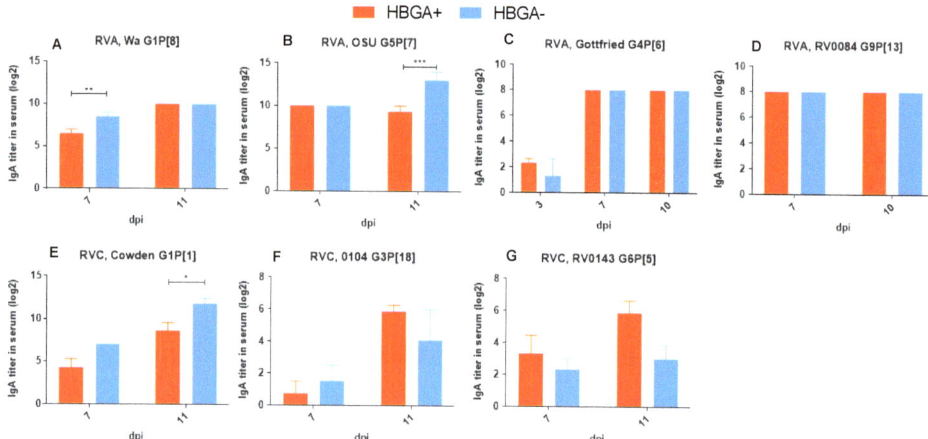

Figure 3. IgA Ab titers in blood samples collected on dpi 0, 3, 7, 11 following RVA (**A–D**), and RVC (**E–G**) infection. ELISA IgA Ab titers were analyzed using two-way ANOVA followed by Duncan's multiple comparisons test (* $p < 0.05$, ** $p < 0.01$, *** $p < 0.001$).

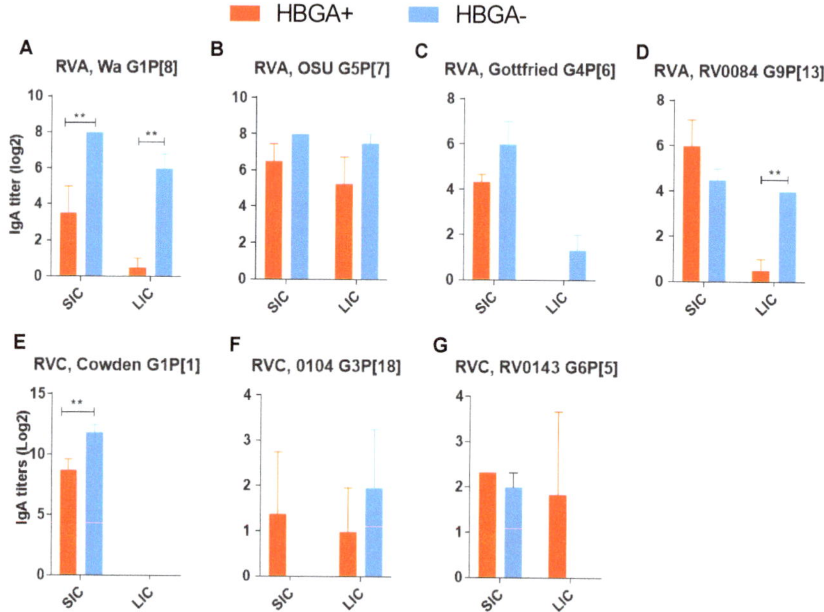

Figure 4. IgA Ab titers in piglet small/large intestinal content (SIC/LIC) samples collected at necropsy 11 days after RVA (**A–D**), and RVC (**E–G**) infection. ELISA IgG Ab titers were analyzed using two-way ANOVA followed by Duncan's multiple comparisons test (** $p < 0.01$).

We also evaluated the cytokine profiles in blood samples collected before (dpi 0) and after (dpi 3) virus inoculation (Figures S2 and S3). In piglets colonized with HBGA$^-$ bacteria, we observed a higher IFN-α concentration compared to those colonized with HBGA$^+$ bacteria after infection with RVA G5P[7] (Figure S2B, $p < 0.05$). The TNF-α concentrations were higher following RVC G3P[18] infection in the piglets colonized with HBGA$^+$ vs. HBGA$^-$ bacteria (Figure S3G, $p < 0.05$). However, the rest of the data on TNF-α and IL-22/IL-10 responses did not allow for discrimination between piglets colonized with HBGA$^+$ vs. HBGA$^-$ bacterial cocktails.

3.3. The Protective Effect of HBGA+ Bacteria Did Not Vary with the Piglet HBGA Phenotype

Although previous studies suggested that the host HBGA phenotype plays an important role in RV infection and evolution [61,62], there are no in vivo data for porcine RVs [6,7]. Here, we aimed to establish whether the protective effect of the bacterial cocktail was independent of the piglet HBGA phenotype. In our study, the data for piglets with the A^{strong} (A$^+$H$^-$) and A^{weak} (A$^+$H$^+$) phenotypes were combined and compared with the O phenotype (A$^-$H$^+$). Our analysis demonstrated that there were no significant differences in diarrhea severity and virus shedding associated with the A$^+$ vs. A$^-$ phenotypes (Table S3). This indicates that the protective effect of HBGA$^+$ bacteria on RV infection was not affected by piglet HBGA phenotype.

4. Discussion

The tripartite RV–host–commensal bacteria interactions have been demonstrated to have profound impacts on RV infection and disease [13,18,63]. In addition to the known mechanisms of protection, such as immunomodulation [64], metabolic and enzymatic support [65], and improved barrier function [66] utilized by probiotics, studies have shown that certain non-pathogenic bacteria possess the ability to directly bind certain viruses [67], including RVs [13,68]. For RVs, this phenomenon has been shown to be associated with the ability of bacteria to express structures similar to what RV uses as attachment sites on IECs, such as HBGAs [16]. However, there is no consensus opinion on the significance of these interactions in vivo [13,69,70]. The current study aimed to evaluate whether direct binding of RV by HBGA$^+$ bacteria is associated with reduced or enhanced RV infection and disease.

Our current data suggest that HBGA$^+$ bacteria improved protection against RVA, RVB, and RVC infection compared to HBGA$^-$ bacteria [16]. Although several studies have shown that the presence of bacterial HBGA-like structures enhanced viral replication [71,72], our study demonstrated that the direct binding of RV virions may represent an additional mechanism of antiviral protection conferred by probiotic/commensal bacteria. These contrasting findings may be attributed to the use of in vitro models (cell culture) in other studies, which lack a key component of the RV–host–bacteria interactions, the mucus, and, thus, may not be physiologically relevant. While RV binding by bacteria can facilitate RV particle delivery to target cells in vitro, the intestinal mucus in the gut may significantly restrict direct contact between bacteria and IECs [73,74], thus limiting the ability of bacteria to serve as a "Trojan horse" for viruses.

There are multiple studies on the use of probiotics against RV infection in which the beneficial/protective effects of certain bacteria are linked to the immunomodulatory effects of bacteria [75–78]. Our findings indicated that in most cases, the IgA Ab titers and IFN-α concentrations following RV infection were either similar or lower in the piglets colonized with HBGA$^+$ bacteria. This suggests that the beneficial effect of HBGA$^+$ bacteria was not associated with improved immune responses. In contrast, increased IgA Ab titers and cytokine concentrations in most cases could be associated with increased virus replication in the piglets colonized with HBGA$^-$ bacteria.

Our study has several limitations related to the inability to control all the aspects of the bacteria–RV interactions between the two different bacterial cocktails. First, in this study, we utilized a different number of bacterial strains in probiotic cocktails (five in the HBGA$^+$ and two in the HBGA$^-$). However, our previous study demonstrated that an increased number of probiotics in treatment does not result in superior anti-RV protection compared to an individual probiotic treatment [20]. In our current study, both HBGA$^+$ and HBGA$^-$

bacteria colonized GF piglets effectively, with higher total bacterial counts observed in piglets colonized with HBGA⁻ bacteria, which could be due to the individual growing characteristics of bacterial strains. This indicates that the higher number of bacterial strains used in the HBGA⁺ bacterial cocktail did not result in a higher bacterial load in the colonized piglets. Next, while we did not have any evidence that the protective effect of the HBGA⁺ probiotic cocktail was immune-mediated, there are other bacteria-mediated effects on the host and RV infection that were not evaluated in this study. For example, bacteria can regulate the mucus composition by stimulating the production or degradation of mucin-type glycans, altering the ability of mucus to provide decoy epitopes for RV attachment. Several probiotics have been shown to up- (*Lactobacilli*) and down-regulate (*Bifidobacteria* and *Streptococci*) mucin secretion [79]. In addition, *B. thetaiotaomicron*, which was used in this study as a component of the HBGA⁺ bacterial cocktail, was previously shown to stimulate mucin secretion in vitro [80]. In addition, various commensal bacteria, including *B. thetaiotaomicron* and *B. longum* were shown to produce sialidases, a group of enzymes responsible for sialic acid removal and mucin degradation, thus affecting the protective role of the intestinal mucus [81,82]. Thus, more studies are needed to dissect the roles of individual/combined HBGA⁺ bacteria in RV infection.

Nevertheless, this is the first study that evaluated a probiotic cocktail broadly protective against genetically distinct RVA, RVB, and RVC strains and demonstrated the role of the HBGA-mediated interactions in this protection in vivo. Thus, our data provide a proof-of-concept that probiotic/commensal bacteria can act as decoy receptors reducing the severity of RV infection and disease in vivo. Further studies are needed to validate the feasibility of this concept in conventional animals.

5. Conclusions

In conclusion, we demonstrated the ability of the HBGA⁺ bacteria to reduce species A, B, and C RV infection in vivo. This may represent a novel mechanism for protection against RV-associated diarrhea. However, the protective effects of the HBGA⁺ bacteria in conventional animals, where HBGA⁺ bacteria would need to compete with already established microbiota, remain to be evaluated. In addition, other aspects of RV–host–bacteria interactions, such as enzymatic and metabolic alterations associated with the HBGA⁺ bacterial cocktail and their effects on IEC integrity, must be further investigated.

Supplementary Materials: The following supporting information can be downloaded at: https://www.mdpi.com/article/10.3390/v16050660/s1, Table S1: Data on the ability of individual HBGA+ and HBGA- bacterial strains to bind virulent RVs; Figure S1: Fecal probiotic bacterial shedding from probiotic colonized piglets. An asterisk indicates a significant difference ($p < 0.05$) in fecal probiotic counts among treatment groups; Figure S2: IFN-α concentrations in piglet blood samples collected on dpi 0 and dpi 3 following RVA (A–D), RVB (E), and RVC (F–H) infection. Significant differences (* $p < 0.05$) are indicated as calculated by two-way ANOVA followed by Sidak's multiple comparisons test; Figure S3: TNF-α concentrations in piglet blood samples collected on dpi 0 and 11 following RVA (A–D), RVB (E), and RVC (F–H) infection. Significant differences (* $p < 0.05$) are indicated as calculated by two-way ANOVA followed by Duncan's multiple comparisons test; Table S2: Primers used in this study; Table S3: Diarrhea and virus shedding in gnotobiotic piglets after oral inoculation with virulent RVs.

Author Contributions: Conceptualization, A.N.V.; methodology, S.A.R. and A.N.V.; formal analysis, S.A.R.; investigation, S.A.R.; technical support, J.O.A., M.K.K. and M.C.; resources, A.N.V.; writing—original draft preparation, S.A.R.; revised draft writing, review, and editing, A.N.V., J.O.A. and S.A.R.; critical review, A.N.V. and L.J.S.; visualization, S.A.R.; supervision, A.N.V.; project administration, A.N.V.; funding acquisition, A.N.V. All authors have read and agreed to the published version of the manuscript.

Funding: This work was supported by the International Development Research Centre (IDRC, Canada), grant number 109053, to Dr. Vlasova and Dr. Vlasova's startup funding.

Institutional Review Board Statement: All our animal experiments were approved by the Institutional Animal Care and Use Committee (IACUC) at Ohio State University) (protocol #2009A0146, approved on 10 December 2021 and #2010A00000088 approved on 23 February 2021).

Data Availability Statement: Data are contained within the article.

Acknowledgments: We gratefully acknowledge the technical assistance of Molly Raque, Ronna Wood, Juliette Hanson, Sara Tallmadge, Megan Strother, Alfred Mainga, and Yusheng Guo. This study was supported by a grant from the IDRC, 109053 (Anastasia N. Vlasova, PI, and Linda J. Saif, co-PI).

Conflicts of Interest: Author J.O.A is employed by the company Givax. The remaining authors declare that the research was conducted in the absence of any commercial or financial relationships that could be construed as a potential conflict of interest. The funders had no role in the design of the study; in the collection, analyses, or interpretation of data; in the writing of the manuscript; or in the decision to publish the results.

References

1. Grimprel, E.; Rodrigo, C.; Desselberger, U. Rotavirus Disease: Impact of Coinfections. *Pediatr. Infect. Dis. J.* **2008**, *27*, S3. [CrossRef]
2. Ma, L.; El Khoury, A.C.; Itzler, R.F. The Burden of Rotavirus Hospitalizations Among Medicaid and Non-Medicaid Children Younger Than 5 Years Old. *Am. J. Public Health* **2009**, *99* (Suppl. S2), S398–S404. [CrossRef] [PubMed]
3. Du, Y.; Chen, C.; Zhang, X.; Yan, D.; Jiang, D.; Liu, X.; Yang, M.; Ding, C.; Lan, L.; Hecht, R.; et al. Global burden and trends of rotavirus infection-associated deaths from 1990 to 2019: An observational trend study. *Virol. J.* **2022**, *19*, 166. [CrossRef]
4. Ishizuka, T.; Kanmani, P.; Kobayashi, H.; Miyazaki, A.; Soma, J.; Suda, Y.; Aso, H.; Nochi, T.; Iwabuchi, N.; Xiao, J.Z.; et al. Immunobiotic Bifidobacteria Strains Modulate Rotavirus Immune Response in Porcine Intestinal Epitheliocytes via Pattern Recognition Receptor Signaling. *PLoS ONE* **2016**, *11*, e0152416. [CrossRef]
5. Amimo, J.O.; Raev, S.A.; Chepngeno, J.; Mainga, A.O.; Guo, Y.; Saif, L.; Vlasova, A.N. Rotavirus Interactions with Host Intestinal Epithelial Cells. *Front. Immunol.* **2021**, *12*, 793841. [CrossRef] [PubMed]
6. Guo, Y.; Candelero-Rueda, R.A.; Saif, L.J.; Vlasova, A.N. Infection of porcine small intestinal enteroids with human and pig rotavirus A strains reveals contrasting roles for histo-blood group antigens and terminal sialic acids. *PLoS Pathog.* **2021**, *17*, e1009237. [CrossRef] [PubMed]
7. Guo, Y.; Raev, S.; Kick, M.K.; Raque, M.; Saif, L.J.; Vlasova, A.N. Rotavirus C Replication in Porcine Intestinal Enteroids Reveals Roles for Cellular Cholesterol and Sialic Acids. *Viruses* **2022**, *14*, 1825. Available online: https://www.ncbi.nlm.nih.gov/pmc/articles/PMC9416568/ (accessed on 15 September 2022). [CrossRef] [PubMed]
8. Hu, L.; Crawford, S.E.; Czako, R.; Cortes-Penfield, N.W.; Smith, D.F.; Le Pendu, J.; Estes, M.K.; Prasad, B.V. Cell attachment protein VP8* of a human rotavirus specifically interacts with A-type histo-blood group antigen. *Nature* **2012**, *485*, 256–259. [CrossRef]
9. Martínez-Alarcón, L.; Ramis, G.; Majado, M.J.; Quereda, J.J.; Herrero-Medrano, J.M.; Ríos, A.; Ramírez, P.; Muñoz, A. ABO and RH1 Blood Group Phenotyping in Pigs (*Sus scrofa*) Using Microtyping Cards. *Transplant. Proc.* **2010**, *42*, 2146–2148. [CrossRef]
10. Smith, D.M.; Newhouse, M.; Naziruddin, B.; Kresie, L. Blood groups and transfusions in pigs. *Xenotransplantation* **2006**, *13*, 186–194. [CrossRef]
11. Bolner, M.; Bertolini, F.; Bovo, S.; Schiavo, G.; Fontanesi, L. Investigation of ABO Gene Variants across More Than 60 Pig Breeds and Populations and Other Suidae Species Using Whole-Genome Sequencing Datasets. *Animals* **2023**, *14*, 5. [CrossRef] [PubMed]
12. Hampton, C.; Dehghanpir, S.; Armstrong, C.; Scully, C.; Baker, R.E.; Mitchell, M. Prevalence of AO blood group and level of agreement for AO blood-typing methods in pet pigs from Louisiana. *J. Vet. Emerg. Crit. Care* **2023**, *33*, 549–558. [CrossRef] [PubMed]
13. Raev, S.A.; Amimo, J.O.; Saif, L.J.; Vlasova, A.N. Intestinal mucin-type O-glycans: The major players in the host-bacteria-rotavirus interactions. *Gut Microbes* **2023**, *15*, 2197833. [CrossRef] [PubMed]
14. Li, H.; Limenitakis, J.P.; Fuhrer, T.; Geuking, M.B.; Lawson, M.A.; Wyss, M.; Brugiroux, S.; Keller, I.; Macpherson, J.A.; Rupp, S.; et al. The outer mucus layer hosts a distinct intestinal microbial niche. *Nat. Commun.* **2015**, *6*, 8292. [CrossRef] [PubMed]
15. Miura, T.; Sano, D.; Suenaga, A.; Yoshimura, T.; Fuzawa, M.; Nakagomi, T.; Nakagomi, O.; Okabe, S. Histo-Blood Group Antigen-Like Substances of Human Enteric Bacteria as Specific Adsorbents for Human Noroviruses. *J. Virol.* **2013**, *87*, 9441–9451. [CrossRef] [PubMed]
16. Raev, S.; Omwando, A.; Guo, Y.; Raque, M.; Amimo, J.; Saif, L.; Vlasova, A. Glycan-mediated interactions between bacteria, rotavirus and the host cells provide an additional mechanism of antiviral defence. *Benef. Microbes* **2022**, *13*, 383–395. [CrossRef] [PubMed]
17. Comstock, L.E.; Kasper, D.L. Bacterial Glycans: Key Mediators of Diverse Host Immune Responses. *Cell* **2006**, *126*, 847–850. [CrossRef] [PubMed]
18. Gozalbo-Rovira, R.; Rubio-Del-Campo, A.; Santiso-Bellón, C.; Vila-Vicent, S.; Buesa, J.; Delgado, S.; Molinero, N.; Margolles, A.; Yebra, M.J.; Collado, M.C.; et al. Interaction of Intestinal Bacteria with Human Rotavirus during Infection in Children. *Int. J. Mol. Sci.* **2021**, *22*, 1010. [CrossRef]

19. Almand, E.A.; Moore, M.D.; Jaykus, L.A. Characterization of human norovirus binding to gut-associated bacterial ligands. *BMC Res. Notes* **2019**, *12*, 607. [CrossRef]

20. Kandasamy, S.; Vlasova, A.N.; Fischer, D.; Kumar, A.; Chattha, K.S.; Rauf, A.; Shao, L.; Langel, S.N.; Rajashekara, G.; Saif, L.J. Differential Effects of Escherichia coli Nissle and Lactobacillus rhamnosus Strain GG on Human Rotavirus Binding, Infection, and B Cell Immunity. *J. Immunol.* **2016**, *196*, 1780–1789. [CrossRef]

21. Liao, S.F.; Nyachoti, M. Using probiotics to improve swine gut health and nutrient utilization. *Anim. Nutr.* **2017**, *3*, 331–343. [CrossRef] [PubMed]

22. Zhao, P.Y.; Kim, I.H. Effect of direct-fed microbial on growth performance, nutrient digestibility, fecal noxious gas emission, fecal microbial flora and diarrhea score in weanling pigs. *Anim. Feed Sci. Technol.* **2015**, *200*, 86–92. [CrossRef]

23. Bekaert, H.; Moermans, R.; Eeckhout, W. Effect of a live yeast culture in diets of weanling piglets on performances and fre-quency of diarrhoea. *Anim. Res.* **1996**, *45*, 369–376. [CrossRef]

24. Harper, A.F.; Kornegay, E.T.; Bryant, K.L.; Thomas, H.R. Efficacy of virginiamycin and a commercially-available lactobacillus probiotic in swine diets. *Anim. Feed Sci. Technol.* **1983**, *8*, 69–76. [CrossRef]

25. Jeong, J.; Kim, J.K.; Lee, S.; Kim, I. Evaluation of Bacillus subtilis and Lactobacillus acidophilus probiotic supplementation on reproductive performance and noxious gas emission in sows. *Ann. Anim. Sci.* **2015**, *15*, 699–710. [CrossRef]

26. Walsh, M.; Saddoris, K.; Sholly, D.; Hinson, R.; Sutton, A.; Applegate, T.; Richert, B.; Radcliffe, J. The effects of direct fed microbials delivered through the feed and/or in a bolus at weaning on growth performance and gut health. *Livest. Sci.* **2007**, *108*, 254–257. [CrossRef]

27. Das, S.; Gupta, P.K.; Das, R.R. Efficacy and Safety of Saccharomyces boulardii in Acute Rotavirus Diarrhea: Double Blind Randomized Controlled Trial from a Developing Country. *J. Trop. Pediatr.* **2016**, *62*, 464–470. [PubMed]

28. Lee, D.K.; Park, J.E.; Kim, M.J.; Seo, J.G.; Lee, J.H.; Ha, N.J. Probiotic bacteria, B. longum and L. acidophilus inhibit infection by rotavirus in vitro and decrease the duration of diarrhea in pediatric patients. *Clin. Res. Hepatol. Gastroenterol.* **2015**, *39*, 237–244. [CrossRef]

29. Maragkoudakis, P.A.; Chingwaru, W.; Gradisnik, L.; Tsakalidou, E.; Cencic, A. Lactic acid bacteria efficiently protect human and animal intestinal epithelial and immune cells from enteric virus infection. *Int. J. Food Microbiol.* **2010**, *141* (Suppl. S1), S91–S97. [CrossRef]

30. Huang, H.C.; Vlasova, A.N.; Kumar, A.; Kandasamy, S.; Fischer, D.D.; Deblais, L.; Paim, F.C.; Langel, S.N.; Alhamo, M.A.; Rauf, A.; et al. Effect of antibiotic, probiotic, and human rotavirus infection on colonisation dynamics of defined commensal microbiota in a gnotobiotic pig model. *Benef. Microbes* **2018**, *9*, 71–86. [CrossRef]

31. Wyatt, R.G.; James, W.D.; Bohl, E.H.; Theil, K.W.; Saif, L.J.; Kalica, A.R.; Greenberg, H.B.; Kapikian, A.Z.; Chanock, R.M. Human rotavirus type 2: Cultivation in vitro. *Science* **1980**, *207*, 189–191. [CrossRef] [PubMed]

32. Amimo, J.O.; Vlasova, A.N.; Saif, L.J. Detection and genetic diversity of porcine group A rotaviruses in historic (2004) and recent (2011 and 2012) swine fecal samples in Ohio: Predominance of the G9P[13] genotype in nursing piglets. *J. Clin. Microbiol.* **2013**, *51*, 1142–1151. [CrossRef] [PubMed]

33. Bohl, E.H.; Theil, K.W.; Saif, L.J. Isolation and serotyping of porcine rotaviruses and antigenic comparison with other rotaviruses. *J. Clin. Microbiol.* **1984**, *19*, 105. [CrossRef] [PubMed]

34. Theil, K.W.; Saif, L.J.; Moorhead, P.D.; Whitmoyer, R.E. Porcine rotavirus-like virus (group B rotavirus): Characterization and pathogenicity for gnotobiotic pigs. *J. Clin. Microbiol.* **1985**, *21*, 340–345. [CrossRef] [PubMed]

35. Tsunemitsu, H.; Saif, L.J.; Jiang, B.M.; Shimizu, M.; Hiro, M.; Yamaguchi, H.; Ishiyama, T.; Hirai, T. Isolation, characterization, and serial propagation of a bovine group C rotavirus in a monkey kidney cell line (MA104). *J. Clin. Microbiol.* **1991**, *29*, 2609–2613. [CrossRef] [PubMed]

36. Meyer, R.C.; Bohl, E.H.; Kohler, E.M. Procurement and Maintenance of Germ-Free Swine for Microbiological Investigations. *Appl. Microbiol.* **1964**, *12*, 295–300. [CrossRef] [PubMed]

37. Krajmalnik-Brown, R.; Ilhan, Z.E.; Kang, D.W.; DiBaise, J.K. Effects of Gut Microbes on Nutrient Absorption and Energy Regulation. *Nutr. Clin. Pract.* **2012**, *27*, 201–214. [CrossRef] [PubMed]

38. Wu, H.J.; Wu, E. The role of gut microbiota in immune homeostasis and autoimmunity. *Gut Microbes* **2012**, *3*, 4–14. [CrossRef] [PubMed]

39. Kayama, H.; Takeda, K. Manipulation of epithelial integrity and mucosal immunity by host and microbiota-derived metabolites. *Eur. J. Immunol.* **2020**, *50*, 921–931. [CrossRef]

40. Johansson, M.E.; Jakobsson, H.E.; Holmén-Larsson, J.; Schütte, A.; Ermund, A.; Rodríguez-Piñeiro, A.M.; Arike, L.; Wising, C.; Svensson, F.; Bäckhed, F.; et al. Normalization of Host Intestinal Mucus Layers Requires Long-Term Microbial Colonization. *Cell Host Microbe* **2015**, *18*, 582–592. [CrossRef]

41. Engevik, M.A.; Banks, L.D.; Engevik, K.A.; Chang-Graham, A.L.; Perry, J.L.; Hutchinson, D.S.; Ajami, N.J.; Petrosino, J.F.; Hyser, J.M. Rotavirus infection induces glycan availability to promote ileum-specific changes in the microbiome aiding rotavirus virulence. *Gut Microbes* **2020**, *11*, 1324–1347. [CrossRef] [PubMed]

42. Ghosh, S.; Whitley, C.S.; Haribabu, B.; Jala, V.R. Regulation of Intestinal Barrier Function by Microbial Metabolites. *Cell. Mol. Gastroenterol. Hepatol.* **2021**, *11*, 1463–1482. [CrossRef] [PubMed]

43. Helmy, Y.A.; Closs, G.; Jung, K.; Kathayat, D.; Vlasova, A.; Rajashekara, G. Effect of Probiotic *E. coli* Nissle 1917 Supplementation on the Growth Performance, Immune Responses, Intestinal Morphology, and Gut Microbes of Campylobacter jejuni Infected Chickens. *Infect. Immun.* **2022**, *90*, e00337-22. [CrossRef] [PubMed]

44. Zhu, W.; Wei, Z.; Xu, N.; Yang, F.; Yoon, I.; Chung, Y.; Liu, J.; Wang, J. Effects of Saccharomyces cerevisiae fermentation products on performance and rumen fermentation and microbiota in dairy cows fed a diet containing low quality forage. *J. Anim. Sci. Biotechnol.* **2017**, *8*, 36. [CrossRef] [PubMed]

45. Whitaker, W.R.; Shepherd, E.S.; Sonnenburg, J.L. Tunable expression tools enable single-cell strain distinction in the gut microbiome. *Cell* **2017**, *169*, 538. [CrossRef] [PubMed]

46. Bartosch, S.; Fite, A.; Macfarlane, G.T.; McMurdo, M.E.T. Characterization of Bacterial Communities in Feces from Healthy Elderly Volunteers and Hospitalized Elderly Patients by Using Real-Time PCR and Effects of Antibiotic Treatment on the Fecal Microbiota. *Appl. Environ. Microbiol.* **2004**, *70*, 3575–3581. [CrossRef] [PubMed]

47. Monteagudo, L.V.; Benito, A.A.; Lázaro-Gaspar, S.; Arnal, J.L.; Martin-Jurado, D.; Menjon, R.; Quílez, J. Occurrence of Rotavirus A Genotypes and Other Enteric Pathogens in Diarrheic Suckling Piglets from Spanish Swine Farms. *Animals* **2022**, *12*, 251. [CrossRef] [PubMed]

48. Robert, H.; Gabriel, V.; Fontagné-Faucher, C. Biodiversity of lactic acid bacteria in French wheat sourdough as determined by molecular characterization using species-specific PCR. *Int. J. Food Microbiol.* **2009**, *135*, 53–59. [CrossRef]

49. Zeng, W.; Wu, J.; Xie, H.; Xu, H.; Liang, D.; He, Q.; Yang, X.; Liu, C.; Gong, J.; Zhang, Q.; et al. Enteral nutrition promotes the remission of colitis by gut bacteria-mediated histidine biosynthesis. *eBioMedicine* **2024**, *100*, 104959. Available online: https://www.thelancet.com/journals/ebiom/article/PIIS2352-3964(23)00525-X/fulltext (accessed on 8 March 2024). [CrossRef]

50. Yuan, L.; Kang, S.Y.; Ward, L.A.; To, T.L.; Saif, L.J. Antibody-Secreting Cell Responses and Protective Immunity Assessed in Gnotobiotic Pigs Inoculated Orally or Intramuscularly with Inactivated Human Rotavirus. *J. Virol.* **1998**, *72*, 330–338. [CrossRef]

51. Hakim, M.S.; Ding, S.; Chen, S.; Yin, Y.; Su, J.; van der Woude, C.J.; Fuhler, G.M.; Peppelenbosch, M.P.; Pan, Q.; Wang, W. TNF-α exerts potent anti-rotavirus effects via the activation of classical NF-κB pathway. *Virus Res.* **2018**, *253*, 28–37. [CrossRef]

52. Jiang, B.; Snipes-Magaldi, L.; Dennehy, P.; Keyserling, H.; Holman, R.C.; Bresee, J.; Gentsch, J.; Glass, R.I. Cytokines as Mediators for or Effectors against Rotavirus Disease in Children. *Clin. Vaccine Immunol.* **2003**, *10*, 995–1001. [CrossRef]

53. Kumar, D.; Shepherd, F.K.; Springer, N.L.; Mwangi, W.; Marthaler, D.G. Rotavirus Infection in Swine: Genotypic Diversity, Immune Responses, and Role of Gut Microbiome in Rotavirus Immunity. *Pathogens* **2022**, *11*, 1078. [CrossRef]

54. Bohl, E.H.; Saif, L.J.; Theil, K.W.; Agnes, A.G.; Cross, R.F. Porcine pararotavirus: Detection, differentiation from rotavirus, and pathogenesis in gnotobiotic pigs. *J. Clin. Microbiol.* **1982**, *15*, 312–319. [CrossRef] [PubMed]

55. Chepngeno, J.; Diaz, A.; Paim, F.C.; Saif, L.J.; Vlasova, A.N. Rotavirus C: Prevalence in suckling piglets and development of virus-like particles to assess the influence of maternal immunity on the disease development. *Vet. Res.* **2019**, *50*, 84. [CrossRef] [PubMed]

56. Marthaler, D.; Homwong, N.; Rossow, K.; Culhane, M.; Goyal, S.; Collins, J.; Matthijnssens, J.; Ciarlet, M. Rapid detection and high occurrence of porcine rotavirus A, B, and C by RT-qPCR in diagnostic samples. *J. Virol. Methods* **2014**, *209*, 30–34. [CrossRef] [PubMed]

57. Arnold, M.; Patton, J.T.; McDonald, S.M. Culturing, Storage, and Quantification of Rotaviruses. *Curr. Protoc. Microbiol.* **2009**, *15*, 15C-3. [CrossRef]

58. Parreno, V.; Hodgins, D.C.; De Arriba, L.; Kang, S.Y.; Yuan, L.; Ward, L.A.; To, T.L.; Saif, L.J. Serum and intestinal isotype antibody responses to Wa human rotavirus in gnotobiotic pigs are modulated by maternal antibodies. *J. Gen. Virol.* **1999**, *80 Pt 6*, 1417–1428. [CrossRef]

59. Vlasova, A.N.; Paim, F.C.; Kandasamy, S.; Alhamo, M.A.; Fischer, D.D.; Langel, S.N.; Deblais, L.; Kumar, A.; Chepngeno, J.; Shao, L.; et al. Protein Malnutrition Modifies Innate Immunity and Gene Expression by Intestinal Epithelial Cells and Human Rotavirus Infection in Neonatal Gnotobiotic Pigs. *mSphere* **2017**, *2*, e00046-17. [CrossRef] [PubMed]

60. Chattha, K.S.; Vlasova, A.N.; Kandasamy, S.; Rajashekara, G.; Saif, L.J. Divergent Immunomodulating Effects of Probiotics on T Cell Responses to Oral Attenuated Human Rotavirus Vaccine and Virulent Human Rotavirus Infection in a Neonatal Gnotobiotic Piglet Disease Model. *J. Immunol.* **2013**, *191*, 2446–2456. [CrossRef]

61. Liu, Y.; Huang, P.; Tan, M.; Liu, Y.; Biesiada, J.; Meller, J.; Castello, A.A.; Jiang, B.; Jiang, X. Rotavirus VP8*: Phylogeny, Host Range, and Interaction with Histo-Blood Group Antigens. *J. Virol.* **2012**, *86*, 9899–9910. [CrossRef]

62. Huang, P.; Xia, M.; Tan, M.; Zhong, W.; Wei, C.; Wang, L.; Morrow, A.; Jiang, X. Spike Protein VP8* of Human Rotavirus Recognizes Histo-Blood Group Antigens in a Type-Specific Manner. *J. Virol.* **2012**, *86*, 4833–4843. Available online: https://journals.asm.org/doi/abs/10.1128/JVI.05507-11 (accessed on 21 January 2022). [CrossRef] [PubMed]

63. Kim, A.H.; Hogarty, M.P.; Harris, V.C.; Baldridge, M.T. The Complex Interactions between Rotavirus and the Gut Microbiota. *Front. Cell. Infect. Microbiol.* **2021**, *10*, 586751. [CrossRef]

64. Mazziotta, C.; Tognon, M.; Martini, F.; Torreggiani, E.; Rotondo, J.C. Probiotics Mechanism of Action on Immune Cells and Beneficial Effects on Human Health. *Cells* **2023**, *12*, 184. [CrossRef] [PubMed]

65. Indira, M.; Venkateswarulu, T.C.; Abraham Peele, K.; Nazneen Bobby Md Krupanidhi, S. Bioactive molecules of probiotic bacteria and their mechanism of action: A review. *3 Biotech* **2019**, *9*, 306. [CrossRef]

66. Ohland, C.L.; Macnaughton, W.K. Probiotic bacteria and intestinal epithelial barrier function. *Am. J. Physiol. Gastrointest. Liver Physiol.* **2010**, *298*, G807–G819. [CrossRef]

67. Almand, E.A.; Moore, M.D.; Jaykus, L.A. Virus-Bacteria Interactions: An Emerging Topic in Human Infection. *Viruses* **2017**, *9*, 58. [CrossRef]
68. Engevik, K.; Banks, L.; Petrosino, J.; Engevik, M.; Hyser, J. Exploring the interaction between rotavirus and Lactobacillus. *FASEB J.* **2021**, *35*. Available online: https://onlinelibrary.wiley.com/doi/abs/10.1096/fasebj.2021.35.S1.04505 (accessed on 24 October 2023). [CrossRef]
69. Robinson, C.M.; Jesudhasan, P.R.; Pfeiffer, J.K. Bacterial lipopolysaccharide binding enhances virion stability and promotes environmental fitness of an enteric virus. *Cell Host Microbe* **2014**, *15*, 36–46. [CrossRef]
70. Kuss, S.K.; Best, G.T.; Etheredge, C.A.; Pruijssers, A.J.; Frierson, J.M.; Hooper, L.V.; Dermody, T.S.; Pfeiffer, J.K. Intestinal microbiota promote enteric virus replication and systemic pathogenesis. *Science* **2011**, *334*, 249–252. [CrossRef]
71. Jones, M.K.; Watanabe, M.; Zhu, S.; Graves, C.L.; Keyes, L.R.; Grau, K.R.; Gonzalez-Hernandez, M.B.; Iovine, N.M.; Wobus, C.E.; Vinjé, J.; et al. Enteric bacteria promote human and mouse norovirus infection of B cells. *Science* **2014**, *346*, 755–759. [CrossRef] [PubMed]
72. Karst, S.M. Identification of a novel cellular target and a co-factor for norovirus infection—B cells & commensal bacteria. *Gut Microbes* **2015**, *6*, 266–271. [PubMed]
73. Johansson, M.E.V.; Hansson, G.C. Immunological aspects of intestinal mucus and mucins. *Nat. Rev. Immunol.* **2016**, *16*, 639–649. [CrossRef] [PubMed]
74. Johansson, M.E.V.; Phillipson, M.; Petersson, J.; Velcich, A.; Holm, L.; Hansson, G.C. The inner of the two Muc2 mucin-dependent mucus layers in colon is devoid of bacteria. *Proc. Natl. Acad. Sci. USA* **2008**, *105*, 15064–15069. [CrossRef]
75. Ahmadi, E.; Alizadeh-Navaei, R.; Rezai, M.S. Efficacy of probiotic use in acute rotavirus diarrhea in children: A systematic review and meta-analysis. *Caspian J. Intern. Med.* **2015**, *6*, 187–195. [PubMed]
76. Gonzalez-Ochoa, G.; Flores-Mendoza, L.K.; Icedo-Garcia, R.; Gomez-Flores, R.; Tamez-Guerra, P. Modulation of rotavirus severe gastroenteritis by the combination of probiotics and prebiotics. *Arch. Microbiol.* **2017**, *199*, 953–961. [CrossRef] [PubMed]
77. Vlasova, A.N.; Kandasamy, S.; Chattha, K.S.; Rajashekara, G.; Saif, L.J. Comparison of probiotic lactobacilli and bifidobacteria effects, immune responses and rotavirus vaccines and infection in different host species. *Vet. Immunol. Immunopathol.* **2016**, *172*, 72–84. [CrossRef] [PubMed]
78. Kandasamy, S.; Vlasova, A.N.; Fischer, D.D.; Chattha, K.S.; Shao, L.; Kumar, A.; Langel, S.N.; Rauf, A.; Huang, H.-C.; Rajashekara, G.; et al. Unraveling the Differences between Gram-Positive and Gram-Negative Probiotics in Modulating Protective Immunity to Enteric Infections. *Front. Immunol.* **2017**, *8*, 334. [CrossRef]
79. Caballero-Franco, C.; Keller, K.; De Simone, C.; Chadee, K. The VSL#3 probiotic formula induces mucin gene expression and secretion in colonic epithelial cells. *Am. J. Physiol.-Gastrointest. Liver Physiol.* **2007**, *292*, G315–G322.
80. Varyukhina, S.; Freitas, M.; Bardin, S.; Robillard, E.; Tavan, E.; Sapin, C.; Grill, J.-P.; Trugnan, G. Glycan-modifying bacteria-derived soluble factors from Bacteroides thetaiotaomicron and Lactobacillus casei inhibit rotavirus infection in human intestinal cells. *Microbes Infect.* **2012**, *14*, 273–278. [CrossRef]
81. Odamaki, T.; Bottacini, F.; Kato, K.; Mitsuyama, E.; Yoshida, K.; Horigome, A.; Xiao, J.-Z.; van Sinderen, D. Genomic diversity and distribution of Bifidobacterium longum subsp. longum across the human lifespan. *Sci. Rep.* **2018**, *8*, 85. [CrossRef] [PubMed]
82. Park, K.-H.; Kim, M.-G.; Ahn, H.-J.; Lee, D.-H.; Kim, J.-H.; Kim, Y.-W.; Woo, E.-J. Structural and biochemical characterization of the broad substrate specificity of Bacteroides thetaiotaomicron commensal sialidase. *Biochim. Biophys. Acta* **2013**, *1834*, 1510–1519. [CrossRef] [PubMed]

Article

Phylogenetic Analyses of Rotavirus A, B and C Detected on a Porcine Farm in South Africa

Amy Strydom [1], Neo Segone [1], Roelof Coertze [1,2], Nikita Barron [1], Muller Strydom [3] and Hester G. O'Neill [1,*]

[1] Department of Microbiology and Biochemistry, University of the Free State, Bloemfontein 9300, South Africa; aimster.strydom@gmail.com (A.S.); neozagonene@gmail.com (N.S.); roelof.coertze@gu.se (R.C.); barronnikita@gmail.com (N.B.)

[2] Department of Infectious Diseases, University of Gothenburg, Guldhedsgatan 10, SE-413 46 Göteborg, Sweden

[3] George Animal Hospital, George 6530, South Africa; vet@georgevet.co.za

* Correspondence: oneillhg@ufs.ac.za

Abstract: Rotaviruses (RVs) are known to infect various avian and mammalian hosts, including swine. The most common RVs associated with infection in pigs are A, B, C and H (RVA-C; RVH). In this study we analysed rotavirus strains circulating on a porcine farm in the Western Cape province of South Africa over a two-year period. Whole genomes were determined by sequencing using Illumina MiSeq without prior genome amplification. Fifteen RVA genomes, one RVB genome and a partial RVC genome were identified. Phylogenetic analyses of the RVA data suggested circulation of one dominant strain (G5-P[6]/P[13]/P[23]-I5-R1-C1-M1-A8-N1-T7-E1-H1), typical of South African porcine strains, although not closely related to previously detected South African porcine strains. Reassortment with three VP4-encoding P genotypes was detected. The study also reports the first complete RVB genome (G14-P[5]-I13-R4-C4-M4-A10-T4-E4-H7) from Africa. The partial RVC (G6-P[5]-IX-R1-C1-MX-A9-N6-T6-EX-H7) strain also grouped with porcine strains. The study shows the continued circulation of an RVA strain, with a high reassortment rate of the VP4-encoding segment, on the porcine farm. Furthermore, incidents of RVB and RVC on this farm emphasize the complex epidemiology of rotavirus in pigs.

Keywords: porcine rotavirus; rotavirus A, B, C; P-type reassortment; South Africa

Citation: Strydom, A.; Segone, N.; Coertze, R.; Barron, N.; Strydom, M.; O'Neill, H.G. Phylogenetic Analyses of Rotavirus A, B and C Detected on a Porcine Farm in South Africa. *Viruses* **2024**, *16*, 934. https://doi.org/10.3390/v16060934

Academic Editors: Ulrich Desselberger and John T. Patton

Received: 10 May 2024
Revised: 3 June 2024
Accepted: 6 June 2024
Published: 8 June 2024

1. Introduction

Rotavirus (RV) causes acute gastroenteritis in various mammalian, including humans and livestock, and avian species. Nine species of rotaviruses have been classified based on VP6 variation, namely A–D and F–J, according to the International Committee on Taxonomy of Viruses (ICTV) ([1]; https://talk.ictvonline.org/taxonomy/ (accessed on 9 March 2024)). RVA has by far the biggest public health impact and is therefore the best studied of all RVs. The virus contains a segmented, double-strand RNA genome, consisting of 11 segments and encoding 6 structural proteins and 5/6 non-structural proteins [2]. These proteins have been classified according to genotype: VP7 (G)—VP4 (P)—VP6 (I)—VP1 (R)—VP2 (C)—VP3 (M)—NSP1 (A)—NSP2 (N)—NSP3 (T)—NSP4 (E)—NSP5/6 (H) with 42G, 58P, 32I, 28R, 24C, 24M, 39A, 28N, 28T, 32E and 28H genotypes assigned to RVA (Rotavirus Classification Working Group: (RCWG). Available online: https://rega.kuleuven.be/cev/viralmetagenomics/virus-classification/rcwg (accessed on 9 March 2024)).

Compared to RVA, knowledge of RVB and RVC is limited despite the impact of these viruses on mortality rates and economic loses in the agricultural sector [3–7]. Advancements in next-generation sequencing initiatives have, however, led to proposals for whole-genome classification of RVB and RVC similar to that for RVA. An RVB classification system was first described in 2018, while revised genotyping was proposed in 2023. Currently, 27G, 6P, 13I, 7R, 6C, 5M, 8A, 10N, 6T, 4E, and 7H types have been identified [8,9]. These genotypes are, for the most part, host–specific for porcine, human, bovine, caprine and murine hosts. Similarly, genotypes described for RVC are also host-specific with only a few exceptions. In 2021, 31G, 26P, 13I, 5R, 5C, 5M, 12A, 10N, 9T, 8E and 4H types were described for RVC strains detected in porcine, human, bovine, canine and ferret [10].

Rotaviruses are endemic in pig populations and varying detection rates have been reported [11–14]. Infected pigs can have clinical or subclinical symptoms with neonatal and suckling piglets worst affected. Rotaviruses A, B, C and H have all been detected in pigs. RVA and RVC are associated with diarrhoea in piglets and weaning animals, whereas RVB has been more associated with older animals [11,14]. Twelve G and 16 P RVA types have been detected in pigs, of which G5P [7] was reported to be the most frequently detected genotype combination [15]. Similarly, 15 G and 16 P RVC types and 27 G and 3 P RVB types have been detected in pigs [9,11]. The probability that porcine populations can act as reservoirs for human infection has been discussed before, and multiple studies have reported evidence of zoonotic transmission and reassortment events of RVA strains [16–19].

Porcine rotavirus was first identified in 1977 in South Africa in faecal samples using electron microscopy [20]. Between 1992 and 1993, rotavirus A, B, and C were identified in porcine faecal samples throughout South Africa [21–23]. This was also the first detection of any non-RVA in Africa [24]. African human RVA strains exhibit a high degree of genotype diversity. Proximity to livestock and the frequency of co-infections with bovine or porcine strains leading to human–animal reassortment events contribute to the diversity [25–27]. However, very little is known about the animal RVA strains in Africa and no surveillance is performed in South Africa. Even less is known about RVB and RVC, and it remains to be seen how Africa compares to developed countries. At the beginning of 2018, we were approached to confirm the occurrence of rotavirus on a porcine farm in the Western Cape of South Africa. The first sample we received tested positive for rotavirus, which prompted further sample testing. Since samples were received infrequently, rotavirus prevalence could not be determined; rather, the study aimed to determine genetic variation in rotavirus on the porcine farm over the course of two years.

2. Materials and Methods

2.1. Sampling and Rotavirus Detection

This animal study was conducted with the approval of the Animal Research Ethics Committee at the University of the Free State (UFS) (UFS-AED2018/0030). One hundred and twenty-one samples were collected on a porcine farm in the Western Cape province of South Africa between January 2018 and February 2020. The farm is a born-to-finish and all in–all out system with a 370-sow unit. At the time of sampling, the average born alive per sow was about 12 piglets. Piglets were weaned at 28 days, moved to a weaner house where they were housed up to 60 days, in their respective groups, after which they were moved to a porker house and stayed there until 84 days, before finally being moved to the grower house until slaughter at 154 days. Samples, both symptomatic (liquid) and asymptomatic (solid), were collected directly from the surface of porcine pens of the farrowing (>28 days) and weaner houses (28–60 days). Total RNA was extracted from the 121 stool samples using Tri-reagent (Sigma Aldrich, St. Louis, MO, USA), and single-stranded RNA was precipitated with 1 M lithium chloride [28]. Extracted RNA was examined by gel electrophoresis on 1% agarose gel and samples with typical rotavirus migration patterns were recorded as positive.

2.2. cDNA Synthesis and Next-Generation Sequencing

The dsRNA of samples with rotavirus profiles were treated with 9 U of DNase I (Sigma Aldrich, St. Louis, MO, USA). An anchor primer was annealed to the dsRNA before sequencing in order to obtain full-length gene segments as previously described [28]. Complementary DNA was synthesized with the Maxima H Minus Double Stranded cDNA kit (Thermo Fisher Scientific, Waltham, MA, USA) using random hexamers. Minor modifications to the manufacturer's instructions included denaturing of the dsRNA at 95 °C for

5 min and first-strand synthesis for two hours at 50 °C [29]. Purified cDNA was submitted for sequencing at the UFS Next-Generation Sequencing Unit (UFS-NGS, Bloemfontein, South Africa). To perform whole-genome sequencing, an Illumina Miseq sequencer (Illumina, Inc., San Diego, CA, USA) was used. Sequencing was performed using a Miseq Reagent kit V2 (500 cycles) with 251 × 2 paired end reads.

2.3. Data Assembly and Analysis

All paired-end reads were screened for poor-quality nucleotides, which were removed using Trimmomatic [30]. Sequencing adapters were clipped and reads shorter than 50 bp were discarded. Additionally, a sliding window of four base pairs was used to remove flanking nucleotides if the average quality score dropped below 20. The overall quality of the reads was assessed before and after trimming using FastQC. De novo assembly of the high-quality paired-end reads was carried out using SPAdes and its default parameters [31]. The identities of the assembled contigs were determined by comparing them to the nucleotide BLAST database, specifying rotavirus A, B, and C. Database matches were used to select possible reference sequences to perform reference mapping for more reliable consensus sequences (Table S1). Reference mapping against the reference sequences was performed locally using Bowtie 2 [32]. Strict mismatch parameters were selected to ensure high-accuracy reference mapping. Calculation of mapping coverage and extraction of consensus sequences was performed using Samtools [33].

Consensus sequences were analysed in BLASTn and RVA genotypes were identified with the Virus Pathogen Database and Analysis Resource (ViPR) [34]. Reference sequences for RVA, B, and C were obtained from GenBank for phylogenetic analyses. The sequences of each segment were aligned with the appropriate reference sequences using MUSCLE in MEGA X [35]. Maximum-likelihood trees were generated using IQtree using the optimal substitution model and ultrafast bootstrap approximation approach [36,37]. Nucleotide distance matrixes were calculated using the p-distance algorithm in MEGA X. Genotypes for RVB and RVC were assigned based on the most recent classification [8,10]. The nucleotide sequence data presented have been deposited in GenBank under the following accession numbers PP669365-PP669534 (RVA), PP669283-PP669293 (RVB) and PP669294-PP669301 (RVC).

3. Results

3.1. Sequencing of Rotavirus-Positive Samples

Electrophoretic analysis of the extracted dsRNA suggested the presence of rotavirus in 16 of the 121 samples collected (Table 1). All positive samples detected in the farrowing and weaner houses were diarrhetic and detected during each of the five sampling dates. These 16 samples were subjected to next-generation sequencing (NGS), which revealed that 15 samples contained RVA strains. Two of the 15 samples had a co-infection with RVC (UFS-BOC009 and UFS-BOC035) and one with RVB (UFS-BOC124) (Table 1). The remaining sample contained rotavirus B (UFS-BOC050). The co-infected samples were detected during separate sampling trips, indicating circulation of different RV groups over time. The RVB sample detected in December 2019 was also detected in the same pen where RVA was detected (Table 1).

Complete open reading frames (ORFs) were obtained for all RVA genome segments. Average coverage (sequence depth) for the RVA consensus sequences ranged from 107.1 to 6562.86 (Table S2). A complete genome was determined for an RVB strain in sample UFS-BOC050. All genome segments were full length except segments 1 and 10 (99.97% and 88.3%, respectively). Average coverage for the RVB segments ranged from 1932.9 to 4688.0. Although all the segments were detected for an RVB strain in UFS-BOC124, with genome segment lengths ranging from 66.3% to 99.9%, a very low number of sequencing reads (average coverage ranged between 4.5 to 11.1) was obtained (Table S2). Sample UFS-BOC009 contained an RVA strain as well as an RVC strain. The average coverage for the RVA strain ranged from 886.7 to 3687.9, whereas the RVC coverage ranged from 31.8 to

131.3. RVC was also identified in sample UFS-BOC035, but similarly to the RVB strain in UFS-BOC124, the average coverage for the RVC strain in UFS-BOC035 was low (5.1 to 18.9) (Table S2).

Table 1. Detection of rotavirus A, B, and C in porcine samples.

Collection Date	Samples Collected	Positive Samples	Age of Positive Piglets	Sample	Rotavirus
10 January 2018	1	1	>28 days [#]	UFS-BOC001	RVA
24 December 2018	11	1	>28 days [#]	UFS-BOC009	RVA RVC
19 February 2019	25	1	>28 days [#]	UFS-BOC035	RVA RVC *
24 December 2019	34	5	5 days	UFS-BOC050	RVB
				UFS-BOC060	RVA
				UFS-BOC063	RVA
				UFS-BOC064	RVA
				UFS-BOC071	RVA
20 February 2020	50	8	28 days	UFS-BOC076	RVA
				UFS-BOC077	RVA
				UFS-BOC078	RVA
			28 days	UFS-BOC079	RVA
			27 days	UFS-BOC081	RVA
				UFS-BOC082	RVA
				UFS-BOC083	RVA
			30 days	UFS-BOC124	RVA RVB *

[#] Exact date-of-birth unknown, but samples obtained from farrowing houses. * Insufficient number of reads to genotype (Table S2).

3.2. Genotyping and Phylogenetic Analyses

3.2.1. Rotavirus A

The rotavirus A strains were identified as G5-I5-R1-C1-M1-A8-N1-T7-E1-H1 in combination with P[6], P[13] or P[23] (Table 2). All the sequences for each segment, apart from those encoding VP4, are nearly identical to each other (Figures 1 and S1). The closest relatives to VP7, VP6, VP1, VP3, NSP1, NSP2, NSP3 and NSP4 encoding genome segments are all derived from South African RVA strains detected in porcine samples (Figure 1). These strains also grouped together in the phylogenetic trees, and in most instances with previously described South African strains (Figure S1). The exception was the NSP5-encoding sequences determined in this study, which grouped separately from previously described South African strains with strains from non-South African countries. The strains were, however, still closely related to the previously described South African porcine strains with nucleotide identities ranging from 96.63% to 99.83% (Figure 1; Table S3).

The variation in the VP4-encoding genotypes detected on the farm is an interesting observation. Two different P[13] (a and b) sequences were detected during the study. The first sample collected in January 2018 contained a P[13]ᵃ genotype (UFS-BOC001). A highly similar P[13]ᵃ was detected almost two years later in December 2019 in four samples sourced from the same pen (Table 2; Figure 1). These samples had co-infections with P[23]. The P[13]ᵃ sequences had approximate 99.5% nucleotide identity with those detected in January 2018 (Table S3). During December 2018 and February 2019, another two P[13]ᵇ strains were detected on the farm with co-infections with P[6] (Table 2). However, these sequences grouped separately from the P[13]ᵃ sequences in the phylogenetic tree and only shared an approximate 83.5% nucleotide identity with these strains (Figure 1; Table S3). Interestingly, the closest relative to the P[13]ᵃ sequences was from Canada—RVA/Pig-wt/CAN/F7P4-A/2006/GXP[13], with only 90.12% nucleotide identity. Similarly, the closest relative to the P[13][b] sequences, RVA/Pig-wt/CHN/SCYA-C7/2019/G9P[13], was from China, with 90.5% nucleotide identity (Table S3).

Table 2. Genome constellations of South African porcine rotavirus A strains.

Collection Date	Pen	Strain	VP7	VP4	VP6	VP1	VP2	VP3	NSP1	NSP2	NSP3	NSP4	NSP5
10 January 2018	unknown	UFS-BOC001	G5	P[13][a]	I5	R1	C1	M1	A8	N1	T7	E1	H1
24 December 2018	unknown	UFS-BOC009	G5	P[6]P[13][b]	I5	R1	C1	M1	A8	N1	T7	E1	H1
19 February 2019	unknown	UFS-BOC035	G5	P[6]P[13][b]	I5	R1	C1	M1	A8	N1	T7	E1	H1
24 December 2019	19134	UFS-BOC060	G5	P[13][a]P[23]	I5	R1	C1	M1	A8	N1	T7	E1	H1
		UFS-BOC063	G5	P[13][a]P[23]	I5	R1	C1	M1	A8	N1	T7	E1	H1
		UFS-BOC064	G5	P[23]	I5	R1	C1	M1	A8	N1	T7	E1	H1
		UFS-BOC071	G5	P[13][a]P[23]	I5	R1	C1	M1	A8	N1	T7	E1	H1
20 February 2020	18202	UFS-BOC076	G5	P[23]	I5	R1	C1	M1	A8	N1	T7	E1	H1
		UFS-BOC077	G5	P[23]	I5	R1	C1	M1	A8	N1	T7	E1	H1
		UFS-B0C078	G5	P[23]	I5	R1	C1	M1	A8	N1	T7	E1	H1
	18212	UFS-BOC079	G5	P[23]	I5	R1	C1	M1	A8	N1	T7	E1	H1
	18119	UFS-BOC081	G5	P[23]	I5	R1	C1	M1	A8	N1	T7	E1	H1
		UFS-BOC082	G5	P[23]	I5	R1	C1	M1	A8	N1	T7	E1	H1
		UFS-BOC083	G5	P[23]	I5	R1	C1	M1	A8	N1	T7	E1	H1
	18197	UFS-BOC124	G5	P[23]	I5	R1	C1	M1	A8	N1	T7	E1	H1

[a] and [b] represent two different P[13] sequences.

The P[6] sequences detected in December 2018 and February 2019 (UFS-BOC009 and UFS-BOC035) were identical and clustered in a group with both human and porcine strains from Asia (Figure 1C). The closest relative was a strain from China: RVA/sewage/CHN/B24-R2/2019/GXP[6], with 95.5% nucleotide identity (Table S3). The two P[6] sequences shared only 91.4% nucleotide identity with a P[6]-containing porcine strain from Mozambique (RVA/Pig-wt/MOZ/MZ-MPT-115/2016/G4P[6]), approximately 89% nucleotide identity with porcine strains from South Africa, and approximately 91% nucleotide identity with South African human strains (Table S3).

Twelve P[23] sequences were detected in December 2019 and February 2020 and were all identical. The closest relative was another South African porcine strain, RVA/Pig-wt/ZAF/MRC-DPRU1487/2007/G3G5P[23], with a nucleotide identity of 95.19%.

3.2.2. Rotavirus B

Due to the low sequence coverage obtained for UFS-BOC124 only UFS-BOC050 was genotyped. The genome constellation was identified as G14-P[5]-I13-R4-C4-M4-A8-T4-E4-H7 (RVB/Pig-wt/ZAF/UFS-BOC050/2019/G14P[5]) using distance matrices and phylogenetic trees for each segment (Figures 2 and S2; Table S4). The pairwise identity of the closest relatives fell within the ranges for the segments as described in 2018 and updated in 2023 (Figure 2) [8,9]. These genotypes are typically associated with RVB detected in porcine samples, and the closest relatives were all of porcine origin. Segments encoding for VP7, VP4, VP6, VP1 and VP3 were all related to strains from the USA [8,9] detected between 2009 and 2015. Segments encoding for VP2, NSP1 and NSP3 were related to Spanish strains [38], and the remaining segments (NSP2, 4 and 5) were related to Asian strains [16]. The nucleotide identities ranged between 82.1% and 88.5%, indicating that the South African strain is diverse from the previously sequenced RVB strains.

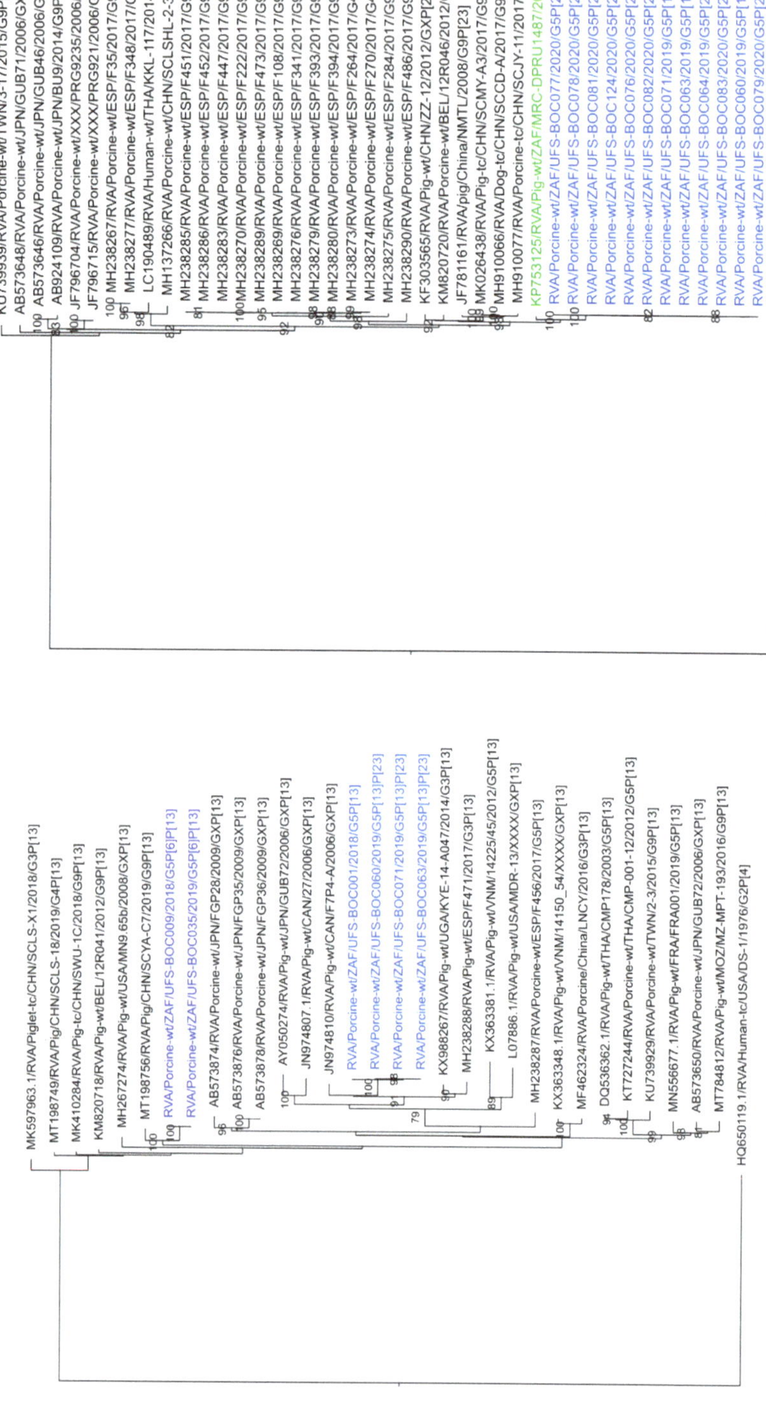

Figure 1. *Cont.*

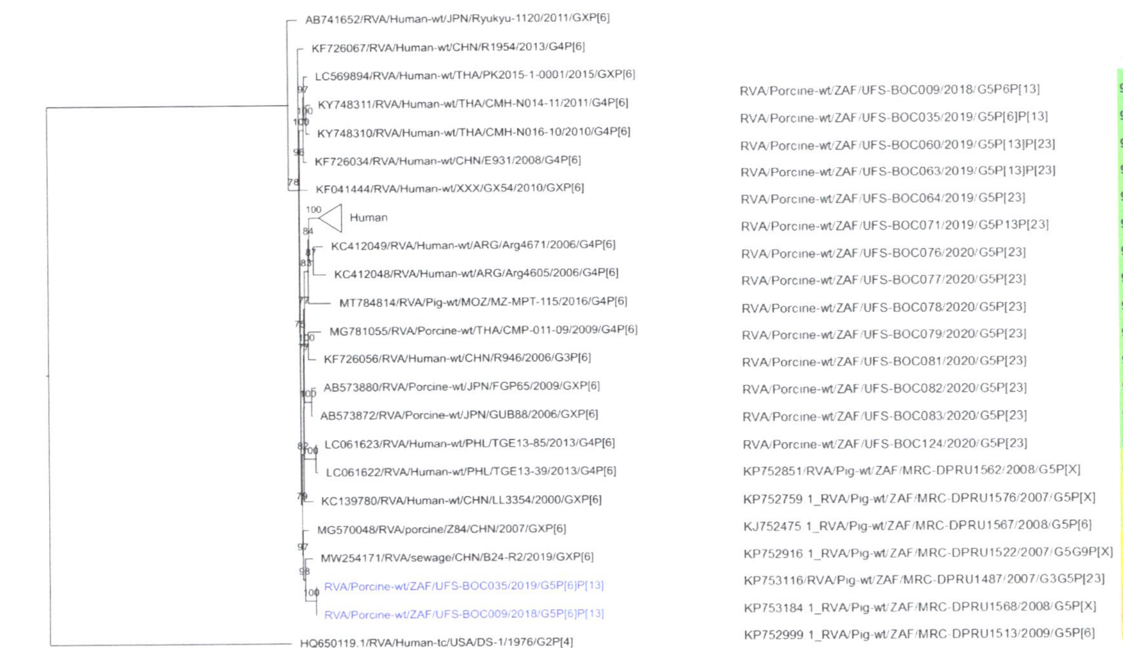

Strain	VP7	VP6	VP1	VP2	VP3	NSP1	NSP2	NSP3	NSP4	NSP5
RVA/Porcine-wt/ZAF/UFS-BOC009/2018/G5P6P[13]	99,69	99,83	99,72	99,81	99,44	99,52	99,68	99,89	99,05	99,83
RVA/Porcine-wt/ZAF/UFS-BOC035/2019/G5P[6]P[13]	99,69	99,83	99,72	99,81	99,44	99,52	99,68	99,89	99,05	99,83
RVA/Porcine-wt/ZAF/UFS-BOC060/2019/G5P[13]P[23]	99,18	99,50	99,60	99,36	99,40	99,52	99,37	99,68	98,48	99,49
RVA/Porcine-wt/ZAF/UFS-BOC063/2019/G5P[13]P[23]	99,18	99,50	99,60	99,36	99,40	99,45	99,37	99,68	98,48	99,49
RVA/Porcine-wt/ZAF/UFS-BOC064/2019/G5P[23]	99,18	99,50	99,60	99,36	99,40	99,45	99,37	99,68	98,48	99,49
RVA/Porcine-wt/ZAF/UFS-BOC071/2019/G5P13P[23]	99,69	99,58	99,63	99,47	99,44	99,52	99,58	99,79	98,67	99,49
RVA/Porcine-wt/ZAF/UFS-BOC076/2020/G5P[23]	99,18	99,50	99,42	99,29	99,24	99,38	99,37	99,68	98,48	99,66
RVA/Porcine-wt/ZAF/UFS-BOC077/2020/G5P[23]	99,18	99,50	99,42	99,29	99,24	99,38	99,37	99,68	98,48	99,66
RVA/Porcine-wt/ZAF/UFS-BOC078/2020/G5P[23]	99,18	99,50	99,42	99,29	99,24	99,38	99,37	99,68	98,48	99,66
RVA/Porcine-wt/ZAF/UFS-BOC079/2020/G5P[23]	99,18	99,50	99,45	99,25	99,44	99,38	99,16	99,68	98,30	99,49
RVA/Porcine-wt/ZAF/UFS-BOC081/2020/G5P[23]	99,18	99,50	99,42	99,29	99,24	99,38	99,37	99,68	98,48	99,66
RVA/Porcine-wt/ZAF/UFS-BOC082/2020/G5P[23]	99,18	99,50	99,42	99,29	99,24	99,38	99,37	99,68	98,48	99,66
RVA/Porcine-wt/ZAF/UFS-BOC083/2020/G5P[23]	99,18	99,50	99,42	99,29	99,24	99,38	99,37	99,68	98,48	99,66
RVA/Porcine-wt/ZAF/UFS-BOC124/2020/G5P[23]	99,18	99,41	99,57	99,33	99,40	99,52	99,48	99,68	98,48	99,49
KP752851/RVA/Pig-wt/ZAF/MRC-DPRU1562/2008/G5P[X]	96,73	93,79	92,59	91,92	90,75	90,35	87,53	93,60	93,56	97,64
KP752759 1_RVA/Pig-wt/ZAF/MRC-DPRU1576/2007/G5P[X]	95,12	94,63	92,68	91,99	90,27	88,02	93,50	93,80	93,37	97,47
KJ752475 1_RVA/Pig-wt/ZAF/MRC-DPRU1567/2008/G5P[6]	95,12	94,37	92,59	91,69	90,55	88,16	87,84	93,90	92,23	97,31
KP752916 1_RVA/Pig-wt/ZAF/MRC-DPRU1522/2007/G5G9P[X]	94,59	93,20	92,93	92,63	89,95	80,90	87,84	93,70	88,07	97,14
KP753116/RVA/Pig-wt/ZAF/MRC-DPRU1487/2007/G3G5P[23]	92,04	96,81	91,74	90,68	90,75	90,42	98,11	83,90	91,86	97,14
KP753184 1_RVA/Pig-wt/ZAF/MRC-DPRU1568/2008/G5P[X]	86,52	94,04	92,44	92,07	90,51	80,77	95,81	84,00	83,90	96,80
KP752999 1_RVA/Pig-wt/ZAF/MRC-DPRU1513/2009/G5P[6]	86,52	92,61	92,50	92,11	90,43	81,59	87,95	93,70	83,90	96,80

Figure 1. Rotavirus A phylogenetic analyses. (**A**): VP4 (P[13]), (**B**): VP4 (P[23]), (**C**): VP4 (P[6]); (**D**): comparison of BOC001 with other study strains and other South African strains. The heatmap indicates the similarity between the study strains (green) and closely related study strains (yellow to red). The South African study strains in the phylogenetic trees are indicated in blue and previously described South African strains are indicated in green. Each gene was compared with sequences available in GenBank and nucleotide alignments were constructed using the MUSCLE algorithm in MEGA X [35]. Phylogenetic trees were generated using IQtree implementing the maximum-likelihood method with ModelFinder, and the trees were statistically supported using 1000 ultrafast bootstrap runs. For P[23] and P[6], K3Pu + F + I + G4 was used, and for P[13], GTR + F + I + G4 was used. The trees are drawn to scale, with branch lengths in the same units as those of the evolutionary distances used to infer the phylogenetic tree.

Sampling date	Pen	Sample	VP7	VP4	VP6	VP1	VP2	VP3	NSP1	NSP2	NSP3	NSP4	NSP5
24 December 2019	19134	UFS-BOC050	G14	P[5]	I13	R4	C4	M4	A8	N10	T4	E4	H7

(A)

Encoding gene segment	Genotype	Nucleotide Cut-off %	%ID	Closest relative
VP7	G14	80	84.58	MF522401.1/RVB/Pig-wt/USA/MT139/2015/G14P[X]
VP4	P[5]	80	83.06	MG272151.1/RVB/Pig-wt/USA/IL14/2013/G16P[5]
VP6	I13	81	88.34	KF882539.1/RVB/Pig-wt/USA/IA09-67/2009/G16P[X]
VP1	R4	79	82.20	MG272093.1/RVB/Pig-wt/USA/KS2/2012/GXP[X]
VP2	C4	79	84.06	MK953186.1/RVB/Pig-wt/ESP/P2B/2017/GXP[4]
VP3	M4	77	83.46	MG272135.1/RVB/Pig-wt/USA/KS2/2012/GXP[X]
NSP1	A8	76	85.46	MK953232.1/RVB/Pig-wt/ESP/B378/2017/GXP[X]
NSP2	N10	83	88.46	KX362385.1/RVB/Pig-wt/VNM/14151_62/2012/GXP[X]
NSP3	T4	78	82.06	MK953213.1_RVB/Pigwt/ESP/B304/2017/G12P[X]
NSP4	E4	76	83.73	MK379206.1/RVB/Pig-wt/CHN/GZ04/2017/GXP[X]
NSP5	H7	79	84.00	AB713981.1/RVB/Pig-wt/JPN/PB-68-C17/2007/GXP[X]

(B)

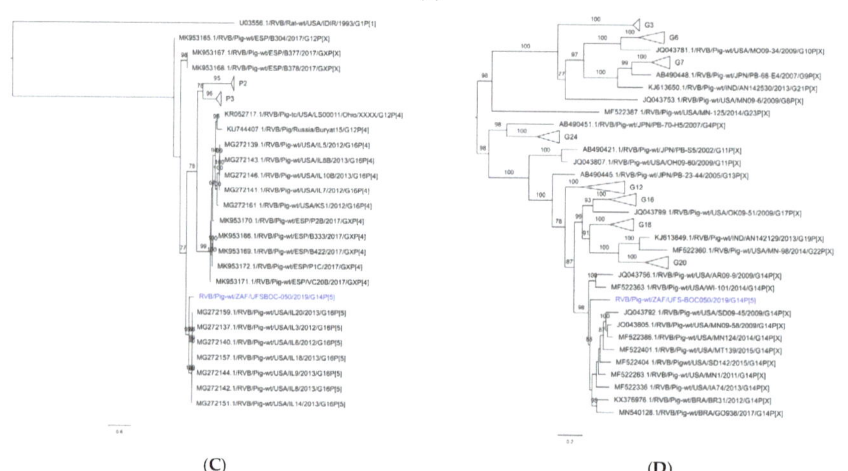

(C)   (D)

Figure 2. Rotavirus B phylogenetic analyses. (**A**): Genome constellation of UFS-BOC050, (**B**): nucleotide identities of the closest relatives of RVB genes, (**C,D**): phylogenetic trees based on the RVB VP4 (**C**) and VP7 (**D**) genes. The South African study strain in the phylogenetic tree is indicated in blue. The sequence was compared with sequences available in GenBank and nucleotide alignments were constructed using the MUSCLE algorithm in MEGA X [35]. The phylogenetic tree was generated using IQtree implementing the maximum-likelihood method with ModelFinder (VP4: TIM3 + F + I + G4; VP7: GTR + F + I + G4) and statistically supported using 1000 ultrafast bootstrap runs. The trees are drawn to scale, with branch lengths in the same units as those of the evolutionary distances used to infer the phylogenetic tree.

3.2.3. Rotavirus C

Similarly to RVB, the sequence data obtained for UFS-BOC035 were deemed insufficient for genotyping. In addition, the average coverage for VP6-, VP3- and NSP4-encoding sequences of UFS-BOC009 ranged between 31 and 43 and was therefore also excluded from further analysis (Table S2). The partial genome constellation of the RVC strain (RVC/Pig-wt/ZAF/UFS-BOC009/2018/G6P[5]) was identified as G6-P[5]-IX-R1-C1-MX-A9-N6-T6-EX-H7 (Figures 3 and S3; Table S5). The nucleotide identities of the remaining segments were in range with the most recently described cut-off levels [10]. The closest relatives to the study strain segments were all detected in pigs from Asia and the USA. The only exception was NSP5 (nucleotide identity: 90.41%), which was related to a South African strain (RVC/Pig-wt/ZAF/BSF3/2021/GXP[X]) detected in the oral virome of a pig [39]. The oral virome study was conducted in KwaZulu Natal province in South Africa in 2021 and two partial RVC strains were identified. However, most of the sequences were too short to include in the phylogenetic analyses of the present study, since only 19–50% of the sequences were determined [39]. Significant diversity from known RVC strains was again observed, with only the VP1-encoding gene of the study strain exhibiting a comparatively high nucleotide identity of 95.84% with RVC/Pig-wt/CHN/VIRES_HeB02_C/2017/GXP[X].

Sampling date	Pen	Sample	VP7	VP4	VP6	VP1	VP2	VP3	NSP1	NSP2	NSP3	NSP4	NSP5
24 December 2018	Unknown	UFS-BOC009	G6	P[5]	IX	R1	C1	MX	A9	N6	T6	EX	H1

(A)

Encoding gene segment	Genotype	Nucleotide Cut-off %	%ID	Closest relative
VP7	G6	85	88.65	MF522701.1_RVC/Pig-wt/USA/MN-265/201/GXP[X]
VP4	P[5]	85	87.50	MG451617.1_RVC/Pig-wt/USA/MN29/2012/G6P[5]
VP1	R1	85	95.84	MK379289.1_RVC/Pig-wt/CHN/VIRES_HeB02_C2/2017/GXP[X]
VP2	C1	84	89.48	LC307108.1_RVC/Pig-wt/JPN/CJ59-32/2003/G5P[4]
NSP1	A9	84	86.01	KX362451.1_RVC/Pig-wt/VNM/12129_51/GXP[X]
NSP2	N6	87	93.38	MG451167.1_RVC/Pig-wt/USA/MO36/2012/G5P[4]
NSP3	T6	85	89.27	LC307026.1_RVC/Pig-wt/JPN/87-G2/2008/GXP[X]
NSP5	H1	79	90.41	OM104995.1_RVC/Pig-wt/ZAF/BSF3/2021/GXP[X]

(B)

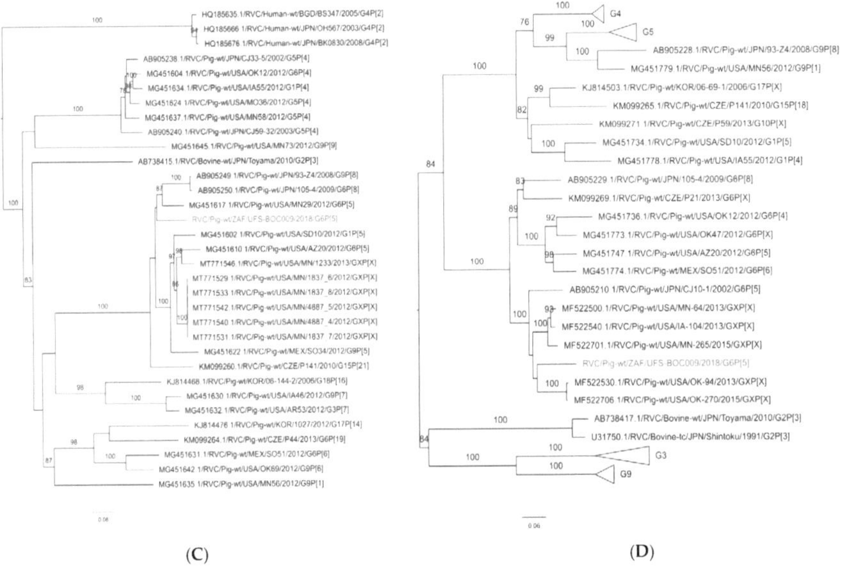

(C)

(D)

Figure 3. Rotavirus C phylogenetic analyses. (**A**): Genome constellation of UFS-BOC009, (**B**): nucleotide identities of the closest relatives of the RVC genes, (**C,D**): phylogenetic trees based on the RVC VP4 (**C**) and VP7 (**D**) genes. The South African study strain in the phylogenetic tree is indicated in blue. The sequence was compared with sequences available in GenBank and nucleotide alignments were constructed using the MUSCLE algorithm in MEGA X [35]. The phylogenetic tree was generated using IQtree implementing the maximum-likelihood method, with ModelFinder (VP7: TIM2 + F + I + G4; VP4: GTR + F + G4) and statistically supported using 1000 ultrafast bootstrap runs. The trees are drawn to scale, with branch lengths in the same units as those of the evolutionary distances used to infer the phylogenetic tree.

4. Discussion

The characterisation and genetic surveillance of porcine rotavirus strains are important on two fronts: firstly, to determine the presence and diversity of the virus causing economic loses in the pork industry for risk analyses and management strategies; and secondly, to understand the influence that porcine rotaviruses might have on the genetic diversity of human strains and their impact on public health. This study describes the genetic diversity of rotavirus strains detected over a two-year period on a porcine farm in the Western Cape province of South Africa. Study limitations include low sample numbers, inconsistent sample sizes, infrequent sampling dates and the fact that samples could only be linked to a pen and not a specific animal. We therefore did not set out to systematically analyse the prevalence of rotavirus, but rather determine the genetic variance of the rotavirus population.

The ability of pigs to harbour species A, B, C, and H (not detected in this study) is well known [11], and the simultaneous detection of multiple species on a farm is not unique either [14]. In a study conducted in Eastern Australia, species A, B, and C were detected in piggeries as mono-infections, but also co-infections. The study reported a higher prevalence for RVA in young pigs (piglets and weaners) whereas RVB and RVC were also detected in older animals (>11 weeks) [14]. Since most of the samples analysed in the current study were obtained from pens housing >28-day-old piglets and weaners, it could explain the higher detection rate for RVA compared to RVB and RVC.

The similarity and consistency between the sequences of the VP7-encoding segment and those of the backbones indicate that there was one dominant RVA strain circulating on the farm for at least two years. Phylogenetic and distance matrix analysis suggested that this porcine RVA strain was similar but not closely related to other South African porcine strains, suggesting the detection of a new porcine strain. The NSP5-encoding segment was the only segment that grouped with non-South African strains, which indicates a possible historical reassortment event for genome segment 11.

The variation in the VP4-encoding segments is of great interest. The origin of the various P types detected in the study is unclear and could indicate that additional RVA strains were present on the farm during the two-year study period. Inconsistency in sample size and infrequent sampling are possible factors that could have contributed to the non-detection of such strains. In the mature virion, the VP4 spike protein fits into a pocket created by VP7 and VP6, and multiple protein–protein interactions ensure the stability of the spike protein [40]. Therefore, the variation in VP4 in the presence of the same VP7 and VP6 proteins warrants further investigation to understand the ability of the porcine strain to harbour different P types.

The P types detected in this study are often reported in porcine RVA studies [11]. It is important to note, though, that in Africa the P[6] genotype is also frequently detected in humans [24,41]. Human and porcine P[6] sequences often cluster together during phylogenetic analysis [41], as was also the case for the P[6] detected in this study, which highlights the zoonotic potential of this genotype. The P[13] sequences identified in the study were not only diverse from each other but also from previously described African P[13] sequences from Uganda and Mozambique [42,43]. Genetic diversity among P[13] sequences has previously been reported for strains detected in piglets in the USA [44]. The only P type detected in the study that was relatively closely related to a South African strain was the P[23], with 95% identity to RVA/Pig-wt/ZAF/MRC-DPRU1487/2007/G3G5P[23]. The P[23] genotype was first detected in December 2019. The P[13]a sequence, first detected in January 2018, was also detected in December 2019, but by February 2020, only the P[23] genotype was detected. It is possible that the P[23] sequences were only introduced to the farm in 2019 and outcompeted the other P genotypes. However, only 12 samples were collected during 2018, and it is therefore possible that an earlier P[23] introduction could have been missed.

This study reports the first full-length genome sequence for a porcine RVB strain from Africa. RVB/Pig-wt/ZAF/UFS-BOC050/2019/G14P[5] was detected in the same pen with various P[13]- and P[23]-containing RVA strains in December 2019, whereas traces of RVB was detected in the RVA/Porcine-wt/ZAF/UFS-BOC124/2020/G5P[23]-containing sample collected two months later, in February 2020. The partial RVC/Pig-wt/ZAF/UFS-BOC009/2018/G6P[5] was also co-detected in a sample containing RVA/Porcine-wt/ZAF/UFS-BOC009/2018/G5P[6]P[13]b during December 2018. Traces of RVC were detected two months later during the February 2019 sampling collection. The low number of reads for the RVB and RVC sequences in samples UFS-BOC124 and UFS-BOC035 could be due to ineffective virus replication.

However, since the traces of both these RVB and RVC strains were detected within two months of RVB/Pig-wt/ZAF/UFS-BOC050/2019/G14P[5] and RVC/Pig-wt/ZAF/UFS-BOC009/2018/G6P[5], it is possible that these infections were clearing. The low nucleotide identities to closest relatives observed for most of the RVB and RVC genome sequences emphasises the lack of sequence data for these strains not only from Africa but also globally.

5. Conclusions

The detection of three rotavirus species (A, B, and C) during a two-year period on a porcine farm in the Western Cape province of South Africa highlights the complex epidemiology of rotavirus in porcine populations. The phylogenetic analyses revealed that the RVB and RVC sequences represent unknown strains and will contribute to the little genetic information available for these groups.

Supplementary Materials: The following supporting information can be downloaded at https://www.mdpi.com/article/10.3390/v16060934/s1. Figure S1: Rotavirus A phylogenetic analysis; Figure S2: Rotavirus B phylogenetic analysis; Figure S3: Rotavirus C phylogenetic analysis; Table S1: Sequences used for reference mapping; Table S2: Evaluation of genome assembly; Table S3: Rotavirus A distance matrix; Table S4: Rotavirus B distance matrix; Table S5: Rotavirus C distance matrix.

Author Contributions: Conceptualization, A.S. and H.G.O.; methodology, investigation, A.S, N.S., R.C., N.B. and M.S.; validation, A.S., R.C. and H.G.O.; formal analysis, A.S., N.S., R.C. and N.B.; resources, M.S.; data curation, A.S. and H.G.O.; writing—original draft preparation, A.S.; writing—review and editing, H.G.O.; funding acquisition, H.G.O. All authors have read and agreed to the published version of the manuscript.

Funding: This research was funded by Deutsche Forschungsgemeinschaft (DFG; JO369/5-1 and JO369/5-2) to H.G.O. which supported A.S., R.C. and N.B. N.S. was supported by scholarships from the National Research Foundation of South Africa and the Poliomyelitis Research Foundation (18/90).

Institutional Review Board Statement: The animal study protocol was approved by the Animal Research Ethics Committee at the University of the Free State (UFS) (UFS-AED2018/0030).

Informed Consent Statement: Not applicable.

Data Availability Statement: The nucleotide sequences generated in this study were submitted to GenBank, and accession numbers PP669365-PP669534 (RVA), PP669283-PP669293 (RVB) and PP669294-PP669301 (RVC) were assigned.

Acknowledgments: The authors would like to express their sincere gratitude to the porcine farm for allowing access and sampling. We would like to thank Martin M Nyaga and the Next-Generation Sequencing Unit at the University of the Free State for excellent assistance with NGS.

Conflicts of Interest: The authors declare no conflicts of interest.

References

1. Matthijnssens, J.; Otto, P.H.; Ciarlet, M.; Desselberger, U.; Van Ranst, M.; Johne, R. VP6-sequence-based cutoff values as a criterion for rotavirus species demarcation. *Arch. Virol.* **2012**, *157*, 1177–1182. [CrossRef]
2. Estes, M.K.; Greenberg, H.B. Rotaviruses. In *Fields Virology*, 6th ed.; Knipe, D.M., Howley, P.M., Cohen, J.I., Griffin, D.E., Lamb, R.A., Martin, M.A., Racaniello, V.R., Roizman, B., Eds.; Wolters Kluwer Health/Lippincott Williams & Wilkins: Philadelphia, PA, USA, 2013; pp. 1347–1401.
3. Chepngeno, J.; Diaz, A.; Paim, F.C.; Saif, L.J.; Vlasova, A.N. Rotavirus C: Prevalence in suckling piglets and development of virus-like particles to assess the influence of maternal immunity on the disease development. *Vet. Res.* **2019**, *50*, 84. [CrossRef]

4. Baumann, S.; Sydler, T.; Rosato, G.; Hilbe, M.; Kümmerlen, D.; Sidler, X. Frequent Occurrence of Simultaneous Infection with Multiple Rotaviruses in Swiss Pigs. *Viruses* **2022**, *14*, 1117. [CrossRef]

5. Amimo, J.O.; Vlasova, A.N.; Saif, L.J. Prevalence and genetic heterogeneity of porcine group C rotaviruses in nursing and weaned piglets in Ohio, USA and identification of a potential new VP4 genotype. *Vet. Microbiol.* **2013**, *164*, 27–38. [CrossRef]

6. Miyabe, F.M.; Dall Agnol, A.M.; Leme, R.A.; Oliveira, T.E.S.; Headley, S.A.; Fernandes, T.; de Oliveira, A.G.; Alfieri, A.F.; Alfieri, A.A. Porcine rotavirus B as primary causative agent of diarrhea outbreaks in newborn piglets. *Sci. Rep.* **2020**, *10*, 22002. [CrossRef]

7. Li, Q.; Wang, Z.; Jiang, J.; He, B.; He, S.; Tu, C.; Guo, Y.; Gong, W. Outbreak of piglet diarrhea associated with a new reassortant porcine rotavirus B. *Vet. Microbiol.* **2024**, *288*, 109947. [CrossRef]

8. Shepherd, K.; Herrera-Ibata, D.M.; Porter, E.; Homwong, N.; Hesse, R.; Bai, J. Whole genome classification and phylogenetic analyses of rotavirus B strains from the United States. *Pathogens* **2018**, *7*, 44. [CrossRef]

9. Brnić, D.; Vlahović, D.; Gudan Kurilj, A.; Maltar-Strmečki, N.; Lojkić, I.; Kunić, V.; Jemeršić, L.; Bačani, I.; Kompes, G.; Beck, R.; et al. The impact and complete genome characterisation of viruses involved in outbreaks of gastroenteritis in a farrow-to-finish holding. *Sci. Rep.* **2023**, *13*, 18780. [CrossRef]

10. Wang, Y.; Porter, E.P.; Lu, N.; Zhu, C.; Noll, L.W.; Hamill, V. Whole-genome classification of rotavirus C and genetic diversity of porcine strains in the USA. *J. Gen Virol.* **2021**, *102*, 001598. [CrossRef]

11. Kumar, D.; Shepherd, F.K.; Springer, N.L.; Mwangi, W.; Marthaler, D.G. Rotavirus Infection in Swine: Genotypic Diversity, Immune Responses, and Role of Gut Microbiome in Rotavirus Immunity. *Pathogens* **2022**, *11*, 1078. [CrossRef]

12. Joshi, M.S.; Arya, S.A.; Shinde, M.S.; Ingle, V.C.; Birade, H.S.; Gopalkrishna, V. Rotavirus C infections in asymptomatic piglets in India, 2009–2013: Genotyping and phylogenetic analysis of all genomic segments. *Arch. Virol.* **2022**, *167*, 2665–2675. [CrossRef]

13. Collins, P.J.; Martella, V.; O'Shea, H. Detection and characterization of group C rotaviruses in asymptomatic piglets in Ireland. *J. Clin. Microbiol.* **2008**, *46*, 2973–2979. [CrossRef]

14. Genz, B.; Gerszon, J.; Pollock, Y.; Gleeson, B.; Shankar, R.; Sellars, M.J.; Moser, R.J. Detection and genetic diversity of porcine rotavirus A, B and C in eastern Australian piggeries. *Aust. Vet. J.* **2023**, *101*, 153–163. [CrossRef]

15. Papp, H.; László, B.; Jakab, F.; Ganesh, B.; De Grazia, S.; Matthijnssens, J.; Ciarlet, M.; Martella, V.; Bányai, K. Review of group A rotavirus strains reported in swine and cattle. *Vet. Microbiol.* **2013**, *165*, 190–199. [CrossRef]

16. Phan, M.V.; Anh, P.H.; Van Cuong, N.; Munnink, B.B.O.; van der Hoek, L.; Tri, T.N. Unbiased whole-genome deep sequencing of human and porcine stool samples reveals circulation of multiple groups of rotaviruses and a putative zoonotic infection. *Virus Evol.* **2016**, *2*, 58875. [CrossRef]

17. Wu, F.T.; Liu, L.T.C.; Jiang, B.; Kuo, T.Y.; Wu, C.Y.; Liao, M.H. Prevalence and diversity of rotavirus A in pigs: Evidence for a possible reservoir in human infection. *Infect. Genet. Evol.* **2022**, *98*, 105198. [CrossRef]

18. Papp, H.; Borzák, R.; Farkas, S.; Kisfali, P.; Lengyel, G.; Molnár, P. Zoonotic transmission of reassortant porcine G4P[6] rotaviruses in Hungarian pediatric patients identified sporadically over a 15year period. *Infect. Genet. Evol.* **2013**, *19*, 71–80. [CrossRef]

19. Suzuki, T.; Hasebe, A.; Miyazaki, A.; Tsunemitsu, H. Analysis of genetic divergence among strains of porcine rotavirus C, with focus on VP4 and VP7 genotypes in Japan. *Virus Res.* **2015**, *197*, 26–34. [CrossRef]

20. Prozesky, L.; Theodoridis, A. Diarrhoea in pigs induced by rotavirus. *J. S. Afr. Vet. Assoc.* **1977**, *44*, 275–278.

21. Geyer, A.; Steele, A.D.; Peenze, I.; Lecatsas, G. Astrovirus-like particles, adenoviruses and rotaviruses associated with diarrhoea in piglets. *J. S. Afr. Vet. Assoc.* **1994**, *65*, 164–166.

22. Geyer, A.; Sebata, T.; Peenze, I.; Steele, A.D. A molecular epidemiological study of porcine rotaviruses. *J. S. Afr. Vet. Assoc.* **1995**, *66*, 202–205.

23. Geyer, A.; Sebata, T.; Peenze, I.; Steele, A.D. Group B and C porcine rotaviruses identified for the first time in South Africa. *J. S. Afr. Vet. Assoc.* **1996**, *67*, 115–116.

24. Seheri, L.M.; Magagula, N.B.; Peenze, I.; Rakau, K.; Ndadza, A.; Mwenda, J.M.; Weldegebriel, G.; Steele, A.D.; Mphahlele, M.J. Rotavirus strain diversity in Eastern and Southern African countries before and after vaccine introduction. *Vaccine* **2018**, *36*, 7222–7230. [CrossRef]

25. Nyaga, M.M.; Jere, K.C.; Esona, M.D.; Seheri, M.L.; Stucker, K.M.; Halpin, R.A.; Akopov, A.; Stockwell, T.B.; Peenze, I.; Diop, A.; et al. Whole genome detection of rotavirus mixed infections in human, porcine and bovine samples co-infected with various rotavirus strains collected from sub-Saharan Africa. *Infect. Genet. Evol.* **2015**, *31*, 321–334. [CrossRef]

26. Strydom, A.; Motanyane, L.; Nyaga, M.M.; João, E.D.; Cuamba, A.; Mandomando, I.; Cassocera, M.; de Deus, N.; O'Neill, H. Whole-genome characterization of G12 rotavirus strains detected in Mozambique reveals a co-infection with a GXP[14] strain of possible animal origin. *J. Gen. Virol.* **2019**, *100*, 932–937. [CrossRef]

27. Mokoena, F.; Esona, M.D.; Seheri, L.M.; Nyaga, M.M.; Magagula, N.B.; Mukaratirwa, A.; Mulindwa, A.; Abebe, A.; Boula, A.; Tsolenyanu, E.; et al. African Rotavirus Surveillance Network. Whole Genome Analysis of African G12P[6] and G12P[8] Rotaviruses Provides Evidence of Porcine-Human Reassortment at NSP2, NSP3, and NSP4. *Front. Microbiol.* **2021**, *12*, 604444. [CrossRef]

28. Potgieter, A.C.; Page, N.A.; Liebenberg, J.; Wright, I.M.; Landt, O.; van Dijk, A.A. Improved strategies for sequence-independent amplification and sequencing of viral double-stranded RNA genomes. *J. Gen. Virol.* **2009**, *90*, 1423–1432. [CrossRef]

29. Strydom, A.; João, E.D.; Motanyane, L.; Nyaga, M.M.; Potgieter, C.A.; Cuamba, A.; Mandomando, I.; Cassocera, M.; de Deus, N.; O'Neill, H.G. Whole genome analyses of DS-1-like Rotavirus A strains detected in children with acute diarrhoea in southern Mozambique suggest several reassortment events. *Infect. Genet. Evol.* **2019**, *69*, 68–75. [CrossRef]

30. Bolger, A.M.; Lohse, M.; Usadel, B. Trimmomatic: A flexible trimmer for Illumina sequence data. *Bioinformatics* **2014**, *30*, 2114–2120. [CrossRef]

31. Bankevich, A.; Nurk, S.; Antipov, D.; Gurevich, A.A.; Dvorkin, M.; Kulikov, A.S.; Lesin, V.M.; Nikolenko, S.I.; Pham, S.; Prjibelski, A.D.; et al. SPAdes: A new genome assembly algorithm and its applications to single-cell sequencing. *J. Comput. Biol.* **2012**, *19*, 455–477. [CrossRef]

32. Langmead, B.; Salzberg, S.L. Fast gapped-read alignment with Bowtie 2. *Nat. Methods* **2012**, *9*, 357–359. [CrossRef]

33. Danecek, P.; Bonfield, J.K.; Liddle, J.; Marshall, J.; Ohan, V.; Pollard, M.O.; Whitwham, A.; Keane, T.; McCarthy, S.A.; Davies, R.M.; et al. Twelve years of SAMtools and BCFtools. *GigaScience* **2021**, *10*, giab008. [CrossRef]

34. Pickett, B.E.; Sadat, E.L.; Zhang, Y.; Noronha, J.M.; Squires, R.B.; Hunt, V. ViPR: An open bioinformatics database and analysis resource for virology research. *Nucleic Acids Res.* **2012**, *40*, 593–598. [CrossRef]

35. Kumar, S.; Stecher, G.; Li, M.; Knyaz, C.; Tamura, K. MEGA X: Molecular evolutionary genetics analysis across computing platforms. *Mol. Biol. Evol.* **2018**, *35*, 1547–1549. [CrossRef]

36. Trifinopoulos, J.; Nguyen, L.T.; von Haeseler, A.; Minh, B.Q. W-IQ-TREE: A fast online phylogenetic tool for maximum likelihood analysis. *Nucleic Acids Res.* **2016**, *44*, W232–W235. [CrossRef]

37. Minh, B.Q.; Nguyen, M.A.T.; Von Haeseler, A. Ultrafast approximation for phylogenetic bootstrap. *Mol. Biol. Evol.* **2013**, *30*, 1188–1195. [CrossRef]

38. Cortey, M.; Díaz, I.; Vidal, A.; Martín-Valls, G.; Franzo, G.; Gómez De Nova, P.J.; Darwich, L.; Puente, H.; Carvajal, A.; Martin, M.; et al. High levels of unreported intraspecific diversity among RNA viruses in faeces of neonatal piglets with diarrhoea. *BMC Vet. Res.* **2019**, *15*, 441. [CrossRef]

39. Chauhan, R.P.; San, J.E.; Gordon, M.L. Metagenomic Analysis of RNA Fraction Reveals the Diversity of Swine Oral Virome on South African Backyard Swine Farms in the uMgungundlovu District of KwaZulu-Natal Province. *Pathogens* **2022**, *11*, 927. [CrossRef]

40. Settembre, E.C.; Chen, J.Z.; Dormitzer, P.R.; Grigorieff, N.; Harrison, S.C. Atomic model of an infectious rotavirus particle. *EMBO J.* **2011**, *30*, 408–416. [CrossRef]

41. Nyaga, M.M.; Tan, Y.; Seheri, M.L.; Halpin, R.A.; Akopov, A.; Stucker, K.M.; Fedorova, N.B.; Shrivastava, S.; Steele, D.A.; Mwenda, J.M.; et al. Whole-genome sequencing and analyses identify high genetic heterogeneity, diversity and endemicity of rotavirus genotype P[6] strains circulating in Africa. *Infect. Genet. Evol.* **2018**, *63*, 79–88. [CrossRef]

42. Bwogi, J.; Jere, K.C.; Karamagi, C.; Byarugaba, D.K.; Namuwulya, P.; Baliraine, F.N.; Desselberger, U.; Iturriza-Gomara, M. Whole genome analysis of selected human and animal rotaviruses identified in Uganda from 2012 to 2014 reveals complex genome reassortment events between human, bovine, caprine and porcine strains. *PLoS ONE* **2017**, *12*, e0178855. [CrossRef]

43. Boene, S.S.; João, E.D.; Strydom, A.; Munlela, B.; Chissaque, A.; Bauhofer, A.F.L.; Nabetse, E.; Latifo, D.; Cala, A.; Mapaco, L.; et al. Prevalence and genome characterization of porcine rotavirus A in southern Mozambique. *Infect. Genet. Evol.* **2021**, *87*, 104637. [CrossRef]

44. Amimo, J.O.; Vlasova, A.N.; Saif, L.J. Detection and genetic diversity of porcine group A rotaviruses in historic (2004) and recent (2011 and 2012) swine fecal samples in Ohio: Predominance of the G9P[13] genotype in nursing piglets. *J. Clin. Microbiol.* **2013**, *51*, 1142–1151. [CrossRef]

Article

Whole-Genome Characterization of Rotavirus G9P[6] and G9P[4] Strains That Emerged after Rotavirus Vaccine Introduction in Mozambique

Benilde Munlela [1,2,*], Eva D. João [1], Amy Strydom [3], Adilson Fernando Loforte Bauhofer [1,2], Assucênio Chissaque [1,2], Jorfélia J. Chilaúle [1], Isabel L. Maurício [4], Celeste M. Donato [5], Hester G. O'Neill [3] and Nilsa de Deus [1,6]

[1] Instituto Nacional de Saúde (INS), Parcela 3943, Vila de Marracuene, Maputo 0205-02, Mozambique; evajoao29@gmail.com (E.D.J.); adilson.bauhofer@ins.gov.mz (A.F.L.B.); assucenio.chissaque@ins.gov.mz (A.C.); jorfelia.chilaule@ins.gov.mz (J.J.C.); nilsa.dedeus@ins.gov.mz (N.d.D.)

[2] Instituto de Higiene e Medicina Tropical (IHMT), Universidade NOVA de Lisboa (UNL), Rua da Junqueira 100, 1349-008 Lisboa, Portugal

[3] Department of Microbiology and Biochemistry, University of the Free State, 205 Nelson Mandela Avenue, Bloemfontein 9301, South Africa; aimster.strydom@gmail.com (A.S.); oneillhg@ufs.ac.za (H.G.O.)

[4] Global Health and Tropical Medicine (GHTM), Associate Laboratory in Translation and Innovation towards Global Health (LA-REAL), Instituto de Higiene e Medicina Tropical (IHMT), Universidade NOVA de Lisboa (UNL), Rua da Junqueira 100, 1349-008 Lisboa, Portugal; isabel.mauricio@ihmt.unl.pt

[5] The Peter Doherty Institute for Infection and Immunity, 792 Elizabeth Street, Melbourne, VIC 3000, Australia; celeste.donato@unimelb.edu.au

[6] Departamento de Ciências Biológicas, Universidade Eduardo Mondlane, Julius Nyerere Avenue, Maputo 3453, Mozambique

[*] Correspondence: benilde.munlela@ins.gov.mz or benildeantnio@gmail.com

Citation: Munlela, B.; João, E.D.; Strydom, A.; Bauhofer, A.F.L.; Chissaque, A.; Chilaúle, J.J.; Maurício, I.L.; Donato, C.M.; O'Neill, H.G.; de Deus, N. Whole-Genome Characterization of Rotavirus G9P[6] and G9P[4] Strains That Emerged after Rotavirus Vaccine Introduction in Mozambique. *Viruses* **2024**, *16*, 1140. https://doi.org/10.3390/v16071140

Academic Editors: Ulrich Desselberger and John T. Patton

Received: 15 May 2024
Revised: 9 July 2024
Accepted: 11 July 2024
Published: 16 July 2024

Abstract: Mozambique introduced the Rotarix® vaccine into the National Immunization Program in September 2015. Following vaccine introduction, rotavirus A (RVA) genotypes, G9P[4] and G9P[6], were detected for the first time since rotavirus surveillance programs were implemented in the country. To understand the emergence of these strains, the whole genomes of 47 ELISA RVA positive strains detected between 2015 and 2018 were characterized using an Illumina MiSeq-based sequencing pipeline. Of the 29 G9 strains characterized, 14 exhibited a typical Wa-like genome constellation and 15 a DS-1-like genome constellation. Mostly, the G9P[4] and G9P[6] strains clustered consistently for most of the genome segments, except the G- and P-genotypes. For the G9 genotype, the strains formed three different conserved clades, separated by the P type (P[4], P[6] and P[8]), suggesting different origins for this genotype. Analysis of the VP6-encoding gene revealed that seven G9P[6] strains clustered close to antelope and bovine strains. A rare E6 NSP4 genotype was detected for strain RVA/Human-wt/MOZ/HCN1595/2017/G9P[4] and a genetically distinct lineage IV or OP354-like P[8] was identified for RVA/Human-wt/MOZ/HGJM0644/2015/G9P[8] strain. These results highlight the need for genomic surveillance of RVA strains detected in Mozambique and the importance of following a One Health approach to identify and characterize potential zoonotic strains causing acute gastroenteritis in Mozambican children.

Keywords: rotavirus A; G9P[4], G9P[6], NSP4 E6 genotype; Mozambique

1. Introduction

Rotavirus remains one of the primary causative agents of gastroenteritis in children under five years of age, exerting a substantial global health burden [1,2]. It is estimated that rotavirus infections resulted in 128,500 deaths in 2016, of which 104,733 occurred in Sub-Saharan Africa [3].

Rotavirus is a member of the Sedoreoviridae family [4]. The virus has an icosahedral capsid formed by three concentric protein layers and a genome comprising 11 double-stranded ribonucleic acid (dsRNA) segments, encoding six viral and structural proteins (VP) and five or six non-structural proteins (NSP) [5]. The gene segments encoding the external capsid proteins, VP7 and VP4, of rotavirus group A (RVA) are used in a binary classification system defining G and P genotypes, respectively [2,5]. Currently, 42 G and 58 P genotypes have been described [6]. Globally, G1P[8], G2P[4], G3P[8], G4P[8], G9P[8] and G12P[8] are the most frequently detected genotype combinations, with varying prevalence observed across different countries [7–10].

A whole-genome classification system (based on nucleotide percent cut-off values) allows for the classification of all 11 RVA genes into genotype constellations designated as Gx-P[x]-Ix-Rx-Cx-Mx-Ax-Nx-Tx-Ex-Hx, with "x" indicating the number of genotypes assigned. These genotypes correspond to genome segments encoding VP7-VP4-VP6-VP1-VP2-VP3-NSP1-NSP2-NSP3-NSP4-NSP5/6 proteins [11]. To date 32I, 28R, 24C, 24M, 39A, 28N, 28T, 32E and 28H genotypes have been described [6]. The most prevalent genotype constellations in humans are the Wa-like (I1-R1-C1-M1-A1-N1-T1-E1-H1) and DS-1-like (I2-R2-C2-M2-A2-N2-T2-E2-H2) constellations. A third group known as AU-1-like (I3-R3-C3-M3-A3-N3-T3-E3-H3) is also detected in humans, albeit at a lower frequency [11–13].

Four rotavirus vaccines have received prequalification from the World Health Organization (WHO): Rotarix®, RotaTeq®, Rotavac® and Rotasiil®. These vaccines have demonstrated significant efficacy in reducing diarrheal morbidity and mortality on a global scale [8]. In September 2015, Mozambique introduced the Rotarix® vaccine into the National Immunization Program. Since then, the prevalence of rotavirus infection in Mozambique has decreased from 40.6% to 19.1% [14,15], the vaccine effectiveness has been estimated to be lower than what was reported in many other African countries, with an effectiveness of 30% against G1P[8] strains and 35% against non-G1P[8] strains [16].

Prior to vaccine implementation, G9P[8] and G1P[8] had been the most predominant genotypes in Mozambique [15,17]. Whole-genome analyses (WGA) of human Mozambican RVA strains before vaccine introduction have suggested genetic diversity was partially driven by reassortment events between animal and human strains [18,19]. However, post-vaccine introduction, G1P[8] became the predominant genotype nationwide [15,17], despite the use of a G1P[8]-based vaccine, although WGA of these G1P[8] strains indicated no significant mutations in epitope regions that might lead to vaccine escape, and no distinct clustering was observed between pre- and post-vaccine strains [20].

During the post-vaccine period, G9P[4], G9P[6], G3P[8] and G3P[4] have emerged as predominant genotype combinations. However, the origin of these strains, as well as their relation to the G9P[8] strains reported before vaccine introduction, remains unclear [14]. Therefore, the whole genomes of strains detected between 2015 and 2018 were determined and analyzed in the current study in order to elucidate the origin of the G9 genotype detected following vaccine introduction in Mozambique.

2. Materials and Methods

2.1. Sample Collection

Fifty-seven fecal samples collected between 2015 and 2018 as part of ongoing hospital-based sampling within the National Diarrhea Surveillance System (ViNaDia) in Mozambique [15] were selected for WGA. The samples collected in 2015 represent the pre-vaccine period and those collected in 2016–2018, the post-vaccine period (Table S1). These samples had previously tested positive for RVA by ELISA (Prospect EIA rotavirus, Basingstoke, United Kingdom) and the binary genotype combination was determined by multiplex Reverse Transcriptase (RT)-PCR [21–23].

2.2. Viral Genomic dsRNA Extraction, cDNA Library Building and Illumina MiSeq Sequencing

Total RNA was extracted from stool samples with TRI-reagent (Sigma, Darmstadt, Germany), and single-stranded RNA was precipitated with lithium chloride. The self-priming

PC3-T7 loop primer (Integrated DNA Technologies, Coralville, IA, USA) was ligated to dsRNA to obtain full-length sequences. Complementary DNA (cDNA) was synthesized using the Maxima H Minus double-stranded cDNA kit (Thermo Fisher Scientific, Massachusetts, MA, USA) as previously described [19]. The cDNA was synthesized at the Next Generation Sequencing Unit at the University of the Free State in Bloemfontein, South Africa. In brief, the cDNA library was made by NEBNext Ultra RNA Library Prep Kit for Illumina v1.2 (New England Biolabs, Ipswich, MA, USA), and NEBNext Multiplex Oligos for Illumina (New England Biolabs, Ipswich, MA, USA) according to the manufacturer's instructions and purified using Agencourt AMPure XP magnetic beads (Beckman Coulter, Brea, CA, USA). Nucleotide sequencing was performed using an Illumina MiSeq sequencer (Illumina, San Diego, CA, USA) using a MiSeq Reagent Kit V3 (Illumina, San Diego, CA, USA) [19].

2.3. Data Analysis
2.3.1. Genome Assembly

A de novo assembly was performed for all samples using CLC Bio Genomics Workbench (12.0.3; Qiagen, Aarhus, Denmark); all contigs with an average coverage above 100 were identified on the Nucleotide Basic Local Alignment Search Tool (BLASTn at the National Center for Biotechnology Information—NCBI). Reference sequences were chosen based on the BLASTn results for reference mapping and extraction of consensus sequences for each segment [19].

2.3.2. Determination of RVA Genotypes

The genotype of each of the 11 genes for each strain was determined using the Virus Pathogenic database and analysis resource (ViPR) according to the guidelines proposed by the Rotavirus Classification Working Group [6,13].

2.3.3. Phylogenetic Analysis

Multiple sequence alignment of each gene was carried out using Multiple Sequence Comparison by Log Expectation (MUSCLE) alignment available in Molecular Evolutionary Genetic Analysis X (MEGA X) [24].

The best nucleotide substitution model, considered as having the lowest Bayesian Information Criterion, was calculated through Maximum Likelihood, as implemented in Mega X, for phylogenetic analysis and the models selected for each gene were: Tamura-3-parameter (T92+G+I) for VP7-G9, VP4-P[6], VP6-I2, VP3-M2, VP7-G1 and VP2-C1, T92+G for VP4-P[8], VP7-G3, VP7-G2, NSP1-A2 and NSP3-T2, T92+I for VP4-P[4], NSP4-E2 and NSP4-E1, T92 for NSP4-E6, NSP1-A1 and NSP2-N1, General Time Reversible (GTR+G+I) for VP1-R2, NSP2-N2 and VP3-M1, GTR+G for VP2-C1 and NSP5/6-H2, GTR+I for VP1-R1 and NSP5/6-H1. Maximum likelihood gene trees based on phylogenetic analysis of the complete ORF of the 11 genome segments for all strains were constructed using MEGA X using 1000 bootstrap replicates to estimate branch support. Pairwise distance matrix nucleotides were obtained in MEGA X using the p-distance algorithm [24]. Mozambican strains previously characterized as DS-1-like and Wa-like constellations were included in the analyses, as well as other genetically similar reference strains obtained from GenBank. The lineages were defined by previously published designations [25–32].

The computational tools for comparative genomics (mVISTA) online platform were used to visualize the sequence similarities of concatenated full genomes of the G9P[4] and G9P[6] exhibiting a DS-1 backbone using RVA/Human-wt/MOZ/HCN1347/2016/G9P[6] strain as reference [33].

2.3.4. Nucleotide Sequence Accession Numbers

The nucleotide sequence data presented were deposited in GenBank under the following accession numbers: PP585813-PP586043 and PP848501-PP848786.

3. Results

3.1. Genome Constellations

Of the 47 successfully sequenced samples, 61.7% (29/4) were collected in males, 53.3% (25/47) were from children between 0–11 months old and 38.3% (18/47) were from children aged between 12–23 months old; 29.7% (14/47) were from unvaccinated children, 27.7%(13/47) from children who received two doses of the vaccine and 4.3% (2/47) were from children that received a single dose of the Rotarix®; 91.5% (43/47) were hospitalized due to their clinical presentation (Table S1).

Twenty-nine strains were identified as G9 strains (Table 1). The G9P[6] (n = 9) and G9P[4] (n = 6) strains presented a typical DS-1-like constellation (-R2-C2-M2-I2-A2-N2-T2-E2-H2), except the strain RVA/Human-wt/MOZ/HCN1595/2017/G9P[4] which contained an E6 NSP4 gene (Table 1). The 14 G9P[8] strains contained a Wa-like constellation (-R1-C1-M1-I1-A1-N1-T1-E1-H1) (Table 1). The remaining strains (n = 18) characterized as G2P[4] (n = 2), G2P[6] (n = 7) G3P[4] (n = 4), G3P[8] (n = 3) and G1P[8] (n = 2) presented typical DS-1-like and Wa-like constellations (Table S2).

3.2. Phylogenetic Analysis

3.2.1. VP7 Encoding Gene

G9 Genotype

The G9 strains, in combination with P[4], P[6] and P[8], formed three different conserved clades within lineage III (Figure 1). The G9P[4] strains from the post-vaccine period grouped with G9P[4] strains from India detected in 2013 and 2014. The nine G9P[6] strains detected in the post-vaccine introduction period clustered with strains from Zimbabwe (2009 and 2011) and South Africa (2008 and 2010), but in combination with P[8] and no other G9P[6] strains previously detected in Southern Africa. The 13 G9P[8] strains, detected before the introduction of the rotavirus vaccine, clustered together with G9P[8] strains from Japan that circulated in 2013. Seven of the nine G9P[6] strains were detected in unvaccinated children. One G9P[8] strain (RVA/Human-wt/MOZ/HGJM0644/2015/G9P[8]), clustered distinctly from other Mozambican strains and shared 99.1–99.2% nucleotide (nt) identity and 98.8–99.1% amino acid (aa) identity with the other 12 G9P[8] Mozambican strains (Table S3). This Mozambican strain clustered with a Japanese strain from 2016 (Figure 1).

G3, G2 and G1 Genotypes

The seven G3 strains clustered in lineage III (Figure S1) and shared an identity of 98.7–99.9% (98.2–100%) nt (aa). The three G3P[8] detected in vaccinated children were closely related to G3P[8] strains from Japan and Kenya detected in 2017 and 2019 (RVA/Human-wt/JPN/Tokyo17-21/2017/G3P[8] and RVA/Human-wt/KEN/KLF0929/2019/G3P[8]). The four G3P[4] were closely related to RVA/Human-wt/PAK/PAK663/2016/G3P[4] from Pakistan (Figure S1). Two Mozambican G3P[8] strains previously described from a rural site in Mozambique (RVA/Human-wt/MOZ/MAN1811450.8/2021/G3P[8] and RVA/Human-wt/MOZ/MAN1811463.8/2021/G3P[8]) clustered between the G3P[8] and G3P[4] study strains and shared an identity of 97.5–98.5% (98.4–99.0%) nt (aa).

All G2 strains from pre (2015) and post-vaccine introduction (2016) grouped together in lineage IV sub lineage a-3 III (Figure S1). These strains were closely related to G2P[4] strains from Southern Africa, including Malawi (RVA/Human-wt/MWI/BID115/2012/G2P[4]) as well as three Mozambican G2P[4] strains that circulated in 2013 (Figure S1).

The G1P[8] strains, RVA/Human-wt/MOZ/HGJM0408/2015/G1P[8] (pre-vaccine period) and RVA/Human-wt/MOZ/HCN1556/2017/G1P[8] (detected in a vaccinated child), clustered in a highly conserved clade of previously characterized G1P[8] Mozambican strains in lineage II (Figure S1).

Table 1. Genotype constellation of Mozambican G9P[4], G9P[6] and G9P[8] strains.

Strain Name	VP7	VP4	VP6	VP1	VP2	VP3	NSP1	NSP2	NSP3	NSP4	NSP5/6
Segment	9	4	6	1	2	3	5	8	7	10	11
Wa-like	G1	P[8]	I1	R1	C1	M1	A1	N1	T1	E1	H1
DS1-like	G2	P[4]	I2	R2	C2	M2	A2	N2	T2	E2	H2
RVA/Human-wt/MOZ/HCN1357/2016/G9P[6]	G9	P[6]	I2	R2	C2	M2	A2	N2	T2	E2	H2
RVA/Human-wt/MOZ/HCN1359/2016/G9P[6]	G9	P[6]	I2	R2	C2	M2	A2	N2	T2	E2	H2
RVA/Human-wt/MOZ/HCN1369/2017/G9P[6]	G9	P[6]	I2	R2	C2	M2	A2	N2	T2	E2	H2
RVA/Human-wt/MOZ/HCN1371/2017/G9P[6]	G9	P[6]	I2	R2	C2	M2	A2	N2	T2	E2	H2
RVA/Human-wt/MOZ/HCN1597/2017/G9P[6]	G9	P[6]	I2	R2	C2	M2	A2	N2	T2	E2	H2
RVA/Human-wt/MOZ/HGJM1782/2017/G9P[6]	G9	P[6]	I2	R2	C2	M2	A2	N2	T2	E2	H2
RVA/Human-wt/MOZ/HGQ1296/2016/G9P[6]	G9	P[6]	I2	R2	C2	M2	A2	N2	T2	E2	H2
RVA/Human-wt/MOZ/HCN1347/2016/G9P[6]	G9	P[6]	I2	R2	C2	M2	A2	N2	T2	E2	H2
RVA/Human-wt/MOZ/HGM1883/2018/G9P[6]	G9	P[6]	I2	R2	C2	M2	A2	N2	T2	E2	H2
RVA/Human-wt/MOZ/HCN1595/2017/G9P[4]	G9	P[4]	I2	R2	C2	M2	A2	N2	T2	E6	H2
RVA/Human-wt/MOZ/HCN1855/2017/G9P[4]	G9	P[4]	I2	R2	C2	M2	A2	N2	T2	E2	H2
RVA/Human-wt/MOZ/HCN1598/2017/G9P[4]	G9	P[4]	I2	R2	C2	M2	A2	N2	T2	E2	H2
RVA/Human-wt/MOZ/HGJM1647/2017/G9P[4]	G9	P[4]	I2	R2	C2	M2	A2	N2	T2	E2	H2
RVA/Human-wt/MOZ/HCN1600/2017/G9P[4]	G9	P[4]	I2	R2	C2	M2	A2	N2	T2	E2	H2
RVA/Human-wt/MOZ/HCN1604/2017/G9P[4]	G9	P[4]	I2	R2	C2	M2	A2	N2	T2	E2	H2
RVA/Human-wt/MOZ/HGM483/2015/G9P[8]	G9	P[8]	I1	R1	C1	M1	A1	N1	T1	E1	H1
RVA/Human-wt/MOZ/HGJM0318/2015/G9P[8]	G9	P[8]	I1	R1	C1	M1	A1	N1	T1	E1	H1
RVA/Human-wt/MOZ/HGJM0334/2015/G9P[8]	G9	P[8]	I1	R1	C1	M1	A1	N1	T1	E1	H1
RVA/Human-wt/MOZ/HGM0355/2015/G9P[8]	G9	P[8]	I1	R1	C1	M1	A1	N1	T1	E1	H1
RVA/Human-wt/MOZ/HCN0370/2015/G9P[8]	G9	P[8]	I1	R1	C1	M1	A1	N1	T1	E1	H1
RVA/Human-wt/MOZ/HGJM0407/2015/G9P[8]	G9	P[8]	I1	R1	C1	M1	A1	N1	T1	E1	H1
RVA/Human-wt/MOZ/HGJM0385/2015/G9P[8]	G9	P[8]	I1	R1	C1	M1	A1	N1	T1	E1	H1
RVA/Human-wt/MOZ/HGM0322/2015/G9P[8]	G9	P[8]	I1	R1	C1	M1	A1	N1	T1	E1	H1
RVA/Human-wt/MOZ/HGJM0497/2015/G9P[8]	G9	P[8]	I1	R1	C1	M1	A1	N1	T1	E1	H1
RVA/Human-wt/MOZ/HGJM0413/2015/G9P[8]	G9	P[8]	I1	R1	C1	M1	A1	N1	T1	E1	H1
RVA/Human-wt/MOZ/HGM0353/2015/G9P[8]	G9	P[8]	I1	R1	C1	M1	A1	N1	T1	E1	H1
RVA/Human-wt/MOZ/HGM0389/2015/G9P[8]	G9	P[8]	I1	R1	C1	M1	A1	N1	T1	E1	H1
RVA/Human-wt/MOZ/HGJM0643/2015/G9P[8]	G9	P[8]	I1	R1	C1	M1	A1	N1	T1	E1	H1
RVA/Human-wt/MOZ/HGJM0644/2015/G9P[8]	G9	P[8]	I1	R1	C1	M1	A1	N1	T1	E1	H1

Wa-like and DS-1-like genotypes constellation are shown in green and red respectively. The G9 and E6 genotype are shown in white and P[6] genotype is shown in blue [12].

VP7-G9

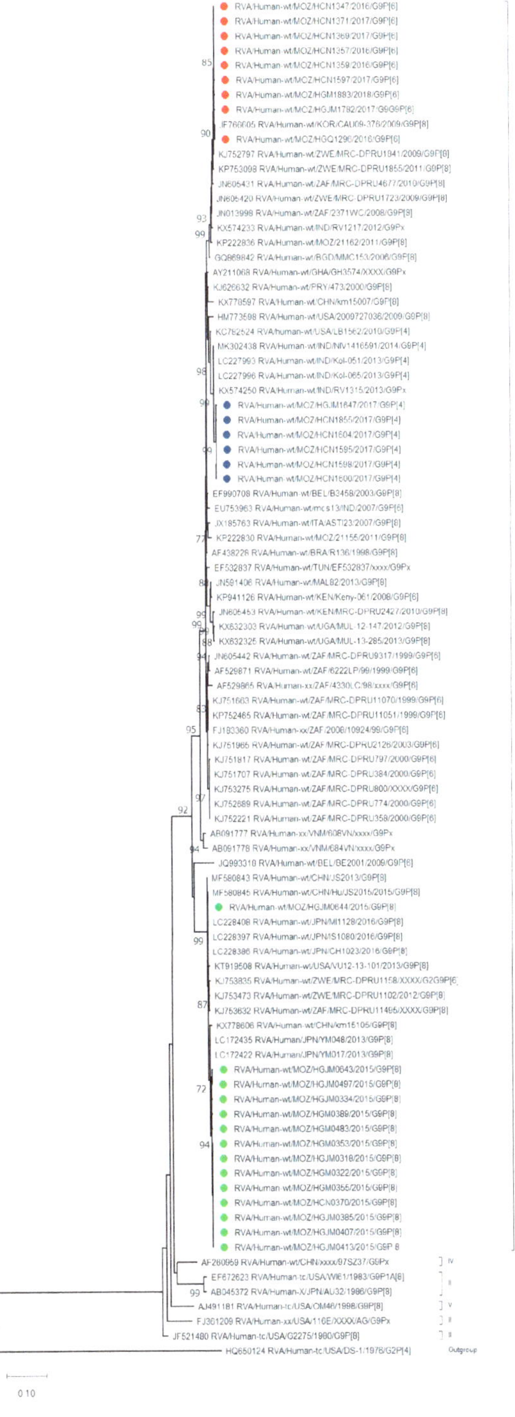

Figure 1. Phylogenetic tree based on the open reading frame (ORF) nucleotide sequence of the VP7-G9 encoding gene of strains circulating in Mozambique compared to global strains obtained from GenBank.

The tree was constructed based on the maximum likelihood method implemented in MEGA X [24], applying Tamura-3-parameter (T92+G+I) as the model. Bootstrap values (1000 replicates) ≥70% are shown with Wa-like strain serving as an out-group. The scale bar indicates genetic distance expressed as the number of nucleotide substitutions per site. G9P[4] Mozambican strains are indicated by blue circles, G9P[6] by red circles and G9P[8] by green circles.

3.2.2. VP4 Encoding Gene
P[4], P[6] and P[8] Genotypes

All P[4] genotypes detected in the study were compared to rotavirus sequences representing the five lineages of the P[4] encoding gene, and the results showed that all sequences were grouped in lineage IV. It was observed that five of the G9P[4] and one G3P[4] strains grouped with the G2P[4] strains from Kenya that circulated in 2012 (RVA/Human-wt/KEN/KLF0569/2012/G2P[4] and RVA/Human-wt/KEN/KLF0593/2012/G2P[4]) and G9P[4] strains from India reported in 2011 (RVA/Human-wt/IND/RV11/2011/G9P[4]) and 2013 (RVA/Human-wt/IND/Kol-047/2013/G9P[4]) and a G3P[4] strain from Pakistan (Figure 2). Interestingly, the remaining G9P[4] strain, RVA/Human-wt/MOZ/HCN1598/2017/G9P[4], detected in northern Mozambique, clustered with G3P[4] Mozambican strains detected in southern Mozambique and shared a % nt (aa) identity of 99.7–100% (99.6–100%) (Figure 2).

All nine P[6] strains in combination with G9 clustered closely together in lineage I with the G2P[6] strains described in this study and with other G12P[6] and G2P[6] strains previously described from Mozambique and other Southern and Eastern African countries. The exception was RVA/Human-wt/MOZ/HCN1328/2016/G2P[6] that grouped separately from the rest of the study strains with G12P[6] Mozambican strains detected in 2012 (Figure 3).

Thirteen G9P[8] strains from 2015 (pre-vaccine period) formed a conserved clade in lineage III and grouped close to G9P[8] strains from Japan and China, as well as G1P[8] strains from Australia and the USA. The three G3P[8] strains were also grouped in lineage III, although in a separate cluster from the G9P[8] strains. The two G1P[8] (RVA/Human-wt/MOZ/HGJM0408/2015/G1P[8] and RVA/Human-wt/MOZ/HCN1556/2017/G1P[8]) strains formed clusters with Mozambican G1P[8] strains reported between 2012 and 2017. One G9P[8] strain, RVA/Human-wt/MOZ/HGJM0644/2015/G9P[8], clustered separately in the rare lineage IV with G1P[8] and G3P[8] strains from Belgium and Russia (2008 and 2009, respectively) (Figure 4).

3.2.3. VP1–VP3 and VP6 Encoding Genes
VP1–VP3

For genotypes R2 (VP1) and C2 (VP2), the G9P[6] and G9P[4] strains grouped into the same cluster with G2P[6] and G3P[4] study strains, and was closely related to Mozambican G2P[4] strains that circulated in 2013. An exception was observed for the C2 genotype, where the G3P[4] strains formed a separate cluster with one of the G9P[6] strains, (RVA/Human-wt/MOZ/HGM1782/2017/G9P[6]) and G3P[4] strains from Pakistan which circulated in 2016 (Figure S1).

Similar groupings were observed for the M2 (VP3) genotype. where the G9P[4] and G9P[6] clustered in the same major clade with the G2P[6] study strains and G2P[4] that circulated in 2013 and 2015. The exception was two study strains, (RVA/Human-wt/MOZ/HGM1782/2017/G9P[6] and RVA/Human-wt/MOZ/HCN1595/2017/G9P[4]), that clustered separately with the study G3P[4] strains from Mozambique (Figure S1).

VP6

The I2 genotype of G9P[4] strains grouped into a conserved cluster in lineage V and were closely related to Malawian (RVA/Human-wt/MWI/BID1JK/2013/G2P[4] and RVA/Human-wt/MWI/BID2DE/2013/G1P[8]) and Indian (RVA/Human-wt/IND/CMC00024/2012/G2Px) strains.

Seven of the G9P[6] strains that circulated between 2016 and 2018 shared an nt (aa) identity of 99.9% (100%). These strains grouped into lineage IX in a cluster that contained animal strains, such as antelope (RVA/Antelope-wt/ZAF/RC-18-08/G6P[14]), with which it shared nt (aa) identity of 98.8% (100%). Also included in this cluster were bovine strains RVA/Cow-wt/ZAF/1604/2007/G8P[1], RVA/Cow-wt/ZAF/MRC-DPRU1604/2007/G6P[1] and RVA/Cow-wt/ZAF/MRC-DPRU3010/2009/G6P[5] with an average nt (aa) identity of 98.3% (99.6%). A VP6-encoding sequence of a human mixed infection strain (RVA/Human-wt/MOZ/0060b/2012/G12P[8]P[14]), previously reported to be of animal origin, also clustered in lineage IX, whereas bovine strains from Mozambique (RVA/Cow-wt/MOZ/MPT-93/2016/G10P[11] and RVA/Cow-wt/MOZ/MPT-307/2016/G10P[11]) clustered in lineages VI and X, respectively, with an average nt (aa) identity of 94.8% (99.6%) to the study strains (Figure 5).

Two G9P[6] strains were grouped in separated clusters; one strain (RVA/Human-wt/MOZ/HGQ1296/2016/G9P[6]) shared an nt identity of only 93.8% with the other seven G9P[6] strains, and formed a cluster with G2P[6] and G2P[4] study strains. Study strain RVA/Human-wt/MOZ/HGJM1782/2017/G9P[6] detected in a vaccinated child shared an average nt identity of 92.5%, with the rest of the G9P[6] strains and grouped with G3P[4] Mozambican study strains (detected in unvaccinated children) and with G3P[4] from Pakistan as in the other segments (Figure 5).

3.2.4. NSP1-NSP5/NSP6 Encoding Genes

The conserved clade observed in the VPs encoding genes, formed by eight G9P[6] and five G9P[4], clustered together with G2P[4] strains from Mozambique that circulated in 2013, 2015 and 2016 for the NSP-encoding genes. The RVA/Human-wt/MOZ/HCN1595/2017/G9P[4], RVA/Human-wt/MOZ/HGM1782/2017/G9P[6] and RVA/Human-wt/MOZ/HGQ1296/2016/G9P[6] strains, continued to show varied clustering patterns across the trees. In the NSP1-encoding gene tree RVA/Human-wt/MOZ/HCN1595/2017/G9P[4] clustered separately with a G9P[4] strains from India, RVA/Human-wt/MOZ/HGM1782/2017/G9P[6] grouped with the four G3P[4] Mozambican strains and RVA/Human-wt/MOZ/HGQ1296/2016/G9P[6] was closely related to the major clade of the G9P[6] and G9P[4] study strains (Figure S1).

In the NSP2, NSP3 and NSP5 trees, only the RVA/Human-wt/MOZ/HCN1595/2017/G9P[4] strain diverged from the group and clustered separately from the major clade. This strain was closely related to G1P[8], G2P[6] and G9P[4] Mozambican study and Asian strains across the trees. The major clade was related to G2P[4] and G2P[6] strains from Mozambique and Kenya (Figure S1).

The E2 NSP4 genotypes were identified in all 15 strains with the exception of RVA/Human-wt/MOZ/HCN1595/2017/G9P[4] strain (Table 1). Five G9P[4] and eight G9P[6] clustered with four G8P[4] and G2P[4] Mozambican strains from 2012 (Figure S1). The rare E6 genotype partial ORF of NSP4 encoding RVA/Human-wt/MOZ/HCN1595/2017/G9P[4] strain clustered with a G9P[4] Indian strain RVA/Human-wt/IND/RV0903/2009/G9P[4] detected in 2009 with nt (aa) identity of 99.6 (98.7)% (Figure S1). This study strain was detected in a vaccinated 8-month-old male child from Nampula province, northern Mozambique (Table S1).

VP4-P[4]

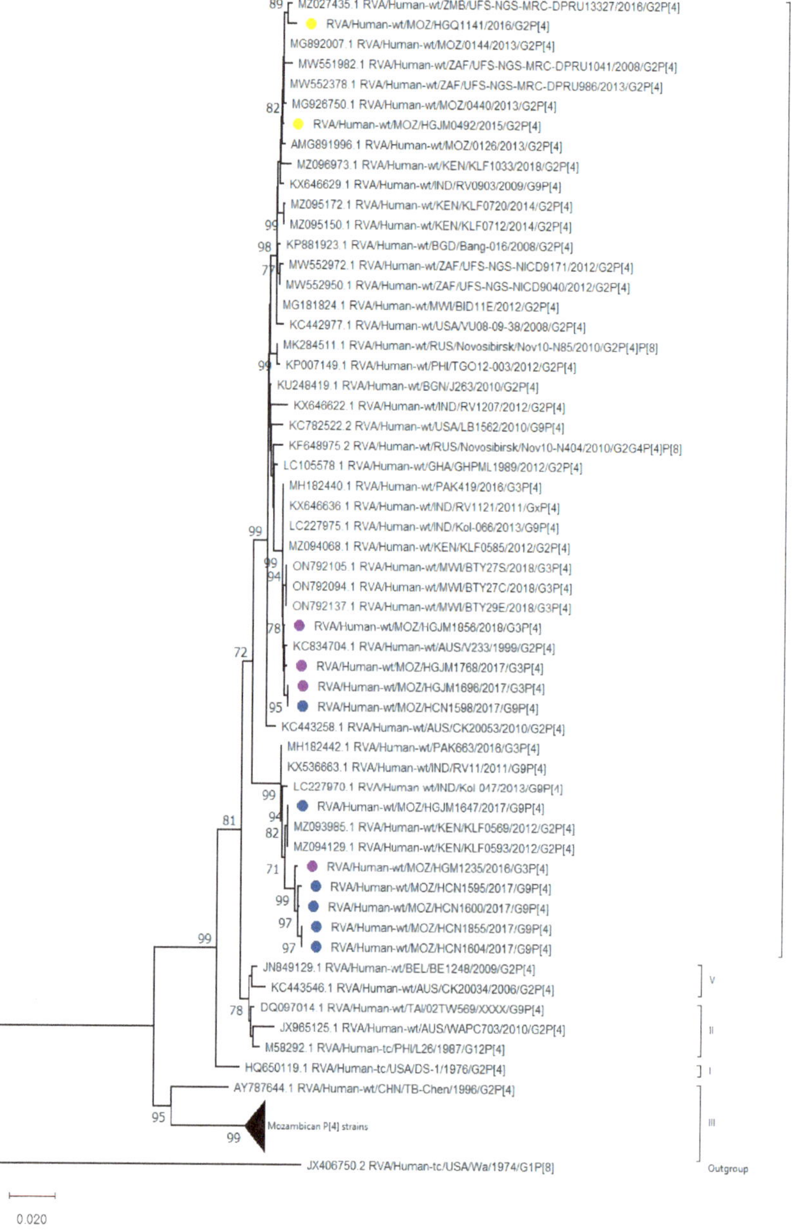

Figure 2. Phylogenetic tree based on the open reading frame (ORF) nucleotide sequence of the VP4-P[4] encoding gene of strains circulating in Mozambique compared to global strains obtained from GenBank. The tree was constructed based on the maximum likelihood method implemented in MEGA X [24], applying Tamura-3-parameter (T92+I) as the model. Bootstrap values (1000 replicates) ≥70% are shown with Wa-like strain serving as an out-group. The scale bar indicates genetic distance expressed as the number of nucleotide substitutions per site. G9P[4] Mozambican strains are indicated by blue circles, G3P[4] by purple circles and G2P[4] by yellow circles.

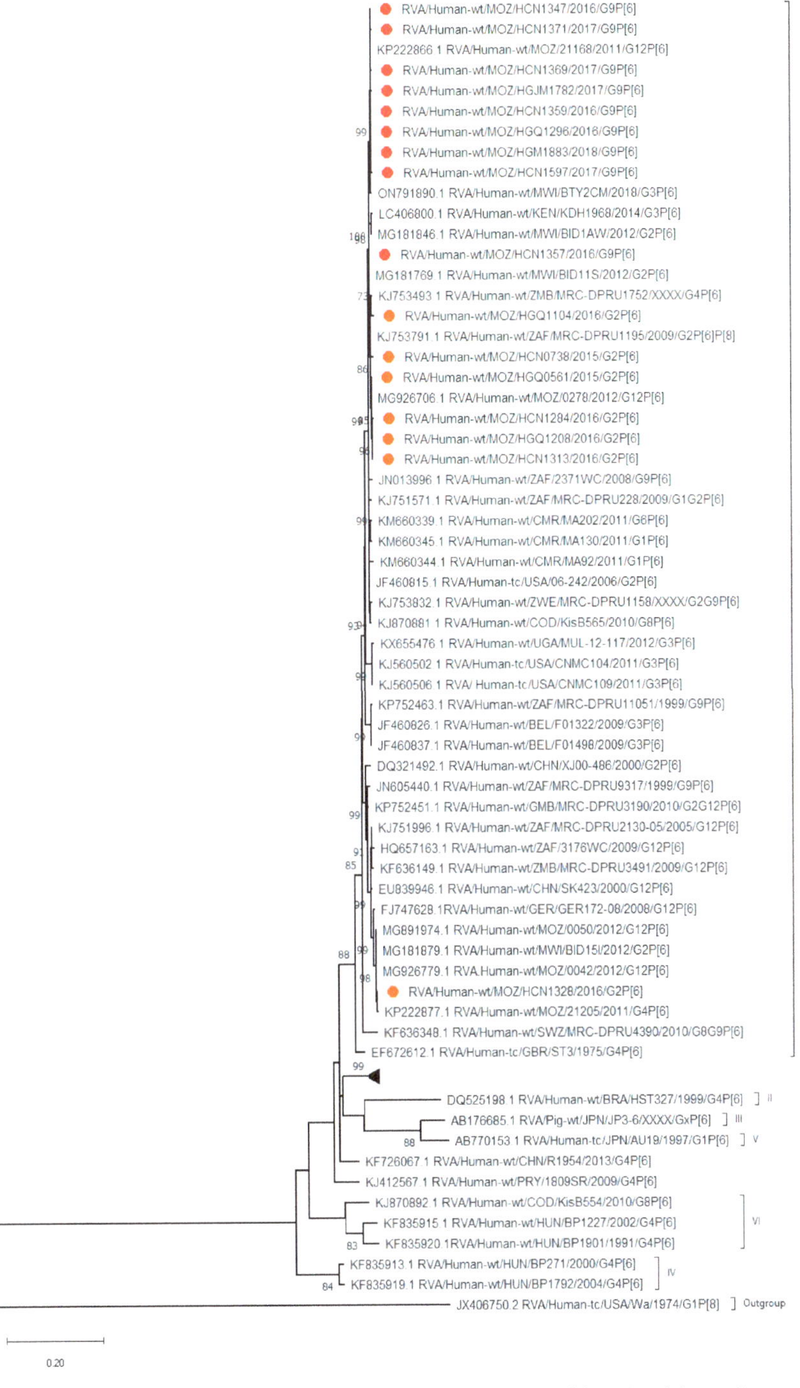

Figure 3. Phylogenetic tree based on the ORF nucleotide sequence of the VP4-P[6] encoding gene of strains circulating in Mozambique compared to global strains obtained from GenBank. The tree was constructed based on the maximum likelihood method implemented in MEGA X [24], applying

T92+G+I as the model. Bootstrap values (1000 replicates) ≥70% are shown with Wa-like strain serving as an out-group. The scale bar indicates genetic distance expressed as the number of nucleotide substitutions per site. G9P[6] Mozambican strains are indicated by red circles and G2P[6] by brown circles.

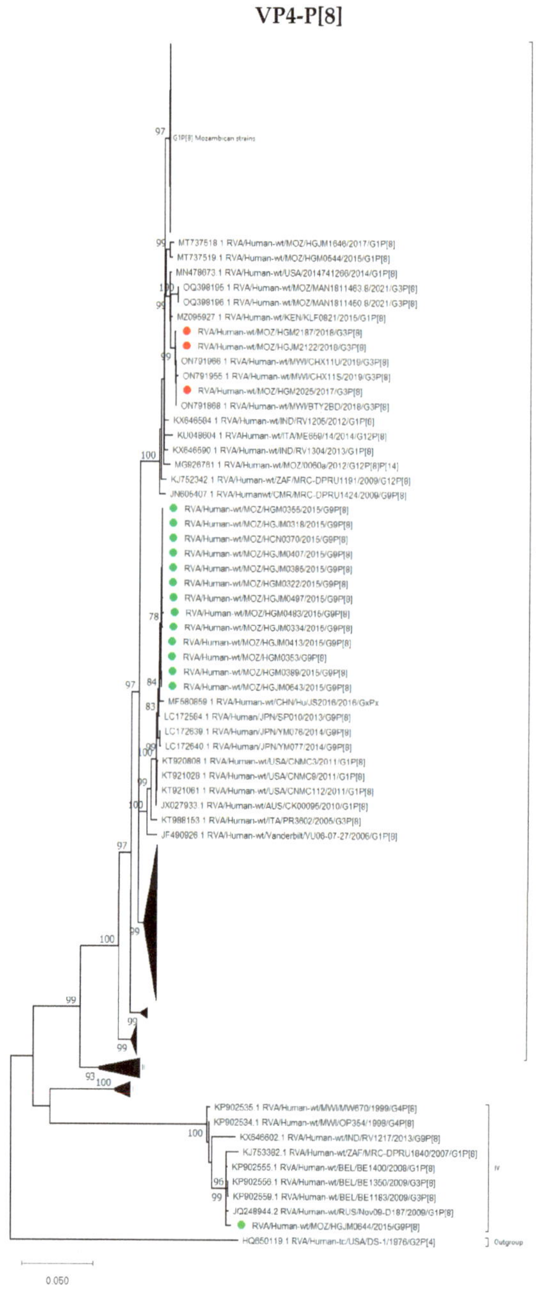

Figure 4. Phylogenetic tree based on the ORF nucleotide sequence of the VP4-encoding genes (P[8]) of strains circulating in Mozambique compared to global strains obtained from GenBank. The tree

was constructed based on the maximum likelihood method implemented in MEGA X [24], applying T92+G as the model. Bootstrap values (1000 replicates) ≥70% are shown with DS-1-like strains serving as an out-group. The scale bar indicates genetic distance expressed as the number of nucleotide substitutions per site. G3P[8] Mozambican strains are indicated by red circles and G9P[8] by green circles.

VP6-I2

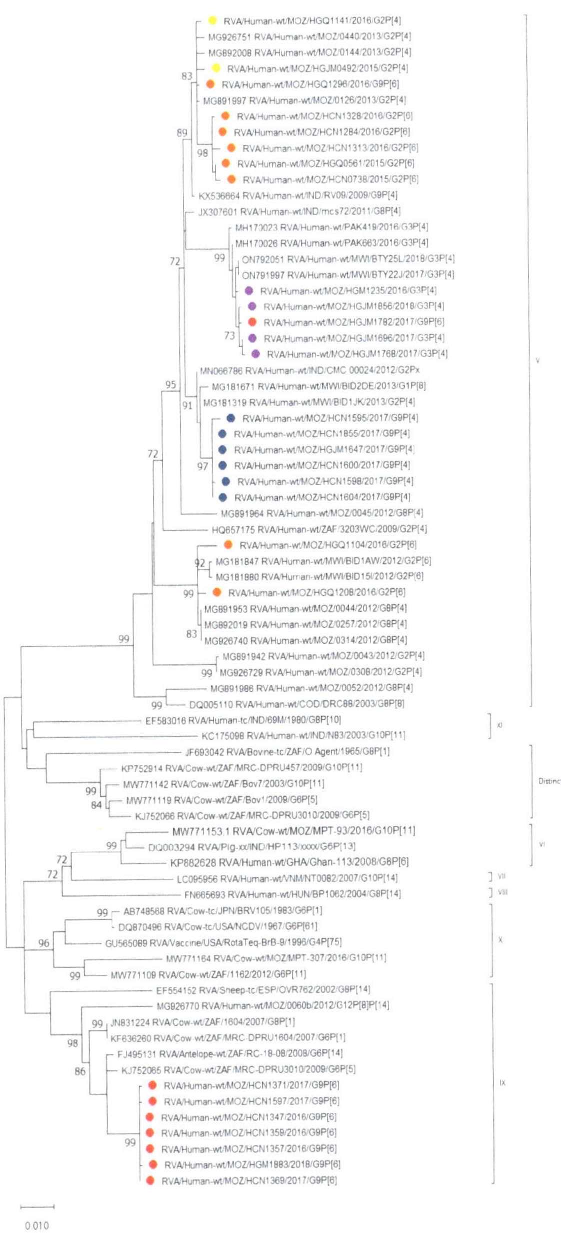

Figure 5. Phylogenetic tree based on the ORF nucleotide sequence of the VP6-encoding I2 genes of strains circulating in Mozambique compared to global strains obtained from GenBank. The best-fit

nucleotide substitution model T92+G+I was used. The tree was constructed based on the maximum likelihood method implemented in MEGA X [24]. Bootstrap values (1000 replicates) ≥70% are shown with Wa-like as out-group. The scale bar indicates genetic distance expressed as the number of nucleotide substitutions per site. G9P[4] Mozambican strains are indicated by blue circles, G9P[6] by red circles, G3P[4] by purple circles, G2P[4] by yellow circles and G2P[6] by brown circles.

3.3. Mvista Analyses

To further study the reassortment events suggested by the phylogenetic analysis, the concatenated DS-1-like genetic backbone (VP6-VP1-VP2-VP3-NSP1-NSP2-NSP3-NSP4 NSP5/6) of the G9P[4] and G9P[6] where aligned by Mvista, using the RVA/Human-wt/MOZ/HCN1347/2016/G9P[6] strain as reference. The results showed that all nine genes of the Mozambican strain exhibited a relatively high degree of conservation, with the exception of the VP6-encoding gene. The VP6 of the G9P[4] strains was conserved in all strains, but showed differences when compared to G9P[6] strains except RVA/Human-wt/MOZ/HGQ1296/2016/G9P[6] and RVA/human-wt/MOZ/HGM1782/2017/G9P[6].

In addition, the VP1–VP3, NSP1, NSP3 and NSP4 encoding genes of the RVA/Human-wt/MOZ/HCN1595/2017/G9P[4] and RVA/Human-wt/MOZ/HGM1782/2017/G9P[6] strains showed a different pattern compared to the other study strains, which likely derived through reassortment events (Figure 6).

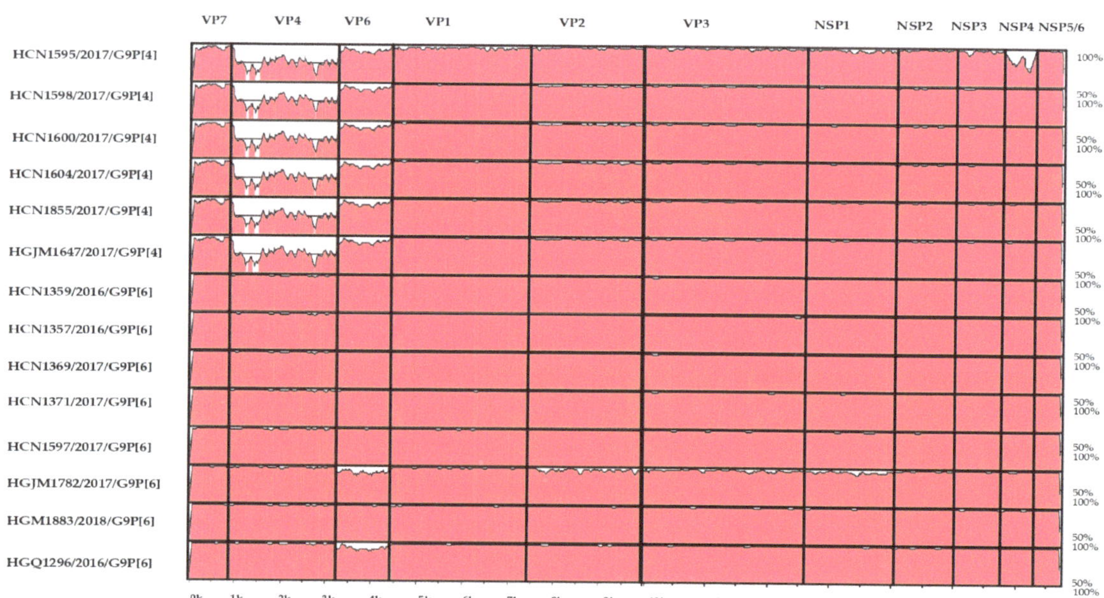

Figure 6. Nucleotide sequence similarities of the G9P[4] and G9P[6] concatenated genomes using the RVA/Human-wt/MOZ/HCN1347/2016/G9P[6] strain as reference. The name of the Mozambican strains is indicated on the left, and the positions of the 11 genes are indicated at the top. The scale indicates the distance in kb.

4. Discussion

In the present study, WGS was performed for 47 strains with specific focus on strains identified as G9P[6], G9P[4] and G9P[8], obtained from Mozambican children with gastroenteritis between 2015 and 2018.

The report of G9P[8] strains as the most predominant genotype in the country before vaccine introduction and the first detection of G9P[4] and G9P[6] genotypes after the vaccine introduction in Mozambique [15], led to the need to monitor changes in strain diversity at gene level. Phylogenetic analysis of the G9 Mozambican strains showed that G9P[8] strains had a Wa-like constellation and the G9P[6] and G9P[4] strains, a DS-1-like constellation with the exception of strain RVA/Human-wt/MOZ/HCN1595/2017/G9P[4] which contained an E6 NSP4 gene. These results highlight the increase of the DS-1 backbone strains after vaccine introduction following the global trend at the time that the strains were detected [10,34–37].

The 13 G9P[8] strains, detected before the introduction of the rotavirus vaccine, clustered together in lineage III. The P[8] lineage III is described as the most common globally [25].

Strain RVA/Human-wt/MOZ/HGJM0644/2015/G9P[8] clustered in a genetically distinct lineage known as lineage IV or OP354-like P[8]. This lineage has been reported in different parts of Europe, Africa and Asia [25,27]. In fact, the Mozambican P[8] lineage IV strain in this study was detected in a 9-month-old female child and could not be detected by RT-PCR, being non-typable for the P genotype. The same observation was reported in Ghana, where 10.4% of non-typeable rotavirus VP4 genes were identified as rare OP354-like P[8] by full-genome sequencing of this rare strain [27,38].

The phylogenetic analysis of the G9 Mozambican strains showed a common pattern between the G9P[4] and G9P[6] strains since most of them clustered together for all gene segments. These results indicate a common ancestral strain with the exception of the RVA/Human-wt/MOZ/HCN1595/2017/G9P[4], RVA/Human-wt/MOZ/HGQ1296/2016/G9P[6], RVA/Human-wt/MOZ/HGM1782/2017/G9P[6] and RVA/Human-wt/MOZ/HCN1598/2017/G9P[4] strains, which clustered distinctly from the larger clade in some segments.

Interestingly, the RVA/Human-wt/MOZ/HGM1782/2017/G9P[6] strain clustered with the G3P[4] strains in segments encoding VP6, VP2, VP3 and NSP1. HGM1782 was detected in the Maputo province, in the same geographic location as the G3P[4] strains, which could explain the similar clustering for the four genome segments. The RVA/Human-wt/MOZ/HCN1598/2017/G9P[4] strain from the northern region of the country, clustered with G3P[4] Mozambican strains from the southern region of Mozambique, for the VP4 encoding gene. These results were confirmed in the Mvista analysis, where these strains were diverse in relation to the others, thus suggesting that they originated through reassortment events. Most of the G9P[4] study strains were detected in vaccinated children, unlike the G9P[6] study strains, which were mostly obtained from unvaccinated children. Regardless, no different clusters were observed between strains detected in vaccinated and unvaccinated children for both genotypes.

The segment encoding for VP6 had the most distinct clustering pattern among the strains. In this segment, seven G9P[6] strains had a higher genetic identity and formed a cluster with animal strains from South Africa and a human Mozambican strain, which had previously been reported as a mixed infection of animal origin [18]. The antelope strain, which was similar to the G9P[6] study strains, was highlighted as having a common origin with the G6P[14] human strains in some segments, excluding the VP6 encoding gene [39]. Three of these Mozambican strains were isolated from children who had contact with animals, including horses, sheep and cats. These results suggest that interspecies transmission occurred.

Unusual G9P[4] RVA strains have been reported in several countries, such as India, Italy, Japan, Benin and Ghana, during pre- and post-vaccine introduction periods [31,32,34–37,40–42]. It has been hypothesized that reassortment events among contemporary human rotavirus strains generated these unusual G9P[4] strains [43]. The E6 NSP4 genotype was first identified in 2000 in India, and the analysis showed that the most common recent ancestor was likely to have been around 1981 in Asia [31]. A recent study in Benin and Ghana described this genotype in Africa for the first time [34,40]. The G9P[4] E6-NSP4 Mozambican strain was identified in the same period as the Benin and Ghanian strains; however, it was related to strains from India. Only one of the characterized strains exhibited the E6 genotype, which indicates a single inter-genotype reassortment event or a sporadic event in Mozambique [37]. Further analyses are necessary to compare diarrhea severity and changes in the NSP4 protein; for example, an evaluation of the

E6 genotype in an animal model could possibly determine if the observed chances are linked to increased pathogenesis of the genotype. In addition, continued genome surveillance is needed to monitor the occurrence of such unusual strains and the possibility of becoming predominant in the country causing severe acute gastroenteritis in children.

Limitations of the study include the short period analyzed (2015–2018) and the inability to calculate the Vesikari score to compare the severity of strains from the post-vaccine period. There is a need to expand the whole-genome analysis to strains detected after 2018 to fully comprehend the genetic diversity of rotavirus strains detected after vaccine introduction.

5. Conclusions

The study results indicate that G9P[4] and G9P[6] strains exhibited a DS-1-like genetic constellation after Rotarix® vaccine introduction in Mozambique. The occurrence of unusual genotypes and close relationship with animal strains, suggesting inter-genotype reassortment and interspecies events, highlight the need for continuous genomic surveillance of RVA strains detected in Mozambique and the importance of following a One Health approach to identify and characterize potential zoonotic strains causing acute gastroenteritis in children.

Supplementary Materials: The following supporting information can be downloaded at: https://www.mdpi.com/article/10.3390/v16071140/s1, Figure S1: Phylogenetic trees based on the ORF nucleotide sequence of the Mozambican strains compared to global strains obtained from GenBank; Table S1: Vaccination status of the Children infected by G2P[6], G2P[6], G3P[4], G3P[8] and G1P[8] Mozambican strains; Table S2: Genotype constellation of Mozambican G2P[6], G2P[6], G3P[4], G3P[8] and G1P[8] strains; Table S3: Nucleotide and amino acid identities of the Mozambicans trains.

Author Contributions: Conceptualization: B.M., E.D.J., H.G.O. and N.d.D.; methodology: B.M., E.D.J., C.M.D. and A.S.; validation: H.G.O., C.M.D. and N.d.D.; formal analysis: B.M. and E.D.J.; investigation: B.M., E.D.J., A.F.L.B. and A.C.; resources: N.d.D. and H.G.O.; data curation: H.G.O. and C.M.D.; writing—original draft preparation: B.M. and E.D.J.; writing—review and editing: B.M., E.D.J., H.G.O., C.M.D., N.d.D., I.L.M., A.S., J.J.C., A.F.L.B. and A.C.; visualization: B.M., E.D.J., C.M.D., N.d.D., A.S., H.G.O. and I.L.M.; supervision: H.G.O., C.M.D., N.d.D. and I.L.M.; project administration: B.M.; and funding acquisition: N.d.D. and H.G.O. All authors have read and agreed to the published version of the manuscript.

Funding: The study was supported by Deutsche Forschungsgemeinschaft (DFG; JO369/5-1 and J0369/5-2) to NdD and HGO (AC, AFLB and AS scholarship). BM Ph.D. is supported by the Calouste Gulbenkian Foundation (270136).

Institutional Review Board Statement: The study was conducted in accordance with the Declaration of Helsinki and approved by the National Ethics Committee of Mozambique (CNBS) under number IRB00002657, reference number 348/CNBS/13.

Informed Consent Statement: Informed consent was obtained from all subjects involved in the study.

Data Availability Statement: The data are available upon request from the corresponding author.

Acknowledgments: The authors would like to thank the caregivers who consented for their children's to be enrolled in the ViNaDia surveillance. We also thank the pediatricians, nurses and all professionals from the sites for their dedication and effort with the children, enrolment and data collection. The Next Generation Sequencing Unit represented by Martin Nyaga. Fundação para a Ciência e Tecnologia for funds to GHTM—UID/04413/2020 and LA-REAL—LA/P/0117/2020.

Conflicts of Interest: The authors declare no conflicts of interest.

References

1. Tate, J.E.; Burton, A.H.; Boschi-Pinto, C.; Parashar, U.D. Global, Regional, and National Estimates of Rotavirus Mortality in Children <5 Years of Age, 2000–2013. *Clin. Infect. Dis.* **2016**, *62*, S96–S105. [CrossRef] [PubMed]
2. Crawford, S.E.; Ramani, S.; Tate, J.E.; Parashar, U.D.; Svensson, L.; Hagbom, M.; Franco, M.A.; Greenberg, H.B.; O'Ryan, M.; Kang, G.; et al. Rotavirus Infection. *Nat. Rev. Dis. Primer* **2017**, *3*, 17083. [CrossRef] [PubMed]

3. Troeger, C.; Blacker, B.F.; Khalil, I.A.; Rao, P.C.; Cao, S.; Zimsen, S.R.; Albertson, S.B.; Stanaway, J.D.; Deshpande, A.; Abebe, Z.; et al. Estimates of the Global, Regional, and National Morbidity, Mortality, and Aetiologies of Diarrhoea in 195 Countries: A Systematic Analysis for the Global Burden of Disease Study 2016. *Lancet Infect. Dis.* **2018**, *18*, 1211–1228. [CrossRef] [PubMed]
4. Matthijnssens, J.; Attoui, H.; Bányai, K.; Brussaard, C.P.D.; Danthi, P.; del Vas, M.; Dermody, T.S.; Duncan, R.; Fāng, Q.; Johne, R.; et al. ICTV Virus Taxonomy Profile: Sedoreoviridae 2022. *J. Gen. Virol.* **2022**, *103*, 001782. [CrossRef] [PubMed]
5. Bányai, K.; Estes, M.K.; Martella, V.; Parashar, U.D. Viral Gastroenteritis. *Lancet* **2018**, *392*, 175–186. [CrossRef] [PubMed]
6. Rotavirus Classification Working Group: RCWG. Available online: https://rega.kuleuven.be/cev/viralmetagenomics/virus-classification/rcwg (accessed on 27 July 2023).
7. Mijatovic-Rustempasic, S.; Jaimes, J.; Perkins, C.; Ward, M.L.; Esona, M.D.; Gautam, R.; Lewis, J.; Sturgeon, M.; Panjwani, J.; Bloom, G.A.; et al. Rotavirus Strain Trends in United States, 2009–2016: Results from the National Rotavirus Strain Surveillance System (NRSSS). *Viruses* **2022**, *14*, 1775. [CrossRef]
8. Burke, R.M.; Tate, J.E.; Kirkwood, C.D.; Steele, A.D.; Parashar, U.D. Current and New Rotavirus Vaccines. *Curr. Opin. Infect. Dis.* **2019**, *32*, 435–444. [CrossRef]
9. Seheri, L.M.; Magagula, N.B.; Peenze, I.; Rakau, K.; Ndadza, A.; Mwenda, J.M.; Weldegebriel, G.; Steele, A.D.; Mphahlele, M.J. Rotavirus Strain Diversity in Eastern and Southern African Countries before and after Vaccine Introduction. *Vaccine* **2018**, *36*, 7222–7230. [CrossRef]
10. Antoni, S.; Nakamura, T.; Cohen, A.L.; Mwenda, J.M.; Weldegebriel, G.; Biey, J.N.M.; Shaba, K.; Rey-Benito, G.; de Oliveira, L.H.; da Costa Oliveira, M.T.; et al. Rotavirus Genotypes in Children under Five Years Hospitalized with Diarrhea in Low and Middle-Income Countries: Results from the WHO-Coordinated Global Rotavirus Surveillance Network. *PLoS Glob. Public Health* **2023**, *3*, e0001358. [CrossRef]
11. Matthijnssens, J.; Ciarlet, M.; McDonald, S.M.; Attoui, H.; Bányai, K.; Brister, J.R.; Buesa, J.; Esona, M.D.; Estes, M.K.; Gentsch, J.R.; et al. Uniformity of Rotavirus Strain Nomenclature Proposed by the Rotavirus Classification Working Group (RCWG). *Arch. Virol.* **2011**, *156*, 1397–1413. [CrossRef]
12. Matthijnssens, J.; Van Ranst, M. Genotype Constellation and Evolution of Group A Rotaviruses Infecting Humans. *Curr. Opin. Virol.* **2012**, *2*, 426–433. [CrossRef]
13. Matthijnssens, J.; Ciarlet, M.; Heiman, E.; Arijs, I.; Delbeke, T.; McDonald, S.M.; Palombo, E.A.; Iturriza-Gómara, M.; Maes, P.; Patton, J.T.; et al. Full Genome-Based Classification of Rotaviruses Reveals a Common Origin between Human Wa-Like and Porcine Rotavirus Strains and Human DS-1-Like and Bovine Rotavirus Strains. *J. Virol.* **2008**, *82*, 3204–3219. [CrossRef]
14. de Deus, N.; Chiláule, J.J.; Cassocera, M.; Bambo, M.; Langa, J.S.; Sitoe, E.; Chissaque, A.; Anapakala, E.; Sambo, J.; Guimarães, E.L.; et al. Early Impact of Rotavirus Vaccination in Children Less than Five Years of Age in Mozambique. *Vaccine* **2018**, *36*, 7205–7209. [CrossRef]
15. João, E.D.; Munlela, B.; Chissaque, A.; Chiláule, J.; Langa, J.; Augusto, O.; Boene, S.S.; Anapakala, E.; Sambo, J.; Guimarães, E.; et al. Molecular Epidemiology of Rotavirus A Strains Pre- and Post-Vaccine (Rotarix®) Introduction in Mozambique, 2012–2019: Emergence of Genotypes G3P[4] and G3P[8]. *Pathogens* **2020**, *9*, 671. [CrossRef]
16. Chissaque, A.; Burke, R.M.; Guimarães, E.L.; Manjate, F.; Nhacolo, A.; Chiláule, J.; Munlela, B.; Chirinda, P.; Langa, J.S.; Cossa-Moiane, I.; et al. Effectiveness of Monovalent Rotavirus Vaccine in Mozambique, a Country with a High Burden of Chronic Malnutrition. *Vaccines* **2022**, *10*, 449. [CrossRef] [PubMed]
17. Manjate, F.; João, E.D.; Chirinda, P.; Garrine, M.; Vubil, D.; Nobela, N.; Kotloff, K.; Nataro, J.P.; Nhampossa, T.; Acácio, S.; et al. Molecular Epidemiology of Rotavirus Strains in Symptomatic and Asymptomatic Children in Manhiça District, Southern Mozambique 2008–2019. *Viruses* **2022**, *14*, 134. [CrossRef]
18. Strydom, A.; Motanyane, L.; Nyaga, M.M.; João, E.D.; Cuamba, A.; Mandomando, I.; Cassocera, M.; de Deus, N.; O'Neill, H. Whole-Genome Characterization of G12 Rotavirus Strains Detected in Mozambique Reveals a Co-Infection with a GXP[14] Strain of Possible Animal Origin. *J. Gen. Virol.* **2019**, *100*, 932–937. [CrossRef]
19. Strydom, A.; João, E.D.; Motanyane, L.; Nyaga, M.M.; Christiaan Potgieter, A.; Cuamba, A.; Mandomando, I.; Cassocera, M.; de Deus, N.; O'Neill, H.G. Whole Genome Analyses of DS-1-like Rotavirus A Strains Detected in Children with Acute Diarrhoea in Southern Mozambique Suggest Several Reassortment Events. *Infect. Genet. Evol.* **2019**, *69*, 68–75. [CrossRef] [PubMed]
20. Munlela, B.; João, E.D.; Donato, C.M.; Strydom, A.; Boene, S.S.; Chissaque, A.; Bauhofer, A.F.L.; Langa, J.; Cassocera, M.; Cossa-Moiane, I.; et al. Whole Genome Characterization and Evolutionary Analysis of G1P[8] Rotavirus A Strains during the Pre- and Post-Vaccine Periods in Mozambique (2012–2017). *Pathogens* **2020**, *9*, 1026. [CrossRef] [PubMed]
21. Gentsch, J.R.; Glass, R.I.; Woods, P.; Gouvea, V.; Gorziglia, M.; Flores, J.; Das, B.K.; Bhan, M.K. Identification of Group A Rotavirus Gene 4 Types by Polymerase Chain Reaction. *J. Clin. Microbiol.* **1992**, *30*, 1365–1373. [CrossRef]
22. Gouvea, V.; Glass, R.I.; Woods, P.; Taniguchi, K.; Clark, H.F.; Forrester, B.; Fang, Z.Y. Polymerase Chain Reaction Amplification and Typing of Rotavirus Nucleic Acid from Stool Specimens. *J. Clin. Microbiol.* **1990**, *28*, 276–282. [CrossRef] [PubMed]
23. Iturriza-Gómara, M.; Kang, G.; Gray, J. Rotavirus Genotyping: Keeping up with an Evolving Population of Human Rotaviruses. *J. Clin. Virol.* **2004**, *31*, 259–265. [CrossRef] [PubMed]
24. Kumar, S.; Stecher, G.; Li, M.; Knyaz, C.; Tamura, K. MEGA X: Molecular Evolutionary Genetics Analysis across Computing Platforms. *Mol. Biol. Evol.* **2018**, *35*, 1547–1549. [CrossRef] [PubMed]

25. Zeller, M.; Heylen, E.; Damanka, S.; Pietsch, C.; Donato, C.; Tamura, T.; Kulkarni, R.; Arora, R.; Cunliffe, N.; Maunula, L.; et al. Emerging OP354-Like P[8] Rotaviruses Have Rapidly Dispersed from Asia to Other Continents. *Mol. Biol. Evol.* **2015**, *32*, 2060–2071. [CrossRef] [PubMed]

26. Zeller, M.; Heylen, E.; Tamim, S.; McAllen, J.K.; Kirkness, E.F.; Akopov, A.; De Coster, S.; Van Ranst, M.; Matthijnssens, J. Comparative Analysis of the Rotarix™ Vaccine Strain and G1P[8] Rotaviruses Detected before and after Vaccine Introduction in Belgium. *PeerJ* **2017**, *5*, e2733. [CrossRef] [PubMed]

27. Damanka, S.A.; Kwofie, S.; Dennis, F.E.; Lartey, B.L.; Agbemabiese, C.A.; Doan, Y.H.; Adiku, T.K.; Katayama, K.; Enweronu-Laryea, C.C.; Armah, G.E. Whole Genome Characterization and Evolutionary Analysis of OP354-like P[8] Rotavirus A Strains Isolated from Ghanaian Children with Diarrhoea. *PLoS ONE* **2019**, *14*, e0218348. [CrossRef] [PubMed]

28. Jere, K.C.; Mlera, L.; O'Neill, H.G.; Potgieter, A.C.; Page, N.A.; Seheri, M.L.; Van Dijk, A.A. Whole Genome Analyses of African G2, G8, G9, and G12 Rotavirus Strains Using Sequence-independent Amplification and 454® Pyrosequencing. *J. Med. Virol.* **2011**, *83*, 2018–2042. [CrossRef] [PubMed]

29. Mwangi, P.N.; Potgieter, R.-L.; Simwaka, J.; Mpabalwani, E.M.; Mwenda, J.M.; Mogotsi, M.T.; Magagula, N.; Esona, M.D.; Steele, A.D.; Seheri, M.L.; et al. Genomic Analysis of G2P[4] Group A Rotaviruses in Zambia Reveals Positive Selection in Amino Acid Site 7 of Viral Protein 3. *Viruses* **2023**, *15*, 501. [CrossRef]

30. Agbemabiese, C.A.; Nakagomi, T.; Damanka, S.A.; Dennis, F.E.; Lartey, B.L.; Armah, G.E.; Nakagomi, O. Sub-Genotype Phylogeny of the Non-G, Non-P Genes of Genotype 2 Rotavirus A Strains. *PLoS ONE* **2019**, *14*, e0217422. [CrossRef]

31. Pradhan, G.N.; Walimbe, A.M.; Chitambar, S.D. Molecular Characterization of Emerging G9P[4] Rotavirus Strains Possessing a Rare E6 NSP4 or T1 NSP3 Genotype on a Genogroup-2 Backbone Using a Refined Classification Framework. *J. Gen. Virol.* **2016**, *97*, 3139–3153. [CrossRef]

32. Tatte, V.S.; Maran, D.; Walimbe, A.M.; Gopalkrishna, V. Rotavirus G9P[4], G9P[6] and G1P[6] Strains Isolated from Children with Acute Gastroenteritis in Pune, Western India, 2013–2015: Evidence for Recombination in Genes Encoding VP3, VP4 and NSP1. *J. Gen. Virol.* **2019**, *100*, 1605–1630. [CrossRef] [PubMed]

33. Frazer, K.; Pachter, L.; Poliakov, A.; Rubin, E.; Dubchak, I. VISTA: Computational Tools for Comparative Genomics. Available online: https://genome.lbl.gov/vista/mvista/mvistacite.shtml (accessed on 20 October 2023).

34. Doan, Y.H.; Dennis, F.E.; Takemae, N.; Haga, K.; Shimizu, H.; Appiah, M.G.; Lartey, B.L.; Damanka, S.A.; Hayashi, T.; Suzuki, T.; et al. Emergence of Intergenogroup Reassortant G9P[4] Strains Following Rotavirus Vaccine Introduction in Ghana. *Viruses* **2023**, *15*, 2453. [CrossRef] [PubMed]

35. Yamamoto, S.P.; Kaida, A.; Ono, A.; Kubo, H.; Iritani, N. Detection and Characterization of a Human G9P[4] Rotavirus Strain in Japan. *J. Med. Virol.* **2015**, *87*, 1311–1318. [CrossRef] [PubMed]

36. Lewis, J.; Roy, S.; Esona, M.D.; Mijatovic-Rustempasic, S.; Hardy, C.; Wang, Y.; Cortese, M.; Bowen, M.D. Full Genome Sequence of a Reassortant Human G9P[4] Rotavirus Strain. *Genome Announc.* **2014**, *2*, e01284-14. [CrossRef] [PubMed]

37. Ianiro, G.; Recanatini, C.; D'Errico, M.M.; Monini, M. Uncommon G9P[4] Group A Rotavirus Strains Causing Dehydrating Diarrhea in Young Children in Italy. *Infect. Genet. Evol.* **2018**, *64*, 57–64. [CrossRef] [PubMed]

38. Damanka, S.; Dennis, F.E.; Agbemabiese, C.; Lartey, B.; Adiku, T.; Nyarko, K.; Enweronu-Laryea, C.C.; Sagoe, K.W.; Ofori, M.; Rodrigues, O.; et al. Identification of OP354-like Human Rotavirus Strains with Subtype P[8]b in Ghanaian Children with Diarrhoea. *Virol. J.* **2016**, *13*, 69. [CrossRef]

39. Matthijnssens, J.; Potgieter, C.A.; Ciarlet, M.; Parreño, V.; Martella, V.; Bányai, K.; Garaicoechea, L.; Palombo, E.A.; Novo, L.; Zeller, M.; et al. Are Human P[14] Rotavirus Strains the Result of Interspecies Transmissions from Sheep or Other Ungulates That Belong to the Mammalian Order Artiodactyla? *J. Virol.* **2009**, *83*, 2917–2929. [CrossRef] [PubMed]

40. Agbla, J.M.; Esona, M.D.; Jaimes, J.; Gautam, R.; Agbankpé, A.J.; Katz, E.; Dougnon, T.V.; Capo-Chichi, A.; Ouedraogo, N.; Razack, O.; et al. Whole Genome Analysis of Rotavirus Strains Circulating in Benin before Vaccine Introduction, 2016–2018. *Virus Res.* **2022**, *313*, 198715. [CrossRef] [PubMed]

41. Zhang, T.; Li, J.; Jiang, Y.-Z.; Xu, J.-Q.; Guan, X.-H.; Wang, L.-Q.; Chen, J.; Liang, Y. Genotype Distribution and Evolutionary Analysis of Rotavirus Associated with Acute Diarrhea Outpatients in Hubei, China, 2013–2016. *Virol. Sin.* **2022**, *37*, 503–512. [CrossRef]

42. Khakha, S.A.; Varghese, T.; Giri, S.; Durbin, A.; Tan, G.S.; Kalaivanan, M.; Prasad, J.H.; Kang, G. Whole-Genome Characterization of Common Rotavirus Strains Circulating in Vellore, India from 2002 to 2017: Emergence of Non-Classical Genomic Constellations. *Gut Pathog.* **2023**, *15*, 44. [CrossRef]

43. Doan, Y.H.; Suzuki, Y.; Fujii, Y.; Haga, K.; Fujimoto, A.; Takai-Todaka, R.; Someya, Y.; Nayak, M.K.; Mukherjee, A.; Imamura, D.; et al. Complex Reassortment Events of Unusual G9P[4] Rotavirus Strains in India between 2011 and 2013. *Infect. Genet. Evol.* **2017**, *54*, 417–428. [CrossRef] [PubMed]

Article

Human Rotaviruses of Multiple Genotypes Acquire Conserved VP4 Mutations during Serial Passage

Maximilian H. Carter [1], Jennifer Gribble [2], Julia R. Diller [1], Mark R. Denison [1,2], Sara A. Mirza [3], James D. Chappell [1], Natasha B. Halasa [1] and Kristen M. Ogden [1,2,*]

[1] Department of Pediatrics, Vanderbilt University Medical Center, Nashville, TN 37232, USA
[2] Department of Pathology, Microbiology, and Immunology, Vanderbilt University Medical Center, Nashville, TN 37232, USA
[3] Centers for Disease Control and Prevention, Atlanta, GA 30329, USA
* Correspondence: kristen.ogden@vumc.org

Abstract: Human rotaviruses exhibit limited tropism and replicate poorly in most cell lines. Attachment protein VP4 is a key rotavirus tropism determinant. Previous studies in which human rotaviruses were adapted to cultured cells identified mutations in VP4. However, most such studies were conducted using only a single human rotavirus genotype. In the current study, we serially passaged 50 human rotavirus clinical specimens representing five of the genotypes most frequently associated with severe human disease, each in triplicate, three to five times in primary monkey kidney cells then ten times in the MA104 monkey kidney cell line. From 13 of the 50 specimens, we obtained 25 rotavirus antigen-positive lineages representing all five genotypes, which tended to replicate more efficiently in MA104 cells at late versus early passage. We used Illumina next-generation sequencing and analysis to identify variants that arose during passage. In VP4, variants encoded 28 mutations that were conserved for all P[8] rotaviruses and 12 mutations that were conserved for all five genotypes. These findings suggest there may be a conserved mechanism of human rotavirus adaptation to MA104 cells. In the future, such a conserved adaptation mechanism could be exploited to study human rotavirus biology or efficiently manufacture vaccines.

Keywords: rotavirus; VP4; genotype; serial passage; culture adaptation

Citation: Carter, M.H.; Gribble, J.; Diller, J.R.; Denison, M.R.; Mirza, S.A.; Chappell, J.D.; Halasa, N.B.; Ogden, K.M. Human Rotaviruses of Multiple Genotypes Acquire Conserved VP4 Mutations during Serial Passage. *Viruses* **2024**, *16*, 978. https://doi.org/10.3390/v16060978

Academic Editors: Ulrich Desselberger and John T. Patton

Received: 14 May 2024
Revised: 6 June 2024
Accepted: 14 June 2024
Published: 18 June 2024

1. Introduction

Rotavirus is the leading cause of diarrheal mortality for children under 5 years of age worldwide, leading to estimates of 130,000 to more than 200,000 infant and child deaths each year [1,2]. Rotaviruses cause diarrheal disease in many animal species, but exhibit narrow host and cell tropism [3,4], which has limited studies in fundamental biology and vaccine design and manufacture for human rotaviruses. A key evolutionary mechanism for RNA viruses, including rotavirus, is misincorporation of nucleotides by the viral RNA polymerase (genetic drift). Like influenza virus, rotavirus has a segmented genome, and can also reassort segments during co-infection (genetic shift) [5]. Evidence suggests there have been frequent rotavirus interspecies transmission events, sometimes with subsequent adaptation [6–9]. Together, genetic drift and shift promote evolution and can potentiate the emergence of antigenically novel rotaviruses and disease outbreaks in naïve populations [10,11]. Understanding adaptive genetic changes for human rotavirus may enhance our capacity to work with these viruses in laboratory settings or to manufacture live-attenuated vaccines at scale.

Rotavirus outer-capsid proteins determine the viral genotype. Rotaviruses are non-enveloped, triple-layered virions with a genome composed of 11 segments of double-stranded RNA [3]. The outer capsid consists of 260 VP7 glycoprotein trimers, with 60 VP4 trimers that project from the surface [12]. VP7 and VP4 determine the viral G and P type, respectively, and are the primary targets of neutralizing antibodies, which may be generated in response to infection or vaccination [13–15]. Although at least 42 G types and 58 P types have been defined to date, only a subset of G/P type combinations infect and cause disease in humans with varying degrees of severity ([16] and https://rega.kuleuven.be/cev/viralmetagenomics/virus-classification/rcwg accessed on 6 June 2024). Predominant human rotavirus genotypes vary by geographic region, but historically G1P[8], G2P[4], G3P[8], G4P[8], G9P[8], and G12P[8] cause the majority of human disease [17]. In recent years, several less common rotavirus genotypes, including G1P[4], G2P[8], G9P[4], G12P[4], G8P[6], G8P[8], and G12P[6], have increasing epidemiological relevance in some parts of Africa, Asia, and South America [7].

While decades of studies have yielded a wealth of knowledge about rotavirus attachment and entry [18], much remains unknown about receptor-dependent cell tropism, especially for human rotaviruses. Trimeric attachment protein VP4 primarily dictates receptor-dependent rotavirus tropism, but major outer-capsid glycoprotein VP7 can interact with integrin coreceptors to mediate internalization [19–24]. VP4 cleavage by intestinal trypsin-like proteases separates the receptor-binding 'head' domain (VP8*) from the stalk domain (VP5*) [23,25]. Glycans serve as attachment receptors and bind VP8* [26,27]. Many animal rotaviruses bind glycans with terminal sialic acid, but most human rotaviruses bind internal sialic acid or histo-blood group antigens, sometimes in a genotype-specific manner [18,26,28–30]. In the monkey kidney epithelial (MA104) cells historically used for rotavirus studies, post-attachment receptors, which interact with VP5* and VP7, contribute to cell specificity and include integrins, heat shock cognate protein hsc70, and in some cases JAM-A and occludin [18,23,24,31,32]. Differences in the capacity of monoclonal antibodies to neutralize human rotaviruses in MA104 cells and human intestinal epithelial cells suggest interactions required for entry differ among these cell types [33]. The VP5* stalk is composed of 'body' and 'foot' domains and helps mediate membrane penetration via conformational rearrangements akin to those of enveloped virus fusion proteins [23,25,34,35]. In the current model, after binding glycan receptors, VP8* head domains separate from VP5*, exposing hydrophobic loops [35]. VP5* rearranges on the surface of the infectious virus particle from an 'upright' to a 'reversed' conformation, which promotes interaction of VP5* hydrophobic loops with target membranes and outward projection of the VP5* foot that was formerly buried in the intermediate capsid layer. VP5* in the reversed information remains tethered to the rotavirus particle, inserts into membranes, and enables Ca^{2+} to cross and promote virus uncoating [35,36]. Thus, interactions of VP4 and VP7 with host molecules and factors that influence VP4 conformational rearrangements may contribute to the types of cells that rotavirus can infect and the efficiency of infection, but factors that limit or enhance human rotavirus infection in specific cell types are incompletely understood.

Human rotaviruses replicate poorly in continuous cell lines unless adapted during serial passage [37]. Consistent with the important role of VP4 in receptor-dependent cell tropism, monoreassortant rotaviruses containing human VP4, engineered using reverse genetics, replicate poorly in most cultured cells [38–40]. Reassortant genetics studies implicate VP4 and VP7 in pathogenesis outcomes in animals, with VP4 specifically linked to tropism [41–46]. While adaptive mutations have somewhat rarely been reported for human rotaviruses, 33 passages in African green monkey kidney (AGMK) cells yielded five non-synonymous changes in the VP4 segment of strain 89-12 [47]. These adaptive changes permitted development of the live, attenuated ROTARIX vaccine. For human rotavirus strains Wa, DC3695, and DC5685, many changes following serial passage in human colonic epithelial HT29 cells or AGMK cells arose in VP7, VP4, and NSP4 segments [48]. Passage of human rotavirus strains Wa and M in MA104 cells yielded attenuation of disease in piglets and more polymorphisms in VP4 than any other segment, including six and eight respective amino acid changes [49]. While 11 or 12 serial passages of human rotavirus vaccine candidate CDC-9 in MA104 cells resulted in no detected nucleotide or amino acid sequence changes, 28 or 44 passages in another monkey kidney epithelial cell line (Vero) resulted in five or six amino acid changes, respectively, in the VP4 gene [50]. These VP4 mutations correlated with both adaptation and attenuation, indicated by increased viral replication in cultured cells, upregulated expression of immunomodulatory cytokines, and reduced virus shedding and diarrhea in neonatal rats. Interestingly, cryo-electron microscopy revealed that at early passages, most VP4 molecules occupied the 'reversed' conformation on CDC-9 virus particles, which is unlikely to be capable of mediating cell entry, whereas at later passages, about half of the VP4 spikes occupied the 'upright' conformation associated with infectious virions [51]. It is hypothesized that an adaptive mutation in VP4 may stabilize the 'upright' conformation. Together, these studies further implicate VP4 in tropism and have advanced rotavirus vaccine candidates. However, all the human rotaviruses for which adaptive mutations have been reported represent a single genotype (G1P[8]), with the exception of the M strain (G3P[8]).

In the current study, we serially passaged supernatants of rotavirus-positive stool samples from pediatric patients treated at Vanderbilt University Medical Center (VUMC) between 2005 and 2013 in cultured cells. The specimens contained rotaviruses with completely sequenced genomes and represented five of the six genotypes most commonly associated with human disease, G1P[8], G2P[4], G3P[8], G9P[8], and G12P[8] [17]. High-passage rotaviruses tended to replicate more efficiently than low-passage rotaviruses, suggesting that passaged virus populations had adapted to the cells. Using next-generation sequencing and variant analysis, we identified sets of VP4 amino acid mutations that were conserved across passaged specimens and genotypes, suggesting a potentially conserved mechanism of cell culture adaptation. Conserved mutations were located primarily near the VP5* hydrophobic loops, which interact with membranes during entry, including in a residue previously identified in several other studies of rotavirus adaptation and attenuation, and near the 'waist,' which is adjacent to the VP7 layer in the reverse conformation. In some cases, these mutations might influence the stability of the upright conformation of VP5* on the particle. These findings help generate new hypotheses about conserved mechanisms by which rotavirus can overcome tropism barriers and replicate efficiently in cultured cells.

2. Materials and Methods

2.1. Rotavirus-Positive Clinical Specimens

Fecal specimens were collected from eligible children presenting with acute gastroenteritis at the Monroe Carell Junior Vanderbilt Children's Hospital and Clinics during the years 2005 to 2013, and complete rotavirus genomes from rotavirus-positive specimens were previously sequenced as described [52–54]. In all cases, collection was performed in accordance with New Vaccine Surveillance Network (NVSN) protocols approved by the Center for Disease Control and Prevention (CDC), VUMC, and the Institutional Review Board. Informed consent, including future specimen use, was provided by a parent or guardian at the time of enrollment. For the current study, we selected 50 specimens representing five genotypes (G1P[8], G2P[4], G3P[8], G9P[8], G12P[8]). The abbreviated strain name, genotype, and year of collection for each specimen used in the current study are listed in Table S1.

2.2. Cells

Primary rhesus (*Macaca mulatta*) monkey kidney (RhMK) cells (Diagnostic Hybrids, Inc., Athens, OH, USA, Cat # 49-0600A) were initially grown in 16 mm glass roller tubes in Eagle's minimum essential medium (EMEM) supplemented with HEPES, fetal bovine serum (FBS), SV5/SV40 antisera, and gentamicin at concentrations proprietary to the manufacturer, as shipped. MA104 and Vero cells were purchased from the American Type Culture Collection (ATCC). We use the lot-specific validation criteria, including cytochrome C oxidase I gene analysis, provided by ATTC, together with visual inspection of cell morphology, culture conditions, and virus susceptibility to validate cell identity. MA104 cells and Vero cells were cultured in EMEM with Earle's salts and L-glutamine (Corning, Corning, NY, USA) supplemented to contain 5% FBS (Gibco, Grand Island, NY, USA). Cells were cultured in serum-free media during serial passages as described in Section 2.4.

All cells were maintained at 37 °C in 5% CO_2. Primary RhMK cells in roller tubes were incubated with slow rotation. Cells were tested for mycoplasma regularly by PCR.

2.3. Enzyme-Linked Immunosorbent Assay (ELISA)

The Rotaclone (Meridian Bioscience, Inc., Cincinnati, OH, USA) ELISA was used to detect rotavirus in lysates from RhMK, MA104, and Vero cells according to manufacturer instructions. Samples with absorbance units (A_{450}) of 0.1 or greater were considered positive.

2.4. Serial Passaging of Clinical Rotavirus Specimens

For an initial passage in primary RhMK cells (P1), 0.1 mL of a 10% (*w/v*) homogenate of stool suspension in Earle's balanced salt solution (Sigma, Burlington, MA, USA) in triplicate for a rotavirus-positive clinical specimen was clarified by centrifugation at $10,000 \times g$ for 15 min. Clarified supernatants, medium alone, and a laboratory stock of SA11-4F rotavirus (0.1 mL at 5×10^6 PFU/mL), were activated with 10 µg/mL trypsin (Worthington Biochemical Corporation, Lakewood, NJ, USA; LS003708) for 1 h at 37 °C then diluted in serum-free EMEM to a final trypsin concentration of <2 µg/mL. Primary RhMK cells were washed three times with serum-free EMEM and adsorbed in roller tubes with each activated specimen for 1 h at 37 °C with slow rotation. Following absorption, inocula were removed, monolayers were washed, and fresh serum-free EMEM containing 0.5 µg/mL of trypsin was added. Cells were incubated with constant rotation at 37 °C for up to 7 days or until cytopathic effect (CPE) was visible, and the cell monolayer was disrupted due to lysis. Cells then were subjected to three rounds of freezing at −80 °C and thawing prior to storage at 4 °C. In two to four subsequent passages, 0.2–1 mL of lysate from each previous lineage and passage was activated with 1 µg/mL trypsin and used as the inoculum for adsorption. The presence of rotavirus in lysates was determined by ELISA following passages one, three, and/or five. For passages subsequent to ELISA, a 1 mL inoculum was used for lysates with values <0.2; 1 mL of a 1:2.5 diluted inoculum was used for lysates with values between 0.2 and 1; and 1 mL of a 1:5 diluted inoculum was used for lysates with values >1, if additional passages were conducted.

After three to five passages in primary RhMK cells, 0.5 mL of rotavirus-positive lysates were activated with 1 µg/mL trypsin for 1 h at 37 °C. Confluent MA104 or Vero cell monolayers in T25 flasks were washed with serum-free EMEM and adsorbed with 0.5 mL (MA104) or 0.3 to 1 mL (Vero) of P3 or P5 RhMK cell lysates for 1 h at 37 °C with occasional rocking. Following absorption, inocula were removed, monolayers were washed, and fresh serum-free EMEM containing 0.5 µg/mL of trypsin was added. Cells were incubated at 37 °C for up to 7 days or until CPE was visible. Cells then were subjected to three rounds of freezing at −80 °C and thawing prior to storage at 4 °C. In up to nine subsequent passages, lysate from each previous passage was activated with 1 µg/mL trypsin and used as the inoculum for adsorption. The presence of rotavirus in lysates was determined by ELISA, typically following P3, P6, and P10. For passages subsequent to ELISA, a 1 mL inoculum was used for lysates with values <1; 1 mL of a 1:10 diluted inoculum was used for lysates with values between 1 and 2; and 1 mL of a 1:100 diluted inoculum was used for lysates with values >2. For rotavirus passages attempted in Vero cells subsequent to passage in MA104 cells, the initial inoculum was 1 mL of rotavirus-positive lysate from MA104 passage 6 or 10 that had been activated with 1 µg/mL trypsin for 1 h at 37 °C.

2.5. Replication Time Course

MA104 cells (~1.7×10^5/well) were seeded in 24-well plates and incubated at 37 °C until confluent. Rotavirus-positive P3 or P10 lysates were diluted to 2.5×10^4 FFU/mL in serum-free EMEM, to achieve a multiplicity of infection (MOI) of 0.01 fluorescent focus units (FFU)/cell with a 0.1 mL inoculum, assuming ~2.5×10^5 MA104 cells per well. Lysates were used neat if titer was less than 2.5×10^4 FFU/mL. Diluted rotavirus-positive lysates were activated with 1 µg/mL trypsin for 1 h at 37 °C. Cells were washed twice and adsorbed with activated viruses for 1 h at 37 °C. Cells were washed to remove unbound virus and incubated with serum-free medium plus 0.5 µg/mL trypsin at 37 °C for 0, 24, or 48 h. Plates were frozen at −80 °C and thawed three times prior to determining virus titer by fluorescent focus assay on MA104 cells. Virus yield was determined by dividing titer at 24 h or 48 h by titer at 0 h.

2.6. Fluorescent Focus Assay (FFA)

MA104 cells (~1 × 10⁵/well) were seeded in black-walled, clear-bottom, 96-well plates and incubated at 37 °C overnight. Virus was activated with 1 µg/mL trypsin for 1 h at 37 °C and serially diluted 1:10 in serum-free EMEM. Following two washes, cells were adsorbed with virus dilutions for 1 h at 37 °C. Cells were washed and incubated at 37 °C for 16–18 h prior to methanol fixation. Cells were stained to detect nuclei using DAPI (Invitrogen, Carlsbad, CA, USA) and rotavirus proteins using sheep α-rotavirus polyclonal serum (Invitrogen) prior to imaging and quantification using an ImageXpress Micro XL Widefield High-Content Analysis System (Applied Biosystems, Waltham, MA, USA). Virus titer was quantified from total and infected cells quantified in four fields of view/well. To determine whether P3 and P10 yield differed at 24 h or 48 h, we used two-way ANOVA followed by Šídák's multiple comparisons test. Statistical analyses were conducted using GraphPad Prism 9.

2.7. RNA Extraction, RT-PCR, and Nucleotide Sequencing

Rotavirus-positive culture lysates were processed for RNA extraction, library preparation, and RNA sequencing. Two × 0.25 mL aliquots of each of MA104 P10 lysate were treated with 1 µL of DNase I (New England Biolabs, Ipswich, MA, USA) for 30 min at 37 °C then with EDTA to a final concentration of 5 mM to inactivate DNase I prior to RNA extraction using TRIzol LS (Invitrogen) according to manufacturer instructions. RNA pellets were resuspended in RNase-free water and incubated in a heat block set at 55 °C for 5–10 min, with small aliquots set aside for RNA quantitation by Qubit. RNA library preparation for Illumina sequencing was conducted using 5 µL of input RNA and the NEBNext Ultra II RNA Library Prep Kit for Illumina (New England Biolabs), according to the manufacturer's instructions. Briefly, RNA was fragmented prior to first-strand and second-strand synthesis, AMPure XP Bead clean up, and end repair. PCR enrichment of adaptor ligated DNA was conducted using NEBNext Multiplex Oligos for Illumina (New England Biolabs). Illumina-ready libraries were sequenced by paired-end sequencing (2 × 150) on a NovaSeq 6000 Sequencing System (Illumina, San Diego, CA, USA). Assistance with quality control and next-generation sequencing was provided by the Vanderbilt Technologies for Advanced Genomics (VANTAGE) research core.

2.8. Illumina RNA-Seq Data Analysis and Variant Calling

Raw reads were processed by first removing the Illumina TruSeq adapter using Trimmomatic default settings [55]. Reads shorter than 36 bp were removed and low-quality bases (Q score < 30) were trimmed from read ends. The raw FASTQ files were aligned to rotavirus reference genome segments (Table S2) using Bowtie2, with the following parameters: bowtie2-p 32-q-x {ref}-1 {sample}_R1_paired.fastq-2 {sample}_R2_paired.fastq-U {sample}_R1_unpaired.fastq,{sample}_R2_unpaired.fastq-S {sample}_bowtie2.sam [56]. The SAMtools [57] suite was used to calculate read depth at each genomic coordinate. LoFreq [58] was used to call single nucleotide variants and indels with the following parameters: lofreq call-parallel--pp-threads 32-f {ref}-d 100000-o {sample}.vcf {sample}_bowtie2.sort.bam. Variants were filtered at a threshold frequency of 0.001, consistent with previous reports [59]. Parsing of variants common across samples was performed via the command line. Identification of encoded amino acid mutations based on published open reading frames and computationally identified variants was performed manually. Putative locations of variants were visualized using UCSF Chimera, developed by the Resource for Biocomputing,

Visualization, and Informatics at the University of California, San Francisco, with support from NIH P41-GM103311 [60].

3. Results

3.1. Human Rotaviruses of Different Genotypes Can Adapt to Replication in Monkey Kidney Cells

To identify polymorphisms that enhance replication in cultured cells, we serially passaged supernatants of rotavirus-positive stool samples from pediatric patients treated at Vanderbilt University Medical Center [61]. Each patient stool sample was considered a 'specimen' and contained rotavirus with a sequenced genome [54]. We attempted to adapt 50 specimens to primary rhesus monkey kidney (RhMK) cells in glass roller tubes. While adaptation historically has been done in primary AGMK cells [48,62], these primary cells were unavailable commercially. The initial inocula were 0.1 mL of trypsin-activated, clarified 10% stool homogenates of rotavirus-positive clinical specimens, each in triplicate parallel 'lineages' (Figure 1). In subsequent passages, we used 0.2–1 mL of lysate as inoculum, based on ELISA score. We also passaged simian laboratory strain SA11 and medium containing trypsin as controls. Following up to five passages in primary RhMK cells, 35 lineages representing 18 distinct specimens tested positive for rotavirus antigen by ELISA, suggesting they had adapted to replication in these cells (Tables S1 and S3). The specimens that tested positive included two (of 15) G1P[8], four (of 11) G3P[8], one (of one) G9P[8], eight (of 14) G12P[8], and three (of 9) G2P[4] rotaviruses.

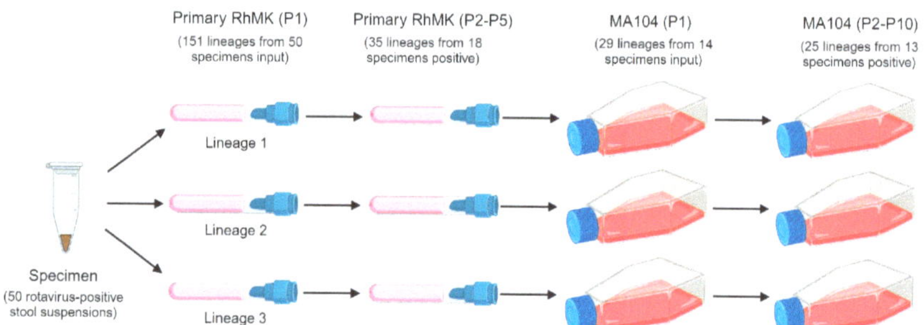

Figure 1. Workflow for rotavirus serial passaging. Each human rotavirus clinical specimen was serially passaged in triplicate lineages in primary RhMK cells in roller tubes three to five times. Most rotavirus-positive lineage lysates from RhMK passage were serially passaged ten times in monkey kidney epithelial MA104 cells in tissue culture flasks. Numbers of input lineages and specimens in each series and of rotavirus positive lineages and specimens at the end of each series are indicated.

To adapt the human rotaviruses to continuous monkey kidney cell culture, we used a subset of 14 of the 18 rotavirus-positive lysates from RhMK cells representing all five genotypes as inocula for a passage series in MA104 cells in T25 culture flasks (Figure 1). We included the 29 rotavirus-positive lineages representing these 14 specimens. The initial inocula were 0.5 mL of trypsin-activated, rotavirus-positive primary RhMK lysates. In subsequent passages, we used 10 µL–1 mL of lysate as inoculum, based on ELISA score. We also passaged simian laboratory strain SA11 and medium containing trypsin as controls. For most specimens and lineages, ELISA scores tended to increase over the passages (Table S4). Following ten passages in MA104 cells, 25 lineages representing 13 distinct specimens tested positive for rotavirus antigen by ELISA, suggesting they had adapted to replication in these cells (Tables S1 and S4). Positive specimens included one (of one) G1P[8], three (of three) G3P[8], one (of one) G9P[8], five (of six) G12P[8], and three (of three) G2P[4] rotaviruses.

We also attempted to adapt the human rotaviruses to Vero cells, which are monkey kidney cells approved for vaccine manufacture. Our reasoning was that identification of sets of mutations that allow human rotaviruses to replicate efficiently in Vero cells could be used in the future to rationally design attenuated vaccine strains that retain many antigenic epitopes of circulating pathogenic human rotaviruses, but replicate more efficiently. To adapt the human rotaviruses, we first used a subset of seven rotavirus-positive lysates from RhMK cells as inocula for a passage series in Vero cells in T25 culture flasks with a methodology identical to that used for MA104 cell passage (Figure 1). Although ELISA scores for about half of the passaged lysates were rotavirus positive after P4, by P10 all were negative (Table S5). We hypothesized that adaptation to MA104 cells might promote adaptation to Vero cells. Therefore, we used MA104 lysates from rotavirus positive specimens and lineages as inocula for a passage series in Vero cells. The initial inocula were 1 mL of trypsin-activated, rotavirus-positive P6 or P10 MA104 lysates. In subsequent passages, we used 10 µL–1 mL of lysate as inoculum, based on the ELISA score. We also passaged simian laboratory strain SA11 and medium containing trypsin. For most specimens and lineages, ELISA scores in early passages were positive, possibly due to the presence of residual rotavirus in diluted inocula (Table S6). ELISA scores tended to decrease over the passages, and none were positive by P10 (Tables S1 and S6). Thus, the human rotavirus clinical specimens in our collection failed to efficiently adapt to Vero cells under the given conditions after primary RhMK passage or primary RhMK and MA104 cell passage.

3.2. Late-Passage Human Rotaviruses Replicate More Efficiently Than Some Early-Passage Viruses

To directly assess whether serially passaged human rotaviruses had adapted to MA104 cells, we compared the replication efficiencies of early- and late-passage viruses for several specimens. We adsorbed MA104 cells with trypsin-activated rotaviruses in P3 or P10 lysates at an MOI of 0.01 FFU/cell. After adsorption, we washed to remove unbound virus, then incubated the cells for 24 or 48 h and quantified virus yield. After 10 passages, the replication efficiency of laboratory strain SA11 was not statistically or appreciably different than that of a pre-passage virus stock at 24 or 48 h (Figure 2). For G1P[8], G9P[8], and G2P[4] human rotaviruses at 48 h, titers of P10 viruses were significantly higher than those of P3 viruses. Although the numbers did not reach statistical significance for G3P[8] and G12P[8] human viruses, likely due to the spread of data points from the many lineages and specimens, at 48 h, titers of P10 viruses were appreciably higher than those of P3 viruses. Consistent with increasing ELISA titers, these observations suggest that human rotaviruses of all genotypes adapted to MA104 cells during serial passage.

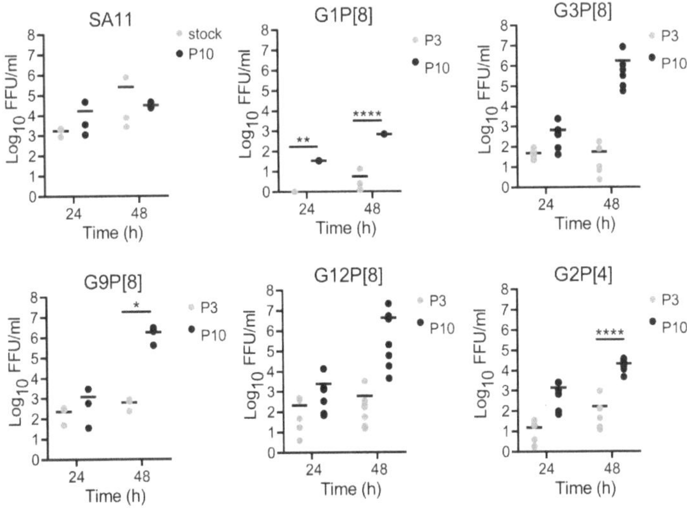

Figure 2. Replication of early- and late-passage rotaviruses in MA104 cells. MA104 cells were adsorbed for 1 h at 37 °C with trypsin-activated, rotavirus-positive P3 or P10 lysates at an MOI of

0.01 FFU/cell or with undiluted lysate if an MOI of 0.01 FFU/cell could not be reached. Cells were washed to remove unbound virus and incubated with serum-free medium plus 0.5 µg/mL trypsin at 37 °C for 0, 24, or 48 h. Plates were frozen at −80 °C and thawed three times prior to determining virus titer by FFA on MA104 cells. The mean and individual data points are shown. Virus yield was determined by dividing titer at 24 h or 48 h by titer at 0 h. *, $p < 0.05$; **, $p < 0.01$; ****, $p < 0.0001$ by two-way ANOVA with Šídák's multiple comparisons.

3.3. Human Rotaviruses Acquire Polymorphisms in VP4 during Serial Passage

In published studies, adaptive mutations in human rotaviruses have been detected following 30–60 passages [47–50]. However, after three to five passages in RhMK cells and 10 passages in MA104 cells, ELISA scores were high, and replication assays suggested the capacity of the viruses to replicate in MA104 cells had substantially improved (Table S4 and Figure 2). Therefore, we decided to determine whether sequence changes had arisen in the genomes of passaged human rotaviruses. To determine the genome sequences of MA104-adapted viruses, we isolated RNA from P10 stocks of 24 lineages representing 13 distinct specimens that tested positive for rotavirus antigen by ELISA, constructed libraries, and used Illumina next-generation sequencing. We also extracted and sequenced RNA from mock-infected MA104 cell lysates and P10 SA11-infected MA104 lysates as controls. When we aligned the resulting viral sequences to the reference genomes (Table S2), we found that for the human rotaviruses, sequence coverage was consistently high for g4, which encodes VP4, and g11, which encodes NSP5 and NSP6 (Figure S1). Coverage for the remaining segments was highly variable and often quite low. However, we obtained high sequence coverage for all segments of SA11, which was prepared using the same method and passaged at similarly high titers (Figure S1). Both g4 and g11 from adapted human rotaviruses contained synonymous and nonsynonymous polymorphisms. However, since g4 sequence coverage was high for all human rotaviruses, and VP4 is an important tropism determinant, we focused our subsequent analyses on this segment and particularly on polymorphisms encoding amino acid mutations.

To identify genetic polymorphisms associated with adaptation to cultured cells, we conducted variant-calling analysis with LoFreq [58]. This approach will identify differences from the reference sequence detected at varying frequencies, not just those that have become fixed in the population. In some cases, we detected polymorphisms at a lower frequency, but the majority were detected at frequencies >95%. While changes were detected throughout the genome, the highest concentration of variants was in g4, with 168 nucleotide polymorphisms encoding 41 amino acid changes in the VP4 protein per sequenced lineage, on average (Tables 1 and S7). G2P[4] virus populations contained the highest numbers of g4 variants, and most P[8] virus populations contained similar numbers of polymorphisms (Tables S7 and S8). While variants were detected in each of the P10 SA11 segments, the number in SA11 g4 (17) was very low relative to the numbers in the human P10 rotavirus g4 segments, and it was unremarkable compared to the number of variants detected in any other P10 SA11 segment. No viral sequences were detected in control P10 cell lysates. Together, these findings suggest that human rotaviruses of all genotypes serially passaged in RhMK cells then MA104 cells acquired polymorphisms in g4, some of which encoded VP4 amino acid changes. We were unable to confidently assess the changes acquired in most other segments.

Table 1. Variant summary.

Genotype	G1P[8]	G3P[8]	G9P[8]	G12P[8]	G2P[4]
Specimens	1	3	1	5	3
Total lineages	1	6	3	7	7
Average total variants	315	224	319	291	837
Average mutation frequency	0.019	0.020	0.027	0.042	0.023
Average g4 variants	152	149	178	176	183
VP4 amino acid changes	35	42	37	39	54

3.4. Putative VP4 Adaptive Polymorphisms Are Conserved across Genotypes

In previous studies, amino acid mutations in the VP4 protein have been associated with tissue culture adaptation [47–50]. We identified several nonsynonymous nucleotide polymorphisms in g4 of P10 MA104-passaged human rotaviruses (Tables 1 and S8). Some of these were located at positions that had been identified in prior studies, although many were not (Tables 2 and S9). We rationalized that if human rotaviruses share a common mechanism of tissue culture adaptation, we might identify VP4 mutations that are conserved for adapted human rotaviruses across genotypes. So, we compared the identities of VP4 mutations in MA104 P10 human rotaviruses and the frequency with which they were detected among the virus lineages. Since we only had a single sequenced rotavirus-positive MA104 P10 G1P[8] rotavirus, all detected VP4 mutations were present in 100% of lineages (Tables 1 and S8). For the other genotypes, we calculated the frequency of mutation detection among individual sequenced lineages. Among the P[8] viruses, we identified 28 VP4 mutations that were conserved in at least half of all sequenced lineages (Table 2). Of these 28 VP8 mutations, 15 were detected in 100% of sequenced lineages. Seven of the 28 mutations in P[8] rotaviruses (S78I, G145S, V390L, V580I, A587I, V604L, and T738I) were encoded by two nucleotide polymorphisms for at least some genotypes, while the rest were encoded by a single nucleotide polymorphism (Tables S10–S26). Among all five rotavirus genotypes, including both P[8] and P[4], we identified 12 VP4 amino acid polymorphisms that were conserved in at least half of all sequenced lineages. Four of the mutations in P[4] rotaviruses (Y295F, D385H, F467L, and V604L) were encoded by two nucleotide polymorphisms, while the rest were encoded by a single nucleotide polymorphism (Tables S10–S33). For each given mutation, the frequency of detection of the variants encoding the mutation among sequenced reads for each lineage was >99% (Tables S10–S33). For each genotype, we identified additional mutations in VP4 encoded by sequence polymorphisms detected in more than 50% of lineages, some of which were conserved among a subset of genotypes (e.g., Y19H) and some of which were unique (e.g., I130V) (Table 2). Polymorphisms conserved among P[8] viruses map to four main regions of VP4, (i) the VP8* head domain, (ii) within or adjacent to the VP5* hydrophobic loops, (iii) clustered around the VP5* 'waist', and (iv) within the VP5* foot (Figure 3A). Polymorphisms conserved among all five rotavirus genotypes map primarily to the three regions of VP5* described above (Figure 3B–D). In the reversed conformation, mutations located in the VP4 waist can be seen adjacent to the VP7 layer of the capsid, and the ring of mutations in and adjacent to the hydrophobic loops is discernable, whereas these clusters are less obvious in the upright conformation of VP4. Together, these observations reveal multiple polymorphisms encoding VP4 amino acid mutations arising during serial passage, a subset of which is conserved across genotypes. Recurrent amino acid changes appearing independently in lineages of different genotypes may contribute to tissue culture adaptation via a common mechanism.

Table 2. VP4 amino acid changes acquired during serial passage by genotype.

G1P[8]	G3P[8]	G9P[8]	G12P[8]	G2P[4]	G1P[8]	G3P[8]	G9P[8]	G12P[8]	G2P[4]
-	Y19H [1]	Y19H	Y19H	-	-	-	A430T	-	-
H52Y *	H52Y *	H52Y *	H52Y *	-	-	-	-	-	I439L
T78I	S78I	S78I	T78I	-	-	-	-	-	M444V
-	-	-	G/N99S	-	-	-	-	-	V463I
I106V	I106V	I106V	I106V	-	**F467L**	**F467L**	**F467L**	**F467L**	**F467L**
V108I	V108I	V108I	V108I	-	-	-	-	-	N498T *
-	D113N	D113N	-	-	S546N	-	S546N	-	-
N120T	N120T	N120T	N120T	-	V560I	V560I	V560I	V560I	-
-	I130V	-	-	-	-	-	-	-	A578V
G145S	G145S	G145S	G145S	-	V580I	V580I	V580I	V580I	-
-	-	-	T149N	-	-	K581R	-	-	-
D150E	D150E	D150E	D150E	-	-	-	-	-	L584I
R162K	R162K	R162K	-	-	-	A586T	A586T	-	-
-	-	-	-	M166I	V587I	V587I	V587I	A587I	-
V173I	-	-	-	V173I	**W590L**	**W590L**	**W590L**	**W590L**	**W590L**
-	-	-	-	S189N	-	-	-	-	S591T
-	-	-	-	D192N	-	-	-	-	D592N
-	S194N	-	-	-	I593V *	-	-	-	A593V *
G195D [2]	**G195D**	**G195D**	**G195D**	**N195D**	-	-	-	-	K595N
T199I	T199I	T199I	T199I	-	-	-	-	-	S596D
-	-	-	-	R245K	-	-	-	-	L598S
D252E	**D252E**	**D252E**	**D252E**	**D252E**	-	S599N	-	-	-
R268S *	**R268S ***	**R268S ***	**R268S ***	**R268S ***	-	-	-	L600V	-
-	-	-	-	V280I	-	-	-	-	D602N
-	I281V	I281V	I281V	I281V	**V604L**	**V604L**	**V604L**	**V604L**	**V604L**
Y295F	**Y295F**	**Y295F**	**Y295F**	**Y295F**	-	A608S	A608S	-	-
-	-	-	-	S303N	-	-	-	R616K	-
-	-	-	-	S305L	N617K	N617K	N617K	S617R	-
V338I	-	-	-	-	**K621R**	**K621R**	**K621R**	**K621R**	**K621R**
-	-	-	-	I352V	-	-	-	-	I629M
A360V	**A360V**	**A360V**	**A360V**	**A360V**	-	A642T	-	-	-
T380A	T380A	T380A	T380A	-	-	-	-	-	V674I
S383R	**S383R**	**S383R**	**S383R**	**S383R**	-	-	-	-	V683I
D385H *	**D385H ***	**D385H ***	**D385H ***	**D385H ***	F689V	F689V	F689V	F689V	-
-	-	-	-	R387S	-	-	-	-	I704V
I388L *	I388L *	I388L *	I388L *	-	T708A	-	T708A	-	-
V390L	**V390L**	**V390L**	**V390L**	**V390A/L**	I711V	-	-	-	-
-	-	-	-	E392A	-	-	-	-	D713N
-	-	-	-	I395V	T738I	T738I	T738I	T738I	-

[1] Amino acid mutations were included in the table if they were detected in a minimum of 50% of sequenced lineages for a given genotype. [2] Bold, underlined text indicates amino acid mutations that are conserved for at least 50% of sequenced lineages per genotype for all five adapted rotavirus genotypes. * Position identified as an adaptive mutation in a prior study [47–50].

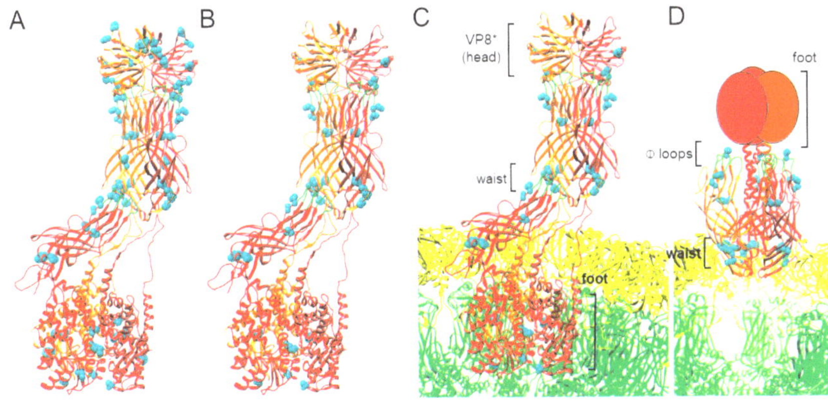

Figure 3. Locations of conserved VP4 amino acid mutations. VP4 of human G1P[8] vaccine candidate CDC-9 is shown in an upright ((**A–C**); PDB ID 7UMS) or reversed ((**D**); PDB ID 7UMT) conformation [51]. In (**C,D**), VP4 is shown relative to VP7 (yellow) and VP6 (green) rotavirus capsid layers. Monomers of trimeric VP4 are shown as red and orange ribbons, with hydrophobic loops colored green, or as red and orange ovals when the structure is unresolved. Locations of VP4 polymorphisms detected following serial passage in MA104 cells that are conserved for P[8] rotaviruses (**A**) or across all tested genotypes (**B–D**) are shown as spheres and colored cyan.

4. Discussion

We serially passaged supernatants of rotavirus-positive stool samples in RhMK cells and MA104 cells. We successfully adapted rotaviruses from genotypes predominantly associated with human disease, including G1P[8], G3P[8], G9P[8], G12P[8], and G2P[4]. For P[8] rotaviruses, genome segments other than those encoding the outer-capsid proteins typically belong to the same genogroup and are more genetically similar to one another than those of G2P[4] rotaviruses, whose segments belong to a distinct genogroup [63]. Thus, it is not surprising that many more mutations were shared among the P[8] rotaviruses in our study than between a given P[8] rotavirus and the G2P[4] rotaviruses (Table 2). Nonetheless, the identification of a subset of VP4 mutations conserved across all five genotypes suggests there may be shared mechanisms of tissue culture adaptation for rotaviruses. We attempted to passage lysates from both RhMK and MA104 serial passages in Vero cells but were unable to adapt the lysates under the conditions used. Thus, in some cases, culture adaptation mechanisms may be cell line specific. Accordingly, in published studies, human rotavirus strain Wa (G1P[8]) acquired different sets of adaptive mutations following serial passage in distinct monkey kidney epithelial cell lines (Table S9) [48,49].

Additional studies are needed to define the steps at which infectivity may be enhanced by VP4 mutations that are conserved across all genotypes. However, their predicted locations based on structures of CDC-9 in upright and reversed conformations on rotavirus particles provide insights that allow some speculation (Figure 3) [35,51]. Mutations in VP5* residue D385 have been identified in multiple culture-adapted human rotaviruses, including 89-12, CDC-9, M, and Wa, all of which are P[8] rotaviruses [47,49,50,64]. D385 is located adjacent to one of the hydrophobic loops that interacts with the lipid bilayer during rotavirus permeabilization of the endosome [36]. While mutations that reduce the hydrophobicity of these loops inhibit viral entry [65], it is unclear how a charge-altering mutation adjacent to one of the loops influences this process. Nonetheless, detection of D385H in VP4 in the majority of successfully passaged G1P[8], G3P[8], G9P[8], G12P[8], and G2P[4] rotaviruses in the current study underscores the importance of this residue for tissue culture adaptation (Table 2). In addition to D385, we identified mutations at conserved amino acid positions Y295, S383, and V390 for most passaged rotaviruses of all genotypes, which also are predicted to reside in or adjacent to the hydrophobic loops (Figure 4B). V390L within one of the loops is predicted to retain its hydrophobicity, and

Y295F will retain its bulkiness. However, S389R will introduce a charged rather than polar residue near the hydrophobic loops, which could potentially influence interactions with membranes. Adaptive mutations in residues 331 and 388 in the hydrophobic loops have been identified for other human rotaviruses, further highlighting the importance of this region (Table S9) [47–50].

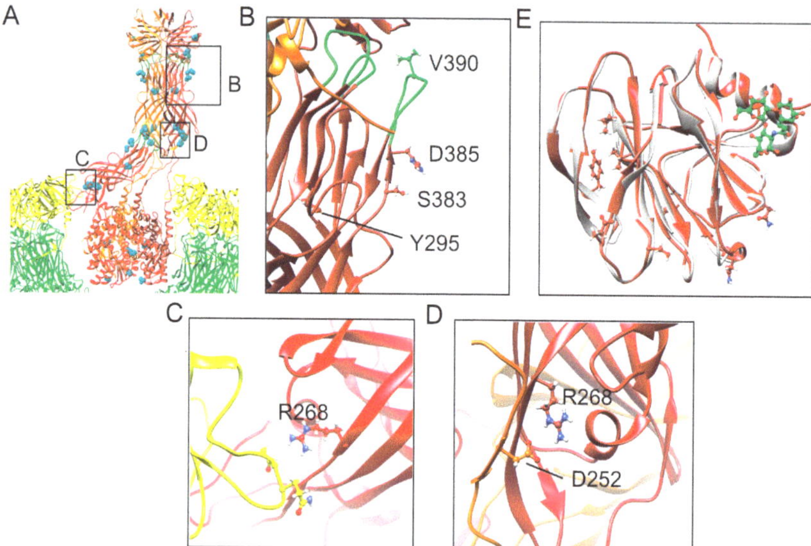

Figure 4. Locations and potential interactions of conserved VP4 amino acid mutations. VP4 of human G1P[8] vaccine candidate CDC-9 is shown in an upright conformation relative to VP7 (yellow) and VP6 (green) rotavirus capsid layers ((**A**); PDB ID 7UMS) [51]. Areas that are magnified in panels (**B–D**) are indicated. Monomers of trimeric VP4 are shown as red and orange ribbons, with hydrophobic loops colored green. Locations of VP4 polymorphisms detected following serial passage in MA104 cells that are conserved across all tested genotypes are shown as cyan spheres in (**A**) or as ball-and-stick representations in (**B–E**). In (**E**), the VP4 head domain of CDC-9 (dark red) onto which locations of conserved VP4 mutations from this study have been mapped, has been aligned with a human P[8] VP8* domain (light gray) bound to the secretory H type-1 antigen (green) (PDB ID 6HA0) [66].

Other conserved VP5* mutations are predicted to localize to the foot and waist regions (Figure 3). The VP5* foot is buried in the VP6 capsid layer in the upright conformation, and its structure in the reverse conformation is unknown, likely because it is unstructured [35,51]. Conserved mutations in the foot domain at V604 and K621 largely maintained their charged or hydrophobic character following adaptation, suggesting only fine-tuning of molecular properties is likely to result from the changes (Table 2). However, it is possible that mutation of K621 alters other functions, such as lysine-linked ubiquitination. Although the specific functions of mutations in the waist region are unknown, several have been detected in prior studies of adapted human rotaviruses, including at residues 262, 267, 268, 364, 368, 471, and 474, suggesting an important role for this region in adaptation (Table S9) [48–50]. For CDC-9, an S331F mutation in VP5* has been proposed to stabilize the upright conformation of VP4, thereby enhancing infectivity [51]. Interestingly, some of the conserved mutations we detected in the VP5* waist might destabilize inter-subunit interactions in the upright conformation. For example, R268 or its equivalent residue in one monomer of VP5* interacts with VP7 in upright RRV and CDC-9 VP4 structures (Figure 4C) [35,51]. In the upright VP4 conformation of CDC-9, D252 and R268 from different VP5* monomers appear to interact ionically (Figure 4D) [51]. In either case, the conserved R268S mutation we detect in adapted human rotaviruses might reduce the stability of interactions with adjacent residues, permitting easier triggering from the upright to reverse conformation (Table 2). It is possible that in addition to having more VP4 molecules in the upright conformation, as proposed by Jenni et al. [51], the capacity to more easily transition to the reversed conformation following interaction of the hydrophobic loops with membranes also increases rotavirus replication efficiency. Adaptation might involve fine adjustments to achieve an optimal metastable state of VP4 that balances binding efficiency with fusogenic conformational transitions.

VP8* is the receptor binding domain of VP4. Although we identified few VP8* polymorphisms that were conserved for both P[4] and P[8] genotypes, several were conserved among P[8] genotypes (Table 2). Mutations we identified in the VP8* head domain of P[8] rotaviruses are not located in known glycan-binding sites [27,66–69]. Alignment of CDC-9 VP8*, onto which locations of conserved P[8] mutations we identified had been mapped, with the structure of a human P[8] VP8* domain bound to the secretory H type-1 antigen, shows that the mutations are distinct from the ligand binding site (Figure 4E) [66]. Residues 145 and 150 are located near the sialic acid binding cleft of some animal rotavirus strains but are not critical for sialic acid binding [27,67]. Thus, detected VP8* mutations are not expected to directly alter the glycan binding capacity of the virus. Conserved TNFR-associated factor (TRAF) binding motifs have been identified in VP4 and VP8* that bind TRAFs and increase in NF-κB activity [70]. Mutations we identified in the VP8* head domain of P[8] rotaviruses are not located in the two conserved TRAF binding motifs that have been identified for human rotaviruses. Previous studies of adapted human rotaviruses have also identified mutations outside of receptor binding regions, including near the base of VP8*, adjacent to VP5*. These include residues 51, 52, 77, 79, 167, 205, and 205 (Table S9) [47–50].

There are several caveats to our findings. In our passage series, we initially inoculated cells with dilutions of stool specimens based on volume rather than virus concentration. Based on the range of ELISA A_{450} values, which are available for stools collected from 2010–2013, specimens varied widely in virus concentration (Table S34). While MOI could influence adaptation, a low ELISA A_{450} value in the inoculum did not preclude adaptation, nor did a high value guarantee successful adaptation in our study. After the initial passages, we made some adjustments to inoculum volume based on the outcome of ELISAs. However, more frequent ELISAs and higher inoculum dilutions between passages might have reduced the effective MOI, allowing for fewer coinfections and increasing selection for individual viruses that replicate efficiently in non-human cells. Reasons for low sequence coverage across regions of the adapted human rotavirus genomes are unclear. We think that problems with our library preparation or the sequencing run are unlikely, since coverage for SA11 is excellent for all segments (Figure S1). Passaged human rotavirus libraries were prepared and sequenced alongside the SA11 library using identical protocols. Adjustments to the SA11 protocol may be needed for efficient library preparation for human rotaviruses. However, differences in %GC content do not necessarily explain virus- or segment-specific differences (Table S35). In many cases, SA11 segments have slightly higher GC content than human rotaviruses. Segments that had good sequence coverage have similar GC content to other segments that had poor coverage. Instead, differences in the quantities of segments present, their stability, or the efficiency of random priming of specific segments or regions might have caused the observed differences in coverage. Our initial rationale for using Illumina sequencing followed by analysis with LoFreq was that we would achieve coverage of the entire genome with depth that would enable sensitive detection of polymorphisms. We reasoned that if identical polymorphisms arose in multiple lineages, even if they had not become fixed in all lineages, their presence might indicate that they were biologically meaningful. In the end, we achieved deep coverage only for g4 and g11, and most conserved polymorphisms in g4 were present in the population at an extremely high frequency. VP4 is a key rotavirus receptor-dependent tropism determinant [23,24,30], but it is possible that synonymous or nonsynonymous changes at other loci contribute to the enhanced replication we observed for human rotaviruses following serial passage (Figure 2). For example, CDC-9 exhibited significantly reduced STAT-1 activation following adaptation to Vero cells and contained amino acid mutations in proteins other than VP4, including a single mutation in innate immune antagonist protein NSP1 [50]. Culture-adapted CDC-9 replicated to higher titers than the parent virus in Caco-2 and Vero cells. While we failed to detect an identical mutation in NSP1 of any of our passaged viruses adapted to MA104 cells, sequence read counts were low in NSP1-encoding g5 for most adapted viruses (Figure S1). Other potential mechanisms of adaptation involve enhanced viral replication or spread. Indeed, the greater difference in replication at 48 h than 24 h for several P10 human rotaviruses compared with P3 (Figure 2) suggests that these viruses may have adapted more efficiently to some aspect of replication or spread in culture following infection, rather than simply enhancing attachment or entry. In the current study, we passaged human rotaviruses three to five times in primary RhMK cells, then 10 times in MA104 cells. Adaptation has historically been initiated in primary AGMK cells, which may yield different results or promote adaptation to Vero cells [48,62]. Nonetheless, our results suggest primary RhMK cells are suitable for human rotavirus adaptation, at least to MA104 cells. For some human rotaviruses, greater passage numbers have been used to identify adaptive polymorphisms or generate vaccine candidates [47,48,50]. With additional passages, a subset of critical adaptive polymorphisms might have become fixed in our virus populations.

Poor replication of contemporary human rotaviruses in cultured cells is an important deterrent to bespoke rotavirus vaccine engineering, and VP4 is a primary rotavirus tropism determinant. In the current study, we were unable to adapt human rotaviruses to Vero cells, which are used for rotavirus vaccine manufacturing. Nonetheless, the detection of conserved polymorphisms upon human rotavirus adaptation to MA104 cells suggests there may be conserved, genotype-independent mechanisms of tissue culture adaptation that can be identified in future studies and used towards this end. The recovery of human rotaviruses and animal rotaviruses containing human outer-capsid antigens by reverse genetics systems underscores their potential as future vaccine platforms [38,39,71–73]. A combination of structural and functional analyses that involve the use of rotavirus reverse genetics systems may allow us to elucidate fundamental mechanisms by which polymorphisms acquired during serial passage enhance rotavirus replication.

Supplementary Materials: The following supporting information can be downloaded at: https://www.mdpi.com/article/10.3390/v16060978/s1, Table S1: Passaged rotavirus-positive fecal specimens; Table S2: GenBank accession numbers for rotavirus reference sequences; Table S3: Clinical specimens and adaptive passages in primary RhMK cells; Table S4: Clinical specimens and adaptive passages in MA104 cells; Table S5: Clinical specimens and adaptive passages in Vero cells; Table S6: Clinical specimens and adaptive passages in Vero cells after passage in MA104 cells; Table S7: Library and analysis statistics; Table S8: Numbers of variants; Table S9: Adaptive mutations from prior studies; Table S10: Conserved, non-synonymous variant frequency for VU12-13-162 (G1P[8]); Table S11: Conserved, non-synonymous variant frequency for VU08-09-16 lineage1 (G3P[8]); Table S12: Conserved, non-synonymous variant frequency for VU08-09-16 lineage2 (G3P[8]); Table S13: Conserved, non-synonymous variant frequency for VU08-09-26 (G3P[8]); Table S14: Conserved, non-synonymous variant frequency for VU11-12-60 lineage 1 (G3P[8]); Table S15: Conserved, non-synonymous variant frequency for VU11-12-60 lineage 2 (G3P[8]); Table S16: Conserved, non-synonymous variant frequency for VU11-12-60 lineage 3 (G3P[8]); Table S17: Conserved, non-synonymous variant frequency for VU12-13-101 lineage 1 (G9P[8]); Table S18: Conserved, non-synonymous variant frequency for VU12-13-101 lineage 2 (G9P[8]); Table S19: Conserved, non-synonymous variant frequency for VU12-13-101 lineage 3 (G9P[8]); Table S20: Conserved, non-synonymous variant frequency for VU11-12-59 lineage 1 (G12P[8]); Table S21: Conserved, non-synonymous variant frequency for VU11-12-59 lineage 2 (G12P[8]); Table S22: Conserved, non-synonymous variant frequency for VU12-13-115 (G12P[8]); Table S23: Conserved, non-synonymous variant frequency for VU12-13-88 (G12P[8]); Table S24: Conserved, non-synonymous variant frequency for VU12-13-31 (G12P[8]); Table S25: Conserved, non-synonymous variant frequency for VU12-13-3 lineage 1 (G12P[8]); Table S26: Conserved, non-synonymous variant frequency for VU12-13-3 lineage 2 (G12P[8]); Table S27: Conserved, non-synonymous variant frequency for VU08-09-11 lineage 1

(G2P[4]); Table S28: Conserved, non-synonymous variant frequency for VU08-09-11 lineage 2 (G2P[4]); Table S29: Conserved, non-synonymous variant frequency for VU12-13-145 lineage 1 (G2P[4]); Table S30: Conserved, non-synonymous variant frequency for VU12-13-145 lineage 2 (G2P[4]); Table S31: Conserved, non-synonymous variant frequency for VU12-13-145 lineage 2 (G2P[4]); Table S32: Conserved, non-synonymous variant frequency for VU12-13-14 lineage 1 (G2P[4]); Table S33: Conserved, non-synonymous variant frequency for VU12-13-14 lineage 2 (G2P[4]); Table S34: ELISA A_{450} values for input inocula; Table S35: Percent GC content of genome segments; Figure S1: Nucleotide coverage by genotype.

Author Contributions: Conceptualization, K.M.O. and M.H.C.; methodology, K.M.O. and M.H.C.; software, J.G.; validation, M.H.C. and J.R.D.; formal analysis, M.H.C. and K.M.O.; investigation, M.H.C., J.G. and J.R.D.; resources, K.M.O., M.R.D., J.D.C. and N.B.H.; data curation, M.H.C. and J.G.; writing—original draft preparation, M.H.C. and K.M.O.; writing—review and editing, J.G., J.R.D., M.R.D., S.A.M., J.D.C. and N.B.H.; visualization, K.M.O.; supervision, K.M.O. and J.R.D.; project administration, K.M.O.; funding acquisition, K.M.O. and N.B.H. All authors have read and agreed to the published version of the manuscript.

Funding: This research was funded by a Turner-Hazinski award from the Department of Pediatrics at VUMC (to K.M.O.) and by the National Institutes of Health (R21 AI146698 to K.M.O.). Collection and sequencing of specimens was supported by the Centers for Disease Control and Prevention (1U01 IP001063-01, 3U01 IP001063-01S1, 5U01 IP000464-05, 3U01 IP000464-05S1, and 3U01 IP000464-05S2 to N.B.H.). The findings and conclusions in this report are those of the authors and do not necessarily represent the official position of the Centers for Disease Control and Prevention or the National Institutes of Health.

Informed Consent Statement: At the time of enrollment for stool specimen collection, informed consent was obtained for future use of samples from all subjects involved in the study [61].

Data Availability Statement: Data generated from Illumina RNA-seq can be accessed at the NCBI Sequence Read Archive (SRA) under BioProject accession number PRJNA1100611. Code utilized in this report can be accessed at https://github.com/ogdenlab1/OrthoreoVariant accessed on 6 June 2024.

Acknowledgments: We acknowledge Matthew B. Scholz and Kimberly R. Drake at Vanderbilt Technologies for Advanced Genomics (VANTAGE) for preliminary analyses of RNA sequencing data. We acknowledge Rendie McHenry and Laura Stewart for project support.

Conflicts of Interest: The authors declare no conflicts of interest.

References

1. Du, Y.; Chen, C.; Zhang, X.; Yan, D.; Jiang, D.; Liu, X.; Yang, M.; Ding, C.; Lan, L.; Hecht, R.; et al. Global burden and trends of rotavirus infection-associated deaths from 1990 to 2019: An observational trend study. *Virol. J.* **2022**, *19*, 166. [CrossRef] [PubMed]
2. Troeger, C.; Khalil, I.A.; Rao, P.C.; Cao, S.; Blacker, B.F.; Ahmed, T.; Armah, G.; Bines, J.E.; Brewer, T.G.; Colombara, D.V.; et al. Rotavirus Vaccination and the Global Burden of Rotavirus Diarrhea among Children Younger Than 5 Years. *JAMA Pediatr.* **2018**, *172*, 958–965. [CrossRef] [PubMed]
3. Desselberger, U. Rotaviruses. *Virus Res.* **2014**, *190*, 75–96. [CrossRef]
4. Greenberg, H.B.; Estes, M.K. Rotaviruses: From pathogenesis to vaccination. *Gastroenterology* **2009**, *136*, 1939–1951. [CrossRef] [PubMed]
5. McDonald, S.M.; Nelson, M.I.; Turner, P.E.; Patton, J.T. Reassortment in segmented RNA viruses: Mechanisms and outcomes. *Nat. Rev. Microbiol.* **2016**, *14*, 448–460. [CrossRef]
6. Carossino, M.; Vissani, M.A.; Barrandeguy, M.E.; Balasuriya, U.B.R.; Parreño, V. Equine Rotavirus A under the One Health Lens: Potential Impacts on Public Health. *Viruses* **2024**, *16*, 130. [CrossRef]
7. Omatola, C.A.; Olaniran, A.O. Genetic heterogeneity of group A rotaviruses: A review of the evolutionary dynamics and implication on vaccination. *Expert. Rev. Anti Infect. Ther.* **2022**, *20*, 1587–1602. [CrossRef] [PubMed]
8. Doro, R.; Farkas, S.L.; Martella, V.; Banyai, K. Zoonotic transmission of rotavirus: Surveillance and control. *Expert. Rev. Anti Infect. Ther.* **2015**, *13*, 1337–1350. [CrossRef]
9. Martella, V.; Banyai, K.; Matthijnssens, J.; Buonavoglia, C.; Ciarlet, M. Zoonotic aspects of rotaviruses. *Vet. Microbiol.* **2010**, *140*, 246–255. [CrossRef]
10. Kirkwood, C.D. Genetic and antigenic diversity of human rotaviruses: Potential impact on vaccination programs. *J. Infect. Dis.* **2010**, *202*, S43–S48. [CrossRef]

11. Desselberger, U.; Iturriza-Gomara, M.; Gray, J.J. Rotavirus epidemiology and surveillance. *Novartis Found. Symp.* **2001**, *238*, 125–147; discussion 147–152.

12. Settembre, E.C.; Chen, J.Z.; Dormitzer, P.R.; Grigorieff, N.; Harrison, S.C. Atomic model of an infectious rotavirus particle. *EMBO J.* **2011**, *30*, 408–416. [CrossRef] [PubMed]

13. Taniguchi, K.; Urasawa, T.; Kobayashi, N.; Ahmed, M.U.; Adachi, N.; Chiba, S.; Urasawa, S. Antibody response to serotype-specific and cross-reactive neutralization epitopes on VP4 and VP7 after rotavirus infection or vaccination. *J. Clin. Microbiol.* **1991**, *29*, 483–487. [CrossRef]

14. Nair, N.; Feng, N.; Blum, L.K.; Sanyal, M.; Ding, S.; Jiang, B.; Sen, A.; Morton, J.M.; He, X.S.; Robinson, W.H.; et al. VP4- and VP7-specific antibodies mediate heterotypic immunity to rotavirus in humans. *Sci. Transl. Med.* **2017**, *9*, eaam5434. [CrossRef] [PubMed]

15. Jiang, B.; Gentsch, J.R.; Glass, R.I. The role of serum antibodies in the protection against rotavirus disease: An overview. *Clin. Infect. Dis.* **2002**, *34*, 1351–1361. [CrossRef] [PubMed]

16. Matthijnssens, J.; Ciarlet, M.; McDonald, S.M.; Attoui, H.; Banyai, K.; Brister, J.R.; Buesa, J.; Esona, M.D.; Estes, M.K.; Gentsch, J.R.; et al. Uniformity of rotavirus strain nomenclature proposed by the Rotavirus Classification Working Group (RCWG). *Arch. Virol.* **2011**, *156*, 1397–1413. [CrossRef] [PubMed]

17. Amin, A.B.; Cates, J.E.; Liu, Z.; Wu, J.; Ali, I.; Rodriguez, A.; Panjwani, J.; Tate, J.E.; Lopman, B.A.; Parashar, U.D. Rotavirus genotypes in the post-vaccine era: A systematic review and meta-analysis of global, regional, and temporal trends in settings with and without rotavirus vaccine introduction. *J. Infect. Dis.* **2023**, *229*, 1460–1469. [CrossRef]

18. Arias, C.F.; López, S. Rotavirus cell entry: Not so simple after all. *Curr. Opin. Virol.* **2021**, *48*, 42–48. [CrossRef]

19. Coulson, B.S.; Londrigan, S.L.; Lee, D.J. Rotavirus contains integrin ligand sequences and a disintegrin-like domain that are implicated in virus entry into cells. *Proc. Natl. Acad. Sci. USA* **1997**, *94*, 5389–5394. [CrossRef]

20. Graham, K.L.; Fleming, F.E.; Halasz, P.; Hewish, M.J.; Nagesha, H.S.; Holmes, I.H.; Takada, Y.; Coulson, B.S. Rotaviruses interact with alpha4beta7 and alpha4beta1 integrins by binding the same integrin domains as natural ligands. *J. Gen. Virol.* **2005**, *86 Pt 12*, 3397–3408. [CrossRef]

21. Guerrero, C.A.; Méndez, E.; Zárate, S.; Isa, P.; López, S.; Arias, C.F. Integrin alpha(v)beta(3) mediates rotavirus cell entry. *Proc. Natl. Acad. Sci. USA* **2000**, *97*, 14644–14649. [CrossRef] [PubMed]

22. Hewish, M.J.; Takada, Y.; Coulson, B.S. Integrins alpha2beta1 and alpha4beta1 can mediate SA11 rotavirus attachment and entry into cells. *J. Virol.* **2000**, *74*, 228–236. [CrossRef]

23. Baker, M.; Prasad, B.V. Rotavirus cell entry. *Curr. Top. Microbiol. Immunol.* **2010**, *343*, 121–148. [CrossRef]

24. Lopez, S.; Arias, C.F. Multistep entry of rotavirus into cells: A Versaillesque dance. *Trends Microbiol.* **2004**, *12*, 271–278. [CrossRef] [PubMed]

25. Estes, M.K.; Greenberg, H.B. Rotaviruses. In *Fields Virology*, 6th ed.; Knipe, D.M., Howley, P.M., Eds.; Lippincott Williams & Wilkins: Philadelphia, PA, USA, 2013; Volume 2, pp. 1347–1401.

26. Coulson, B.S. Expanding diversity of glycan receptor usage by rotaviruses. *Curr. Opin. Virol.* **2015**, *15*, 90–96. [CrossRef] [PubMed]

27. Ramani, S.; Hu, L.; Venkataram Prasad, B.V.; Estes, M.K. Diversity in Rotavirus-Host Glycan Interactions: A "Sweet" Spectrum. *Cell Mol. Gastroenterol. Hepatol.* **2016**, *2*, 263–273. [CrossRef]

28. Marionneau, S.; Cailleau-Thomas, A.; Rocher, J.; Le Moullac-Vaidye, B.; Ruvoën, N.; Clément, M.; Le Pendu, J. ABH and Lewis histo-blood group antigens, a model for the meaning of oligosaccharide diversity in the face of a changing world. *Biochimie* **2001**, *83*, 565–573. [CrossRef]

29. Venkataram Prasad, B.V.; Shanker, S.; Hu, L.; Choi, J.M.; Crawford, S.E.; Ramani, S.; Czako, R.; Atmar, R.L.; Estes, M.K. Structural basis of glycan interaction in gastroenteric viral pathogens. *Curr. Opin. Virol.* **2014**, *7*, 119–127. [CrossRef]

30. Amimo, J.O.; Raev, S.A.; Chepngeno, J.; Mainga, A.O.; Guo, Y.; Saif, L.; Vlasova, A.N. Rotavirus Interactions With Host Intestinal Epithelial Cells. *Front. Immunol.* **2021**, *12*, 793841. [CrossRef]

31. Graham, K.L.; Halasz, P.; Tan, Y.; Hewish, M.J.; Takada, Y.; Mackow, E.R.; Robinson, M.K.; Coulson, B.S. Integrin-using rotaviruses bind alpha2beta1 integrin alpha2 I domain via VP4 DGE sequence and recognize alphaXbeta2 and alphaVbeta3 by using VP7 during cell entry. *J. Virol.* **2003**, *77*, 9969–9978. [CrossRef]

32. Torres-Flores, J.M.; Silva-Ayala, D.; Espinoza, M.A.; Lopez, S.; Arias, C.F. The tight junction protein JAM-A functions as coreceptor for rotavirus entry into MA104 cells. *Virology* **2015**, *475*, 172–178. [CrossRef] [PubMed]

33. Feng, N.; Hu, L.; Ding, S.; Sanyal, M.; Zhao, B.; Sankaran, B.; Ramani, S.; McNeal, M.; Yasukawa, L.L.; Song, Y.; et al. Human VP8* mAbs neutralize rotavirus selectively in human intestinal epithelial cells. *J. Clin. Investig.* **2019**, *129*, 3839–3851. [CrossRef] [PubMed]

34. Dormitzer, P.R.; Nason, E.B.; Prasad, B.V.; Harrison, S.C. Structural rearrangements in the membrane penetration protein of a non-enveloped virus. *Nature* **2004**, *430*, 1053–1058. [CrossRef] [PubMed]

35. Herrmann, T.; Torres, R.; Salgado, E.N.; Berciu, C.; Stoddard, D.; Nicastro, D.; Jenni, S.; Harrison, S.C. Functional refolding of the penetration protein on a non-enveloped virus. *Nature* **2021**, *590*, 666–670. [CrossRef] [PubMed]

36. de Sautu, M.; Herrmann, T.; Jenni, S.; Harrison, S.C. The rotavirus VP5*/VP8* conformational transition permeabilizes membranes to Ca^{2+}. *PLoS Pathog* **2024**, *20*, e1011750. [CrossRef]

37. Ward, R.L.; Knowlton, D.R.; Pierce, M.J. Efficiency of human rotavirus propagation in cell culture. *J. Clin. Microbiol.* **1984**, *19*, 748–753. [CrossRef] [PubMed]

38. Falkenhagen, A.; Patzina-Mehling, C.; Gadicherla, A.K.; Strydom, A.; O'Neill, H.G.; Johne, R. Generation of Simian Rotavirus Reassortants with VP4- and VP7-Encoding Genome Segments from Human Strains Circulating in Africa Using Reverse Genetics. *Viruses* **2020**, *12*, 201. [CrossRef]
39. Kanai, Y.; Onishi, M.; Kawagishi, T.; Pannacha, P.; Nurdin, J.A.; Nouda, R.; Yamasaki, M.; Lusiany, T.; Khamrin, P.; Okitsu, S.; et al. Reverse Genetics Approach for Developing Rotavirus Vaccine Candidates Carrying VP4 and VP7 Genes Cloned from Clinical Isolates of Human Rotavirus. *J. Virol.* **2020**, *95*, e01374-20. [CrossRef]
40. Greenberg, H.B.; Wyatt, R.G.; Kapikian, A.Z.; Kalica, A.R.; Flores, J.; Jones, R. Rescue and serotypic characterization of noncultivable human rotavirus by gene reassortment. *Infect. Immun.* **1982**, *37*, 104–109. [CrossRef]
41. Bridger, J.C.; Tauscher, G.I.; Desselberger, U. Viral determinants of rotavirus pathogenicity in pigs: Evidence that the fourth gene of a porcine rotavirus confers diarrhea in the homologous host. *J. Virol.* **1998**, *72*, 6929–6931. [CrossRef]
42. Feng, N.; Sen, A.; Wolf, M.; Vo, P.; Hoshino, Y.; Greenberg, H.B. Roles of VP4 and NSP1 in determining the distinctive replication capacities of simian rotavirus RRV and bovine rotavirus UK in the mouse biliary tract. *J. Virol.* **2011**, *85*, 2686–2694. [CrossRef] [PubMed]
43. Ijaz, M.K.; Sabara, M.I.; Alkarmi, T.; Frenchick, P.J.; Ready, K.F.; Longson, M.; Dar, F.K.; Babiuk, L.A. Characterization of two rotaviruses differing in their in vitro and in vivo virulence. *J. Vet. Med. Sci.* **1993**, *55*, 963–971. [CrossRef] [PubMed]
44. Kirkwood, C.D.; Bishop, R.F.; Coulson, B.S. Attachment and growth of human rotaviruses RV-3 and S12/85 in Caco-2 cells depend on VP4. *J. Virol.* **1998**, *72*, 9348–9352. [CrossRef] [PubMed]
45. Tsugawa, T.; Tatsumi, M.; Tsutsumi, H. Virulence-associated genome mutations of murine rotavirus identified by alternating serial passages in mice and cell cultures. *J. Virol.* **2014**, *88*, 5543–5558. [CrossRef] [PubMed]
46. Wang, W.; Donnelly, B.; Bondoc, A.; Mohanty, S.K.; McNeal, M.; Ward, R.; Sestak, K.; Zheng, S.; Tiao, G. The rhesus rotavirus gene encoding VP4 is a major determinant in the pathogenesis of biliary atresia in newborn mice. *J. Virol.* **2011**, *85*, 9069–9077. [CrossRef] [PubMed]
47. Ward, R.L.; Kirkwood, C.D.; Sander, D.S.; Smith, V.E.; Shao, M.; Bean, J.A.; Sack, D.A.; Bernstein, D.I. Reductions in cross-neutralizing antibody responses in infants after attenuation of the human rotavirus vaccine candidate 89-12. *J. Infect. Dis.* **2006**, *194*, 1729–1736. [CrossRef] [PubMed]
48. Tsugawa, T.; Tsutsumi, H. Genomic changes detected after serial passages in cell culture of virulent human G1P[8] rotaviruses. *Infect. Genet. Evol.* **2016**, *45*, 6–10. [CrossRef]
49. Guo, Y.; Wentworth, D.E.; Stucker, K.M.; Halpin, R.A.; Lam, H.C.; Marthaler, D.; Saif, L.J.; Vlasova, A.N. Amino Acid Substitutions in Positions 385 and 393 of the Hydrophobic Region of VP4 May Be Associated with Rotavirus Attenuation and Cell Culture Adaptation. *Viruses* **2020**, *12*, 408. [CrossRef]
50. Resch, T.K.; Wang, Y.; Moon, S.; Jiang, B. Serial Passaging of the Human Rotavirus CDC-9 Strain in Cell Culture Leads to Attenuation: Characterization from In Vitro and In Vivo Studies. *J. Virol.* **2020**, *94*, e00889-20. [CrossRef] [PubMed]
51. Jenni, S.; Li, Z.; Wang, Y.; Bessey, T.; Salgado, E.N.; Schmidt, A.G.; Greenberg, H.B.; Jiang, B.; Harrison, S.C. Rotavirus VP4 Epitope of a Broadly Neutralizing Human Antibody Defined by Its Structure Bound with an Attenuated-Strain Virion. *J. Virol.* **2022**, *96*, e0062722. [CrossRef]
52. Dennis, A.F.; McDonald, S.M.; Payne, D.C.; Mijatovic-Rustempasic, S.; Esona, M.D.; Edwards, K.M.; Chappell, J.D.; Patton, J.T. Molecular epidemiology of contemporary G2P[4] human rotaviruses cocirculating in a single U.S. community: Footprints of a globally transitioning genotype. *J. Virol.* **2014**, *88*, 3789–3801. [CrossRef] [PubMed]
53. McDonald, S.M.; McKell, A.O.; Rippinger, C.M.; McAllen, J.K.; Akopov, A.; Kirkness, E.F.; Payne, D.C.; Edwards, K.M.; Chappell, J.D.; Patton, J.T. Diversity and relationships of cocirculating modern human rotaviruses revealed using large-scale comparative genomics. *J. Virol.* **2012**, *86*, 9148–9162. [CrossRef]
54. Ogden, K.M.; Tan, Y.; Akopov, A.; Stewart, L.S.; McHenry, R.; Fonnesbeck, C.J.; Piya, B.; Carter, M.H.; Fedorova, N.B.; Halpin, R.A.; et al. Multiple introductions and antigenic mismatch with vaccines may contribute to increased predominance of G12P[8] rotaviruses in the United States. *J. Virol.* **2018**, *93*, e01476-18. [CrossRef] [PubMed]
55. Bolger, A.M.; Lohse, M.; Usadel, B. Trimmomatic: A flexible trimmer for Illumina sequence data. *Bioinformatics* **2014**, *30*, 2114–2120. [CrossRef]
56. Langmead, B.; Wilks, C.; Antonescu, V.; Charles, R. Scaling read aligners to hundreds of threads on general-purpose processors. *Bioinformatics* **2019**, *35*, 421–432. [CrossRef] [PubMed]
57. Li, H.; Handsaker, B.; Wysoker, A.; Fennell, T.; Ruan, J.; Homer, N.; Marth, G.; Abecasis, G.; Durbin, R. The Sequence Alignment/Map format and SAMtools. *Bioinformatics* **2009**, *25*, 2078–2079. [CrossRef]
58. Wilm, A.; Aw, P.P.; Bertrand, D.; Yeo, G.H.; Ong, S.H.; Wong, C.H.; Khor, C.C.; Petric, R.; Hibberd, M.L.; Nagarajan, N. LoFreq: A sequence-quality aware, ultra-sensitive variant caller for uncovering cell-population heterogeneity from high-throughput sequencing datasets. *Nucleic Acids Res.* **2012**, *40*, 11189–11201. [CrossRef] [PubMed]
59. Nakamura, K.; Oshima, T.; Morimoto, T.; Ikeda, S.; Yoshikawa, H.; Shiwa, Y.; Ishikawa, S.; Linak, M.C.; Hirai, A.; Takahashi, H.; et al. Sequence-specific error profile of Illumina sequencers. *Nucleic Acids Res.* **2011**, *39*, e90. [CrossRef]
60. Pettersen, E.F.; Goddard, T.D.; Huang, C.C.; Couch, G.S.; Greenblatt, D.M.; Meng, E.C.; Ferrin, T.E. UCSF Chimera—A visualization system for exploratory research and analysis. *J. Comput. Chem.* **2004**, *25*, 1605–1612. [CrossRef]

61. Bowen, M.D.; Mijatovic-Rustempasic, S.; Esona, M.D.; Teel, E.N.; Gautam, R.; Sturgeon, M.; Azimi, P.H.; Baker, C.J.; Bernstein, D.I.; Boom, J.A.; et al. Rotavirus Strain Trends During the Postlicensure Vaccine Era: United States, 2008–2013. *J. Infect. Dis.* **2016**, *214*, 732–738. [CrossRef]

62. Arnold, M.; Patton, J.T.; McDonald, S.M. Culturing, storage, and quantification of rotaviruses. *Curr. Protoc. Microbiol.* **2009**, *Unit 15.C.3*, 15C.3.1–15C.3.24. [CrossRef]

63. Matthijnssens, J.; Ciarlet, M.; Heiman, E.; Arijs, I.; Delbeke, T.; McDonald, S.M.; Palombo, E.A.; Iturriza-Gomara, M.; Maes, P.; Patton, J.T.; et al. Full genome-based classification of rotaviruses reveals a common origin between human Wa-Like and porcine rotavirus strains and human DS-1-like and bovine rotavirus strains. *J. Virol.* **2008**, *82*, 3204–3219. [CrossRef] [PubMed]

64. Ward, R.L.; Bernstein, D.I. Rotarix: A rotavirus vaccine for the world. *Clin. Infect. Dis.* **2009**, *48*, 222–228. [CrossRef] [PubMed]

65. Kim, I.S.; Trask, S.D.; Babyonyshev, M.; Dormitzer, P.R.; Harrison, S.C. Effect of mutations in VP5 hydrophobic loops on rotavirus cell entry. *J. Virol.* **2010**, *84*, 6200–6207. [CrossRef]

66. Gozalbo-Rovira, R.; Ciges-Tomas, J.R.; Vila-Vicent, S.; Buesa, J.; Santiso-Bellón, C.; Monedero, V.; Yebra, M.J.; Marina, A.; Rodríguez-Díaz, J. Unraveling the role of the secretor antigen in human rotavirus attachment to histo-blood group antigens. *PLoS Pathog.* **2019**, *15*, e1007865. [CrossRef] [PubMed]

67. Sun, X.; Li, D.; Duan, Z. Structural Basis of Glycan Recognition of Rotavirus. *Front. Mol. Biosci.* **2021**, *8*, 658029. [CrossRef] [PubMed]

68. Hu, L.; Sankaran, B.; Laucirica, D.R.; Patil, K.; Salmen, W.; Ferreon, A.C.M.; Tsoi, P.S.; Lasanajak, Y.; Smith, D.F.; Ramani, S.; et al. Glycan recognition in globally dominant human rotaviruses. *Nat. Commun.* **2018**, *9*, 2631. [CrossRef]

69. Xu, S.; Ahmed, L.U.; Stuckert, M.R.; McGinnis, K.R.; Liu, Y.; Tan, M.; Huang, P.; Zhong, W.; Zhao, D.; Jiang, X.; et al. Molecular basis of P[II] major human rotavirus VP8* domain recognition of histo-blood group antigens. *PLoS Pathog.* **2020**, *16*, e1008386. [CrossRef]

70. LaMonica, R.; Kocer, S.S.; Nazarova, J.; Dowling, W.; Geimonen, E.; Shaw, R.D.; Mackow, E.R. VP4 differentially regulates TRAF2 signaling, disengaging JNK activation while directing NF-kappa B to effect rotavirus-specific cellular responses. *J. Biol. Chem.* **2001**, *276*, 19889–19896. [CrossRef]

71. Hamajima, R.; Lusiany, T.; Minami, S.; Nouda, R.; Nurdin, J.A.; Yamasaki, M.; Kobayashi, N.; Kanai, Y.; Kobayashi, T. A reverse genetics system for human rotavirus G2P[4]. *J. Gen. Virol.* **2022**, *103*, 001816. [CrossRef]

72. Kawagishi, T.; Nurdin, J.A.; Onishi, M.; Nouda, R.; Kanai, Y.; Tajima, T.; Ushijima, H.; Kobayashi, T. Reverse Genetics System for a Human Group A Rotavirus. *J. Virol.* **2020**, *94*, e00963-19. [CrossRef] [PubMed]

73. Sánchez-Tacuba, L.; Feng, N.; Meade, N.J.; Mellits, K.H.; Jaïs, P.H.; Yasukawa, L.L.; Resch, T.K.; Jiang, B.; López, S.; Ding, S.; et al. An Optimized Reverse Genetics System Suitable for Efficient Recovery of Simian, Human, and Murine-Like Rotaviruses. *J. Virol.* **2020**, *94*, e01294-20. [CrossRef] [PubMed]

Safety, Immunogenicity, and Mechanism of a Rotavirus mRNA-LNP Vaccine in Mice

Chenxing Lu [†], Yan Li [†], Rong Chen, Xiaoqing Hu, Qingmei Leng, Xiaopeng Song, Xiaochen Lin, Jun Ye, Jinlan Wang, Jinmei Li, Lida Yao, Xianqiong Tang, Xiangjun Kuang, Guangming Zhang, Maosheng Sun, Yan Zhou *,[‡] and Hongjun Li *,[‡]

Institute of Medical Biology, Chinese Academy of Medical Science & Peking Union Medical College, Yunnan Key Laboratory of Vaccine Research and Development on Severe Infectious Disease, Kunming 650118, China; luchenxing@student.pumc.edu.cn (C.L.); yjlz2314@163.com (Y.L.); chenrong@imbcams.com.cn (R.C.); huxiaoqing@imbcams.com.cn (X.H.); lqm212855240@163.com (Q.L.); igtheshy131@gmail.com (X.S.); linxiaochen@imbcams.com.cn (X.L.); yejun@imbcams.com.cn (J.Y.); lanlingyu@student.pumc.edu.cn (J.W.); lijinmei917@163.com (J.L.); adayao0926@163.com (L.Y.); tangxq8859@163.com (X.T.); kuangxiangjun@imbcams.com.cn (X.K.); zhangguangming@imbcams.com.cn (G.Z.); sunmaosheng@imbcams.com.cn (M.S.)
* Correspondence: zhouxiaobao_850@163.com (Y.Z.); lihongjun@imbcams.com.cn (H.L.); Tel.: +86-13888340684 (Y.Z.); +86-13888918945 (H.L.)
[†] These authors contributed equally to this work.
[‡] These authors contributed equally to this work.

Abstract: Rotaviruses (RVs) are a major cause of diarrhea in young children worldwide. The currently available and licensed vaccines contain live attenuated RVs. Optimization of live attenuated RV vaccines or developing non-replicating RV (e.g., mRNA) vaccines is crucial for reducing the morbidity and mortality from RV infections. Herein, a nucleoside-modified mRNA vaccine encapsulated in lipid nanoparticles (LNP) and encoding the VP7 protein from the G1 type of RV was developed. The $5'$ untranslated region of an isolated human RV was utilized for the mRNA vaccine. After undergoing quality inspection, the VP7-mRNA vaccine was injected by subcutaneous or intramuscular routes into mice. Mice received three injections in 21 d intervals. IgG antibodies, neutralizing antibodies, cellular immunity, and gene expression from peripheral blood mononuclear cells were evaluated. Significant differences in levels of IgG antibodies were not observed in groups with adjuvant but were observed in groups without adjuvant. The vaccine without adjuvant induced the highest antibody titers after intramuscular injection. The vaccine elicited a potent antiviral immune response characterized by antiviral clusters of differentiation CD8[+] T cells. VP7-mRNA induced interferon-γ secretion to mediate cellular immune responses. Chemokine-mediated signaling pathways and immune response were activated by VP7-mRNA vaccine injection. The mRNA LNP vaccine will require testing for protective efficacy, and it is an option for preventing rotavirus infection.

Keywords: rotavirus; mRNA vaccine; structural protein VP7; lipid nanoparticles; neutralizing antibody

Citation: Lu, C.; Li, Y.; Chen, R.; Hu, X.; Leng, Q.; Song, X.; Lin, X.; Ye, J.; Wang, J.; Li, J.; et al. Safety, Immunogenicity, and Mechanism of a Rotavirus mRNA-LNP Vaccine in Mice. *Viruses* **2024**, *16*, 211. https://doi.org/10.3390/v16020211

Academic Editors: Ulrich Desselberger and John T. Patton

Received: 16 December 2023
Revised: 22 January 2024
Accepted: 22 January 2024
Published: 31 January 2024

1. Introduction

Rotaviruses (RVs) are classified as a genus in the family of Reoviridae. RVs are a major cause of diarrhea in young children worldwide [1]. Each year, infection by RVs results in ~114 million cases of acute gastroenteritis in children under 5 years of age. Diarrhea due to RV infection accounts for 5% of all global deaths in this age group, leading to ~200,000 infant fatalities [2,3]. Specific treatment is lacking, but vaccination is an effective means of preventing RV infection and the resulting gastroenteritis.

Seven live RV vaccines are in use: Rotarix (G1P [8]) [4]; RotaTeq (G1P [5], G2P [5], G3P [5], G4P [5], G6P [8]) [5]; Rotavac (G9P [11]) [6]; ROTASIIL (G1P [5], G2P [5], G3P [5], G4P [5], and G9P [5]) [7]; Lanzhou lamb rotavirus vaccine (G10P [12]) [8]; Rotalan (G2P [12], G3P [2], and G4P [12]) [9]; and Rotavin-M1 (G1P [8]) [7]. These vaccines have very important roles in reducing the burden of gastroenteritis caused by RV infection. These vaccines offer partial protection against infection, but their efficacy varies across different regions worldwide [10,11]. The risk of intussusception must also be considered [12]. Therefore, further optimization of live attenuated RV vaccines, or development of non-replicating RV vaccines to replace live attenuated RV vaccines (e.g., inactivated vaccines, recombinant subunit vaccines) is of great importance to further reduce the morbidity and mortality caused by RV infection.

Messenger (m)RNA vaccines are the third generation of nucleic acid vaccines after traditional (inactivated, live attenuated) vaccines and new (subunit, viral vector) vaccines. By introducing mRNA encoding one or more target antigenic proteins into the cytoplasm of host cells, antigenic proteins are expressed in host cells and then presented to the immune system of the host. This action activates the immune system to produce antibodies. This strategy has attracted extensive attention and research [13,14] due to its short development cycle, easy industrialization, simple and controllable production process, easy response to new variants, and better induction of humoral immunity and cellular immunity. Hence, vaccines based on the mRNA of viruses, bacteria, parasites, and tumor cells are being investigated [15].

In the early days of the COVID-19 outbreak, two mRNA vaccines of the novel coronavirus achieved great success, so mRNA vaccine technology has received widespread attention [16,17]. Research and development of mRNA vaccines are active worldwide, with the focus on treatment of infectious diseases and cancer [18]. In addition to severe acute respiratory syndrome coronavirus-2, mRNA technology is being used to create vaccines for influenza viruses [19], Zika virus [20], human immunodeficiency virus [21–24], respiratory syncytial virus [25], herpes simplex virus [26], varicella zoster virus [27], human cytomegalovirus [28], rabies virus [29], and Dengue virus [30]. A recent study showed that monovalent and trivalent LS-P2-VP8* induced superior humoral responses to P2-VP8* in guinea pigs, with encouraging responses detected against the most prevalent P genotypes [31].

RV is an unenveloped double-stranded RNA virus. The genome comprises 11 segmented double-stranded RNAs that encode six structural proteins (VP1–VP7) and six non-structural proteins (NSP1–NSP6) [32–34]. The VP7 protein, encoded by the structural gene VP7, along with the structural protein VP4, constitutes the outermost layer of the RV structure. VP7 and VP4, two capsid proteins, harbor neutralizing epitopes and have vital roles in invading and infecting target cells [35]. Consequently, they are utilized frequently as candidates for genetically engineered RV vaccines [36]. Glycoprotein VP7 is a structurally neutralizing antigen of RVs that can elicit the production of immunoglobulin (Ig)G antibodies, which are associated with protection afforded by the immune system [37].

Herein, we assessed the immunogenicity of an mRNA vaccine for RVs. Our approach involved creating an mRNA vaccine with an encoded G1P [8] RV VP7 protein and enveloping it in lipid nanoparticles (LNP). Subsequently, mice were immunized with doses (2, 5, or 10 µg) through intramuscular (IM) or subcutaneous (SC) routes. Finally, humoral and cellular immunity were assessed following three immunizations. The VP7 mRNA vaccine could elicit production of RV-specific antibodies and activate T-cell immune responses. The mRNA LNP vaccine will require testing for protective efficacy.

2. Materials and Methods

2.1. Ethical Approval of the Study Protocol

The experimental protocol was approved (DWLL202208007) by the Experimental Animal Welfare Ethics Committee of the Institute of Medical Biology within the Chinese Academy of Medical Sciences (Beijing, China). Bodyweight and temperature were monitored daily. Animals exhibiting significant reductions in these parameters (as well as other severe health issues) were killed humanely to enable sample collection.

2.2. Cells and Viruses

The virus named "ZTR-68-A" (G1P [8]) was obtained from a child suffering from diarrhea in Yunnan Province (China). ZTR-68-A (G1P [8]) was preserved by the Molecular Biology Laboratory in the Institute of Medical Biology within the Chinese Academy of Medical Sciences. HEK293 cells were obtained from OBIO Technology (Shanghai, China). MA104 cells were stored in the Laboratory of Molecular Biology, Institute of Medical Biology, Chinese Academy of Medical Sciences (Kunming, China).

2.3. Generation of mRNA and mRNA-LNP

Wild-type rotaviral VP7 (GenBank: JX509940.1) was synthesized by Integrated DNA Technologies (Coralville, IA, USA) and constructed in the pUC57-Kan-SapI-free vector by Genscript Biotech (Piscataway, NJ, USA). The construct contained a T7 promoter site for in vitro transcription of mRNA, a 5' untranslated region (UTR) derived from rotavirus, a full-length sequence of VP7 CDS, a 3'UTR derived from human β-globin [38,39], and a poly A tail with 115 nucleosides. The plasmid was extracted and linearized using BspQI enzyme (DD4302; Vazyme, Nanjing, China). After completion, DNA magnetic beads (N411; Vazyme) were used for purification. RNA was amplified using the T7 High Yield RNA Transcription Kit (N^1-Me-Pseudo UTP) (DD4202; Vazyme). RNA magnetic beads (N412; Vazyme) were used to purify IVT production. Then, the mRNA was capped with Vaccinia Capping Enzyme (DD4109; Vazyme) and 2'-O-Methyltransferase (DD4110; Vazyme). Subsequently, mRNA was purified using RNA magnetic beads (N412; Vazyme) and dissolved in RNase-free water. The mRNA concentration was determined using an ultra-micro-spectrophotometer (Thermo Fisher Technologies, Waltham, MA, USA). After purification, mRNA was stored at −80 °C until use.

After determining the expression effect of VP7, the capped and purified mRNA was diluted with 50 mmol/L sodium acetate buffer (pH5.5) to 200 ng/μL, and then LNP-wrapped. LNP's components can be divided into ionizable lipid, DSPC, cholesterol and polyethylene glycol-lipid (AVT, Shanghai, China). The preparation method is to dissolve the above four lipid components in anhydrous ethanol according to a molar ratio of 50:10:38.5:1.5 and then form LNP after blowing and mixing. mRNA-LNP was obtained using a microfluidic mixer (INano™L; Apenzy Biosciences, Shrewsbury, MA, USA) to complete the process, and the flow rate ratio of LNP and mRNA was 1:3. Meanwhile, empty LNP was also included as a control. The obtained mRNA-LNP vaccine was diluted 100 times with Tris-HCL buffer (PH7.5), and then the vaccine was concentrated to the original volume using an ultrafiltration tube (100 K) to complete the replacement of anhydrous ethanol. Finally, the preparation was passed through a 0.22 μm filter and stored at 4 °C until use. The size and potential of LNP were analyzed with a laser particle-size analyzer (Malvern Instruments, Malvern, UK). A RiboGreen® assay (Thermo Fisher Technologies) was employed to determine the encapsulation and concentration of mRNA. Transmission electron microscopy (JEM-1200EX, JEOL, Tokyo, Japan) was used to observe the morphology and size of LNP.

2.4. Transfection and Viral Protein Expression

mRNA was transfected into HEK293 cells using a transfection reagent (jetMESSENGER™; PolyPlus, Brant, France). After 24 h, total cell protein and whole-cell supernatants were collected. Each lysate sample underwent sodium dodecyl sulfate–polyacrylamide gel electrophoresis on 10% gels using TRIS-HCl. Then, proteins were transferred onto polyvinylidene fluoride (PVDF) membranes, which were enclosed in TBST (Tris-buffered saline containing Tween 20) solution with 5% skimmed milk. PVDF membranes were treated with primary (rabbit anti-VP7) antibody, followed by addition of a secondary antibody (horseradish peroxidase-coupled goat anti-rabbit IgG; Abcam, Cambridge, UK), followed by incubation for 1 h at room temperature. After washing, treated PVDF membranes were exposed to a highly sensitive luminescence solution (PK10003; Proteintech, Chicago, IL, USA) and imaged using an electrochemiluminescence system (Thermo Fisher Scientific, Waltham, MA, USA).

2.5. Mouse Experiments

Vaccines were mixed with/without an equal volume of aluminum hydroxide. Then, vaccines were injected in female Balb/c mice aged 6–8 weeks. Mice were divided into three groups: IM injection with adjuvant (group A); SC injection with adjuvant (group B); and IM injection without adjuvant (group C). Each group had dosage subgroups of 2, 5 and 10 μg. Subgroups were named with their group number–injection method–dose. In the three vaccine groups, sera were collected at days 0, 20, 41, and 56 to evaluate the immunogenicity of the vaccine. The immunization scheme is shown in Figure 1.

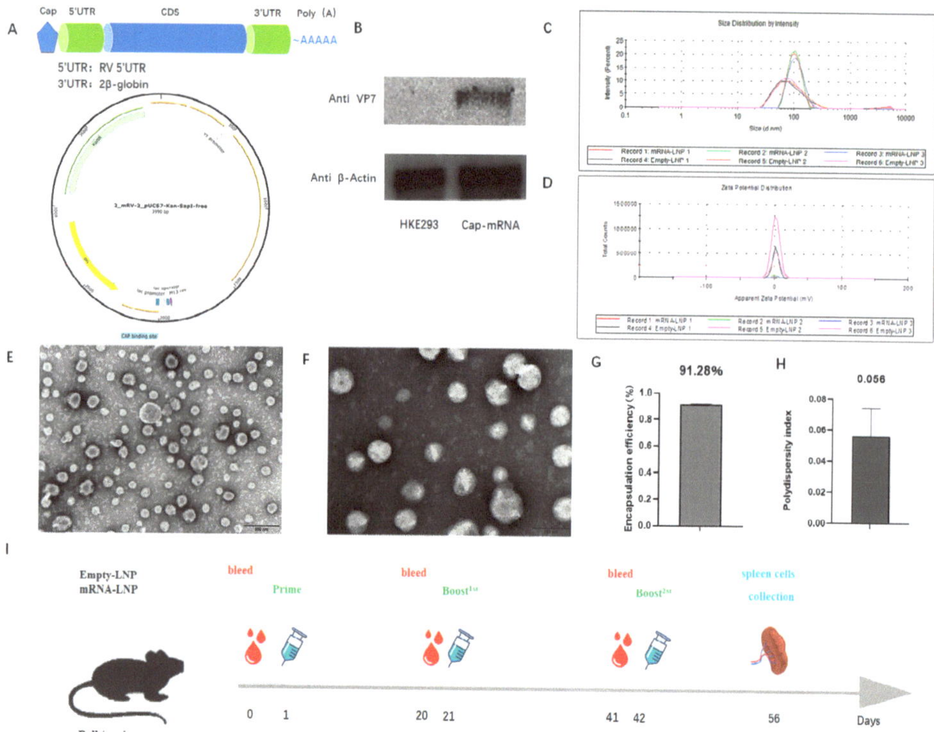

Figure 1. Vaccine characterization and immunization strategy. (**A**) Model of vaccine construction. (**B**) Detection of mRNA expression after transfection of Cap-mRNA to HEK293 cells by Western blotting; β-Actin was used as an internal reference. (**C**) Particle size detection with a Malvin laser granmeter. (**D**) Electric potential of the mRNA-LNP. (**E,F**) Shape of mRNA-LNP particles observed by transmission electron microscopy. The scale of (**E,F**) are 500 nm and 200 nm, respectively. (**G**) Percentage encapsulation of the encapsulated vaccine. Data are represented as mean ± SD. (**H**) PDI of the encapsulated vaccine. Data are represented as mean ± SD. (**I**) Schedule of immunization with the VP7-mRNA vaccine and blood collection. Mice were divided into three groups: IM injection with adjuvant; SC injection with adjuvant; and IM injection without adjuvant. Each group had a dosage subgroup of 2, 5 and 10 μg. The immunization procedure was three doses with an interval of 21 days, and the spleen was removed 14 days after the final immunization.

2.6. Detection of IgG Antibody and Neutralizing Antibody

For IgG antibodies' detection, the RV concentrate was added to an enzyme-linked immunosorbent assay-coated solution (C1050; Solarbio, Beijing, China) at a ratio of 1:100. Next, the mixture was coated onto a 96-well plate (Corning, NY, USA) with a flat bottom at a volume of 100 μL per well. The plate was kept overnight at 4 °C and subsequently washed and sealed with 3% bovine serum albumin for 1 h. Serum samples were diluted in buffer and incubated for 1 h at 37 °C, followed by five washings. Horseradish peroxidase-conjugated goat anti-mouse IgG antibody (HA1006; Huabio, Woburn, MA, USA) was diluted in 3% BSA at 1:20,000 and incubated for 1 h. The plates were evaluated using an EPOCH microplate reader (BioTek, Winooski, VT, USA) at an absorbance of 450 with a reference wavelength of 650 nm. If the A450 values of the serum dilution were higher than 0.105, the IgG/IgA antibody was considered to be positive, while the reciprocal of the highest positive serum dilution was considered as the IgG/IgA titer.

To detect neutralizing antibodies, sera with different dilutions were mixed with RV and incubated at 37 °C for 2 h. The mixture was then added to a 96-well plate filled with MA104 cells and incubated at 37 °C for 7 days. After freezing and thawing twice, the lysate was transferred to a 96-well plate coated with RV antibody, incubated at 37 °C for 1 h, then washed with PBST 5 times and added RV enzyme labeled antibody at 1:3000 dilution. Absorbance was measured by Biotek at 450 nm and 650 nm using an enzyme-labeling instrument (Biotek).

2.7. Flow Cytometry

The whole blood of mice was collected using collection vessels coated with anticoagulant (heparin sodium). Then, surface markers were stained with CD3ε-PerCP, CD4-APC, and CD8-PE (Biolegend, San Diego, CA, USA) for 30 min. Next, red blood cell lysate (R1010; Solarbio) at 3 × volume was added, followed by gentle vortex-mixing or tube inversion. After cooling on ice for 15 min, centrifugation (450× g, 10 min, 4 °C) was undertaken. The supernatant was discarded and red blood cell lysate (2 × volume) was added. The mixture was agitated gently and centrifuged (450× g; 10 min, 4 °C). The supernatant was discarded and the cell pellet resuspended with 500 μL cell-staining buffer (420201; Darco, Syracuse, NY, USA) for analyses on a high-speed flow cytometer (BD LSRFortessa™; BD Biosciences, Franklin Lakes, NJ, USA). The resulting data were analyzed using FlowJo V10 (BD Biosciences).

Mouse spleens were isolated and ground following strict aseptic procedures. Splenic lymphocytes were isolated using Mouse Lymphocyte Isolation Solution (7211011; Darko, Bedford Heights, OH, USA) and fixed with Cyto-Fast Fix/Perm Buffer (426803; Darko) and CD3ε-PerCP. Surface markers were stained for 30 min using CD4-APC and CD8-PE (BioLegend). After centrifuging for 350× g, 5 min, 500 μL of cell-staining buffer (420201; Darco) was added, and the sample was analyzed on a high-speed flow cytometer (BD LSRFortessa). FlowJo V10 was used for data analyses.

2.8. Enzyme-Linked Immunosorbent Spot (ELISpot)

Spleen lymphocytes (2 million cells/well from immunized mice) were cultured in 96-well plates for measurement of IFN-γ expression using an ELISpot assay kit (3321-4AST-2; Mabtech, Stockholm, Sweden) following the manufacturer's instructions. A VP7 peptide (final concentration = 20 μg/mL) was utilized to stimulate specific T-cell responses. An identical volume of PMA + ionomycin was utilized as a positive control. Spots were enumerated using an ELISpot reader system (Autoimmun Diagnostika, Strasbourg, France).

2.9. Detection of Cytokines in Serum

After the slide chip had dried completely, the cytokine standard was prepared. Sample diluent (100 μL) was added to each hole of the chip. The quantitative antibody chip was incubated on a shaker for 1 h at room temperature before being closed. After cleaning, a detection antibody was added to each well followed by incubation overnight on a shaker for 2 h at 4 °C. After cleaning, CY3-streptaavin was added to each well and the slide wrapped in aluminum foil and incubated on a shaker for 1 h at room temperature. After additional cleaning, fluorescence detection was undertaken using a laser scanner (InnoScan 300 Microarray Scanner; Innopsys, Chicago, IL, USA). Data analyses were carried out using QAM-CYT-1 (Raybiotech, Norcross, GA, USA).

2.10. Transcriptome Sequencing of Peripheral Blood Mononuclear Cells (PBMCs)

Samples of PBMC from two groups of mice (mRNA-LNP-immunized and control) were collected 14 days after the third immunization. Total RNA was extracted using TRIzol® Reagent (Invitrogen, Carlsbad, CA, USA) according to the manufacturer's protocols. Then, mRNA libraries were constructed using the VAHTS Universal V6 RNAseq Library Prep Kit according to manufacturer's (Vazyme) instructions. Sequencing and analyses of the transcriptome were conducted by OE Biotech (Shanghai, China). Raw reads in fastq format were processed using fastp' (https://github.com/OpenGene/fastp, accessed on 8 April 2020). Low-quality reads were removed to obtain clean reads. Then, ~6.97 million clean reads for each sample were retained for subsequent analyses. Clean reads were mapped to the reference genome using HISAT22 (https://github.com/DaehwanKimLab/hisat2, accessed on 8 June 2017). The fragments per kilobase million (FPKM)3 of each gene was calculated. The read counts of each gene were obtained by HTSeq-count4 (https://github.com/htseq/htseq/blob/main/doc/htseqcount.rst, accessed on 8 October 2023). Analyses of differential expression were undertaken using DESeq25 (https://github.com/thelovelab/DESeq2, accessed on 8 October 2023). Q < 0.05 and fold change (FC) > 1.5 or <0.67 were set as thresholds for significantly differentially expressed genes (DEGs). Based on the hypergeometric distribution, enrichment analyses of DEGs were carried out based on the Gene Ontology (GO; https://geneontology.org, accessed on 8 October 2023), Kyoto Encyclopedia of Genes and Genomes (KEGG; www.genome.jp accessed on 8 October 2023), Reactome (https://reactome.org, accessed on 8 October 2023), and WikiPathways (https://www.wikipathways.org, accessed on 8 October 2023) databases using R 3.2.0 Institute for Statistical Computing (Vienna, Austria).

2.11. Statistical Analyses

Prism 9.0.2.161 (GraphPad, La Jolla, CA, USA) was used for data analyses and mapping. Experimental results are expressed as the geometric mean ± standard error. Between-group differences were analyzed using a two-tailed Student's *t*-test or Tukey's multiple comparison test. $p < 0.05$ was considered significant.

3. Results

3.1. Construction, Characterization, and Protein Expression of RV VP7 mRNA-LNP

The plasmid was synthesized according to the design strategy for the mRNA vaccine, as shown in Figure 1A. The capped products were transfected with 2 µg of vaccine into HEK293 cells, and mRNA expression was analyzed at the cellular level via Western blotting. The resulting Cap-mRNA was expressed in cells (Figure 1B). After the Cap-RNA had been encapsulated with LNP using microfluidic technology, the particle size and potential were measured and observed under an electron microscope. The particle size of the obtained LNP-mRNA was ~100 nm (Figure 1C) and the electric potential was ~0 mV (Figure 1D). The product demonstrated spherical particles with round edges according to TEM (Figure 1E,F). The percent encapsulation of the vaccine was 91.28% according to the RiboGreen kit (Figure 1G). Furthermore, the average polydispersity index was < 0.105 (Figure 1H).

3.2. RV mRNA Vaccine Elicited Effective Humoral and Cellular Immune Responses

The serum level of IgG antibody in mice was measured to evaluate the immunogenicity of the VP7-mRNA vaccine. Sera from mice were collected at days 0, 20, 41, and 56 for measurement of IgG antibody for the three-dose group. Sera from mice were collected at days 0, 20, and 35 for measurement of IgG antibody for the two-dose group. A four-fold increase in the serum IgG antibody titer induced by doses of 2, 5, and 10 µg indicated that the conversion was 100%. After two immunizations, in group A, the IgG antibody titer ($\log_2$) (GMT) increased in the 2, 5, and 10 µg groups by 8.05, 10.77, and 11.27 (Figure 2A), whereas in group B it increased by 8.89, 10.62, and 12.56 (Figure 2B), respectively. In group C, the IgG antibody titer increased by 8.79, 8.59, and 14.18 (Figure 2C). After three immunizations, in group A, the IgG antibody titer ($\log_2$) (GMT) increased in the 2, 5, and

10 μg groups by 11.23, 12.32, and 11.28; in group B, it increased by 9.21, 10, and 10.91; and and in group C, it increased by 11.8, 13.59, and 11.55, respectively (Figure 2A,C). After comparing the three immunization routes, the IgG antibody level was highest in the intramuscular injection group without adjuvant. There was no statistically significant difference in the IgG antibody level between the two-dose and three-dose immunization groups with adjuvant, while there was a statistically significant difference in the two-dose and three-dose immunization groups without adjuvant (Figure 2D).

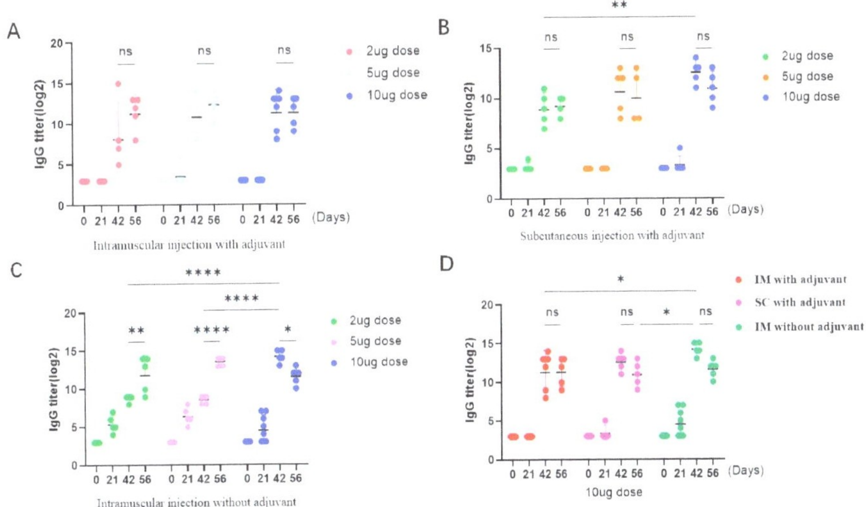

Figure 2. The vaccine triggered humoral immune responses in mice. (**A**) Levels of IgG antibodies that the vaccine triggers in 2, 5, and 10 μg groups after intramuscular injection with adjuvant. (**B**) Levels of IgG antibodies that the vaccine triggers in 2, 5, and 10 μg groups after subcutaneous injection with adjuvant. (**C**) Levels of IgG antibodies that the vaccine triggers in 2, 5, and 10 μg groups by intramuscular injection without adjuvant. (**D**) The effects of the three immunization routes were compared in the 10 μg dose group. Data are presented as geometric mean with geometric SD. Significant differences were determined by a two-way ANOVA (* $p < 0.05$, ** $p < 0.01$ and **** $p < 0.0001$; ns. indicates not significant).

To explore if the VP7-mRNA vaccine could induce a cellular immune response after immunization of mice, splenic lymphocytes were collected for flow cytometry and ELISpot detection. In the group of 5 μg by intramuscular injection without adjuvant, VP7-specific IFN-γ responses could be elicited according to the ELISpot assay (Figure 3A). A statistical chart of the number of spots is displayed (Figure 3B). Hence, in the group with 5 μg by intramuscular injection without adjuvant, the vaccine stimulated a T helper (Th)1 cell immune response in mouse lymphocytes. In the group of 10 μg by intramuscular injection without adjuvant, after the third injection, the neutralizing antibody titer (log_2) (GMT) increased by 4.23 (Figure 3C). For the 10 μg group, flow cytometry revealed that the percentage of $CD8^+$ cells increased after vaccine immunization of mice, which suggested an increase in the number of $CD8^+$ T cells (Figure 3D–F).

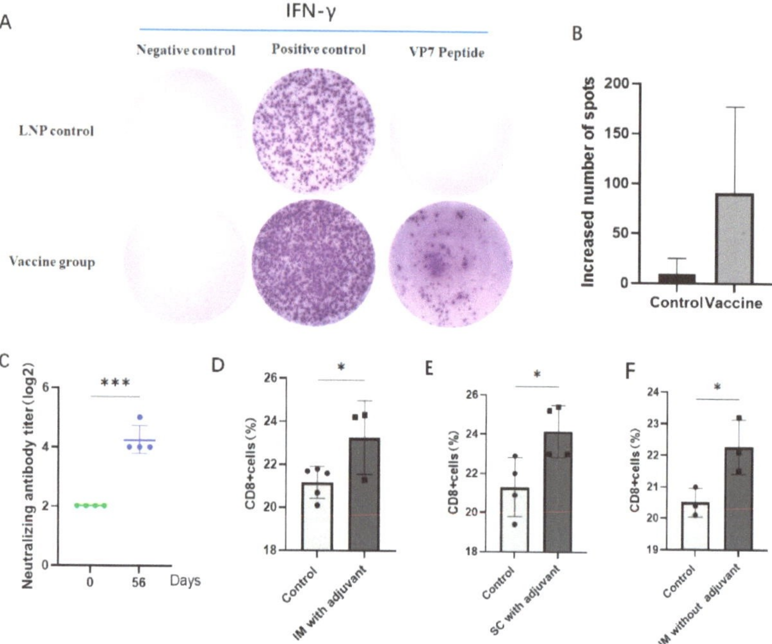

Figure 3. The vaccine triggered cellular immunity and neutralizing antibodies in mice. (**A**) In the group of 5 μg by intramuscular injection without adjuvant, the amount of interferon-γ produced by splenic lymphocytes in the control group and vaccine group was measured by ELISpot. (**B**) The number of spots in the control group and vaccine group was mapped and counted in ELISpot. Data are represented as mean ± SD. (**C**) In the group of 10 μg by intramuscular injection without adjuvant, the vaccine triggered neutralizing antibodies after the third immunization. Data are presented as geometric mean with geometric SD. Significant differences were determined by an unpaired *t* test. (*** $p < 0.001$). (**D**) The percentage of CD8$^+$ cells in total spleen lymphocytes in the group of 10 μg by intramuscular injection with adjuvant. (**E**) The percentage of CD8$^+$ cells in total spleen lymphocytes in the group of 10 μg by subcutaneous injection with adjuvant and (**F**) the group of 10μg by intramuscular injection without adjuvant. Data are presented as mean ± SD. Significant differences were determined by an unpaired *t* test (* $p < 0.05$).

3.3. Transcriptome Sequencing of Peripheral Blood Mononuclear Cells (PBMCs)

To investigate the gene expression changes in PBMCs after vaccine injection, transcriptome sequencing on the splenocytes of mice that received 10 μg of the vaccine without adjuvant was performed. Transcriptome sequencing was carried out 14 days after vaccination. Then, DEGs were identified through FC. The threshold set for genes with upregulated and downregulated expression was FC ≥ 2.0. Compared with the genes in the control group, 65 genes had upregulated expression and 200 genes had downregulated expression after one immunization. The top 10 genes with upregulated expression were *LOC115488350, Nup62cl, Vpreb1, Tspan18, Plxna4os1, Gm36614, Gm39234, Stfa1,* and *Kcnq4.* The top 10 genes with downregulated expression were *Lcn12, 1700122E12Rik, Gm35611, Gm41061, Amy2a1, Gm34771, Pnlip, Gm41061, Ceacam18,* and *4631405J19Rik.* Among these DEGs, *LOC115488350* had the highest upregulation after one dose of vaccine, and *Lcn12* had the highest downregulation (Figure 4A).

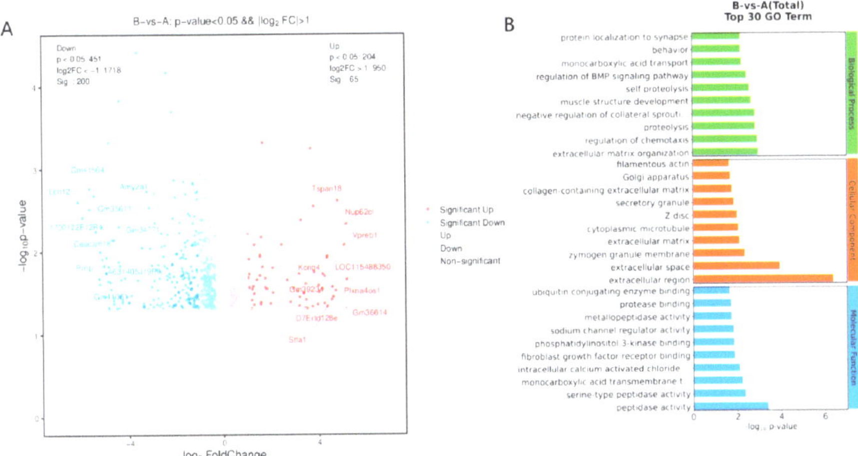

Figure 4. Changes in gene expression of PBMCs. (**A**) The differences generated by the comparison are reflected in the volcano map. Gray shows genes with a non-significant difference in expression. Red and blue are genes with a significant difference in expression. The horizontal axis is log$_2$ fold change. The vertical axis is $-\log_{10}$ of the *p*-value. (**B**) Analyses of functional enrichment (using the GO database) of the top 30 genes (based on selection of GO items corresponding to PopHits $\geq$ 5 in the three categories and ranking 10 items from largest to smallest according to the corresponding $-\log_{10}$ *p*-value of each item).

The DEGs after each immunization were subjected to analyses of functional enrichment using the GO database based on biological process (BP), cellular component (CC), and molecular function (MF). For BP, the DEGs were enriched mainly in "extracellular matrix organization", whereas they were enrieched mainly in "extracellular space" in CC, and "peptidase activity" in MF (Figure 4B). Enrichment of the signaling pathways of DEGs was assessed using the KEGG database. DEGs showed enrichment in "immune system", "infectious disease: viral", "signaling molecules and interaction", and "signal transduction". The "immune system" pathway was clustered in four terms: "chemokine signaling pathway", "complement and coagulation", "NOD-like receptor signaling pathway" and "intestinal immune network for IgA production". *C-C motif chemokine 28* (*CCL28*; FC = 2.9), *serpin f2* (FC = 10.3), and *caspase 12* (FC = 3.0) were involved in the "immune system" pathway. The "infectious disease: viral" pathway was clustered in three terms: "hepatitis B", "human cytomegalovirus infection" and "Kaposi sarcoma-associated herpes virus infection". *Caspase 12* (FC = 3.0), platelet-derived growth factor receptor a (*Pdgfra*) (FC = 9.5), and *Cd200r2* (FC = 2.9) were involved in the "infectious disease: viral" pathway (Figure 5).

3.4. Vaccine Safety

To evaluate the safety of the vaccine, body weight monitoring of mice was performed after immunization. The weight of the mice was measured once a week. The body weight of mice in each group did not decrease significantly, and maintained a stable increase (Figure 6). The results showed that the mice grew normally during immunization. Immunizing mice with the vaccine did not affect body weight, indicating that the vaccine had no serious side effects.

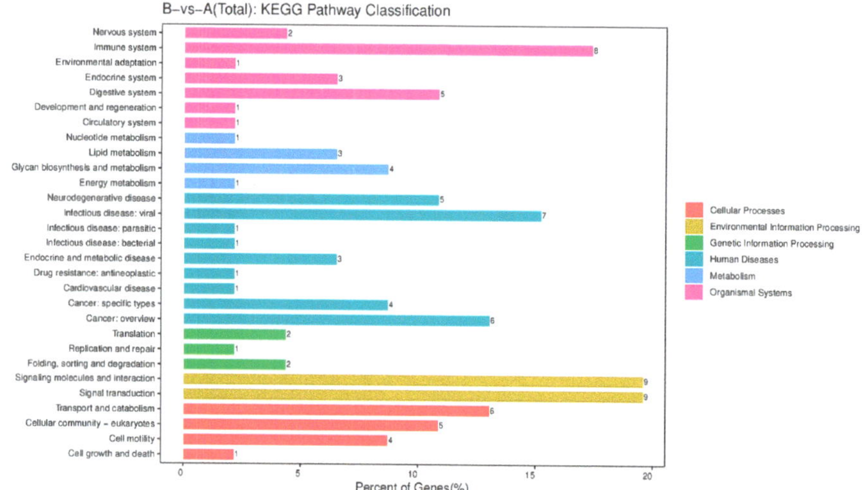

Figure 5. KEGG pathway classification. The horizontal axis is the percentage of the total number of genes annotated to each pathway (differentially expressed genes (DEGs)) and all genes annotated to the KEGG database (DEGs). The vertical axis represents the pathway name, and the number on the right side of the column represents the number of DEGs annotated to this pathway.

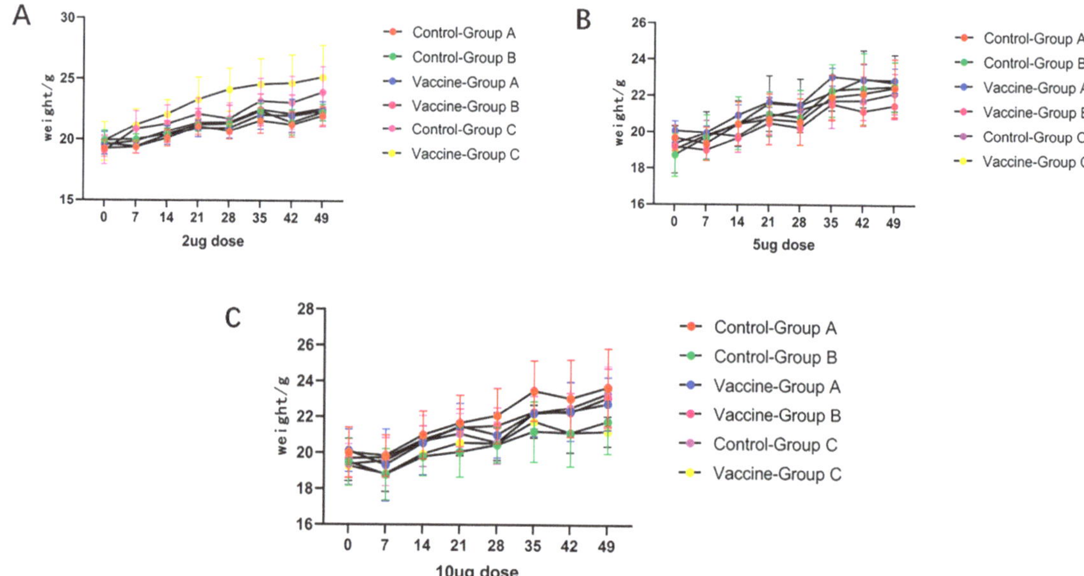

Figure 6. Changes in the body weight of mice. Starting at week 0, weight was measured weekly until the end of week 7. The weight of mice in different dose groups was divided into a (**A**) 2 μg group, (**B**) 5 μg group, and (**C**) 10 μg group. The weight data are presented in the form of mean ± SD.

4. Discussion

mRNA technology has been utilized to develop vaccines for different infectious diseases. The first mRNA vaccine was created for influenza viruses [40]. Moderna completed phase I clinical trials using LNP as a carrier for the influenza vaccines H10N8 and H7N9. This mRNA vaccine prevented and protected against influenza A infection while exhibiting strong immunogenicity, safety, and tolerance in humans [41]. Meyer and colleagues administered two mRNA vaccines that encode Ebola virus (EBOV) glycoproteins to guinea pigs infected with EBOV. Those guinea pigs could produce EBOV glycoprotein-specific IgG antibodies and EBOV-neutralizing antibodies. These phenomena resulted in all the guinea pigs infected with EBOV surviving [42]. Roth et al. developed an LNP-modified mRNA vaccine that encoded the non-structural protein of Dengue virus type I as the target antigen. Then, they immunized mice with human leukocyte antigen class-I molecules. Virus-specific CD8$^+$ T cells (which have a crucial protective role) [43] were generated.

RV infection is a significant contributor to diarrhea among infants and young children worldwide. It accounts for 5% of deaths among children under 5 years of age [2,44]. Live attenuated RV vaccines can prevent RV infection to a certain extent, but are hampered by safety issues (e.g., intestinal adverse reactions), efficacy differences between countries (developed countries have higher efficacy than developing countries), and expenses (e.g., problems in cold-chain transportation). Development of RV vaccines faces significant challenges [45–49]. To further reduce the risk of the morbidity and mortality caused by RV infection, development of a new RV (e.g., mRNA) vaccine is imperative because it could supplement the use of live vaccines.

The VP7 protein serves as the structural protein for, and determines the G type of, RVs. It plays a crucial part in the infection process of RV cells and in the assembly of RV particles. It also contains several neutralizing epitopes, making it a key neutralizing antigen for RVs [1]. We developed and formulated a VP7-mRNA vaccine expressing the UTR of RVs and investigated its immunogenicity in terms of dose response and immune pathway. Immunization with the VP7-mRNA vaccine (2, 5, or 10 µg) induced a humoral immune response and cytotoxic T-cell response. Furthermore, adjuvant-free vaccination with 10 µg of the vaccine resulted in higher IgG antibody levels. The group of mice that did not receive adjuvant consistently outperformed the group receiving adjuvant. Non-replicating vaccines primarily activate the innate immune response at the inoculation site, which highlights the importance of the site and route of inoculation. In the present study, IM injection of the VP7-mRNA vaccine produced greater stimulation of humoral immunity than that by SC injection. This effect occurred because of the abundance of antigen-presenting cells (APCs) in muscle blood vessels. APCs can trigger the immune response rapidly after capturing the antigen, whereas adipose tissue contains fewer APCs, thereby making IM injection more effective than SC injection. Furthermore, it has been reported that IM injection of mRNA-LNP is safer than cortical or intravenous injection for inducing the production of anti-LNP antibodies. Our ELISpot results indicated that interferon-γ could be stimulated by the VP7-mRNA vaccine. This vaccine immunization could also stimulate the proliferation of activated T cells, activate CD4+ T cells to differentiate into Th0 cells, promote the differentiation of Th0 cells to Th1 cells, and secrete interferon-γ. The amount of neutralizing antibody is relatively low. In the future, the VP7 sequence and structural optimization are needed to facilitate the production of more highly neutralizing antibodies.

The total RNA of PBMC samples was extracted for transcriptome sequencing to explore the gene expression changes in PBMCs after vaccine injection. Among the top 10 DEGs, the protein encoded by *Vpreb1* belongs to the Ig superfamily and is expressed selectively at the early stages of B-cell development in pro-B and early pre-B cells. This gene encodes the iota polypeptide chain that is associated with the µ chain of the Ig molecule to form a molecular complex which is expressed on the surface of pre-B cells. This molecular complex is thought to regulate *Ig* rearrangements in the early steps of B-cell differentiation. Immunization by the vaccine can cause changes in levels of chemokines and immune regulation-related genes. Among them, *CCL28* (FC = 2.9) and *Pdgfra* (FC = 9.5) have important regulatory roles in triggered immune system- and infectious disease-related pathways. CCL28 is a β or CC chemokine. The chemokine encoded by this gene displays chemotactic activity for resting CD4 T cells or CD8 T cells and eosinophils and IgA production of the intestinal immune network. *Pdgfra* acts as a cell surface receptor for platelet-derived growth factor (PDGF)A, PDGFB, and PDGFC, and has an essential role in the regulation of embryonic development, cell proliferation, cell survival, and chemotaxis.

5. Conclusions

In conclusion, we designed a novel VP7-mRNA vaccine carrying the viral UTR-encoded VP7 protein of RV. The mRNA vaccine could express the rotaviral VP7 protein and induce humoral and cellular immunity by regulating immune system-related genes and multiple signaling pathways. It is feasible to use VP7 as an immunogen for rotavirus mRNA design. The protective efficacy of the novel RV candidate vaccine should be determined before further development. In addition, multivalent and multi-target vaccines also should be considered.

Author Contributions: C.L.: Conceptualization, Writing—Original draft preparation writing, Investigation, Data curation, Formal analysis; Y.L.: Methodology, Data curation, Formal analysis; R.C. and X.H.: Investigation, Formal analysis; Q.L. and X.S.: Visualization, Data curation; X.L., J.Y., J.W. and J.L.: Investigation; L.Y. and X.T.: Software; X.K.: Investigation; G.Z.: Supervision; M.S.: Supervision, Validation; Y.Z.: Investigation, Conceptualization, Writing—Reviewing and Editing; H.L.: Conceptualization, Writing—Reviewing and Editing. All authors have read and agreed to the published version of the manuscript.

Funding: This work was supported by the Major Science and Technology Special Project of Yunnan Province (Biomedicine) #1 under grant number 202202AA100006; the Science and Technology Project of Yunnan Province—general program #2 under grant number 202201AT070236; the CAMS Innovation Fund for Medical Sciences (CIFMS) #3 under grant number 2021-I2M-1-043; the Science and Technology Project of Yunnan Province—general program #4 under grant number 2019FB020; and the Yunnan Province Innovative Vaccine Technology and Industrial Transformation Platform #5 under grant number 202002AA100009.

Institutional Review Board Statement: The animal study protocol was approved by the Experimental Animal Welfare Ethics Committee of the Institute of Medical Biology within the Chinese Academy of Medical Sciences (DWLL202208007 on 24 August 2022).

Informed Consent Statement: Not applicable.

Data Availability Statement: Data are contained within the article.

Conflicts of Interest: The authors declare that they have no conflicts of interest.

References

1. Sadiq, A.; Bostan, N.; Yinda, K.C.; Naseem, S.; Sattar, S. Rotavirus: Genetics, pathogenesis and vaccine advances. *Rev. Med. Virol.* **2018**, *28*, e2003. [CrossRef] [PubMed]
2. Tate, J.E.; Burton, A.H.; Boschi-Pinto, C.; Steele, A.D.; Duque, J.; Parashar, U.D. 2008 estimate of worldwide rotavirus-associated mortality in children younger than 5 years before the introduction of universal rotavirus vaccination programmes: A systematic review and meta-analysis. *Lancet Infect. Dis.* **2012**, *12*, 136–141. [CrossRef] [PubMed]
3. Wang, H.; Naghavi, M.; Allen, C.; Barber, R.M.; Bhutta, Z.A.; Carter, A.; Casey, D.C.; Charlson, F.J.; Chen, A.Z.; Coates, M.M.; et al. Global, regional, and national life expectancy, all-cause mortality, and cause-specific mortality for 249 causes of death, 1980–2015: A systematic analysis for the Global Burden of Disease Study 2015. *Lancet* **2016**, *388*, 1459–1544. [CrossRef] [PubMed]
4. O'Ryan, M.; Linhares, A. Update on Rotarix: An oral human rotavirus vaccine. *Expert Rev. Vaccines* **2009**, *8*, 1627–1641. [CrossRef]
5. Chandran, A.; Santosham, M. RotaTeq: A three-dose oral pentavalent reassortant rotavirus vaccine. *Expert Rev. Vaccines* **2009**, *7*, 1475–1480. [CrossRef]
6. Changotra, H.; Vij, A. Rotavirus virus-like particles (RV-VLPs) vaccines: An update. *Rev. Med. Virol.* **2017**, *27*, e1954. [CrossRef]
7. Skansberg, A.; Sauer, M.; Tan, M.; Santosham, M.; Jennings, M.C. Product review of the rotavirus vaccines ROTASIIL, ROTAVAC, and Rotavin-M1. *Hum. Vaccines Immunother.* **2020**, *17*, 1223–1234. [CrossRef]
8. Li, J.; Zhang, Y.; Yang, Y.; Liang, Z.; Tian, Y.; Liu, B.; Gao, Z.; Jia, L.; Chen, L.; Wang, Q. Effectiveness of Lanzhou lamb rotavirus vaccine in preventing gastroenteritis among children younger than 5 years of age. *Vaccine* **2019**, *37*, 3611–3616. [CrossRef]
9. Xia, S.; Du, J.; Su, J.; Liu, Y.; Huang, L.; Yu, Q.; Xie, Z.; Gao, J.; Xu, B.; Gao, X.; et al. Efficacy, immunogenicity and safety of a trivalent live human-lamb reassortant rotavirus vaccine (LLR3) in healthy Chinese infants: A randomized, double-blind, placebo-controlled trial. *Vaccine* **2020**, *38*, 7393–7400. [CrossRef]
10. Burnett, E.; Parashar, U.; Tate, J. Rotavirus Vaccines: Effectiveness, Safety, and Future Directions. *Pediatr. Drugs* **2018**, *20*, 223–233. [CrossRef]

11. Weaver, E.A.; Chilengi, R.; Simuyandi, M.; Beach, L.; Mwila, K.; Becker-Dreps, S.; Emperador, D.M.; Velasquez, D.E.; Bosomprah, S.; Jiang, B. Association of Maternal Immunity with Rotavirus Vaccine Immunogenicity in Zambian Infants. *PLoS ONE* **2016**, *11*, e0150100.

12. Murphy, T.V.; Gargiullo, P.M.; Massoudi, M.S.; Nelson, D.B.; Jumaan, A.O.; Okoro, C.A.; Zanardi, L.R.; Setia, S.; Fair, E.; LeBaron, C.W.; et al. Intussusception among infants given an oral rotavirus vaccine. *N. Engl. J. Med.* **2001**, *344*, 564–572. [CrossRef] [PubMed]

13. Youssef, M.; Hitti, C.; Puppin Chaves Fulber, J.; Kamen, A.A. Enabling mRNA Therapeutics: Current Landscape and Challenges in Manufacturing. *Biomolecules* **2023**, *13*, 1497. [CrossRef] [PubMed]

14. Barbier, A.J.; Jiang, A.Y.; Zhang, P.; Wooster, R.; Anderson, D.G. The clinical progress of mRNA vaccines and immunotherapies. *Nat. Biotechnol.* **2022**, *40*, 840–854. [CrossRef] [PubMed]

15. Qin, S.; Tang, X.; Chen, Y.; Chen, K.; Fan, N.; Xiao, W.; Zheng, Q.; Li, G.; Teng, Y.; Wu, M.; et al. mRNA-based therapeutics: Powerful and versatile tools to combat diseases. *Signal Transduct. Target Ther.* **2022**, *7*, 166. [CrossRef] [PubMed]

16. Baden, L.R.; El Sahly, H.M.; Essink, B.; Kotloff, K.; Frey, S.; Novak, R.; Diemert, D.; Spector, S.A.; Rouphael, N.; Creech, C.B.; et al. Efficacy and Safety of the mRNA-1273 SARS-CoV-2 Vaccine. *N. Engl. J. Med.* **2021**, *384*, 403–416. [CrossRef] [PubMed]

17. Polack, F.P.; Thomas, S.J.; Kitchin, N.; Absalon, J.; Gurtman, A.; Lockhart, S.; Perez, J.L.; Pérez Marc, G.; Moreira, E.D.; Zerbini, C.; et al. Safety and Efficacy of the BNT162b2 mRNA COVID-19 Vaccine. *N. Engl. J. Med.* **2020**, *383*, 2603–2615. [CrossRef]

18. Guevara, M.L.; Persano, F.; Persano, S. Advances in Lipid Nanoparticles for mRNA-Based Cancer Immunotherapy. *Front. Chem.* **2020**, *8*, 589959. [CrossRef]

19. Freyn, A.W.; Ramos da Silva, J.; Rosado, V.C.; Bliss, C.M.; Pine, M.; Mui, B.L.; Tam, Y.K.; Madden, T.D.; de Souza Ferreira, L.C.; Weissman, D.; et al. A Multi-Targeting, Nucleoside-Modified mRNA Influenza Virus Vaccine Provides Broad Protection in Mice. *Mol. Ther.* **2020**, *28*, 1569–1584. [CrossRef]

20. Pardi, N.; Hogan, M.J.; Pelc, R.S.; Muramatsu, H.; Andersen, H.; DeMaso, C.R.; Dowd, K.A.; Sutherland, L.L.; Scearce, R.M.; Parks, R.; et al. Zika virus protection by a single low-dose nucleoside-modified mRNA vaccination. *Nature* **2017**, *543*, 248–251. [CrossRef]

21. Zhang, P.; Narayanan, E.; Liu, Q.; Tsybovsky, Y.; Boswell, K.; Ding, S.; Hu, Z.; Follmann, D.; Lin, Y.; Miao, H.; et al. A multiclade env–gag VLP mRNA vaccine elicits tier-2 HIV-1-neutralizing antibodies and reduces the risk of heterologous SHIV infection in macaques. *Nat. Med.* **2021**, *27*, 2234–2245. [CrossRef] [PubMed]

22. Aldon, Y.; McKay, P.F.; Moreno Herrero, J.; Vogel, A.B.; Lévai, R.; Maisonnasse, P.; Dereuddre-Bosquet, N.; Haas, H.; Fábián, K.; Le Grand, R.; et al. Immunogenicity of stabilized HIV-1 Env trimers delivered by self-amplifying mRNA. *Mol. Ther.-Nucleic Acids* **2021**, *25*, 483–493. [CrossRef]

23. Mu, Z.; Wiehe, K.; Saunders, K.O.; Henderson, R.; Cain, D.W.; Parks, R.; Martik, D.; Mansouri, K.; Edwards, R.J.; Newman, A.; et al. mRNA-encoded HIV-1 Env trimer ferritin nanoparticles induce monoclonal antibodies that neutralize heterologous HIV-1 isolates in mice. *Cell Rep.* **2022**, *38*, 110514. [CrossRef] [PubMed]

24. Gomez, C.E.; Perdiguero, B.; Usero, L.; Marcos-Villar, L.; Miralles, L.; Leal, L.; Sorzano, C.O.S.; Sanchez-Corzo, C.; Plana, M.; Garcia, F.; et al. Enhancement of the HIV-1-Specific Immune Response Induced by an mRNA Vaccine through Boosting with a Poxvirus MVA Vector Expressing the Same Antigen. *Vaccines* **2021**, *9*, 959. [CrossRef] [PubMed]

25. Qiu, X.; Xu, S.; Lu, Y.; Luo, Z.; Yan, Y.; Wang, C.; Ji, J. Development of mRNA vaccines against respiratory syncytial virus (RSV). *Cytokine Growth Factor Rev.* **2022**, *68*, 37–53. [CrossRef]

26. Awasthi, S.; Knox, J.J.; Desmond, A.; Alameh, M.G.; Gaudette, B.T.; Lubinski, J.M.; Naughton, A.; Hook, L.M.; Egan, K.P.; Tam, Y.K.; et al. Trivalent nucleoside-modified mRNA vaccine yields durable memory B cell protection against genital herpes in preclinical models. *J. Clin. Investig.* **2021**, *131*, e152310. [CrossRef]

27. Monslow, M.A.; Elbashir, S.; Sullivan, N.L.; Thiriot, D.S.; Ahl, P.; Smith, J.; Miller, E.; Cook, J.; Cosmi, S.; Thoryk, E.; et al. Immunogenicity generated by mRNA vaccine encoding VZV gE antigen is comparable to adjuvanted subunit vaccine and better than live attenuated vaccine in nonhuman primates. *Vaccine* **2020**, *38*, 5793–5802. [CrossRef]

28. Nelson, C.S.; Jenks, J.A.; Pardi, N.; Goodwin, M.; Roark, H.; Edwards, W.; McLellan, J.S.; Pollara, J.; Weissman, D.; Permar, S.R. Human Cytomegalovirus Glycoprotein B Nucleoside-Modified mRNA Vaccine Elicits Antibody Responses with Greater Durability and Breadth than MF59-Adjuvanted gB Protein Immunization. *J. Virol.* **2020**, *94*, e00186-20. [CrossRef]

29. Alberer, M.; Gnad-Vogt, U.; Hong, H.S.; Mehr, K.T.; Backert, L.; Finak, G.; Gottardo, R.; Bica, M.A.; Garofano, A.; Koch, S.D.; et al. Safety and immunogenicity of a mRNA rabies vaccine in healthy adults: An open-label, non-randomised, prospective, first-in-human phase 1 clinical trial. *Lancet* **2017**, *390*, 1511–1520. [CrossRef] [PubMed]

30. Wollner, C.J.; Richner, M.; Hassert, M.A.; Pinto, A.K.; Brien, J.D.; Richner, J.M. A Dengue Virus Serotype 1 mRNA-LNP Vaccine Elicits Protective Immune Responses. *J. Virol.* **2021**, *95*, e02482-20. [CrossRef]

31. Roier, S.; Mangala Prasad, V.; McNeal, M.M.; Lee, K.K.; Petsch, B.; Rauch, S. mRNA-based VP8* nanoparticle vaccines against rotavirus are highly immunogenic in rodents. *npj Vaccines* **2023**, *8*, 190. [CrossRef]

32. Crawford, S.E.; Ramani, S.; Tate, J.E.; Parashar, U.D.; Svensson, L.; Hagbom, M.; Franco, M.A.; Greenberg, H.B.; O'Ryan, M.; Kang, G.; et al. Rotavirus infection. *Nat. Rev. Dis. Prim.* **2017**, *3*, 17083. [CrossRef]

33. Esona, M.D.; Gautam, R. Rotavirus. *Clin. Lab. Med.* **2015**, *35*, 363–391. [CrossRef]

34. Spindler, K.R.; Long, C.P.; McDonald, S.M. Rotavirus genome replication: Some assembly required. *PLoS Pathog.* **2017**, *13*, e1006242.

35. Morozova, O.V.; Sashina, T.A.; Fomina, S.G.; Novikova, N.A. Comparative characteristics of the VP7 and VP4 antigenic epitopes of the rotaviruses circulating in Russia (Nizhny Novgorod) and the Rotarix and RotaTeq vaccines. *Arch. Virol.* **2015**, *160*, 1693–1703. [CrossRef]

36. Rodriguez, J.M.; Luque, D. Structural Insights into Rotavirus Entry. *Adv. Exp. Med. Biol.* **2019**, *1215*, 45–68.

37. Zhao, B.; Pan, X.; Teng, Y.; Xia, W.; Wang, J.; Wen, Y.; Chen, Y. Rotavirus VP7 epitope chimeric proteins elicit cross-immunoreactivity in guinea pigs. *Virol. Sin.* **2015**, *30*, 363–370. [CrossRef]

38. Zhuang, X.; Qi, Y.; Wang, M.; Yu, N.; Nan, F.; Zhang, H.; Tian, M.; Li, C.; Lu, H.; Jin, N. mRNA Vaccines Encoding the HA Protein of Influenza A H1N1 Virus Delivered by Cationic Lipid Nanoparticles Induce Protective Immune Responses in Mice. *Vaccines* **2020**, *8*, 123. [CrossRef] [PubMed]

39. Holtkamp, S.; Kreiter, S.; Selmi, A.; Simon, P.; Koslowski, M.; Huber, C.; Türeci, O.z.; Sahin, U. Modification of antigen-encoding RNA increases stability, translational efficacy, and T-cell stimulatory capacity of dendritic cells. *Blood* **2006**, *108*, 4009–4017. [CrossRef] [PubMed]

40. Hekele, A.; Bertholet, S.; Archer, J.; Gibson, D.G.; Palladino, G.; Brito, L.A.; Otten, G.R.; Brazzoli, M.; Buccato, S.; Bonci, A.; et al. Rapidly produced SAM®vaccine against H7N9 influenza is immunogenic in mice. *Emerg. Microbes Infect.* **2019**, *2*, e52. [CrossRef] [PubMed]

41. Bahl, K.; Senn, J.J.; Yuzhakov, O.; Bulychev, A.; Brito, L.A.; Hassett, K.J.; Laska, M.E.; Smith, M.; Almarsson, Ö.; Thompson, J.; et al. Preclinical and Clinical Demonstration of Immunogenicity by mRNA Vaccines against H10N8 and H7N9 Influenza Viruses. *Mol. Ther.* **2017**, *25*, 1316–1327. [CrossRef]

42. Meyer, M.; Huang, E.; Yuzhakov, O.; Ramanathan, P.; Ciaramella, G.; Bukreyev, A. Modified mRNA-Based Vaccines Elicit Robust Immune Responses and Protect Guinea Pigs from Ebola Virus Disease. *J. Infect. Dis.* **2018**, *217*, 451–455. [CrossRef]

43. Roth, C.; Cantaert, T.; Colas, C.; Prot, M.; Casadémont, I.; Levillayer, L.; Thalmensi, J.; Langlade-Demoyen, P.; Gerke, C.; Bahl, K.; et al. A Modified mRNA Vaccine Targeting Immunodominant NS Epitopes Protects Against Dengue Virus Infection in HLA Class I Transgenic Mice. *Front. Immunol.* **2019**, *10*, 1424. [CrossRef]

44. Tate, J.E.; Burton, A.H.; Boschi-Pinto, C.; Parashar, U.D.; World Health Organization–Coordinated Global Rotavirus Surveillance Network; Agocs, M.; Serhan, F.; de Oliveira, L.; Mwenda, J.M.; Mihigo, R.; et al. Global, Regional, and National Estimates of Rotavirus Mortality in Children <5 Years of Age, 2000–2013. *Clin. Infect. Dis.* **2016**, *62* (Suppl. 2), S96–S105.

45. Beres, L.K.; Tate, J.E.; Njobvu, L.; Chibwe, B.; Rudd, C.; Guffey, M.B.; Stringer, J.S.A.; Parashar, U.D.; Chilengi, R. A Preliminary Assessment of Rotavirus Vaccine Effectiveness in Zambia. *Clin. Infect. Dis.* **2016**, *62*, S175–S182. [CrossRef] [PubMed]

46. Sahakyan, G.; Grigoryan, S.; Wasley, A.; Mosina, L.; Sargsyan, S.; Asoyan, A.; Gevorgyan, Z.; Kocharyan, K.; Avagyan, T.; Lopman, B.; et al. Impact and Effectiveness of Monovalent Rotavirus Vaccine in Armenian Children. *Clin. Infect. Dis.* **2016**, *62*, S147–S154. [CrossRef] [PubMed]

47. Ruiz-Palacios, G.M.; Pérez-Schael, I.; Velázquez, F.R.; Abate, H.; Breuer, T.; Clemens, S.C.; Cheuvart, B.; Espinoza, F.; Gillard, P.; Innis, B.L.; et al. Safety and efficacy of an attenuated vaccine against severe rotavirus gastroenteritis. *N. Engl. J. Med.* **2006**, *354*, 11–22. [CrossRef] [PubMed]

48. Jonesteller, C.L.; Burnett, E.; Yen, C.; Tate, J.E.; Parashar, U.D. Effectiveness of Rotavirus Vaccination: A Systematic Review of the First Decade of Global Postlicensure Data, 2006–2016. *Clin. Infect. Dis.* **2017**, *65*, 840–850. [CrossRef]

49. Carvalho, M.F.; Gill, D. Rotavirus vaccine efficacy: Current status and areas for improvement. *Hum. Vaccines Immunother.* **2019**, *15*, 1237–1250. [CrossRef] [PubMed]

*viruses*

MDPI

Article

Novel Universal Recombinant Rotavirus A Vaccine Candidate: Evaluation of Immunological Properties

Dmitriy L. Granovskiy [1,*,†], Nelli S. Khudainazarova [1,†], Ekaterina A. Evtushenko [1], Ekaterina M. Ryabchevskaya [1], Olga A. Kondakova [1], Marina V. Arkhipenko [1], Marina V. Kovrizhko [2], Elena P. Kolpakova [2], Tatyana I. Tverdokhlebova [2], Nikolai A. Nikitin [1] and Olga V. Karpova [1]

[1] Department of Virology, Faculty of Biology, Lomonosov Moscow State University, 119991 Moscow, Russia; nelly.khudaynazarova@bk.ru (N.S.K.); trifonova.katerina@gmail.com (E.A.E.); eryabchevskaya@gmail.com (E.M.R.); olgakond1@yandex.ru (O.A.K.); armar74@mail.ru (M.V.A.); nikitin@mail.bio.msu.ru (N.A.N.); okar@genebee.msu.ru (O.V.K.)

[2] Rostov Research Institute of Microbiology and Parasitology, 344010 Rostov-On-Don, Russia; npo-kovrizhko@yandex.ru (M.V.K.); kolpako-va@mail.ru (E.P.K.); rostovniimp@rniimp.ru (T.I.T.)

[*] Correspondence: dgran98@gmail.com

[†] These authors contributed equally to this work.

Abstract: Rotavirus infection is a leading cause of severe dehydrating gastroenteritis in children under 5 years of age. Although rotavirus-associated mortality has decreased considerably because of the introduction of the worldwide rotavirus vaccination, the global burden of rotavirus-associated gastroenteritis remains high. Current vaccines have a number of disadvantages; therefore, there is a need for innovative approaches in rotavirus vaccine development. In the current study, a universal recombinant rotavirus antigen (URRA) for a novel recombinant vaccine candidate against rotavirus A was obtained and characterised. This antigen included sequences of the VP8* subunit of rotavirus spike protein VP4. For the URRA, for the first time, two approaches were implemented simultaneously—the application of a highly conserved neutralising epitope and the use of the consensus of the extended protein's fragment. The recognition of URRA by antisera to patient-derived field rotavirus isolates was proven. Plant virus-based spherical particles (SPs), a novel, effective and safe adjuvant, considerably enhanced the immunogenicity of the URRA in a mouse model. Given these facts, a URRA + SPs vaccine candidate is regarded as a prospective basis for a universal vaccine against rotavirus.

Keywords: rotavirus; rotavirus vaccine; recombinant vaccine; recombinant antigen; structurally modified plant viruses; tobacco mosaic virus; plant virus adjuvants; spherical particles

Citation: Granovskiy, D.L.; Khudainazarova, N.S.; Evtushenko, E.A.; Ryabchevskaya, E.M.; Kondakova, O.A.; Arkhipenko, M.V.; Kovrizhko, M.V.; Kolpakova, E.P.; Tverdokhlebova, T.I.; Nikitin, N.A.; et al. Novel Universal Recombinant Rotavirus A Vaccine Candidate: Evaluation of Immunological Properties. *Viruses* **2024**, *16*, 438. https://doi.org/10.3390/v16030438

Academic Editors: Ulrich Desselberger and John T. Patton

Received: 8 February 2024
Revised: 4 March 2024
Accepted: 10 March 2024
Published: 12 March 2024

1. Introduction

Group A rotaviruses (RVA) remain a leading cause of acute gastroenteritis in young children and infants throughout the world. Rotavirus infection is responsible for an estimated 258 million episodes of diarrhoea and 130,000 deaths among children under 5 years of age annually, with a disproportionately high occurrence in low-income countries [1]. The best way to prevent rotavirus infection is vaccination. Since 2006, four attenuated rotavirus vaccines have been licensed in more than 100 countries worldwide. All these vaccines are live-attenuated and require an oral route of administration [2]. Two of these vaccines are used most widely: RotaTeq® (Merck & Co., Rahway, NJ, USA) and Rotarix™ (GlaxoSmithKline, Rixensart, Belgium). RotaTeq® is a live-attenuated pentavalent vaccine based on human–bovine reassortant rotavirus strains (with antigens G1, G2, G3, G4, and P[8]). Rotarix™ is a monovalent live-attenuated vaccine based on the human RVA strain G1P[8]. Both vaccines have been shown to be highly effective in preventing severe rotavirus infection in middle- and high-income countries, but post-licensure studies have demonstrated that existing vaccines have been far less efficacious in low-income countries, where the incidence of rotavirus-associated diarrhoea is already high [3]. Presumably, this is related to the high titres of maternally-derived antibodies, co-infections with other enteropathogens, and the greater diversity of rotavirus circulating in these countries [2]. The disadvantages of existing rotavirus vaccines include some serious side effects, such as intestinal intussusception [4–7], a wide range of contraindications, the risks of chronic infection [8–10], the reversion of the vaccine strain to a virulent phenotype [11], and the recombination of the vaccine strain with wild-type strains [12–14]. The latter poses risks of the emergence of new, more pathogenic rotavirus strains [15]. The extent to which existing live rotavirus vaccines provide protection against non-vaccine genotypes is currently a controversial issue. Several studies reported substantial changes in the composition of circulating RVA strains and an increase in the proportion of heterotypic genotypes in the post-vaccination era [16,17]. However, it is not yet known whether these changes are due to the selective pressure of vaccines or natural evolutionary processes [18,19].

The shortcomings of the existing vaccines highlight the need for newer approaches to rotavirus vaccine development. One of the most promising directions of research in this field is the creation of non-replicating recombinant vaccines with a parenteral route of administration. The use of such vaccines avoids the multiplication of the vaccine strain in the intestine, which minimises the risk of side effects, chronic infection, and reassortment with wild-type RVA. The advantages of these vaccines also include greater safety and purity of the preparation [20,21]. For the development of a recombinant RVA vaccine, it is advisable to use rotavirus structural proteins, which are essential for an effective immune response. The rotavirus virion is composed of three protein shells—an outer capsid, an inner capsid, and an internal core—that enclose 11 segments of double-stranded RNA. Two rotaviral structural proteins of the outer capsid, the spike protein VP4 (protease-cleaved protein, P), and VP7 (glycosylated protein, G), define both the serotype and the genotype of rotavirus strains [22] and are considered to be crucial for vaccine development [23,24]. During infection, the VP4 spike protein is cleaved by intestinal trypsin into two subunits: VP8* and VP5*. Both of them provide a good basis for the development of a recombinant rotavirus vaccine [25–29]. In particular, it has been shown that a VP8*-induced immune response is sufficient for disease prevention [30]. However, studies have revealed that such subunit vaccine candidates generate low cross-reactive immune responses to heterologous strains of RVA [27]. In an attempt to provide broader protection, multivalent vaccines based on antigens from several RVA genotypes have been formulated [31], but further evaluations of efficiency are required to better understand the ability of such vaccines to provide cross-serotype protection.

Peptides corresponding to the neutralising epitopes of rotavirus antigens are also being considered as a potential basis for a recombinant vaccine against rotavirus. It is assumed that the use of peptides that mimic pathogen's epitopes allows the production of antibodies against specific regions of antigens, including sites that may be otherwise inaccessible to the immune system [32,33]. In addition, it is assumed that immunisation with constructs containing highly conserved epitopes induces the production of cross-reactive antibodies, which is essential since strain diversity is one of the most fundamental problems in the development of a vaccine against RVA [34]. Despite such possibilities, the development of peptide-based vaccines remains limited. This is most likely related to the poor immunogenic activity of peptides themselves and the lack of effective and safe adjuvants available for use in humans [32].

The present work is devoted to the development of a novel, broad-spectrum, highly immunogenic RVA vaccine candidate. A new recombinant rotavirus antigen was designed as the basis for the vaccine candidate. For this, two methods were combined: the application of a conserved neutralising epitope and the use of the consensus of the extended protein's fragment. The resulting antigen, named URRA (Universal Recombinant Rotavirus Antigen), consists of a short peptide ep8 corresponding to a neutralising epitope (from 1 to 10 aa VP8*) highly conserved among RVA strains [34,35] and ΔVP8*, the truncated VP8* subunit (from 65 to 223 aa of VP8*), obtained using the consensus approach on the base of a wide range of RVA isolates of genotype P[8] [28]. The coding sequences of ep8 and ΔVP8* were designed previously [28]. The potential impact of vaccines' selective pressure on RVA strains' genetic diversity and distribution, the emergence of previously uncommon RVA strains, and rapid evolutionary changes in the RVA population are all concerns that are widely discussed [14,16–19,36,37]. Because of this, the current research focused on the antigenic properties of the URRA protein. The correspondence of URRA to currently circulating RVA variants was examined using antisera to patient-derived field RVA isolates.

The proposed vaccine candidate contains a special adjuvant, spherical particles (SPs) obtained from the tobacco mosaic virus (TMV). Previous studies have demonstrated the properties of TMV SPs as an adjuvant and a platform for the adsorption and stabilisation of various antigens [38–40]. SPs are also known to be safe in a wide variety of animal models and biodegradable [41–43]. Here, the immunogenicity of the vaccine candidate was evaluated in a murine model and compared with the immunogenicity of the individually formulated URRA. The immunogenicity of SPs was measured separately to estimate the immune response to the adjuvant. The data obtained show that a vaccine candidate based on a URRA + SPs composition provides a possible solution to the fundamental challenges of recombinant RVA vaccine development. Such a vaccine could be a prospective subject for further research.

2. Materials and Methods

2.1. Expression and Purification of Rotavirus Recombinant Antigen

Escherichia coli strain XL1-Blue was applied for rotavirus recombinant antigen expression. The cultures were grown in 3 mL of a 2YT medium containing 1.6% (w/v) tryptone, 1% (w/v) yeast extract, 0.5% (w/v) NaCl, and 100 µg/mL of ampicillin at a temperature of 37 °C with shaking at 180 rpm overnight. The cultures were added to 200 mL of 2YT with the same composition and were grown at 37 °C with shaking at 180 rpm for 3 h. After that, cultures were induced with IPTG to a final concentration of 2 mM and were cultured for 5 h at 37 °C and shaking at 180 rpm. Cell pellets were centrifugated for 10 min at $5000\times g$ (JA-14 rotor, Avanti JXN-30 centrifuge, Beckman Coulter Inc., Brea, CA, USA) at 4 °C, then stored at -20 °C and subsequently used for chromatographic isolation and the purification of recombinant protein. Metal affinity chromatography with Ni^{2+}-NTA resin (Qiagen, Hilden, Germany) under denaturing conditions was applied. For this, sediment cell pellets were resuspended and lysed in 5 mL of a solution containing 6 M GuHCl and 0.2% (w/v) natrium deoxycholate at 25 °C with shaking at 120 rpm for 1 h. The recombinant protein was eluted from the column according to the manufacturer's protocol (Qiagen) and then dialysed against deionised water (for 2 h) and Milli-Q (for 2 h) (Simplicity UV, Merck Millipore, Darmstadt, Germany) in the ratio 1:250, with hourly water replacement, and then stored at -20 °C.

2.2. Obtaining Sera to Untyped Field Rotavirus Isolate

Serum №1 and Serum №2 were obtained from two corresponding groups of outbreed CD-1 mice. Each group consisted of 10 individual males aged 6–8 weeks old. Mice were immunised with RV strains RVV-5 (Serum №1) and RRV-6 (Serum №2). These strains are stored in the collection of the Rostov Research Institute of Microbiology and Parasitology (Rostov-On-Don, Russian Federation). The strains were isolated from the material obtained from children with laboratory-confirmed rotavirus gastroenteritis who were undergoing hospital treatment in the infectious diseases department of the City Hospital of Rostov-on-Don. The bacterial and fungal flora-free material was adapted to growth on continuous mammalian cell cultures VERO and SPEV and purified by high-speed centrifugation. Both strains were assigned to RVA by PCR and ELISA using appropriate test systems (manufactured by Vector-Best, Russian Federation; AmpliSens, Russian Federation). Mice were immunised twice intramuscularly in two pelvic limbs in equal amounts (100 µL in each limb). No adjuvants were used for the immunisation.

2.3. Western Blot Analysis

First, an SDS-PAGE with an 8–20% acrylamide linear gradient was performed. Proteins separated by electrophoresis were then transferred to a PVDF membrane (Invitrogen, TM, ThermoFisher Scientific, Waltham, MA, USA) using a MINI PROTEAN II (Bio-Rad Laboratories Inc., USA) transfer system. The membrane was blocked with 5% (*w/v*) non-fat dry milk in TTBS (0.01 M Tris-HCl (pH 7.4), 0.15 M NaCl, and 0.05% (*v/v*) Tween-20). Then, the membrane was treated with primary mice polyclonal Abs (Serum №1 and Serum №2; a description is given in Section 2.2 of the "Materials and Methods") to untyped patient-derived rotavirus isolates in a 1:500 dilution, and then with secondary Abs to mouse IgG conjugated with horseradish peroxidase (Jackson Immunoresearch Inc., West Grove, PA, USA) in a 1:20,000 dilution. WesternBright ECL substrate (Advansta Inc., San Jose, CA, USA) was applied, and the signal was detected using the ChemiDoc™ XRS documentation system with Image Lab™ Software Version 6.1 (Bio-Rad Laboratories, Hercules, CA, USA).

2.4. Enzyme-Linked Immunosorbent Assay (ELISA) for Qualitative Assessment of Protein-Serum Interaction

For the qualitative assessment of the interaction of URRA with Serum №1 and Serum №2, incubation and washing schedules were consistent with the protocol described by Kovalenko et al. (2022) [39]. URRA or SPs were used as an antigen for coating a 96-well plate in concentrations of 10 µg/mL, 50 µg/mL, 100 µg/mL, or 200 µg/mL. Analyses were conducted in two replicates for each antigen concentration for each serum analysed. Serum №1 or Serum №2 were used in a dilution of 1:100. Anti-mouse total IgG HRP conjugate (#ab6728, Abcam, Cambridge, UK) was used in a dilution of 1:10,000.

2.5. TMV Isolation and Spherical Particles Generation

The TMV and the SPs were obtained according to the protocol described by Trifonova et al. (2015) [44], with some modifications. A TMV solution with a concentration of 2 mg/mL was used for SPs formation. The TMV solution was aliquoted into 1.5 mL polypropylene tubes (Greiner Bio-One GmbH, Frickenhausen, Germany). Aliquots of 500 µL each were heated to 98 °C in a "Termite" thermostat (DNA technology, Moscow, Russia) and incubated at 98 °C for 10 min. After cooling the aliquots at 4 °C for 5 min, the preparations were vortexed and re-incubated at 98 °C for 10 min.

2.6. Immunofluorescence Analysis

URRA + SPs or SPs samples formulated in PBS were loaded onto the coverslips coated with formvar. The loaded samples were incubated for 10 min. Then, the excess of the samples was removed. Then, the coverslips were dried in the air for 10 min. The resulting coverslips with the loaded samples were incubated for 1 h with a blocking solution (1% bovine serum albumin (BSA) and 0.05% Tween-20 in PBS); then, for 1 h with 1:50 dilution of polyclonal anti-URRA serum obtained from mice immunised with URRA twice with a two-week interval between immunisations. During the first immunisation, complete Freund's adjuvant was applied, while during the second, the incomplete version was used. For controls without primary antibodies, the coverslips were incubated with a blocking solution for an additional hour instead. The coverslips were washed three times with a washing solution (0.25% BSA and 0.05% Tween-20 in PBS) and subsequently incubated for 1 h with Alexa 546 fluorophore-conjugated secondary antibodies to mouse IgG (Invitrogen, USA; 1:100 dilution in the blocking solution). After that, the coverslips were washed three times with the washing solution, once with PBS, and finally rinsed with pure water and dried in air. Immediately prior to the examination of the samples, the preparations were treated with a photo-protector 1,4-diazabicyclo[2.2.2]octane and studied under an Axiovert 200 M fluorescence microscope (Carl Zeiss, Oberkochen, Germany) equipped with an ORCAII-ERG2 integrated camera (Hamamatsu Photonics, Shizuoka, Japan).

2.7. Immunisation of Mice for the Immunogenicity Studies

To study the immunogenicity of the vaccine candidate and the individually formulated URRA, four groups of BALB/c mice were used. Each group consisted of 25 individual females. The non-immunised control was represented by group 1. Group 2 served as an adjuvant control, and the mice were immunised with SPs and TMV in an amount of 250 µg per dose. Group 3 mice were immunised with individual rotavirus antigen URRA in an amount of 15 µg per dose. Those in Group 4 were immunised with a URRA + SPs composition. For this group, one dose contained 15 µg of URRA protein and 250 µg of SPs; therefore, the ratio of antigen to SPs, by mass, was 15:250. A description of the immunisation groups and the scheme of the experiment are provided in Section 3.4 of the "Results". All samples administered were prepared in PBS. The final volume of one dose was 260 µL/animal. The mice were immunised intramuscularly in a pelvic limb with 260 µL of the solution. The pelvic limb selected for the immunisation was changed between the immunisations. Blood collection and euthanasia were carried out by decapitation [45].

2.8. Ethical Statement

The immunogenicity evaluation experiments on mice were approved by the Ethics Committee of the Sechenov First Moscow State Medical University (Protocol №111 dated 21 October 2022). The experiments for obtaining Serum №1 and Serum №2 were approved by the Ethics Committee of the Rostov Scientific Research Institute of Microbiology and Parasitology (Protocol №05/17 dated 23 May 2023).

2.9. Statistical Analysis

The Mann–Whitney Test with the Holm–Bonferroni correction was used for multiple comparisons. The Mann–Whitney Test was used for a single pairwise comparison. Comparison results were considered to be significant with a probability value (p-value) of less than 0.05. Statistical processing of the results and the plotting of graphs were carried out using the GraphPadPrism 9.1.0 program (GraphPad Software, La Jolla, San Diego, CA, USA).

2.10. Enzyme-Linked Immunosorbent Assay (ELISA) for Titre Measurement

For the measurement of anti-URRA and anti-SPs antibody titres, an ELISA was performed according to the protocol described by Kovalenko et al. (2022) [39]. URRA or SPs were used as an antigen for coating the 96-well plates with a concentration of 10 µg/mL. All sera samples collected were titrated in three-fold serial dilutions, starting from 1:30. Anti-mouse total IgG HRP conjugate (#ab6728, Abcam, Cambridge, UK), anti-mouse IgG1 HRP conjugate (#ab97240, Abcam, Cambridge, UK), anti-mouse IgG2a HRP conjugate (#ab97245, Abcam, Cambridge, UK), anti-mouse IgG2b HRP conjugate (#ab97250, Abcam, Cambridge, UK), or anti-mouse total IgG3 HRP conjugate (#ab97260, Abcam, Cambridge, UK) was used in a dilution of 1:10,000. The serum titre was defined as the reciprocal of the serum dilution at which A_{450} was equal to the mean of the background signal + 3 SD. The background signal was taken as the mean value obtained from 24 wells for each plate separately, into which no test serum was added (neither experimental nor non-immune). If the A_{450} in a 1:30 dilution was below the mean value of the background signal + 3 SD, the serum titre was considered to be 30. If the sera titre was measured in more than one replicate, the geometric mean of these values was used for further calculations and presentation.

3. Results

3.1. Designing Universal Recombinant Rotavirus Antigen URRA

A Universal recombinant rotavirus antigen (URRA) was designed for the purpose of the current research (Figure 1). The amino acid sequence of the URRA is based on the sequence of the rotavirus VP8* protein (one of two subunits formed following rotavirus spike protein VP4 cleavage by trypsin). It consists of two parts. The N-terminus of the protein is represented by a short peptide ep8 corresponding to a linear neutralising B cell epitope with the sequence ^{1}MASLIYRQLL^{10} (1–10 aa of VP8*), which is highly conserved among the vast majority of RVA strains of all genotypes. The ep8 peptide sequence is followed by the sequence of the ΔVP8*P[8] towards the C-terminus of the protein. The ΔVP8*P[8] is the truncated VP8* subunit (65–223 aa of VP8*) obtained using a consensus approach based on a wide range of RVA isolates of genotype P[8], as described previously [28]. The C-terminus of the URRA contains a hexahistidine tag-coding sequence. The estimated molecular weight of the URRA, calculated by amino acid sequence using the ProtParam EXPaSy proteomics server, Swiss Institute of Bioinformatics, http://expasy.org/ (accessed on 6 February 2024), was 20.397 kDa.

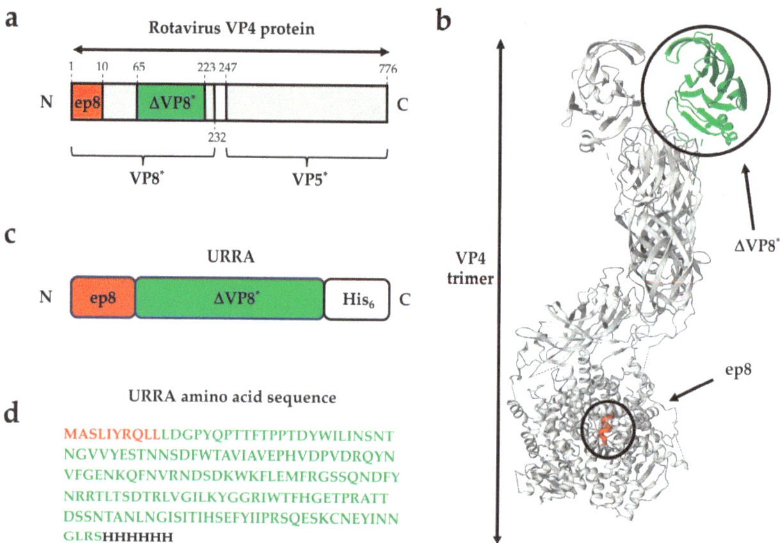

Figure 1. Rotavirus spike protein structure and graphical overview of a URRA. (**a**) Linear diagram of rotavirus VP4 spike protein (1–776 aa) showing the location of fragments VP8* (1–232 aa), VP5* (247–776 aa), ΔVP8* (65–223 aa), epitope ep8 (1–10 aa) (not to scale). VP4 fragment 232–247 aa is being removed during VP4 in vivo proteolysis. (**b**) Structural model of the rotavirus VP4 protein trimer showing fragment ΔVP8* and epitope ep8. The protein VP4 structure was visualised using Mol*, https://molstar.org/ (accessed on 6 February 2024). based on cryo-electron microscopy data (Protein Data Bank [PDB]: 6WXE). (**c**) Schematic representation of the URRA's structure. (**d**) The amino acid sequence of a URRA. All images (**a–d**) use the following colour code: ΔVP8*, green; epitope ep8, red.

3.2. Interaction of URRA with Antisera to Patient-Derived Field Rotavirus Isolates

The ability of the recombinant rotavirus antigen, URRA, to interact with polyclonal antisera to untyped patient-derived field rotavirus isolates circulating in the Russian Federation was evaluated. Serum №1 and Serum №2, described in Section 2.2 of the "Materials and Methods", were used for the analyses. The interaction of the URRA with both antisera was qualitatively assessed by means of an ELISA, as described in Section 2.4 of the "Materials and Methods", and a Western blot analysis. The results of the ELISA and Western blot analysis performed using Serum №1 and the ELISA performed using Serum №2 are presented in Figures 2–4, respectively. For Serum №1, the recognition of the URRA by the serum was demonstrated by both the ELISA (Figure 2, Table S1) and the Western blot analysis (Figure 3a, lane 3). For Serum №2, the recognition of the URRA by the serum was demonstrated by the ELISA (Figure 4, Table S2), but the Western blot analysis did not reveal any interaction between the antigen and the serum.

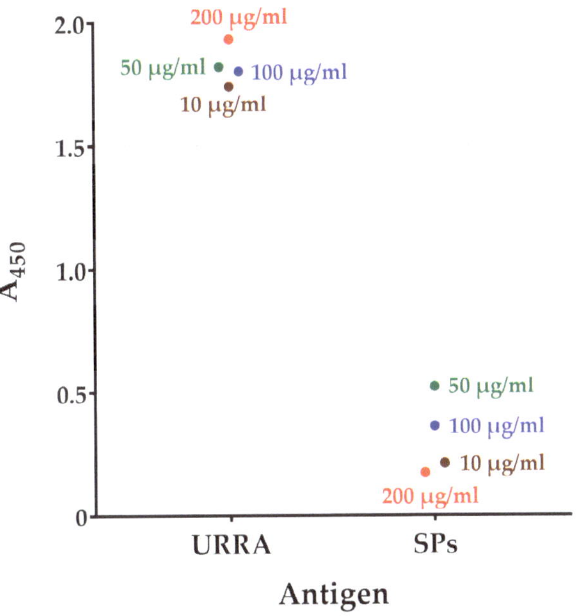

Figure 2. Interaction of URRA with mice antiserum to field rotavirus isolate (Serum №1). The efficiency of interaction was estimated based on absorbance at wavelength 450 nm (A_{450}), as evaluated by indirect ELISA. A parallel experiment with SPs as an antigen was performed as a negative control. Secondary HRP-conjugated antibodies to mouse IgG were used. The analyses were carried out using four different concentrations of antigen for adsorption on a microplate. Antigen concentrations are marked on the figure near the corresponding point (10 µg/mL, 50 µg/mL, 100 µg/mL, and 200 µg/mL, respectively). •, geometric means of A_{450} values for certain antigen concentrations. The analyses were conducted in two replicates for each antigen concentration. The complete data on A_{450}, for each replicate and for all analyses, are presented in Table S1.

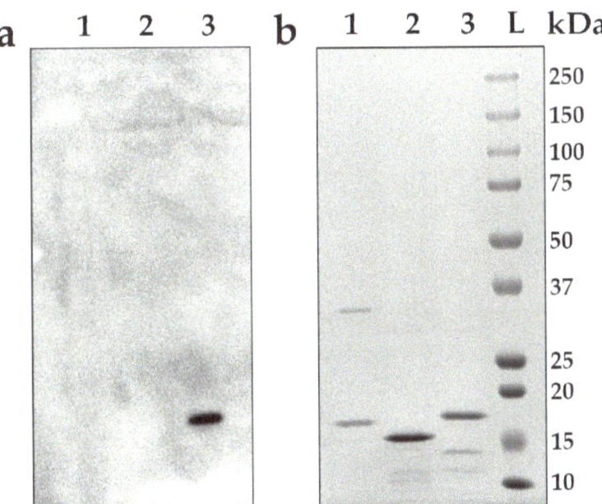

Figure 3. Interaction of URRA with antiserum to field rotavirus isolate (Serum №1). (**a**) Western blot analysis with primary antiserum №1 to field rotavirus isolate (1:500) and secondary HRP-conjugated antibodies (1:10,000). 1—SPs (negative control), 2—heterologous hexahistidine tag-containing recombinant protein (negative control), 3—URRA. (**b**) Electrophoresis analysis in 8–20% SDS-PAGE, staining by Coomassie G-250. 1—SPs (negative control), 2—heterologous hexahistidine tag-containing recombinant protein (negative control), 3—URRA, L—protein molecular weight markers ladder (molecular weights, in kDa, are indicated on the right).

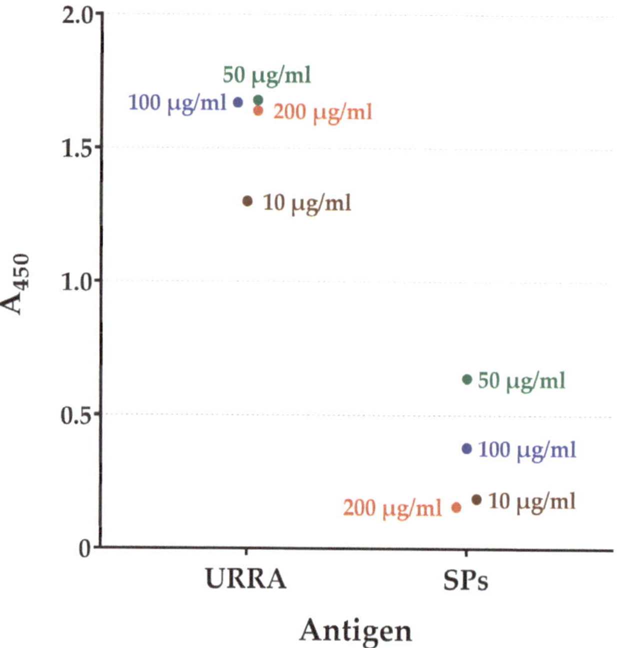

Figure 4. Interaction of URRA with mice antiserum to field rotavirus isolate (Serum №2). The efficiency of interaction was estimated based on absorbance at wavelength 450 nm (A_{450}), as evaluated by indirect ELISA. A parallel experiment with SPs as an antigen was performed as a negative control. Secondary HRP-conjugated antibodies to mouse IgG were used. The analyses were carried out using four different concentrations of antigen for adsorption on a microplate. Antigen concentrations are marked on the figure near the corresponding point (10 µg/mL, 50 µg/mL, 100 µg/mL, and 200 µg/mL, respectively). •, geometric means of A_{450} values for certain antigen concentrations. The analyses were conducted in two replicates for each antigen concentration. The complete data on A_{450}, for each replicate and for all analyses, are presented in Table S2.

3.3. The Adsorption of URRA to SPs

In this study, the authors propose the use of spherical particles (SPs), obtained from the tobacco mosaic virus through heat treatment, as an adjuvant for the URRA in a vaccine candidate formulation. The ability of the URRA to form composition with SPs by adsorbing to their surface was examined by indirect immunofluorescence analysis (Figure 5. Negative controls are presented in Figure S1). The URRA:SPs mass ratio within the composition was 15:250. The results obtained demonstrated that the URRA was able to effectively adsorb to SPs. The presence of the fluorescent signal indicates that the rotavirus antigen URRA maintains its antigenic properties while being adsorbed to SPs.

3.4. The Immunogenicity of Individual URRA and of a Vaccine Candidate (URRA + SPs)

The immunisation of the mice was carried out to evaluate the immunogenicity of the vaccine candidate and the impact of SPs on the immunogenicity of the URRA. The immunisation schedule and a brief description of the groups of mice are presented in Figure 6. Four groups of mice, each consisting of 25 animals, were used in the experiment. Group 1 was not immunised and served as a control group. Groups 2, 3, and 4 were immunised with SPs (which served as an adjuvant control), URRA, or the URRA + SPs composition, respectively. For all groups, ten mice were immunised once, and blood was collected on the 21st day after immunisation; 15 mice were immunised twice, with a 21 day interval between immunisations, and the blood was collected on the 42nd day after the second immunisation.

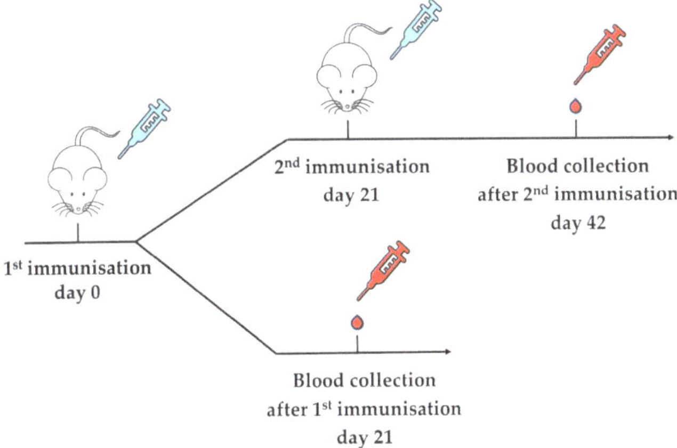

Figure 5. Immunofluorescence analysis of the URRA + SPs composition. (**a**,**b**) are the same image, presented in fluorescence and phase contrast modes, respectively. The URRA + SPs composition was obtained in PBS. The URRA:SPs mass ratio within the composition was 15:250. The URRA + SPs composition was treated with polyclonal mouse anti-URRA serum, obtained using Freund's adjuvants and secondary antibodies conjugated to Alexa Fluor® 546. Scale bars, 5 μm. Negative controls are presented in Figure S1.

Group #	Formulation	Number of animals
1	– (Non-immunised group)	25
2	SPs	25
3	URRA	25
4	URRA + SPs	25*

Figure 6. Immunisation schedule and description of mice groups involved in the experiment to evaluate the immunogenicity of URRA individually and in composition with spherical particles (SPs). The control group (group 1) was not immunised. Other mice groups were immunised intramuscularly. Ten mice in each group were immunised once, and 15 mice were immunised twice, with a 21-day interval between immunisations. Mice in group 2 were immunised with 250 μg of SPs, in group 3

with 15 µg of URRA, and in group 4 with 250 µg of SPs and 15 µg of URRA. All samples were administered with PBS in a total volume of 0.26 mL. SPs, spherical particles obtained by the thermal remodelling of TMV; n, number of mice participating in the corresponding stage of the experiment; *, in group 4, only 14 mice were involved in the second immunisation, while nine mice were involved in blood sampling after the second immunisation.

The sera were obtained from all blood samples collected. Total anti-URRA IgG titres were measured for all sera samples obtained using an ELISA, as described in Section 2.10 of the "Materials and Methods". The results of the ELISA and statistical analyses carried out for sera from mice after the first immunisation are presented in Figure 7 (complete data on titres are presented in Table S3). Those for sera obtained after the second immunisation are presented in Figure 8 (complete data on titres are presented in Table S4). The Wilcoxon–Mann–Whitney Test with the Holm–Bonferroni correction was used for subsequent comparisons for sera obtained from both once- and twice-immunised mice. Anti-URRA total IgG titres elicited by the individually formulated URRA (group 3) or the URRA + SPs composition (group 4, vaccine candidate) were subjected to pairwise comparison with those in the non-immunised group (group 1). Anti-URRA IgG titres elicited by the individual URRA and the URRA + SPs composition were compared with each other to evaluate the impact of SPs on the immunogenicity of the URRA. Finally, titres from groups immunised with SPs-containing formulations (groups 2 and 4) were also compared with each other.

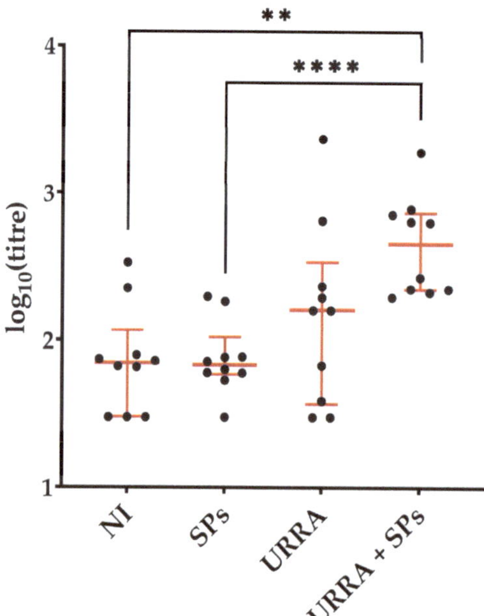

Figure 7. Immunogenicity of URRA individually and in composition with spherical particles (SPs) after the first immunisation. Anti-URRA IgG titres in four groups of mice are presented. The scheme of the study is presented in Figure 6. Sera titres were evaluated using indirect ELISA. The concentration of antigen used for adsorption on a microplate was 10 µg/mL. *p*-values were calculated using the Wilcoxon–Mann–Whitney Test with the Holm–Bonferroni correction. •, IgG titres of individual mice; NI, non-immunised group; **, $p < 0.01$; ****, $p < 0.0001$; —, median. Error bars represent the interquartile range. Formulations used for the immunisation of corresponding groups of mice are marked under the horizontal axis. The complete data on anti-URRA sera titres for corresponding mice are presented in Table S3.

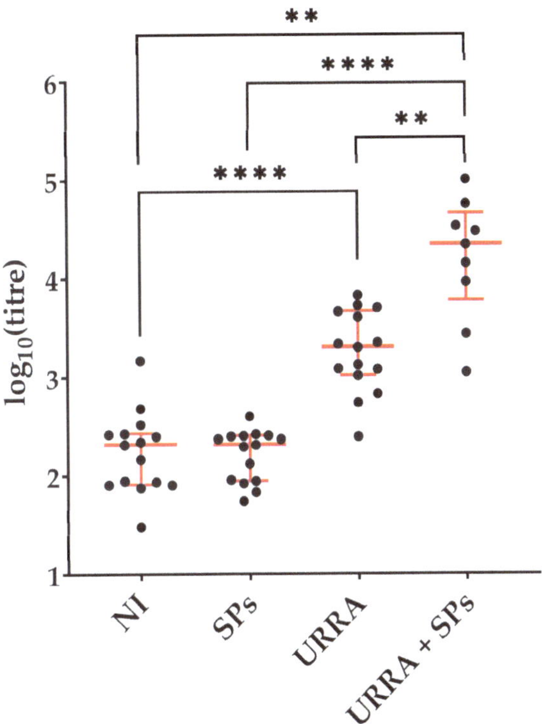

Figure 8. Immunogenicity of URRA individually and in composition with spherical particles (SPs) after the second immunisation. Anti-URRA IgG titres in four groups of mice are presented. The scheme of the study is presented in Figure 6. Sera titres were evaluated using indirect ELISA. The concentration of antigen used for adsorption on a microplate was 10 μg/mL. *p*-values were calculated using the Wilcoxon–Mann–Whitney Test with the Holm–Bonferroni correction. •, IgG titres of individual mice; NI, non-immunised group; **, $p < 0.01$; ****, $p < 0.0001$; —, median. Error bars represent the interquartile range. Formulations used for the immunisation of corresponding groups of mice are marked under the horizontal axis. The complete data on anti-URRA sera titres for corresponding mice are presented in Table S4.

After the first immunisation (Figure 7, Table S3), of all the pairwise comparisons conducted, significant differences in anti-URRA IgG titres were revealed between sera from the group immunised with the URRA + SPs composition (group 4, median titre 4.5×10^2) and two control groups: the non-immunised group (group 1, median titre 6.94×10^1) and the group immunised with SPs (group 2, adjuvant control, median titre 6.73×10^1). There was no significant difference between anti-URRA IgG titres of mice immunised with the individual antigen URRA (group 3, median titre 1.59×10^2) and either the sera titres of mice from the non-immunised group (group 1) or those from the group immunised with the vaccine candidate (group 4).

After the second immunisation (Figure 8, Table S4), the significant difference found in anti-URRA IgG titres between the non-immunised control group (group 1, median titre 2.05×10^2) and the group immunised with the vaccine candidate (group 4, median titre 2.18×10^4) was again demonstrated. In contrast to sera obtained after the first immunisation, a significant difference was demonstrated between anti-URRA IgG titres induced by the URRA (group 3, median titre 1.99×10^3) and those induced by the vaccine candidate, which were 11 times higher. The significant difference between the groups immunised with SPs only (group 2, median titre 2.03×10^2) and those immunised with the URRA + SPs composition was also repeated.

In the present study, the immunogenicity of the vaccine candidate and the individual URRA after the second immunisation was assessed not only by total IgG titres but also separately by IgG1, IgG2a, IgG2b, and IgG3 isotype titres. The ELISA and statistical analyses were carried out in the same manner as they were for the assessment of total IgG titres. The results of the ELISA and statistical analyses are presented in Figure 9 (the complete data on anti-URRA IgG1, IgG2a, IgG2b, and IgG3 sera titres are presented in Tables S5, S6, S7, and S8, respectively). For IgG2a, IgG2b, and IgG3, no significant differences were revealed between any of the groups compared. Both the individual URRA (group 3, median titre 1.5×10^3) and the vaccine candidate (group 4, median titre 4.51×10^4) elicited a significant number of anti-URRA IgG1 antibodies, compared with the non-immunised control group (group 1, median titre 8.65×10^1). The sera IgG1 titres induced by the vaccine candidate were 30 times higher than those induced by the individual URRA. The significant difference in anti-URRA IgG1 titres was also revealed between the group immunised with the vaccine candidate and the group immunised with SPs only (group 2, median titre 1.46×10^2).

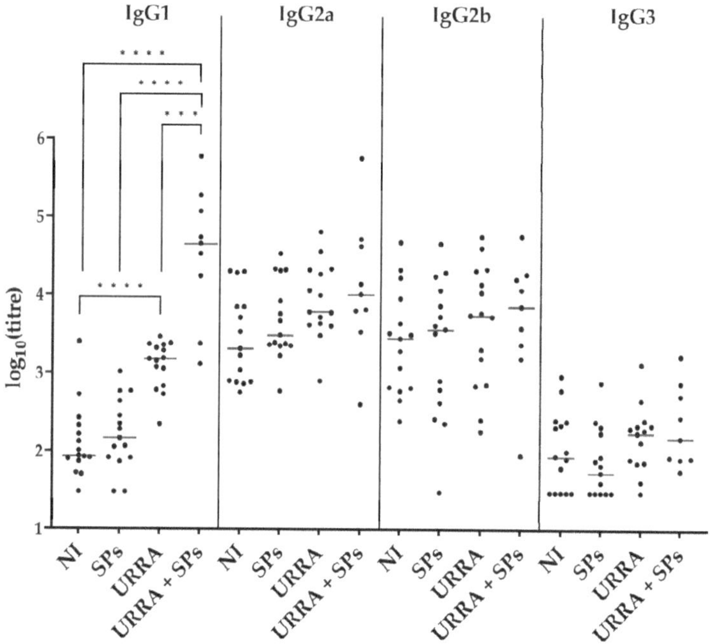

Figure 9. Comparison of anti-URRA IgG isotypes' (IgG1, IgG2a, IgG2b, IgG3) titres in groups after the second immunisation. The scheme of the experiment is presented in Figure 6. Sera titres were evaluated using indirect ELISA. The concentration of antigen used for adsorption on a microplate was 10 µg/mL. *p*-values were calculated using the Wilcoxon–Mann–Whitney Test with the Holm–Bonferroni correction. •, IgG isotype titres of individual mice; NI, non-immunised group; ***, $p < 0.001$; ****, $p < 0.0001$; —, median. Formulations used for the immunisation of corresponding groups of mice are marked under the horizontal axis. The IgG isotypes are marked above the corresponding graphs. The complete data on anti-URRA IgG1, IgG2a, IgG2b and IgG3 sera titres are presented in Tables S5, S6, S7 and S8, respectively.

The ratio of the immune response to an antigen and an adjuvant is an important characteristic of a vaccine. Thus, the anti-URRA and anti-SPs total IgG titres elicited after two immunisations with the vaccine candidate were measured with an ELISA and compared using the Wilcoxon–Mann–Whitney Test. The results are presented in Figure 10 (the complete data on titres are presented in Table S9). It was revealed that the titres of anti-URRA IgG (median titre 2.18×10^4) were 14 times higher than those of anti-SPs IgG (median titre 1.52×10^3).

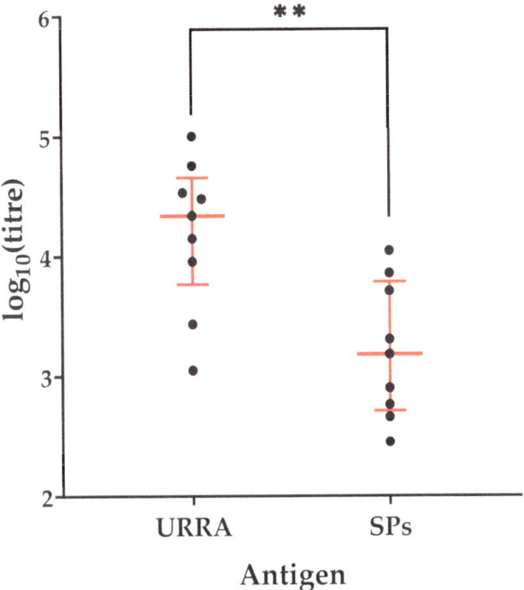

Figure 10. Comparison of anti-URRA and anti-SPs immune responses induced by the URRA + SPs composition after the second immunisation. The scheme of the study is presented in Figure 6. Sera titres were evaluated using indirect ELISA. The concentration of antigen used for adsorption on a microplate was 10 μg/mL. *p*-values were calculated using the Wilcoxon–Mann–Whitney Test. •, IgG titres of individual mice; NI, non-immunised group; **, $p < 0.01$; —, median. Error bars represent the interquartile range. The complete data on anti-SPs sera titres for corresponding mice are presented in Table S9.

4. Discussion

In the present work, a recombinant rotavirus A (RVA) antigen was developed based on the VP8* subunit of the VP4 spike protein. VP4 is one of the components of the rotavirus capsid's outer layer. During infection, VP4 is cleaved by intestinal trypsin into two subunits, VP8* and VP5*, which are rotavirus antigens containing various neutralising epitopes. Here, to obtain a universal RVA antigen URRA, two approaches were combined—the application of the extended protein's fragment and the use of a conserved neutralising epitope. Thus, the antigen URRA is composed of two parts. The larger part is ΔVP8*P[8], the truncated VP8* subunit (65–223 aa of VP8*) obtained using a consensus approach on the base of a wide range of RVA isolates of genotype P[8]. The coding sequence of ΔVP8*P[8] was designed as described previously [28]. The consensus sequence of ΔVP8*P[8] represents the "spike head", the concanavalin-like domain of the VP8* subunit, and contains 159 amino acid residues. Both VP8* and its truncated forms are known to be highly immunogenic and stimulate the production of virus-neutralising antibodies, which makes VP8* a promising basis for a vaccine candidate against RVA [27,31]. One of the most advanced developments in this area is the trivalent subunit vaccine P2-VP8* [31]. It consists of a truncated VP8* segment (65–223 aa) of the three most common RVA genotypes—P[8], P[4], and P[6]—fused to the Th2 epitope of the tetanus toxoid to enhance immunogenicity. This vaccine candidate has successfully passed phase 1 and 2 clinical trials and is currently in phase 3 (ClinicalTrials.gov NCT04010488). The vaccine candidate P2-VP8* was shown to induce a high immune response and elicit the production of neutralising antibodies to the P[4], P[6], and P[8] genotypes of RVA strains [31]. At the same time, 58 P-genotypes of rotavirus have already been described in humans and animals worldwide [46], and more research is required to evaluate the ability of subunit vaccines to provide sufficiently broad cross-serotype protection. A promising approach to extending the vaccine's protection is the application of short peptides that mimic neutralising epitopes, highly conserved among a wide range of RVA strains. For this reason, the lesser, N-terminal part of the URRA is represented by the short peptide ep8 corresponding to a linear neutralising B-cell epitope with the sequence ^{1}MASLIYRQLL^{10} (1–10 aa of VP8*) [34,35], which is highly conserved among the vast majority of RV strains of all genotypes. The sequence of ep8 was obtained as described previously [28]. Peptide vaccines are supposed to induce specific immune responses to pathogens' neutralising epitopes, including those epitopes that are otherwise inaccessible to the immune system. Conversely, owing to the relatively small size of peptides, they are often weakly immunogenic and, therefore, require carrier platforms for delivery and adjuvating [32,33]. In the study by Kovacs-Nolan et al. (2006), the immunogenicity of a peptide with the sequence ^{1}MASLIYRQLL^{10} (1–10 aa VP8*) was evaluated in an animal model [34]. The peptide was covalently linked to a thioredoxin carrier protein fused with the P2 epitope of the tetanus toxin in order to boost immunogenicity. The authors of this paper believe that immunisation with such constructs has the potential to induce broad-spectrum immunity against multiple serological variants of RVA.

The global introduction of rotavirus vaccines has resulted in a considerable reduction in rotavirus-related deaths and hospitalisations. However, recent studies revealed changes in the composition of circulating RVA strains and the emergence of previously uncommon rotavirus strains after the introduction of the vaccines [14,16,17,47]. This may be due to either vaccine selective pressure or natural evolutionary processes [18,19]. One way or another, this situation raises concerns and highlights the necessity to focus on currently relevant RVA strains when developing a new vaccine. For this reason, the authors decided to examine the antigenic specificity of the URRA protein using polyclonal antisera (Serum №1 and Serum №2) to patient-derived field RVA isolates currently circulating in the Russian Federation. The possibility of the URRA interacting with these sera was proven using the Western blot and indirect ELISA analyses. In the ELISA assay, Serum №1 interacted effectively with the URRA (Figure 2). The Western blot analysis performed with Serum №1 also revealed the recognition of the URRA (Figure 3a, lane 3). Thus, the results of the Western blot and ELISA analyses with Serum №1 coincided, revealing that the URRA corresponds to a relevant RVA field isolate. Serum №2 interacted effectively with the URRA in the ELISA assay (Figure 4). However, no recognition of the antigen in the Western blot analysis was observed. In this case, the results of the Western blot analysis and ELISA are contradictory. This may be explained by the fact that Serum №2 recognised some conformational epitopes in the URRA, which are well-preserved under the native conditions of an ELISA but are not preserved under the denaturing conditions of the Western blot analysis. Certainly, the authors' assumption requires further research. Nevertheless, it is concluded that the results of the experiments indicate the ability of the URRA to interact with both antisera and currently circulating field RVA isolates.

The authors proposed a vaccine candidate, URRA + SPs, that represents the composition of URRA and spherical particles (SPs) generated by the thermally-induced rearrangement of the tobacco mosaic virus (TMV). Previous studies have demonstrated the properties of TMV SPs as a prospective safe and biodegradable adjuvant and a platform for the adsorption and stabilisation of various antigens [38–44]. To examine the antigenic properties of the URRA in the URRA + SPs composition, immunofluorescence analysis with primary polyclonal sera to the URRA was carried out, revealing that the URRA adsorbs effectively to SPs while maintaining antigenic specificity. This suggests that SPs could be used as a platform for the URRA-based vaccine candidate. The immunogenicity of the vaccine candidate was evaluated and compared with the immunogenicity of the individual URRA. The experiments were carried out in mice. After the first immunisation, the anti-URRA total IgG titres induced by the URRA + SPs composition were significantly higher than the titres elicited in control groups. In contrast, the differences between anti-URRA IgG titres elicited by individual URRA and the titres elicited in control groups were not significant. This indicates the low immunogenicity of the URRA by itself and points to the necessity of using adjuvants in general and SPs as an adjuvant in particular. After the second immunisation, it was revealed that anti-URRA IgG titres elicited by the URRA + SPs were significantly higher than those elicited by the URRA individually. This indicates that SPs increase the immunogenicity of the URRA considerably when co-administered and that SPs could serve as an appropriate adjuvant for the URRA in the vaccine candidate. Anti-URRA IgG titres induced by vaccine candidate, URRA + SPs, after the second immunisation were 48 times higher than anti-URRA titres in the corresponding group after the first immunisation. The results have revealed that at least double immunisation with the vaccine candidate URRA + SPs is required to provide high immunogenicity. Experiments studying the titres of IgG isotypes separately demonstrated that immune responses to both the vaccine candidate and individual antigens are mostly represented by the IgG1 isotype. This indicates that SPs do not alter the polarisation of the immune response to rotavirus antigen URRA. According to various estimates, a natural RVA infection in mice (EDIM, epizootic diarrhoea of infant mice) induces an IgG1 predominant [48], or IgG1/IgG2 balanced [49], immune response.

Protein adjuvants and carriers are known to be able to activate a self-immune response. In the current study, the immunogenicity of SPs was evaluated after double immunisation with the URRA + SPs vaccine candidate. Total IgG titres to SPs were shown to be 14 times less than to the URRA after the second immunisation with the URRA + SPs composition. The prevalence of anti-URRA antibodies confirms the possibility of applying SPs as an adjuvant for a rotavirus vaccine candidate. These results are consistent with previous studies on SPs-based vaccines against rubella, COVID-19, and anthrax [38–40]. In all these cases, IgG titres to SPs were significantly lower than to the antigen of interest. A possible limitation of using URRA + SPs as a vaccine composition is the immunity to SPs adjuvant, which is induced during the first immunisation and can potentially reduce the effectiveness of each subsequent one. Pre-existing immunity to a platform or an adjuvant is a serious issue in the vaccine research field. However, it is known that the problem with the effectiveness of booster immunisations does not always arise. Some studies on protein immunopreparations based on plant viruses demonstrated that pre-existing immunity did not reduce the effectiveness of such drugs [50,51], or even increased it [52]. In the current research, a considerable increase in anti-URRA antibody titres was detected after the second immunisation with the URRA + SPs composition compared to the first immunisation (Tables S4 and S3, respectively). Presumably, these data indicate that in the case of SPs, the antibodies to a platform might not affect the effectiveness of further immunisations.

5. Conclusions

In this research, a sequence of the universal recombinant rotavirus antigen, URRA, was designed based on a wide variety of rotavirus strains, combining two approaches to achieve the goal of creating an antigen able to provide an effective immune response.

In serological studies, URRA demonstrated consistency with rotavirus strains circulating in the Russian Federation. This makes this protein a promising basis for a recombinant rotavirus vaccine. The immunogenicity of the URRA was assessed in individual form and when combined with spherical particles (SPs) obtained from the tobacco mosaic virus. Individual URRA was only able to elicit anti-URRA titres after two immunisations. At the same time, when paired with SPs, the URRA induced a significant immune response, even after a single immunisation. Moreover, SPs were able to enhance the immunogenicity of the URRA after two immunisations. Combined with the fact that the immune response to spherical particles themselves was shown to be significantly lower than that to the rotavirus antigen, the results obtained indicate that they may be considered an appropriate adjuvant for the URRA. Therefore, the recombinant rotavirus antigen, URRA, paired with spherical particles obtained from the tobacco mosaic virus in a 15:250 mass ratio might provide an elegant solution to the challenge of recombinant rotavirus vaccine development.

6. Limitations of the Current Study

The present study mainly focuses on the assessment of the immunological properties of the vaccine candidate, including the ability to interact with the antisera to existing rotavirus strains and immunogenicity. In this regard, this research has several strengths and limitations. The main strengths of the study are the demonstrated ability of the URRA to interact with two patient-derived strains of rotavirus circulating in the Russian Federation and the detailed analyses of the total IgG and all IgG subclasses titres elicited after the two-step immunisation of mice with the individual URRA and with a vaccine candidate. In terms of limitations, first and foremost, the immunogenicity data has to be additionally supported by the protectiveness assessment, which is the subject of further investigation. Secondly, comparing the adjuvant effect of SPs on the immunogenicity of the URRA to the effect of other adjuvants might provide an overall picture for the further rotavirus vaccine design. Finally, the assessment of IgA titres in the vaccinated mice may enable us to draw deeper conclusions about the effectiveness of the vaccine candidate.

Supplementary Materials: The following supporting information can be downloaded at https://www.mdpi.com/article/10.3390/v16030438/s1, Figure S1: Controls for immunofluorescence analysis of the URRA + SPs composition. (a–f) are the same image, presented in fluorescence and phase contrast modes, respectively. All samples were formulated in PBS. (a,b). The URRA + SPs composition. URRA:SPs mass ratio within the composition was 15:250. The URRA + SPs composition was only treated with secondary antibodies conjugated to Alexa Fluor® 546. (c,d). SPs. SPs were treated with polyclonal mouse anti-URRA serum, obtained using Freund's adjuvant, and secondary antibodies conjugated to Alexa Fluor® 546. (e,f). SPs. SPs were only treated with secondary antibodies conjugated to Alexa Fluor® 546. Scale bars: 5 μm; Table S1: Absorbance at wavelength 450 nm (A_{450}), evaluated by indirect ELISA performed with mice antiserum to field rotavirus isolate (Serum №1) and URRA as an antigen. A parallel experiment with SPs antigen was used as a negative control. The analyses were carried out using four different concentrations of antigen for adsorption on a microplate (10 μg/mL, 50 μg/mL, 100 μg/mL and 200 μg/mL). The analyses were conducted in two replicates for each antigen concentration; Table S2: Absorbance at wavelength 450 nm (A_{450}), evaluated by indirect ELISA performed with mice antiserum to field rotavirus isolate (Serum №2) and URRA as an antigen. A parallel experiment with SPs antigen was used as a negative control. The analyses were carried out using four different concentrations of antigen for adsorption on a microplate (10 μg/mL, 50 μg/mL, 100 μg/mL and 200 μg/mL). The analyses were conducted in two replicates for each antigen concentration; Table S3: Total IgG titres to URRA in blood sera of mice after the first immunisation. Groups of mice were immunised with SPs (group 2), URRA (group 3) and URRA + SPs (group 4). The control group (group 1) was not immunised. The scheme of the experiment is presented in Figure 6. Sera titres were evaluated using indirect ELISA. The concentration of antigen used for adsorption on a microplate was 10 μg/mL; Table S4: Total IgG titres of URRA in blood sera of mice after the second immunisation. Groups of mice were immunised with SPs (group 2), URRA (group 3) and URRA + SPs (group 4). The control group (group 1) was not immunised. The scheme of the experiment is presented in Figure 6. Sera titres were evaluated using indirect ELISA. The concentration of antigen used for adsorption on a microplate was 10 μg/mL; Table S5: IgG1 titres of URRA in blood sera of mice after the second immunisation. 6 Groups of mice were immunised with SPs (group 2), URRA (group 3) and URRA + SPs (group 4). The control group (group 1) was not immunised. The scheme of the experiment is presented in Figure 6. Sera titres were evaluated using indirect ELISA. The concentration of antigen used for adsorption on a microplate was 10 μg/mL; Table S6: IgG2a titres of URRA in blood sera of mice after the second immunisation. Groups of mice were immunised with SPs (group 2), URRA (group 3) and URRA + SPs (group 4). The control group (group 1) was not immunised. The scheme of the experiment is presented in Figure 6. Sera titres were evaluated using indirect ELISA. The concentration of antigen used for adsorption on a microplate was 10 μg/mL; Table S7: IgG2b titres of URRA in blood sera of mice after the second immunisation. Groups of mice were immunised with SPs (group 2), URRA (group 3) and URRA + SPs (group 4). The control group (group 1) was not immunised. The scheme of the experiment is presented in Figure 6. Sera titres were evaluated using indirect ELISA. The concentration of antigen used for adsorption on a microplate was 10 μg/mL; Table S8: IgG3 titres of URRA in blood sera of mice after the second immunisation. Groups of mice were immunised with

SPs (group 2), URRA (group 3) and URRA + SPs (group 4). The control group (group 1) was not immunised. The scheme of the experiment is presented in Figure 6. Sera titres were evaluated using indirect ELISA. The concentration of antigen used for adsorption on a microplate was 10 µg/mL; Table S9: Total IgG titres of SPs and of URRA in blood sera of mice immunised with the URRA + SPs composition (group 4) after the second immunisation. The scheme of the experiment is presented in Figure 6. Sera titres were evaluated using indirect ELISA. The concentration of antigen used for adsorption on a microplate was 10 µg/mL.

Author Contributions: Conceptualization, D.L.G., N.S.K., E.A.E., E.M.R. and O.V.K.; Data curation, D.L.G., N.S.K. and E.A.E.; Formal analysis, D.L.G. and N.S.K.; Funding acquisition, O.V.K.; Investigation, D.L.G., N.S.K., E.M.R., M.V.A., M.V.K., E.P.K. and N.A.N.; Methodology, D.L.G., N.S.K., E.A.E., E.M.R., O.A.K., M.V.A., M.V.K., E.P.K., T.I.T., N.A.N. and O.V.K.; Project administration, E.A.E., N.A.N. and O.V.K.; Resources, T.I.T. and O.V.K.; Supervision, O.V.K.; Visualization, D.L.G. and N.S.K.; Writing—original draft, D.L.G. and N.S.K.; Writing—review & editing, E.A.E., E.M.R., O.A.K., M.V.A., M.V.K., E.P.K., T.I.T., N.A.N. and O.V.K. All authors have read and agreed to the published version of the manuscript.

Funding: This work was supported by the Russian Science Foundation (Grant number 23-74-01008).

Institutional Review Board Statement: Not applicable.

Informed Consent Statement: Not applicable.

Data Availability Statement: All the relevant data are provided in this paper and in Supplementary Materials.

Conflicts of Interest: The authors declare no conflict of interest. The funders had no role in the design of the study, in the collection, analyses, or interpretation of data, in the writing of the manuscript, or in the decision to publish the results.

References

1. Troeger, C.; Khalil, I.A.; Rao, P.C.; Cao, S.; Blacker, B.F.; Ahmed, T.; Kang, G. Rotavirus vaccination and the global burden of rotavirus diarrhea among children younger than 5 years. *JAMA Pediatr.* **2018**, *172*, 958–965. [CrossRef]
2. Varghese, T.; Gagandeep, K.; Steele, A. Understanding Rotavirus Vaccine Efficacy and Effectiveness in Countries with High Child Mortality. *Vaccines* **2022**, *10*, 346. [CrossRef]
3. Burnett, E.; Parashar, U.; Tate, J.E. Real-world effectiveness of rotavirus vaccines, 2006–2019: A literature review and meta-analysis. *Lancet Glob. Health* **2020**, *8*, 1195–1202. [CrossRef]
4. Weintraub, E.S.; Baggs, J.; Duffy, J.; Vellozzi, C.; Belongia, E.A.; Irving, S.; Jackson, L.A. Risk of intussusception after monovalent rotavirus vaccination. *N. Engl. J. Med.* **2014**, *370*, 513–519. [CrossRef] [PubMed]
5. Carlin, J.B.; Macartney, K.K.; Lee, K.J.; Quinn, H.E.; Buttery, J.; Lopert, R.; Bines, J.; McIntyre, P.B. Intussusception risk and disease prevention associated with rotavirus vaccines in Australia's National Immunization Program. *Clin. Infect. Dis.* **2013**, *57*, 1427–1434. [CrossRef]
6. Patel, M.M.; Lopez-Collada, V.R.; Bulhoes, M.M.; De Oliveira, L.H.; Bautista Marquez, A.; Flannery, B. Intussusception risk and health benefits of rotavirus vaccination in Mexico and Brazil. *N. Engl. J. Med.* **2011**, *364*, 2283–2292. [CrossRef] [PubMed]
7. Stowe, J.; Andrews, N.; Ladhani, S.; Miller, E. The risk of intussusception following monovalent rotavirus vaccination in England: A self-controlled case- series evaluation. *Vaccine* **2016**, *34*, 3684–3689. [CrossRef]
8. Kaplon, J.; Cros, G.; Ambert-Balay, K.; Leruez-Ville, M.; Chomton, M.; Fremy, C. Rotavirus vaccine virus shedding; viremia and clearance in infants with severe combined immune deficiency. *Pediatr. Infect. Dis.* **2015**, *34*, 326–328. [CrossRef]
9. Klinkenberg, D.; Blohm, M.; Hoehne, M.; Mas Marques, A.; Malecki, M.; Schildgen, V. Risk of Rotavirus Vaccination for Children with SCID. *Pediatr. Infect. Dis.* **2015**, *34*, 114–115. [CrossRef]
10. Palau, M.J.; Vescina, C.M.; Regairaz, L.; Cabanillas, D.; Stupka, J.A.; Degiuseppe, J.I. Persistent infection with a rotavirus vaccine strain in a child suffering from Severe Combined Immunodeficiency in Argentina. *Rev. Argent. Microbiol.* **2021**, *53*, 216–219. [CrossRef]
11. Simsek, C.; Bloemen, M.; Jansen, D.; Descheemaeker, P.; Reynders, M.; Van Ranst, M.; Matthijnssens, J. Rotavirus vaccine-derived cases in Belgium: Evidence for reversion of attenuating mutations and alternative causes of gastroenteritis. *Vaccine* **2022**, *35*, 5114–5125. [CrossRef]
12. Boom, J.A.; Sahni, L.C.; Payne, D.C.; Gautam, R.; Lyde, F.; Mijatovic- Rustempasic, S.; Bowen, M.D.; Tate, J.E.; Rench, M.A.; Gentsch, J.R.; et al. Symptomatic infection and detection of vaccine and vaccine- reassortant rotavirus strains in 5 children: A case series. *J. Infect. Dis.* **2012**, *206*, 1275–1279. [CrossRef]
13. Bucardo, F.; Rippinger, C.M.; Svensson, L.; Patton, J.T. Vaccine-derived NSP2 segment in rotaviruses from vaccinated children with gastroenteritis in Nicaragua. *Infect. Genet. Evol.* **2012**, *12*, 1282–1294. [CrossRef]

14. Dóró, R.; László, B.; Martella, V.; Leshem, E.; Gentsch, J.; Parashar, U.; Banyai, K. Review of global rotavirus strain prevalence data from six years post vaccine licensure surveillance: Is there evidence of strain selection from vaccine pressure? *Infect. Genet. Evol.* **2014**, *28*, 446–461. [CrossRef]

15. Payne, D.C.; Edwards, K.M.; Bowen, M.D.; Keckley, E.; Peters, J.; Esona, M.D.; Teel, E.N.; Kent, D.; Parashar, U.D.; Gentsch, J.R. Sibling transmission of vaccine-derived rotavirus (RotaTeq) associated with rotavirus gastroenteritis. *J. Pediatr.* **2010**, *125*, 438–441. [CrossRef] [PubMed]

16. Lucien, M.A.B.; Esona, M.D.; Pierre, M.; Joseph, G.; Rivière, C.; Leshem, E.; Aliabadi, N.; Desormeaux, A.M.; Andre-Alboth, J.; Fitter, D.L.; et al. Diversity of rotavirus strains circulating in Haiti before and after introduction of monovalent vaccine. *IJID Reg.* **2022**, *14*, 146–151. [CrossRef]

17. Hungerford, D.; Allen, D.J.; Nawaz, S.; Collins, S.; Ladhani, S.; Vivancos, R.; Iturriza-Gómara, M. Impact of rotavirus vaccination on rotavirus genotype distribution and diversity in England, September 2006 to August 2016. *Euro Surveill.* **2019**, *24*, 1700774. [CrossRef]

18. Cates, J.E.; Amin, A.B.; Tate, J.E.; Lopman, B.; Parashar, U. Do Rotavirus Strains Affect Vaccine Effectiveness? A Systematic Review and Meta-analysis. *J. Pediatr. Infect. Dis.* **2021**, *40*, 1135–1143. [CrossRef] [PubMed]

19. Amin, A.B.; Cates, J.E.; Liu, Z.; Wu, J.; Ali, I.; Rodriguez, A.; Panjwani, J.; Tate, J.E.; Lopman, B.A.; Parashar, U.D. Rotavirus genotypes in the post-vaccine era: A systematic review and meta-analysis of global, regional, and temporal trends in settings with and without rotavirus vaccine introduction. *J. Infect. Dis.* **2023**, jiad403. [CrossRef] [PubMed]

20. Kirkwood, C.D.; Ma, L.F.; Carey, M.E.; Steele, A.D. The rotavirus vaccine development pipeline. *Vaccine* **2019**, *37*, 7328–7335. [CrossRef]

21. Song, J.M. Parenteral, non-live rotavirus vaccine: Recent history and future perspective. *Clin. Exp. Vaccine Res.* **2021**, *10*, 203–210. [CrossRef]

22. Matthijnssens, J.; Ciarlet, M.; Rahman, M.; Attoui, H.; Bányai, K.; Estes, M.K.; Gentsch, J.R.; Iturriza-Gómara, M.; Kirkwood, C.D.; Martella, V.; et al. Recommendations for the classification of group A rotaviruses using all 11 genomic RNA segments. *Arch. Virol.* **2008**, *153*, 1621–1629. [CrossRef]

23. Li, Y.; Xue, M.; Yu, L.; Luo, G.; Yang, H.; Jia, L.; Zeng, Y.; Li, T.; Ge, S.; Xia, N. Expression and characterization of a novel truncated rotavirus VP4 for the development of a recombinant rotavirus vaccine. *Vaccine* **2018**, *36*, 2086–2092. [CrossRef] [PubMed]

24. Khodabandehloo, M.; Shahrabadi, M.S.; Keyvani, H.; Bambai, B.; Sadigh, Z. Recombinant outer capsid glycoprotein (VP7) of rotavirus expressed in insect cells induces neutralizing antibodies in rabbit. *Iran. J. Public Health* **2012**, *41*, 73–84.

25. Wen, X.; Cao, D.; Jones, R.W.; Li, J.; Szu, S.; Hoshino, Y. Construction and characterization of human rotavirus recombinant VP8* subunit parenteral vaccine candidates. *Vaccine* **2012**, *30*, 6121–6126. [CrossRef] [PubMed]

26. Xia, M.; Huang, P.; Jiang, X.; Tan, M. Immune response and protective efficacy of the S particle presented rotavirus VP8* vaccine in mice. *Vaccine* **2019**, *37*, 4103–4110. [CrossRef]

27. Groome, M.J.; Koen, A.; Fix, A.; Page, N.; Jose, L.; Madhi, S.A.; McNeal, M.; Dally, L.; Cho, I.; Power, M.; et al. Safety and immunogenicity of a parenteral P2-VP8-P[8] subunit rotavirus vaccine in toddlers and infants in South Africa: A randomised; double-blind; placebo-controlled trial. *Lancet Infect. Dis.* **2017**, *17*, 843–853. [CrossRef]

28. Kondakova, O.A.; Ivanov, P.A.; Baranov, O.A.; Ryabchevskaya, E.M.; Arkhipenko, M.V.; Skurat, E.V.; Evtushenko, E.A.; Nikitin, N.A.; Karpova, O.V. Novel antigen panel for modern broad-spectrum recombinant rotavirus A vaccine. *Clin. Exp. Vaccine Res.* **2021**, *10*, 123–131. [CrossRef]

29. Nair, N.; Feng, N.; Blum, L.K.; Sanyal, M.; Ding, S.; Jiang, B.; Sen, A.; Morton, J.M.; He, X.S.; Robinson, W.H.; et al. VP4-and VP7-specific antibodies mediate heterotypic immunity to rotavirus in humans. *Sci. Transl. Med.* **2017**, *9*, 5434. [CrossRef]

30. Xue, M.; Yu, L.; Che, Y.; Lin, H.; Zeng, Y.; Fang, M.; Li, T.; Ge, S.; Xia, N. Characterization and protective efficacy in an animal model of a novel truncated rotavirus VP8 subunit parenteral vaccine candidate. *Vaccine* **2015**, *33*, 2606–2613. [CrossRef]

31. Groome, M.J.; Fairlie, L.; Morrison, J.; Fix, A.; Koen, A.; Masenya, M.; Jose, L.; Shabir, M.; Page, N.; McNeal, M.; et al. Safety and immunogenicity of a parenteral trivalent P2-VP8 subunit rotavirus vaccine: A multisite; randomised; double-blind; placebo-controlled trial. *Lancet Infect. Dis.* **2020**, *20*, 851–863. [CrossRef]

32. Li, W.; Joshi, M.D.; Singhania, S.; Ramsey, K.H.; Murthy, A.K. Peptide Vaccine: Progress and Challenges. *Vaccines* **2014**, *2*, 515–536. [CrossRef]

33. Malonis, R.J.; Lai, J.R.; Vergnolle, O. Peptide-Based Vaccines: Current Progress and Future Challenges. *Chem. Rev.* **2020**, *120*, 3210–3229. [CrossRef]

34. Kovacs-Nolan, J.; Mine, Y. Tandem copies of a human rotavirus VP8 epitope can induce specific neutralizing anti- bodies in BALB/c mice. *Biochim. Et Biophys. Acta (BBA)-Gen. Subj.* **2006**, *1760*, 1884–1893. [CrossRef]

35. Kovacs-Nolan, J.; Yoo, D.; Mine, Y. Fine mapping of sequential neutralization epitopes on the subunit protein VP8 of human rotavirus. *Biochem. J.* **2003**, *376*, 269–275. [CrossRef]

36. Sadiq, A.; Khan, J. Rotavirus in developing countries: Molecular diversity, epidemiological insights, and strategies for effective vaccination. *Front. Microbiol.* **2024**, *5*, 1297269. [CrossRef]

37. Omatola, C.A.; Ogunsakin, R.E.; Olaniran, A.O. Prevalence, Pattern and Genetic Diversity of Rotaviruses among Children under 5 Years of Age with Acute Gastroenteritis in South Africa: A Systematic Review and Meta-Analysis. *Viruses* **2021**, *13*, 1905. [CrossRef]

38. Trifonova, E.A.; Zenin, V.A.; Nikitin, N.A.; Yurkova, M.S.; Ryabchevskaya, E.M.; Putlyaev, E.V.; Donchenko, E.K.; Kondakova, O.A.; Fedorov, A.N.; Atabekov, J.G.; et al. Study of rubella candidate vaccine based on a structurally modified plant virus. *Antivir. Res.* **2017**, *144*, 27–33. [CrossRef] [PubMed]

39. Kovalenko, A.O.; Ryabchevskaya, E.M.; Evtushenko, E.A.; Manukhova, T.I.; Kondakova, O.A.; Ivanov, P.A.; Arkhipenko, M.V.; Gushchin, V.A.; Nikitin, N.A.; Karpova, O.V. Vaccine Candidate Against COVID-19 Based on Structurally Modified Plant Virus as an Adjuvant. *Front. Microbiol.* **2022**, *13*, 845316. [CrossRef] [PubMed]

40. Granovskiy, D.L.; Ryabchevskaya, E.M.; Evtushenko, E.A.; Kondakova, O.A.; Arkhipenko, M.V.; Kravchenko, T.B.; Bakhteeva, I.V.; Timofeev, V.S.; Nikitin, N.A.; Karpova, O.V. New formulation of a recombinant anthrax vaccine stabilised with structurally modified plant viruses. *Front. Microbiol.* **2022**, *13*, 1003969. [CrossRef] [PubMed]

41. Bruckman, M.A.; Randolph, L.N.; VanMeter, A.; Hern, S.; Shoffstall, A.J.; Taurog, R.E.; Steinmetz, N.F. Biodistribution, pharmacokinetics, and blood compatibility of native and PEGylated tobacco mosaic virus nano-rods and -spheres in mice. *Virology* **2014**, *449*, 163–173. [CrossRef] [PubMed]

42. Nikitin, N.A.; Zenin, V.A.; Trifonova, E.A.; Ryabchevskaya, E.M.; Kondakova, O.A.; Fedorov, A.N.; Atabekov, J.G.; Karpova, O.V. Assessment of structurally modified plant virus as a novel adjuvant in toxicity studies. *Regul. Toxicol. Pharmacol.* **2018**, *97*, 127–133. [CrossRef] [PubMed]

43. Nikitin, N.A.; Malinin, A.S.; Rakhnyanskaya, A.A.; Trifonova, E.A.; Karpova, O.V.; Yaroslavov, A.A.; Atabekov, J.G. Use of a polycation spacer for noncovalent immobilization of albumin on thermally modified virus particles. *Polym. Sci. Ser. A* **2011**, *53*, 1026–1031. [CrossRef]

44. Trifonova, E.A.; Nikitin, N.A.; Kirpichnikov, M.P.; Karpova, O.V.; Atabekov, J.G. Obtaining and characterization of spherical particles—New biogenic platforms. *Moscow Univ. Biol. Sci. Bull.* **2015**, *70*, 194–197. [CrossRef]

45. Society of Laboratory Animal Science. Specialist information from the Committee of Animal Welfare Officers (GV-SOLAS) and Working Group 4 in TVT Recommendation for Blood Sampling in Laboratory Animals, Especially Small Laboratory Animals. 2017. Available online: https://www.gv-solas.de/wp-content/uploads/2017/03/tie_blutentnahme17_e.pdf (accessed on 6 February 2024).

46. Rotavirus Classification Working Group: RCWG. Available online: https://rega.kuleuven.be/cev/viralmetagenomics/virus-classification/rcwg (accessed on 6 February 2024).

47. João, E.D.; Munlela, B.; Chissaque, A.; Chilaúle, J.; Langa, J.; Augusto, O.; Boene, S.S.; Anapakala, E.; Sambo, J.; Guimarães, E.; et al. Molecular Epidemiology of Rotavirus A Strains Pre- and Post-Vaccine (Rotarix®) Introduction in Mozambique, 2012–2019: Emergence of Genotypes G3P[4] and G3P[8]. *Pathogens* **2020**, *9*, 671. [CrossRef]

48. Knipping, K.; McNeal, M.M.; Crienen, A.; van Amerongen, G.; Garssen, J.; van't Land, B. A gastrointestinal rotavirus infection mouse model for immune modulation studies. *Virol. J.* **2011**, *8*, 109. [CrossRef]

49. Reimerink, J.H.J.; Boshuizen, J.A.; Einerhand, A.W.C.; Duizer, E.; van Amerongen, G.; Schmidt, N.; Koopmans, M.P.G. Systemic immune response after rotavirus inoculation of neonatal mice depends on source and level of purification of the virus: Implications for the use of heterologous vaccine candidates. *J. Gen. Virol.* **2007**, *88*, 604–612. [CrossRef]

50. Rioux, G.; Babin, C.; Majeau, N.; Leclerc, D. Engineering of papaya mosaic virus (PapMV) nanoparticles through fusion of the HA11 peptide to several putative surface-exposed sites. *PLoS ONE* **2012**, *7*, 31925. [CrossRef]

51. Koo, M.; Bendahmane, M.; Lettieri, G.A.; Paoletti, A.D.; Lane, T.E.; Fitchen, J.H.; Buchmeier, M.J.; Beachy, R.N. Protective immunity against murine hepatitis virus (MHV) induced by intranasal or subcutaneous administration of hybrids of tobacco mosaic virus that carries an MHV epitope. *Proc. Natl. Acad. Sci. USA* **1999**, *96*, 7774–7779. [CrossRef]

52. Shukla, S.; Wang, C.; Beiss, V.; Steinmetz, N.F. Antibody Response against Cowpea Mosaic Viral Nanoparticles Improves In Situ Vaccine Efficacy in Ovarian Cancer. *ACS Nano* **2020**, *14*, 2994–3003. [CrossRef] [PubMed]

Review

Update on Early-Life T Cells: Impact on Oral Rotavirus Vaccines

Catherine Montenegro, Federico Perdomo-Celis and Manuel A. Franco *

Instituto de Genética Humana, Facultad de Medicina, Pontificia Universidad Javeriana, Bogotá 110221, Colombia; ca.montenegro@javeriana.edu.co (C.M.); perdomo_federico@javeriana.edu.co (F.P.-C.)
* Correspondence: mafranco@javeriana.edu.co

Abstract: Rotavirus infection continues to be a significant public health problem in developing countries, despite the availability of several vaccines. The efficacy of oral rotavirus vaccines in young children may be affected by significant immunological differences between individuals in early life and adults. Therefore, understanding the dynamics of early-life systemic and mucosal immune responses and the factors that affect them is essential to improve the current rotavirus vaccines and develop the next generation of mucosal vaccines. This review focuses on the advances in T-cell development during early life in mice and humans, discussing how immune homeostasis and response to pathogens is established in this period compared to adults. Finally, the review explores how this knowledge of early-life T-cell immunity could be utilized to enhance current and novel rotavirus vaccines.

Keywords: rotavirus; vaccine; T cell; regulatory T cell; early life; layered immunity

1. Introduction

Despite having a very important impact on human health, oral rotavirus (RV) vaccines need improvement [1], especially in developing countries where they underperform and where RV diarrhea represents a high burden of disease [2,3]. In these settings, protective immune responses against pathogens are influenced, amongst other factors, by malnutrition and chronic gut inflammation (associated with microbial dysbiosis) that manifest as environmental enteric dysfunction (EED). In turn, EED is thought to affect oral vaccine immunogenicity and efficacy, and we currently do not have validated EED biomarkers [4–8]. In addition, the heterogeneity in the results of EED studies in children receiving oral RV vaccines highlights the need for more preclinical animal studies to support clinical research [8]. Oral RV vaccines face an important challenge related to the age of the intended recipients they must protect (infants and young children). In adult mice, CD8$^+$ T cells are important in mediating short-term protection against reinfection, while antibodies (whose production largely depends on CD4$^+$ T-cell help) mediate long-term protection [9]. Thus, T cells are key players in antiviral protection. Very little is known about immunity and T-cell responses to RV in early-life mice, particularly in local intestinal responses [10,11], and recent advances in this area suggest that early-life T cells are different from adult T cells [12,13]. Here, we will review recent advances in early-life T-cell responses in mice (a tractable model in which most T-cell studies have been performed) and humans that may impact the improvement or development of new RV vaccines. We will focus on T cells expressing the αβ antigen receptor (TCR), both effector and regulatory (Treg) cells, and, when available, we will highlight specific aspects of mucosal/intestinal T cells in this age group. We will refer to the neonatal and infant periods in humans, while we will use the term early life to refer to the neonatal and pre-weaning periods in mice.

Excellent recent reviews on T-cell responses to RV [14], immunity to intestinal viruses including RV [15], immunity in neonates [16,17], CD8$^+$ T-cell immunity in early life [12], intestinal immunity in early life [13], and intestinal Treg cells [18] have recently been

Citation: Montenegro, C.; Perdomo-Celis, F.; Franco, M.A. Update on Early-Life T Cells: Impact on Oral Rotavirus Vaccines. *Viruses* **2024**, *16*, 818. https://doi.org/10.3390/v16060818

Academic Editors: Ulrich Desselberger and John T. Patton

Received: 8 April 2024
Revised: 21 May 2024
Accepted: 21 May 2024
Published: 22 May 2024

published. Here, we will review recent advances in the three aspects of T-cell development that are shared by both mice and humans: (a) "Layered immunity" and "neonatal window of opportunity" theories that aim to explain how immune homeostasis is established in early life (Figure 1); (b) the critical role of Treg cells to maintain tolerance to dietary and microbial components; (c) the phenotypic and functional differences in early-life effector T cells relative to adult cells. We will focus on studies that may have an impact on our understanding of T-cell responses to RV vaccines.

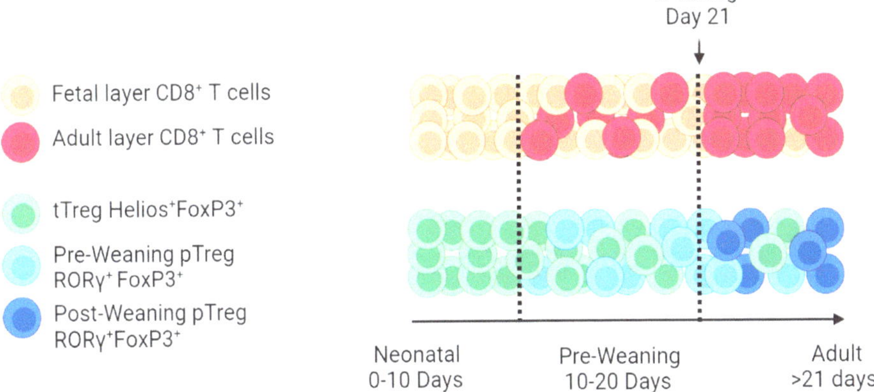

Figure 1. Layered development of CD8+ T cells and CD4+ Treg cells in early-life mice. Many cells of the immune system appear progressively as layers of cells with distinct functions [19]. A small fraction of each layer persists in adults [19]. Recently, this has been shown to occur for early-life CD8+ T cells, which express Lyn28b, the master regulator of fetal lymphopoiesis [12]. Particularly, early-life CD8+ T cells are characterized by bystander activation by innate cytokines and the acquisition of an effector-like profile. On the other hand, in early life, CD4+ T cells are biased towards the development of Treg cells that also seem to develop in layers [18]. The first layer of Treg cells to appear seems to be enriched in Helios+ thymus-derived Treg (tTreg) cells that are mostly selected to mediate self-tolerance [18]. Between day 10 of life and the moment of weaning (day 21 of life), a second layer of RORγ+ peripheral Treg (pTreg) cells develops, and they mediate tolerance mostly to gut bacteria [20]. Figure designed with Biorender.

2. Early-Life T Cells in Mice

2.1. Layered T-Cell Immunity in Early-Life Mice

Neonates and infants suffer more infections compared to adults [13], and have poorer responses to vaccines [21], suggesting they have an impaired ability to develop long-lasting protective immunity. Historically, it was assumed that this was due to an "immature" immune system. However, recent studies suggest that neonates can develop robust immune responses in epithelial barriers to tolerate food antigens, to permit colonization by commensal microorganisms, or for pathogen elimination. However, early-life T cells have a lower capacity to generate long-term memory compared to adult T cells [13,22,23]. These functions adapted to early life are thought to be mediated by multiple populations of immune cells that appear progressively as layers with different phenotypes and functions according to the environments and challenges to which they must adapt [17,19,24]. The existence of layers of immune cells like B cells [25] and γδ T cells [26] has been known for some time but has only recently become appreciated for αβ TCR CD8+ T cells [12,24] and Treg cells [18,27] (Figure 1). Although the concept of a specific lineage of early-life CD8+ T cells is not completely clear [28], in CD8+ T cells (analogous to B cells [25]), the expression of Lyn28b (the master regulator of fetal lymphopoiesis [29]) is associated with features of early-life CD8+ T cells in mice [30]. In addition, these cells are prone to respond in a bystander, non-antigen-specific manner [28]. In humans, Lyn28b was also associated with the development of early-life Treg cells [31]. An important feature of the immune cell layers is that they can persist and modulate the response in adulthood [12]. Thus, intrinsic and environmental factors that occur in early life determine a "window of opportunity" that may influence tolerogenic versus effector-type responses that may persist and determine immune responses in the adult.

2.2. Mouse Treg Cells in Early Life

Early-life CD4$^+$ T cells have a propensity to develop into Treg cells, and these account for the tolerogenic milieu necessary for self-discrimination and sensing of commensal microbiota [12,13]. Broadly speaking, two major Treg cell subsets can be identified in mice: those that develop in the thymus (tTreg) and those that develop in the periphery (pTreg). These subsets are difficult to differentiate phenotypically. However, tTreg cells mainly express the transcription factor Helios, while pTreg cells express the transcription factor RORγ [18]. RORγ$^+$ Treg cells are strongly modulated by gut bacteria, mediate tolerance to commensals, are partially dependent on FoxP3, and are predominant in the colon relative to the small intestine. In contrast, Helios$^+$ Treg cells require FoxP3, are not modulated by microorganisms, and are dominant in the small intestine [18]. The following three phases of early-life tolerance induction and development of Treg cells have been proposed [20]:

1. Neonatal phase (from 0 to 10 days of life): During this period, gut antigens and bacteria are poorly translocated into the small intestine or colon, and the development of self-tolerance probably relies on tTreg cells. Thymectomy performed during the first 2–3 days of life (but not later) in mice leads to overt multi-organ autoimmunity [27]. Self-tolerance generated at this period is mediated by a subset of Treg cells that develop in the thymus and remain in adult mice [18,27]. These Treg cells have a distinct transcriptome, activation profile, and TCR repertoire compared to adult cells [27]. Moreover, they may also play an active role in suppressing effector responses to pathogens such as herpes simplex virus more efficiently than adult Tregs [32].

2. The pre-weaning period (from 10 to 21 days of life), during which, bacteria and luminal antigens transiently traverse the intestinal epithelium. In this phase, pTreg cells (RORγ$^+$) begin to generate, and they are responsible for tolerance to gut bacteria [20]. If bacteria are not encountered during this period, antigen-specific effector responses will occur in inflammatory settings [20]. During the pre-weaning period, the generation of Treg cells that mediate tolerance to the microbiota is influenced by coordinated internal and external factors in the mouse pup. The principal external factor is maternal milk: during this period, maternal milk decreases the concentration of epidermal growth factor that indirectly regulates the capacity of bacteria to traverse the colon and induce pTreg cells [20,33]. Maternal milk antibodies produced by B cells stimulated in the intestine that have migrated to the breast recognize specific subsets of bacteria and are key in the development of Treg cells, that in turn are linked to the frequency of naive (CD44$^-$) intestinal T cells [23,34]. In addition, while the proportion of gut RORγ$^+$ versus Helios$^+$ Treg cells is determined genetically, it is also influenced by maternal antibodies, and this effect can be transmitted through multiple generations of mice [18,35]. Another factor regulating the development of RORγ$^+$ Treg cells is the interplay with intestinal IgA-producing B cells during weaning. Indeed, RORγ$^+$ Treg cells and IgA-producing B cells have a reciprocal regulation: RORγ$^+$ Treg cells inhibit the development of intestinal IgA-producing B cells and thus the coating of the microbiota by secretory IgA. In turn, the secretory IgA coating of intestinal bacteria inhibits the development of RORγ$^+$ Treg cells [35]. Mammary gland IgA is produced by cells that originally develop in the intestine, thus generating a feedback loop between RORγ$^+$ Treg cells and the mother's IgA-producing B cells specific for intestinal bacteria [35].

Synchronously with these external factors in mouse pups, during the pre-weaning period, a specialized subset of antigen-presenting cells (denominated Thetis cells) appears in the colon. A subset of these cells (of unknown function) resembles thymic antigen-presenting cells, as the cells also express the transcription factor autoimmune regulator

(AIRE) [36]. Another subset of Thetis cells expresses the integrin $\alpha v \beta 8$, which, together with $\alpha v \beta 6$, activates latent extracellular transforming growth factor (TGF)-β. Importantly, the induction of pTreg cells by Thetis cells via TGF-β signaling is a critical process to avoid autoimmunity and colitis [36]. Other populations of tolerogenic antigen-presenting cells may also participate in the process [13]. Finally, a critical event that occurs during the pre-weaning period is the appearance of mature microfold (M) cells in gut-associated lymphoid tissues, which facilitates antigen transport and the promotion of pTreg cells [13].

3. The post-weaning phase (day 21 of life onwards) is the phase in which the translocation of intestinal bacteria and the capacity to develop tolerance to gut microbiota is stopped. After weaning, mice start to eat solid foods, and this event promotes the diversification of the microbiota, maturation of the mucosal immune cell composition, and the formation of germinal centers in lymphoid tissues [13].

The role of microbiota in the development of tolerogenic versus effector responses has important implications in the settings of enteric viral infections and EED. For instance, antibiotic microbiota ablation in mothers and their mouse pups resulted in reduced homologous RV infection/diarrhea in the pups, which was associated with a more pronounced antiviral antibody response [37]. However, a delicate balance must be maintained between commensal and non-commensal bacteria in the gut. In keeping with this notion, antibiotic exposure in early-life mice induces intestinal microbiota alterations (dysbiosis), reducing antibody responses to five parenteral human vaccines. This vaccine hyporesponse can be rescued by the restoration of commensal microbiota, but this effect is not seen in adult mice [38]. In addition, the mice that received antibiotics in early life had enhanced T-cell cytokine recall responses in vitro [38]. Thus, the level and composition of gut microbiota in early life influence the quality of antigen-specific adaptive immune responses, likely related to changes in Treg cell populations.

Malnutrition and the dysbiosis of microbiota are associated with EED, which could be associated with the reduced immunogenicity of oral RV vaccines in infants [4–7]. Recently, an EED mouse model was developed by colonizing malnourished 21-day-old mice with a single adherent-invasive *E. coli* isolate [39]. When these mice are vaccinated orally with *E. coli*-labile toxin, antigen-specific $CD4^+$ T-cell responses are dampened in the small intestine but not the mesenteric lymph nodes, and vaccine efficiency is reduced. Moreover, EED mice exhibit increased frequencies of small-intestine $ROR\gamma t^+$ $FoxP3^+$ Treg cells. The deletion of this Treg subset restores small-intestine $CD4^+$ T-cell responses and the capacity of the vaccine to protect from challenge. However, the removal of Treg cells results in increased EED-related stunting, suggesting a fine equilibrium between the capacity of the organism to thrive and defend itself from microorganisms. This model illustrates that the decreased efficacy of oral vaccines in EED may depend on $ROR\gamma t^+$ $FoxP3^+$ Treg cells, an effect that is restricted to local but not systemic immunity [39].

2.3. Effector/Memory T Cells in Early Life

Early-life effector memory $CD4^+$ and $CD8^+$ T cells differ from those of adults in phenotype, function, and TCR repertoire [17]. Mouse T cells begin to express the terminal deoxynucleotidyl transferase enzyme (that augments TCR variability) at one week of age, making neonatal TCRs shorter and more cross-reactive than those of adults [12]. Moreover, they express more innate Toll-like receptors (TLRs) and have increased bystander activation by innate cytokines such as interleukin (IL)-12 and IL-18 [17,28]. Consistently, early-life effector T cells have a greater capacity to proliferate to pathogens compared to adult cells [22,40,41]. However, compared to adult mice, neonatal mice have a reduced capacity to generate long-lived memory T-cell responses [12].

In the case of $CD4^+$ T cells, enhanced TCR-mediated signaling enables them to respond to low antigen doses [41]. Moreover, early-life $CD4^+$ T cells have a propensity to develop into Treg or Th2 cells [12]. In contrast, early-life $CD8^+$ T cells can mount fast, short-lived effector responses [22]. As mentioned above, neonatal $CD8^+$ T cells express Lyn28b and persist and modulate immune responses in adults [24,30]. In addition, a population of naïve (non-antigen-experienced) cells that express some memory markers (CD122$^+$ CD44$^+$ CD49d$^-$) has been characterized in early life, and these cells also persist in adult mice and modulate responses to pathogens [24].

In addition, metabolism is an important regulator of T-cell function and differentiation. It is known that, upon activation, naïve T cells switch their metabolism infection from oxidative phosphorylation to glycolysis in response to infection to mobilize their transcriptional and translational machinery and to undergo clonal expansion. However, when infection is cleared, T cells must decrease anabolic activity to become a more quiescent memory cell, switching from glycolysis back to fatty acid oxidation [42]. Nonetheless, it was demonstrated that neonatal CD8$^+$ T cells are biased to exhibit higher glycolytic activity than their adult counterparts after infection, which limits the formation of memory cells [43]. Most likely, other metabolic pathways may be implicated in the regulation of neonatal T-cell responses, which could be promising targets to improve memory formation in early life [44].

T-cell development in the intestinal mucosa differs from that in peripheral blood and non-mucosal lymphoid organs and is thought to be coordinated with the abovementioned weaning factors [13]. Very few studies have addressed the development of intestinal T cells in early-life mice [23,45,46]. At birth, intestinal T cells are infrequent, and their differential isolation from intestinal lamina propria and intraepithelial compartments is a technical challenge [23]. Thus, we do not have a clear picture of the frequencies and numbers of CD4$^+$ and CD8$^+$ T cells in these individual compartments in early life [23]. Before weaning, TCR $\alpha\beta^+$ cells seem to be enriched in CD4$^+$ over CD8$\alpha\beta^+$ cells in the small intestine (evaluated as a whole) [23]. In addition, T cells in lamina propria, and to some degree in the intraepithelial compartment, have low expression of typical markers of tissue-resident memory T cells (T$_{RM}$), like CD69. Indeed, this marker is only fully acquired post-weaning [45]. The expression of the memory marker CD44 is also reduced in lamina propria CD4$^+$ T cells [23,34], but it is increased in CD8$^+$ T cells due to the presence of atypical naïve cells [24].

In general, two types of effector/memory T-cell subsets exist, circulating T cells and non-circulating T$_{RM}$ cells, that are difficult to differentiate phenotypically [47]. As an alternative, the two cell subsets are differentiated using in vivo intravascular staining (T$_{RM}$ cells are not stained by the intravascular antibody) [47,48]. Importantly, T$_{RM}$ cells are critical in the mucosal response against several types of pathogens [47,48]. However, while multiple studies of antiviral intestinal T cells [15,49] and studies of intestinal T$_{RM}$ cells with model microorganisms have revealed a protective role of these subsets in mucosal tissues [50,51], very few studies have explored the role of intestinal T$_{RM}$ cells specific for a natural intestinal pathogen. In this regard, experiments in adult mice showed that intestinal CD8$^+$ T$_{RM}$ cells modulate norovirus persistence [52]. In addition, compared to intestinal RV-specific T cells of adult mice, those of neonatal mice differ in kinetics and fine specificity, while both have a relatively short persistence [53]. Furthermore, previous studies in neonatal mouse T cells have shown that these cells may play a role in protection after RV vaccination [10] and that Treg cells expressing the latency-associated peptide (precursor of TGF-β) may modulate vaccine-induced protection [11]. The formal characterization of intestinal RV-specific T$_{RM}$ cells in neonatal mice and their capacity to mediate protection has not been performed.

Currently, early-life mucosal T$_{RM}$ responses have only been evaluated in the lung. As such, it has been described that, compared to adult mice, neonatal mice respond to respiratory syncytial virus (RSV) infection with variable levels of CD8$^+$ T$_{RM}$ cells (intravascular negative cells) in the lungs. In addition, neonatal mice fail to maintain lung CD8$^+$ T$_{RM}$ cells at 40 days post-infection and are less protected upon viral rechallenge [54]. However, if neonatal mice are primed and boosted with RSV in the presence of CpG (TLR9 ligand), an adult-like induction of CD8$^+$ T$_{RM}$ cells is observed, and protection is established [54]. In a second model, it was observed that, after influenza virus infection, two-week-old mice developed virus-specific CD4$^+$ and CD8$^+$ effector responses comparable to those in adult mice [40]. However, six weeks after primary influenza infection, the frequencies of lung virus-specific T_{RM} cells were reduced in neonatal mice relative to adults, and they were less protected upon viral challenge [40]. Interestingly, the propensity to generate fewer lung T_{RM} cell responses in early-life mice was associated with an increased expression of the transcription factor T-bet (associated with effector-like T-cell responses), and the reduction in T-bet levels in infant mice increased lung T_{RM} development [40]. These data are in line with the tendency of neonatal T cells to develop short-lived effector responses, which comes at the expense of the generation of long-term memory. In addition, these studies indicate that the limitation to establishing protective T_{RM} cells in neonates can be overcome by augmenting both innate immune activation and antigen exposure, as well as by modulating T-bet expression.

3. T Cells in Human Neonates and Infants

Given the difficulty of sampling blood and tissues from neonatal humans and infants, our knowledge of T-cell dynamics in early life has been scarce. However, recent studies using samples from pediatric and adult tissues have revealed how age, location, and several other variables shape T-cell function [13,48]. In humans, the neonatal period goes up to the first 28 days of life. Although there are no equivalent stages between neonatal mice and humans, and important anatomical, physiological, and microbial differences are found between both species [13], they do share some mechanisms of immune development. In this section, we will discuss key insights from neonatal and infant T-cell responses in humans.

Similar to mouse T cells, neonatal human T cells have intrinsic properties that evolved to exert a specialized role for the host. As previously discussed, the layered immune system hypothesis postulates that the temporal emergence of hematopoietic stem cells (HSCs) gives rise to diverse cell populations at different stages of life [19]. Supporting this hypothesis in humans, it has been shown that fetal CD4$^+$ T cells are functionally and transcriptionally different from adult cells [55–57]. Specifically, fetal cells are prone to proliferation and preferentially become Treg cells [55]. In addition, fetal and adult HSCs also have different gene signatures that determine the matured T-cell profile [56]. Interestingly, preliminary findings indicate that T-cell tolerance induced in utero may be maintained until early adulthood through the establishment of long-lived Treg cells [55]. More recently, a transcriptomic analysis of naive CD4$^+$ T cells revealed that cells derived from cord blood are closely related to fetal cells but distinct from circulating adult cells [58]. Moreover, cord blood cells were enriched in genes associated with a rapid proliferative response [58]. In line with these data, the single-cell transcriptomic analysis of naïve-phenotype CD8$^+$ T cells from fetal, neonatal, and adult individuals revealed that multiple and distinct innate-like fetal clusters are present at birth, and some of them disappear with age [28]. These findings suggest that the human T-cell compartment is highly diverse and evolves with age. In contrast, it has been proposed that T-cell composition may progressively change from fetal-like properties towards a more adult phenotype with age. This "gradual change model" is supported by the recent single-cell transcriptomic profiling of fetal, neonatal, and adult human T cells [59]. As such, newborn cells appeared to be relatively homogeneous, and their developmental stage was placed intermediate between fetal and adult cells. Moreover, the authors propose that this divergence between the three life stages was not explained by distinct waves of HSC progenitors [59].

Relative to adults, increased frequencies of human Treg cells are found in fetal blood and lymphoid tissues, and they play an important role in suppressing effector CD4$^+$ and CD8$^+$ T-cell proliferation and cytokine secretion [60]. Importantly, within the first two years of life, this 6-to-10-fold higher frequency of Treg cells is also observed in multiple lymphoid and mucosal tissues (including the gut), compared with adult tissues [61], consistent with the maintenance of a tolerogenic environment during early human life. In addition, in infant blood and most of the tissues, the predominant T-cell subset exhibits a naïve CD31$^+$ phenotype, consistent with recent thymic emigrants [61]. In contrast, effector memory T cells can only be found in the lungs and intestine [61]. These data illustrate the in situ control of immune responses by regulatory mechanisms in early life. In addition, the intestinal mucosa, with lower Treg/effector cell ratios, seems to constitute a hotspot of effector immune surveillance for the rapid response against exogenous antigens.

In addition to varying grades of diversity, similar to what occurs in mice [16], neonatal human T cells are biased toward broadly cross-reactive (and self-reactive) TCRs. As such, human naïve T cells in cord blood exhibit higher CD5 expression than adult cells, a marker associated with the strength of self-peptide major histocompatibility complex reactivity [62]. In line with this, T cells specific for leukemia-associated self-antigens are found in higher frequency in cord blood than in adult blood [63]. This TCR cross-reactivity might facilitate a more rapid response against different types of antigens during early life. Importantly, in addition to TCR-mediated stimulation, neonatal human T cells express several pattern recognition receptors, including TLR 2, 3, and 5, as well as complement receptors, which induce their activation, proliferation, and cytokine production [64,65]. In keeping with these observations, naïve CD8$^+$ T cells from human neonates have a transcriptional and chromatin landscape that predispose them to innate-like functions [66], such as the production of interleukin-8 (CXCL8) [67]. Overall, these characteristics of neonatal T cells are reminiscent of other innate-like populations, such as B1 cells [25] and $\gamma\delta$ T cells [26].

The above studies indicate that neonatal T cells are armed with innate immune features that could contribute to a rapid host defense upon infection. Nonetheless, a consequence of this lower activation threshold in human neonatal T cells is increased proliferation and more rapid differentiation towards short-lived cells [40,41,68], as has been described for neonatal mice cells [22]. These intrinsic properties could affect the development of long-term memory, with important consequences for secondary immune responses. On the one hand, it has been described that this different program of human newborn T cells facilitates Treg cell differentiation [31]. In addition, it was shown that CXCL8$^+$ CD4$^+$ T cells from neonates are direct precursors of Th1 cells upon sustained proliferation [69]. On the other hand, recent studies have demonstrated that, similar to mouse cells, human infant T cells are intrinsically programmed for short-term responses, driven by the increased expression of the transcription factor T-bet [40] and other transcription factors [28,70]. As such, infant cells exhibit higher levels of T-bet relative to adult cells in vivo and after in vitro activation, which negatively correlates with the expression of the long-lived memory marker CD127 [40]. In agreement with these findings, CD8$^+$ T cell responses of children (less than 4 years old) to an inactivated vaccine at day 10 after vaccination are similar to those of adults; however, contrary to adults, the response had practically disappeared by day 28 after vaccination [71]. Also, in keeping with these data, a recent study reported that children maintain a novel CD8$^+$ T cell subset epigenetically poised for rapid effector responses, and this subset is lost with age [72]. Of note, the predisposition for effector-like responses in the neonatal period also impacts the generation of CD69$^+$ CD103$^+$ T_{RM} cells in the respiratory tract. Indeed, the frequency of this tissue-resident population increases with age, suggesting that the aforementioned transcriptional program of neonatal T cells directly regulates the generation of this memory population [73]. A similar dynamic seems to occur for T_{RM} cells in the human neonatal gut [74]. In the early weeks of human life, there is a low frequency of $\alpha\beta$ T_{RM}-phenotype cells in the gut, and this population accumulates progressively with age, peaking at around 1.4 years and remaining stable thereafter [75,76].

In summary, neonatal human T cells are intrinsically programmed for rapid differentiation to effector-like responses. Enhanced TCR signaling and localized effector-like responses could protect newborns from newly encountered pathogens. However, a higher proportion of effector-phenotype T cells in infants could contribute to tissue injury during viral infection [77]. Moreover, short-lived effector responses in neonates are generated at the expense of the induction of long-term memory. As a result, the generation of T_{RM} cells at the site of infection during early life may undermine the immune response, resulting in recurrent symptomatic disease. Likely, the promotion of localized effector-like immune responses, such as in bronchus- and gut-associated lymphoid tissues [78], could be a beneficial vaccine strategy in early life to prevent infection or illness, waiting for the developmental maturation and acquisition of adult immune traits.

4. Conclusions and Future Directions: How May Early-Life T-Cell Immunity Impact RV Vaccines?

Recent studies have provided important insights into the influence of the neonatal period on the establishment of enduring immune responses and the implications for health and disease. Beyond the inherent attributes of mouse and human T cells discussed above, various additional elements modulate and shape the neonatal immune system. Within this "neonatal window of opportunity", environmental factors, such as commensal microbiota or some specific pathogens, may shift immune responses towards more tolerizing or effector-type responses and thus predispose an individual to immune-mediated and other diseases in adult life [79]. For instance, it has recently been shown that early microbial exposure modulates and programs mouse $CD8^+$ T-cell development, and fetal-derived cells in a dirty environment are more responsive to stimulation, acquiring an effector-like profile [80]. Considering that RV vaccines provide non-sterilizing immunity in children and they are less effective in low-income countries [9], it is possible that poor environmental conditions in these settings influence early-life immunity. A potential outcome of these processes is the generation of monofunctional effector-like responses [81] and low frequencies of long-lived memory RV-specific cells [82], thus impairing long-term immunity in adult life [83]. Likely, poor tissue-resident memory responses are also affected.

As mentioned previously, protection against RV reinfection in adult mice is mediated by $CD4^+$ T cell-dependent antibodies and by short-lived $CD8^+$ T cells [9], and the role of early-life T cells is much less well known. The following studies of early-life T cells may be envisioned for preclinical basic research or clinical studies that could have an impact on the improvement/development of new RV vaccines:

1. Promotion of T_{RM}-cell responses: The improvement of antigen delivery to the gut tissue by oral vaccines [78] or using parenteral vaccines to boost after priming with an oral vaccine could be beneficial strategies in early life.

2. Innate immune activation simultaneously with mucosal antigen administration [54]. Double-stranded RNA and dmLT are two mucosal adjuvants that are being tested in ongoing clinical trials with orally administered vaccines [84].

3. The modulation of transcriptional circuits to promote long-lived memory T cells in early life, like reducing T-bet expression [40]. Moreover, other transcriptional or metabolic pathways could be modulated to promote memory-like cells. In this regard, previous studies have shown that inhibition of the mTORC1 (mammalian target of rapamycin complex 1) pathway promotes memory T-cell differentiation in mice [85], as well as enhances immune function and the response to influenza vaccination while reducing infections in the elderly [86].

4. The study of Treg cells in EED and RV-vaccinated children [39]. In this regard, since the evaluation of $ROR\gamma t^+$ Treg cells (or their elimination seeking a benefit) is difficult given the need for tissue isolation [35], measuring antibody-coated bacteria in stool samples could be a more accessible and useful correlate of Treg cells [85], as well as a potential biomarker in children [87]. Given the impact of maternal milk on the thymus and T cells in general [88], these studies would need to be performed with breast-fed and formula-fed children as separate groups, but in both groups, levels of IgA-covered bacteria may correlate with Treg cells.

5. Phenotypic and functional innate-like features of T cells may be used as biomarkers of anti-RV immune responses and vaccine responses.

6. Our lack of understanding of basic aspects of early-life T cells highlights our need for more basic studies in neonatal mouse models of RV infection/vaccination [11]. As observed in this review (and summarized in Table 1), several basic aspects of the early-life immune response of mice also seem to occur in human neonates and infants, justifying these studies.

Table 1. Summary of characteristics shared by human and mouse early-life T cells.

T cells are functionally and transcriptionally different from adult T cells, supporting the hypothesis of the layered immunity theory.
T cells exhibit increased expression levels of innate-like receptors and respond in a bystander fashion.
TCRs are cross-reactive/autoreactive with limited diversity due to a lack of expression of TdT.
$CD8^+$ T cells are prone to rapidly proliferating and differentiating in short-lived cells depending on the expression of T-bet.
Long-lived memory T-cell responses are diminished.
$CD4^+$ T cells are prone to become Treg cells, and both humans and mice have comparable $Helios^+$ and $ROR\gamma^+$ Treg subsets.
Both seem to have reduced capacity for the generation of mucosal T_{RM}.

However, some important differences between human and mouse early-life T cells exist; for example, the diversification of the T-cell repertoire occurs earlier in humans than in mice and the post-thymic maturation of T cells is shorter in mice [17]. For this reason, the extrapolation of results from mice to humans should be carried out with caution.

In conclusion, we are just starting to understand how the gradual diversification of the T-cell compartment and functions in early life may determine the clinical outcome after infection. To improve RV vaccines, it is crucial to consider environmental factors such as diet, microbiome, and the level of exogenous antigen exposure, which can alter T-cell immune ontogeny, along with the regulatory mechanisms that govern this layered immunity. The use of advanced molecular techniques for global immune monitoring could provide valuable tools for developing next-generation RV vaccines. These vaccines should not only target antigen-specific responses but also promote effective tissue-localized and long-lived memory responses.

Author Contributions: C.M., F.P.-C. and M.A.F. conceived and wrote the manuscript. C.M. conceived and designed the figure. All authors have read and agreed to the published version of the manuscript.

Funding: This work was funded by grants from Vicerrectoría de Investigación, Pontificia Universidad Javeriana grant number 000010013, and Sistema General de Regalías, Colombia, grant number BPIN 2020000100360. Catherine Montenegro was funded with a partial scholarship by Pontificia Universidad Javeriana.

Acknowledgments: We thank Immunogenetics and Immunomodulation lab members, Pontificia Universidad Javeriana for critical discussion.

Conflicts of Interest: The authors declare no conflicts of interest.

References

1. Chen, J.; Grow, S.; Iturriza-Gómara, M.; Hausdorff, W.P.; Fix, A.; Kirkwood, C.D. The Challenges and Opportunities of Next-Generation Rotavirus Vaccines: Summary of an Expert Meeting with Vaccine Developers. *Viruses* **2022**, *14*, 2565. [CrossRef] [PubMed]
2. Hallowell, B.D.; Chavers, T.; Parashar, U.; Tate, J.E. Global Estimates of Rotavirus Hospitalizations Among Children Below 5 Years in 2019 and Current and Projected Impacts of Rotavirus Vaccination. *J. Pediatr. Infect. Dis. Soc.* **2022**, *11*, 149–158. [CrossRef]
3. Keita, A.M.; Doh, S.; Sow, S.O.; Powell, H.; Omore, R.; Jahangir Hossain, M.; Ogwel, B.; Ochieng, J.B.; Jones, J.C.M.; Zaman, S.M.A.; et al. Prevalence, Clinical Severity, and Seasonality of Adenovirus 40/41, Astrovirus, Sapovirus, and Rotavirus Among Young Children With Moderate-to-Severe Diarrhea: Results From the Vaccine Impact on Diarrhea in Africa (VIDA) Study. *Clin. Infect. Dis.* **2023**, *76*, S123–S131. [CrossRef] [PubMed]
4. Ahmed, S.; Iqbal, J.; Sadiq, K.; Umrani, F.; Rizvi, A.; Kabir, F.; Jamil, Z.; Syed, S.; Ehsan, L.; Zulqarnain, F.; et al. Association of Anti-Rotavirus IgA Seroconversion with Growth, Environmental Enteric Dysfunction and Enteropathogens in Rural Pakistani Infants. *Vaccine* **2022**, *40*, 3444–3451. [CrossRef] [PubMed]
5. Kim, A.H.; Armah, G.; Dennis, F.; Wang, L.; Rodgers, R.; Droit, L.; Baldridge, M.T.; Handley, S.A.; Harris, V.C. Enteric Virome Negatively Affects Seroconversion Following Oral Rotavirus Vaccination in a Longitudinally Sampled Cohort of Ghanaian Infants. *Cell Host Microbe* **2022**, *30*, 110–123.e5. [CrossRef] [PubMed]

6. Parker, E.P.K.; Bronowski, C.; Sindhu, K.N.C.; Babji, S.; Benny, B.; Carmona-Vicente, N.; Chasweka, N.; Chinyama, E.; Cunliffe, N.A.; Dube, Q.; et al. Impact of Maternal Antibodies and Microbiota Development on the Immunogenicity of Oral Rotavirus Vaccine in African, Indian, and European Infants. *Nat. Commun.* **2021**, *12*, 7288. [CrossRef]

7. Robertson, R.C.; Church, J.A.; Edens, T.J.; Mutasa, K.; Min Geum, H.; Baharmand, I.; Gill, S.K.; Ntozini, R.; Chasekwa, B.; Carr, L.; et al. The Fecal Microbiome and Rotavirus Vaccine Immunogenicity in Rural Zimbabwean Infants. *Vaccine* **2021**, *39*, 5391–5400. [CrossRef]

8. Lauer, J.M.; Kirby, M.A.; Muhihi, A.; Ulenga, N.; Aboud, S.; Liu, E.; Choy, R.K.M.; Arndt, M.B.; Kou, J.; Fawzi, W.; et al. Assessing Environmental Enteric Dysfunction via Multiplex Assay and Its Relation to Growth and Development among HIV-Exposed Uninfected Tanzanian Infants. *PLoS Negl. Trop. Dis.* **2023**, *17*, e0011181. [CrossRef]

9. Crawford, S.E.; Ramani, S.; Tate, J.E.; Parashar, U.D.; Svensson, L.; Hagbom, M.; Franco, M.A.; Greenberg, H.B.; O'Ryan, M.; Kang, G.; et al. Rotavirus Infection. *Nat. Rev. Dis. Primers* **2017**, *3*, 17083. [CrossRef]

10. VanCott, J.L.; Prada, A.E.; McNeal, M.M.; Stone, S.C.; Basu, M.; Huffer, B., Jr.; Smiley, K.L.; Shao, M.; Bean, J.A.; Clements, J.D.; et al. Mice Develop Effective but Delayed Protective Immune Responses When Immunized as Neonates Either Intranasally with Nonliving VP6/LT(R192G) or Orally with Live Rhesus Rotavirus Vaccine Candidates. *J. Virol.* **2006**, *80*, 4949–4961. [CrossRef]

11. Rey, L.M.; Gil, J.Á.; Mateus, J.; Rodríguez, L.-S.; Rondón, M.A.; Ángel, J.; Franco, M.A. LAP+ Cells Modulate Protection Induced by Oral Vaccination with Rhesus Rotavirus in a Neonatal Mouse Model. *J. Virol.* **2019**, *93*, 10–1128. [CrossRef] [PubMed]

12. Tabilas, C.; Smith, N.L.; Rudd, B.D. Shaping Immunity for Life: Layered Development of CD8+ T Cells. *Immunol. Rev.* **2023**, *315*, 108–125. [CrossRef]

13. Torow, N.; Hand, T.W.; Hornef, M.W. Programmed and Environmental Determinants Driving Neonatal Mucosal Immune Development. *Immunity* **2023**, *56*, 485–499. [CrossRef] [PubMed]

14. Laban, N.M.; Goodier, M.R.; Bosomprah, S.; Simuyandi, M.; Chisenga, C.; Chilyabanyama, O.N.; Chilengi, R. T-Cell Responses after Rotavirus Infection or Vaccination in Children: A Systematic Review. *Viruses* **2022**, *14*, 459. [CrossRef]

15. Lockhart, A.; Mucida, D.; Parsa, R. Immunity to Enteric Viruses. *Immunity* **2022**, *55*, 800–818. [CrossRef] [PubMed]

16. Rudd, B.D. Neonatal T Cells: A Reinterpretation. *Annu. Rev. Immunol.* **2020**, *38*, 229–247. [CrossRef]

17. Davenport, M.P.; Smith, N.L.; Rudd, B.D. Building a T Cell Compartment: How Immune Cell Development Shapes Function. *Nat. Rev. Immunol.* **2020**, *20*, 499–506. [CrossRef]

18. Ramanan, D.; Pratama, A.; Zhu, Y.; Venezia, O.; Sassone-Corsi, M.; Chowdhary, K.; Galván-Peña, S.; Sefik, E.; Brown, C.; Gélineau, A.; et al. Regulatory T Cells in the Face of the Intestinal Microbiota. *Nat. Rev. Immunol.* **2023**, *23*, 749–762. [CrossRef]

19. Herzenberg, L.A.; Herzenberg, L.A. Toward a Layered Immune System. *Cell* **1989**, *59*, 953–954. [CrossRef]

20. Knoop, K.A.; Gustafsson, J.K.; McDonald, K.G.; Kulkarni, D.H.; Coughlin, P.E.; McCrate, S.; Kim, D.; Hsieh, C.-S.; Hogan, S.P.; Elson, C.O.; et al. Microbial Antigen Encounter during a Preweaning Interval Is Critical for Tolerance to Gut Bacteria. *Sci. Immunol.* **2017**, *2*, eaao1314. [CrossRef]

21. Lynn, D.J.; Benson, S.C.; Lynn, M.A.; Pulendran, B. Modulation of Immune Responses to Vaccination by the Microbiota: Implications and Potential Mechanisms. *Nat. Rev. Immunol.* **2022**, *22*, 33–46. [CrossRef] [PubMed]

22. Smith, N.L.; Wissink, E.; Wang, J.; Pinello, J.F.; Davenport, M.P.; Grimson, A.; Rudd, B.D. Rapid Proliferation and Differentiation Impairs the Development of Memory CD8+ T Cells in Early Life. *J. Immunol.* **2014**, *193*, 177–184. [CrossRef]

23. Torow, N.; Yu, K.; Hassani, K.; Freitag, J.; Schulz, O.; Basic, M.; Brennecke, A.; Sparwasser, T.; Wagner, N.; Bleich, A.; et al. Active Suppression of Intestinal CD4(+)TCRαβ(+) T-Lymphocyte Maturation during the Postnatal Period. *Nat. Commun.* **2015**, *6*, 7725. [CrossRef] [PubMed]

24. Smith, N.L.; Patel, R.K.; Reynaldi, A.; Grenier, J.K.; Wang, J.; Watson, N.B.; Nzingha, K.; Yee Mon, K.J.; Peng, S.A.; Grimson, A.; et al. Developmental Origin Governs CD8+ T Cell Fate Decisions during Infection. *Cell* **2018**, *174*, 117–130.e14. [CrossRef]

25. Baumgarth, N. The Shaping of a B Cell Pool Maximally Responsive to Infections. *Annu. Rev. Immunol.* **2021**, *39*, 103–129. [CrossRef]

26. Hu, Y.; Hu, Q.; Li, Y.; Lu, L.; Xiang, Z.; Yin, Z.; Kabelitz, D.; Wu, Y. Γδ T Cells: Origin and Fate, Subsets, Diseases and Immunotherapy. *Signal Transduct. Target Ther.* **2023**, *8*, 434. [PubMed]

27. Yang, S.; Fujikado, N.; Kolodin, D.; Benoist, C.; Mathis, D. Immune Tolerance. Regulatory T Cells Generated Early in Life Play a Distinct Role in Maintaining Self-Tolerance. *Science* **2015**, *348*, 589–594. [CrossRef] [PubMed]

28. Watson, N.B.; Patel, R.K.; Kean, C.; Veazey, J.; Oyesola, O.O.; Laniewski, N.; Grenier, J.K.; Wang, J.; Tabilas, C.; Yee Mon, K.J.; et al. The Gene Regulatory Basis of Bystander Activation in CD8+ T Cells. *Sci. Immunol.* **2024**, *9*, eadf8776. [CrossRef]

29. Yuan, J.; Nguyen, C.K.; Liu, X.; Kanellopoulou, C.; Muljo, S.A. Lin28b Reprograms Adult Bone Marrow Hematopoietic Progenitors to Mediate Fetal-like Lymphopoiesis. *Science* **2012**, *335*, 1195–1200. [CrossRef]

30. Wang, J.; Wissink, E.M.; Watson, N.B.; Smith, N.L.; Grimson, A.; Rudd, B.D. Fetal and Adult Progenitors Give Rise to Unique Populations of CD8+ T Cells. *Blood* **2016**, *128*, 3073–3082. [CrossRef]

31. Bronevetsky, Y.; Burt, T.D.; McCune, J.M. Lin28b Regulates Fetal Regulatory T Cell Differentiation through Modulation of TGF-β Signaling. *J. Immunol.* **2016**, *197*, 4344–4350. [CrossRef]

32. Fernandez, M.A.; Puttur, F.K.; Wang, Y.M.; Howden, W.; Alexander, S.I.; Jones, C.A. T Regulatory Cells Contribute to the Attenuated Primary CD8+ and CD4+ T Cell Responses to Herpes Simplex Virus Type 2 in Neonatal Mice. *J. Immunol.* **2008**, *180*, 1556–1564. [CrossRef]

33. Knoop, K.A.; McDonald, K.G.; Coughlin, P.E.; Kulkarni, D.H.; Gustafsson, J.K.; Rusconi, B.; John, V.; Ndao, I.M.; Beigelman, A.; Good, M.; et al. Synchronization of Mothers and Offspring Promotes Tolerance and Limits Allergy. *JCI Insight* **2020**, *5*, 137943. [CrossRef] [PubMed]

34. Torow, N.; Li, R.; Hitch, T.C.A.; Mingels, C.; Al Bounny, S.; van Best, N.; Stange, E.-L.; Simons, B.; Maié, T.; Rüttger, L.; et al. M Cell Maturation and CDC Activation Determine the Onset of Adaptive Immune Priming in the Neonatal Peyer's Patch. *Immunity* **2023**, *56*, 1220–1238.e7. [CrossRef] [PubMed]

35. Ramanan, D.; Sefik, E.; Galván-Peña, S.; Wu, M.; Yang, L.; Yang, Z.; Kostic, A.; Golovkina, T.V.; Kasper, D.L.; Mathis, D.; et al. An Immunologic Mode of Multigenerational Transmission Governs a Gut Treg Setpoint. *Cell* **2020**, *181*, 1276–1290.e13. [CrossRef]

36. Akagbosu, B.; Tayyebi, Z.; Shibu, G.; Paucar Iza, Y.A.; Deep, D.; Parisotto, Y.F.; Fisher, L.; Pasolli, H.A.; Thevin, V.; Elmentaite, R.; et al. Novel Antigen-Presenting Cell Imparts Treg-Dependent Tolerance to Gut Microbiota. *Nature* **2022**, *610*, 752–760. [CrossRef]

37. Uchiyama, R.; Chassaing, B.; Zhang, B.; Gewirtz, A.T. Antibiotic Treatment Suppresses Rotavirus Infection and Enhances Specific Humoral Immunity. *J. Infect. Dis.* **2014**, *210*, 171–182. [CrossRef] [PubMed]

38. Lynn, M.A.; Tumes, D.J.; Choo, J.M.; Sribnaia, A.; Blake, S.J.; Leong, L.E.X.; Young, G.P.; Marshall, H.S.; Wesselingh, S.L.; Rogers, G.B.; et al. Early-Life Antibiotic-Driven Dysbiosis Leads to Dysregulated Vaccine Immune Responses in Mice. *Cell Host Microbe* **2018**, *23*, 653–660.e5. [CrossRef] [PubMed]

39. Bhattacharjee, A.; Burr, A.H.P.; Overacre-Delgoffe, A.E.; Tometich, J.T.; Yang, D.; Huckestein, B.R.; Linehan, J.L.; Spencer, S.P.; Hall, J.A.; Harrison, O.J.; et al. Environmental Enteric Dysfunction Induces Regulatory T Cells That Inhibit Local CD4+ T Cell Responses and Impair Oral Vaccine Efficacy. *Immunity* **2021**, *54*, 1745–1757.e7. [CrossRef]

40. Zens, K.D.; Chen, J.K.; Guyer, R.S.; Wu, F.L.; Cvetkovski, F.; Miron, M.; Farber, D.L. Reduced Generation of Lung Tissue-Resident Memory T Cells during Infancy. *J. Exp. Med.* **2017**, *214*, 2915–2932. [CrossRef]

41. Thapa, P.; Guyer, R.S.; Yang, A.Y.; Parks, C.A.; Brusko, T.M.; Brusko, M.; Connors, T.J.; Farber, D.L. Infant T Cells Are Developmentally Adapted for Robust Lung Immune Responses through Enhanced T Cell Receptor Signaling. *Sci. Immunol.* **2021**, *6*, eabj0789. [CrossRef] [PubMed]

42. Reina-Campos, M.; Scharping, N.E.; Goldrath, A.W. CD8+ T Cell Metabolism in Infection and Cancer. *Nat. Rev. Immunol.* **2021**, *21*, 718–738. [CrossRef] [PubMed]

43. Tabilas, C.; Wang, J.; Liu, X.; Locasale, J.W.; Smith, N.L.; Rudd, B.D. Cutting Edge: Elevated Glycolytic Metabolism Limits the Formation of Memory CD8+ T Cells in Early Life. *J. Immunol.* **2019**, *203*, 2571–2576. [CrossRef] [PubMed]

44. Holm, S.R.; Jenkins, B.J.; Cronin, J.G.; Jones, N.; Thornton, C.A. A Role for Metabolism in Determining Neonatal Immune Function. *Pediatr. Allergy Immunol.* **2021**, *32*, 1616–1628. [CrossRef] [PubMed]

45. Kuo, S.; El Guindy, A.; Panwala, C.M.; Hagan, P.M.; Camerini, V. Differential Appearance of T Cell Subsets in the Large and Small Intestine of Neonatal Mice. *Pediatr. Res.* **2001**, *49*, 543–551. [CrossRef] [PubMed]

46. Steege, J.C.; Buurman, W.A.; Forget, P.P. The Neonatal Development of Intraepithelial and Lamina Propria Lymphocytes in the Murine Small Intestine. *Dev. Immunol.* **1997**, *5*, 121–128. [CrossRef]

47. Masopust, D.; Soerens, A.G. Tissue-Resident T Cells and Other Resident Leukocytes. *Annu. Rev. Immunol.* **2019**, *37*, 521–546. [CrossRef] [PubMed]

48. Paik, D.H.; Farber, D.L. Anti-Viral Protective Capacity of Tissue Resident Memory T Cells. *Curr. Opin. Virol.* **2021**, *46*, 20–26. [CrossRef] [PubMed]

49. Parsa, R.; London, M.; Rezende de Castro, T.B.; Reis, B.; Buissant des Amorie, J.; Smith, J.G.; Mucida, D. Newly Recruited Intraepithelial Ly6A+CCR9+CD4+ T Cells Protect against Enteric Viral Infection. *Immunity* **2022**, *55*, 1234–1249.e6. [CrossRef]

50. von Hoesslin, M.; Kuhlmann, M.; de Almeida, G.P.; Kanev, K.; Wurmser, C.; Gerullis, A.-K.; Roelli, P.; Berner, J.; Zehn, D. Secondary Infections Rejuvenate the Intestinal CD103 $^+$ Tissue-Resident Memory T Cell Pool. *Sci. Immunol.* **2022**, *7*, eabp9553. [CrossRef]

51. Fung, H.Y.; Teryek, M.; Lemenze, A.D.; Bergsbaken, T. CD103 Fate Mapping Reveals That Intestinal CD103- Tissue-Resident Memory T Cells Are the Primary Responders to Secondary Infection. *Sci. Immunol.* **2022**, *7*, eabl9925. [CrossRef] [PubMed]

52. Tomov, V.T.; Palko, O.; Lau, C.W.; Pattekar, A.; Sun, Y.; Tacheva, R.; Bengsch, B.; Manne, S.; Cosma, G.L.; Eisenlohr, L.C.; et al. Differentiation and Protective Capacity of Virus-Specific CD8+ T Cells Suggest Murine Norovirus Persistence in an Immune-Privileged Enteric Niche. *Immunity* **2017**, *47*, 723–738.e5. [CrossRef] [PubMed]

53. Jaimes, M.C.; Feng, N.; Greenberg, H.B. Characterization of Homologous and Heterologous Rotavirus-Specific T-Cell Responses in Infant and Adult Mice. *J. Virol.* **2005**, *79*, 4568–4579. [CrossRef] [PubMed]

54. Malloy, A.M.W.; Lu, Z.; Kehl, M.; Pena DaMata, J.; Lau-Kilby, A.W.; Turfkruyer, M. Increased Innate Immune Activation Induces Protective RSV-Specific Lung-Resident Memory T Cells in Neonatal Mice. *Mucosal Immunol.* **2023**, *16*, 593–605. [CrossRef] [PubMed]

55. Mold, J.E.; Michaëlsson, J.; Burt, T.D.; Muench, M.O.; Beckerman, K.P.; Busch, M.P.; Lee, T.-H.; Nixon, D.F.; McCune, J.M. Maternal Alloantigens Promote the Development of Tolerogenic Fetal Regulatory T Cells in Utero. *Science* **2008**, *322*, 1562–1565. [CrossRef] [PubMed]

56. Mold, J.E.; Venkatasubrahmanyam, S.; Burt, T.D.; Michaëlsson, J.; Rivera, J.M.; Galkina, S.A.; Weinberg, K.; Stoddart, C.A.; McCune, J.M. Fetal and Adult Hematopoietic Stem Cells Give Rise to Distinct T Cell Lineages in Humans. *Science* **2010**, *330*, 1695–1699. [CrossRef]

57. Li, N.; van Unen, V.; Abdelaal, T.; Guo, N.; Kasatskaya, S.A.; Ladell, K.; McLaren, J.E.; Egorov, E.S.; Izraelson, M.; Chuva de Sousa Lopes, S.M.; et al. Memory CD4+ T Cells Are Generated in the Human Fetal Intestine. *Nat. Immunol.* **2019**, *20*, 301–312. [CrossRef]
58. Hiwarkar, P.; Hubank, M.; Qasim, W.; Chiesa, R.; Gilmour, K.C.; Saudemont, A.; Amrolia, P.J.; Veys, P. Cord Blood Transplantation Recapitulates Fetal Ontogeny with a Distinct Molecular Signature That Supports CD4+ T-Cell Reconstitution. *Blood Adv.* **2017**, *1*, 2206–2216. [CrossRef]
59. Bunis, D.G.; Bronevetsky, Y.; Krow-Lucal, E.; Bhakta, N.R.; Kim, C.C.; Nerella, S.; Jones, N.; Mendoza, V.F.; Bryson, Y.J.; Gern, J.E.; et al. Single-Cell Mapping of Progressive Fetal-to-Adult Transition in Human Naive T Cells. *Cell Rep.* **2021**, *34*, 108573. [CrossRef]
60. Michaëlsson, J.; Mold, J.E.; McCune, J.M.; Nixon, D.F. Regulation of T Cell Responses in the Developing Human Fetus. *J. Immunol.* **2006**, *176*, 5741–5748. [CrossRef]
61. Thome, J.J.; Bickham, K.L.; Ohmura, Y.; Kubota, M.; Matsuoka, N.; Gordon, C.; Granot, T.; Griesemer, A.; Lerner, H.; Kato, T.; et al. Early-Life Compartmentalization of Human T Cell Differentiation and Regulatory Function in Mucosal and Lymphoid Tissues. *Nat. Med.* **2016**, *22*, 72–77. [CrossRef] [PubMed]
62. Mandl, J.N.; Monteiro, J.P.; Vrisekoop, N.; Germain, R.N. T Cell-Positive Selection Uses Self-Ligand Binding Strength to Optimize Repertoire Recognition of Foreign Antigens. *Immunity* **2013**, *38*, 263–274. [CrossRef]
63. St John, L.S.; Wan, L.; He, H.; Garber, H.R.; Clise-Dwyer, K.; Alatrash, G.; Rezvani, K.; Shpall, E.J.; Bollard, C.M.; Ma, Q.; et al. PR1-Specific Cytotoxic T Lymphocytes Are Relatively Frequent in Umbilical Cord Blood and Can Be Effectively Expanded to Target Myeloid Leukemia. *Cytotherapy* **2016**, *18*, 995–1001. [CrossRef] [PubMed]
64. McCarron, M.; Reen, D.J. Activated Human Neonatal CD8+ T Cells Are Subject to Immunomodulation by Direct TLR2 or TLR5 Stimulation. *J. Immunol.* **2009**, *182*, 55–62. [CrossRef] [PubMed]
65. Pekalski, M.L.; García, A.R.; Ferreira, R.C.; Rainbow, D.B.; Smyth, D.J.; Mashar, M.; Brady, J.; Savinykh, N.; Dopico, X.C.; Mahmood, S.; et al. Neonatal and Adult Recent Thymic Emigrants Produce IL-8 and Express Complement Receptors CR1 and CR2. *JCI Insight* **2017**, *2*, 93739. [CrossRef] [PubMed]
66. Galindo-Albarrán, A.O.; López-Portales, O.H.; Gutiérrez-Reyna, D.Y.; Rodríguez-Jorge, O.; Sánchez-Villanueva, J.A.; Ramírez-Pliego, O.; Bergon, A.; Loriod, B.; Holota, H.; Imbert, J.; et al. CD8+ T Cells from Human Neonates Are Biased toward an Innate Immune Response. *Cell Rep.* **2016**, *17*, 2151–2160. [CrossRef] [PubMed]
67. Gibbons, D.; Fleming, P.; Virasami, A.; Michel, M.L.; Sebire, N.J.; Costeloe, K.; Carr, R.; Klein, N.; Hayday, A. Interleukin-8 (CXCL8) Production Is a Signatory T Cell Effector Function of Human Newborn Infants. *Nat. Med.* **2014**, *20*, 1206–1210. [CrossRef] [PubMed]
68. Schönland, S.O.; Zimmer, J.K.; Lopez-Benitez, C.M.; Widmann, T.; Ramin, K.D.; Goronzy, J.J.; Weyand, C.M. Homeostatic Control of T-Cell Generation in Neonates. *Blood* **2003**, *102*, 1428–1434. [CrossRef] [PubMed]
69. Das, A.; Rouault-Pierre, K.; Kamdar, S.; Gomez-Tourino, I.; Wood, K.; Donaldson, I.; Mein, C.A.; Bonnet, D.; Hayday, A.C.; Gibbons, D.L. Adaptive from Innate: Human IFN-Γ+CD4+ T Cells Can Arise Directly from CXCL8-Producing Recent Thymic Emigrants in Babies and Adults. *J. Immunol.* **2017**, *199*, 1696–1705. [CrossRef]
70. Montecino-Rodriguez, E.; Casero, D.; Fice, M.; Le, J.; Dorshkind, K. Differential Expression of PU.1 and Key T Lineage Transcription Factors Distinguishes Fetal and Adult T Cell Development. *J. Immunol.* **2018**, *200*, 2046–2056. [CrossRef]
71. He, X.-S.; Holmes, T.H.; Mahmood, K.; Kemble, G.W.; Dekker, C.L.; Arvin, A.M.; Greenberg, H.B. Phenotypic Changes in Influenza-Specific CD8+ T Cells after Immunization of Children and Adults with Influenza Vaccines. *J. Infect. Dis.* **2008**, *197*, 803–811. [CrossRef]
72. Thomson, Z.; He, Z.; Swanson, E.; Henderson, K.; Phalen, C.; Zaim, S.R.; Pebworth, M.-P.; Okada, L.Y.; Heubeck, A.T.; Roll, C.R.; et al. Trimodal Single-Cell Profiling Reveals a Novel Pediatric CD8αα+ T Cell Subset and Broad Age-Related Molecular Reprogramming across the T Cell Compartment. *Nat. Immunol.* **2023**, *24*, 1947–1959. [CrossRef]
73. Connors, T.J.; Baird, J.S.; Yopes, M.C.; Zens, K.D.; Pethe, K.; Ravindranath, T.M.; Ho, S.-H.; Farber, D.L. Developmental Regulation of Effector and Resident Memory T Cell Generation during Pediatric Viral Respiratory Tract Infection. *J. Immunol.* **2018**, *201*, 432–439. [CrossRef]
74. Senda, T.; Dogra, P.; Granot, T.; Furuhashi, K.; Snyder, M.E.; Carpenter, D.J.; Szabo, P.A.; Thapa, P.; Miron, M.; Farber, D.L. Microanatomical Dissection of Human Intestinal T-Cell Immunity Reveals Site-Specific Changes in Gut-Associated Lymphoid Tissues over Life. *Mucosal Immunol.* **2019**, *12*, 378–389. [CrossRef]
75. Connors, T.J.; Matsumoto, R.; Verma, S.; Szabo, P.A.; Guyer, R.; Gray, J.; Wang, Z.; Thapa, P.; Dogra, P.; Poon, M.M.L.; et al. Site-Specific Development and Progressive Maturation of Human Tissue-Resident Memory T Cells over Infancy and Childhood. *Immunity* **2023**, *56*, 1894–1909.e5. [CrossRef]
76. Poon, M.M.L.; Caron, D.P.; Wang, Z.; Wells, S.B.; Chen, D.; Meng, W.; Szabo, P.A.; Lam, N.; Kubota, M.; Matsumoto, R.; et al. Tissue Adaptation and Clonal Segregation of Human Memory T Cells in Barrier Sites. *Nat. Immunol.* **2023**, *24*, 309–319. [CrossRef]
77. Connors, T.J.; Ravindranath, T.M.; Bickham, K.L.; Gordon, C.L.; Zhang, F.; Levin, B.; Baird, J.S.; Farber, D.L. Airway CD8+ T Cells Are Associated with Lung Injury during Infant Viral Respiratory Tract Infection. *Am. J. Respir. Cell Mol. Biol.* **2016**, *54*, 822–830. [CrossRef]
78. Matsumoto, R.; Gray, J.; Rybkina, K.; Oppenheimer, H.; Levy, L.; Friedman, L.M.; Khamaisi, M.; Meng, W.; Rosenfeld, A.M.; Guyer, R.S.; et al. Induction of Bronchus-Associated Lymphoid Tissue Is an Early Life Adaptation for Promoting Human B Cell Immunity. *Nat. Immunol.* **2023**, *24*, 1370–1381. [CrossRef]

79. Renz, H.; Adkins, B.D.; Bartfeld, S.; Blumberg, R.S.; Farber, D.L.; Garssen, J.; Ghazal, P.; Hackam, D.J.; Marsland, B.J.; McCoy, K.D.; et al. The Neonatal Window of Opportunity-Early Priming for Life. *J. Allergy Clin. Immunol.* **2018**, *141*, 1212–1214. [CrossRef]

80. Tabilas, C.; Iu, D.S.; Daly, C.W.P.; Mon, K.J.Y.; Reynaldi, A.; Wesnak, S.P.; Grenier, J.K.; Davenport, M.P.; Smith, N.L.; Grimson, A.; et al. Early Microbial Exposure Shapes Adult Immunity by Altering CD8+ T Cell Development. *Proc. Natl. Acad. Sci. USA* **2022**, *119*, e2212548119. [CrossRef]

81. Parra, M.; Herrera, D.; Jácome, M.F.; Mesa, M.C.; Rodríguez, L.-S.; Guzmán, C.; Angel, J.; Franco, M.A. Circulating Rotavirus-Specific T Cells Have a Poor Functional Profile. *Virology* **2014**, *468–470*, 340–350. [CrossRef]

82. Jaimes, M.C.; Rojas, O.L.; González, A.M.; Cajiao, I.; Charpilienne, A.; Pothier, P.; Kohli, E.; Greenberg, H.B.; Franco, M.A.; Angel, J. Frequencies of Virus-Specific CD4(+) and CD8(+) T Lymphocytes Secreting Gamma Interferon after Acute Natural Rotavirus Infection in Children and Adults. *J. Virol.* **2002**, *76*, 4741–4749. [CrossRef]

83. Angel, J.; Steele, A.D.; Franco, M.A. Correlates of Protection for Rotavirus Vaccines: Possible Alternative Trial Endpoints, Opportunities, and Challenges. *Hum. Vaccin. Immunother.* **2014**, *10*, 3659–3671. [CrossRef]

84. Lavelle, E.C.; Ward, R.W. Mucosal Vaccines—Fortifying the Frontiers. *Nat. Rev. Immunol.* **2022**, *22*, 236–250. [CrossRef]

85. Araki, K.; Turner, A.P.; Shaffer, V.O.; Gangappa, S.; Keller, S.A.; Bachmann, M.F.; Larsen, C.P.; Ahmed, R. MTOR Regulates Memory CD8 T-Cell Differentiation. *Nature* **2009**, *460*, 108–112. [CrossRef] [PubMed]

86. Mannick, J.B.; Del Giudice, G.; Lattanzi, M.; Valiante, N.M.; Praestgaard, J.; Huang, B.; Lonetto, M.A.; Maecker, H.T.; Kovarik, J.; Carson, S.; et al. MTOR Inhibition Improves Immune Function in the Elderly. *Sci. Transl. Med.* **2014**, *6*, 268ra179. [CrossRef] [PubMed]

87. DuPont, H.L.; Jiang, Z.-D.; Alexander, A.S.; DuPont, A.W.; Brown, E.L. Intestinal IgA-Coated Bacteria in Healthy- and Altered-Microbiomes (Dysbiosis) and Predictive Value in Successful Fecal Microbiota Transplantation. *Microorganisms* **2022**, *11*, 93. [CrossRef]

88. Hsu, P.S.; Nanan, R. Does Breast Milk Nurture T Lymphocytes in Their Cradle? *Front. Pediatr.* **2018**, *6*, 268. [CrossRef]

Article

Rotavirus-Specific Maternal Serum Antibodies and Vaccine Responses to RV3-BB Rotavirus Vaccine Administered in a Neonatal or Infant Schedule in Malawi

Benjamin Morgan [1,†], Eleanor A. Lyons [1,†], Amanda Handley [1,2], Nada Bogdanovic-Sakran [1], Daniel Pavlic [1], Desiree Witte [3,4], Jonathan Mandolo [3,5], Ann Turner [4], Khuzwayo C. Jere [3,4], Frances Justice [1], Darren Suryawijaya Ong [1], Rhian Bonnici [1], Karen Boniface [1], Celeste M. Donato [1], Ashley Mpakiza [3], Anell Meyer [1,6], Naor Bar-Zeev [4], Miren Iturriza-Gomara [4,7], Nigel A. Cunliffe [4], Margaret Danchin [1,7,8] and Julie E. Bines [1,6,9,*]

1. Enteric Diseases, Murdoch Children's Research Institute, Parkville, VIC 3052, Australia; amanda.handley@mcri.edu.au (A.H.); nada.bogdanovic@mcri.edu.au (N.B.-S.); daniel.pavlic@mcri.edu.au (D.P.); fran.justice@mcri.edu.au (F.J.); darren.ong@mcri.edu.au (D.S.O.); rhian.bonnici@mcri.edu.au (R.B.); celeste.donato@mcri.edu.au (C.M.D.); margie.danchin@mcri.edu.au (M.D.)
2. Medicines Development for Global Health, Melbourne, VIC 3001, Australia
3. Malawi Liverpool Welcome Trust Programme, Blantyre P.O. Box 30096, Chichi, Malawi; dwitte@mlw.mw (D.W.); jmandolo@mlw.mw (J.M.); khuzwayo.jere@liverpool.ac.uk (K.C.J.)
4. Institute of Infection, Veterinary and Ecological Sciences, University of Liverpool, Liverpool L69 7ZX, UK; amturner@liverpool.ac.uk (A.T.); m.iturriza-gomara@liverpool.ac.uk (M.I.-G.); n.a.cunliffe@liverpool.ac.uk (N.A.C.)
5. Department of Clinical Sciences, Liverpool School of Tropical Medicine, Liverpool L3 5QA, UK
6. Department of Gastroenterology and Clinical Nutrition, Royal Children's Hospital, Parkville, VIC 3052, Australia
7. GSK Vaccines for Global Health Institute, 53100 Sienna, Italy
8. Department of General Medicine, Royal Children's Hospital, Parkville, VIC 3052, Australia
9. Department of Paediatrics, The University of Melbourne, Parkville, VIC 3052, Australia
* Correspondence: julie.bines@mcri.edu.au; Tel.: +61-38341 6451
† These authors contributed equally to this work.

Citation: Morgan, B.; Lyons, E.A.; Handley, A.; Bogdanovic-Sakran, N.; Pavlic, D.; Witte, D.; Mandolo, J.; Turner, A.; Jere, K.C.; Justice, F.; et al. Rotavirus-Specific Maternal Serum Antibodies and Vaccine Responses to RV3-BB Rotavirus Vaccine Administered in a Neonatal or Infant Schedule in Malawi. *Viruses* 2024, 16, 1488. https://doi.org/10.3390/v16091488

Academic Editors: Ulrich Desselberger and John T. Patton

Received: 16 August 2024
Revised: 11 September 2024
Accepted: 13 September 2024
Published: 19 September 2024

Abstract: High titres of rotavirus-specific maternal antibodies may contribute to lower rotavirus vaccine efficacy in low- and middle-income countries (LMICs). RV3-BB vaccine (G3P[6]) is based on a neonatal rotavirus strain that replicates well in the newborn gut in the presence of breast milk. This study investigated the association between maternal serum antibodies and vaccine response in infants administered the RV3-BB vaccine. Serum was collected antenatally from mothers of 561 infants enrolled in the RV3-BB Phase II study conducted in Blantyre, Malawi, and analysed for rotavirus-specific serum IgA and IgG antibodies using enzyme-linked immunosorbent assay. Infant vaccine take was defined as cumulative IgA seroconversion (≥ 3 fold increase) and/or stool vaccine shedding. Maternal IgA or IgG antibody titres did not have a negative impact on vaccine-like stool shedding at any timepoint. Maternal IgG (but not IgA) titres were associated with reduced take post dose 1 ($p < 0.005$) and 3 ($p < 0.05$) in the neonatal vaccine schedule group but not at study completion (week 18). In LMICs where high maternal antibodies are associated with low rotavirus vaccine efficacy, RV3-BB in a neonatal or infant vaccine schedule has the potential to provide protection against severe rotavirus disease.

Keywords: rotavirus vaccine; maternal antibodies; RV3-BB vaccine

1. Introduction

Rotavirus vaccines are recommended for all children and have been introduced in 126 countries worldwide, providing a cost-effective pathway to reduce hospitalization and death among young children [1,2]. However, despite this significant achievement, rotavirus gastroenteritis remains the most common cause and still accounts for a quarter of global diarrheal deaths in children less than 5 years of age [3]. Barriers to the success of rotavirus vaccines include challenges to vaccine access (such as cost, supply, health prioritization, and service delivery), timely administration, and vaccine safety concerns [1,4]. A significant disparity in vaccine efficacy and effectiveness has been observed for the licensed vaccines Rotarix® (GlaxoSmithKline, Rixensart, Belgium) and RotaTeq® (Merck, Whitehouse Station, NJ, USA) when rotavirus vaccines have been implemented in low- and middle-income countries (LMICs) compared to high-income countries (HICs) [5]. The median vaccine effectiveness of the Rotarix® vaccine is estimated at 57% when implemented in LMICs compared to 84% in HICs. A similar pattern has been reported with the RotaTeq® vaccine at 45% and 90% for LMICs and HICs, respectively [6]. It has been postulated that co-administration with oral polio vaccine, interference in replication by other enteric pathogens, immaturity of the immune system, and malnutrition could contribute to this disparity [5–7].

Maternal serum (IgG) and breast milk (IgA) antibodies are critical to provide passive immunity while infants establish their own immune defences and have been proposed as a major factor contributing to lower rotavirus vaccine efficacy in LMICs [5,7–10]. Serum IgG antibodies are transferred across the placenta and provide systemic passive immunity for a newborn during their first few months of life, with high titres commonly observed at birth followed by a decline over time, with a half-life of 3 to 4 weeks (Figure 1) [8]. The rate of this decline is influenced by the initial antibody titre, maternal health and nutrition, and exposure to wild-type infection [8,11]. High titres of maternal serum rotavirus-specific IgG antibodies have consistently been associated with reduced immunogenicity after administration of rotavirus vaccines and have been proposed to be associated with reduced vaccine efficacy in LMICs [8,9,12,13]. Serum IgA antibodies are detected in maternal serum but not transmitted transplacentally. Infants rely on maternal breast milk IgA to provide local immunity to the gut mucosa and support protection from enteric pathogenic bacteria and viruses [8]. Reduced immunogenicity in response to a rotavirus vaccine was reported in infants of mothers with high rotavirus-specific IgA titres in breast milk in Vietnam and Zambia but not in Nicaragua [12–14]. However, clinical trials of withholding breastfeeding (30 min to 60 min) prior to administration of the Rotarix® vaccine in South Africa, Pakistan, and India failed to show a difference in seroconversion in those infants who had breastfeeding withheld compared to those who did not [15–17].

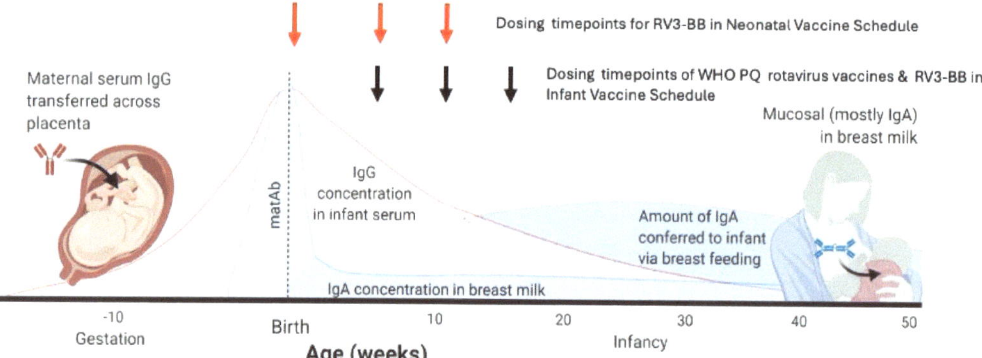

Figure 1. Timing of administration of rotavirus vaccines in association with age-related changes in titres of maternal serum and breastmilk antibodies and infant serum IgG and IgA titres. Neonatal vaccine schedule (first dose at birth) (red arrows) with RV3-BB vaccine; infant vaccine schedule (first dose at 6–8 weeks of age) (black arrows): WHO-prequalified rotavirus vaccines (Rotarix®, RotaTeq®, Rotavac® (Bharat Biotech, Hyderabad, India) and Rotasiil® (Serum Institute of India, Pune, India) and RV3-BB vaccine administered in the infant schedule). Adapted from: Otero CE, Langel SN, Blasi M, Permar SR. PLoS Pathogens 2020 16(11):e1009010 [8].

Administration of a rotavirus vaccine at birth or soon after birth has the potential to address some of the current challenges to vaccine implementation and to improve the performance and safety of a rotavirus vaccine [18]. However, administration of an oral rotavirus vaccine from birth maximises exposure to high levels of maternal serum IgG antibodies and breast milk IgA [8] (Figure 1). The RV3-BB rotavirus vaccine is based on an asymptomatic neonatal rotavirus strain (RV3: G3P[6]) that replicates well in the newborn gut despite the presence of breast milk [19,20]. Clinical trials have shown that RV3-BB is well tolerated in a neonatal vaccine schedule, is immunogenic, and was associated with a vaccine efficacy of 75% against severe rotavirus disease at 18 months in Indonesia [21,22]. In clinical trials in New Zealand and Indonesia, there was no association between maternal rotavirus-specific IgG antibody titre, colostrum or breast milk IgA antibodies, and vaccine take, serum IgA response, or stool vaccine virus shedding after three doses of RV3-BB vaccine [23,24].

Infants living in Malawi are often exposed to rotavirus infection early in life [25]. High titres of maternal rotavirus-specific antibodies have been reported in mothers in Malawi and were negatively correlated with response to Rotarix® vaccine, in particular vaccine virus shedding [7]. The primary aim of this study was to determine whether maternal serum rotavirus-specific IgA and IgG antibody titres in mothers in Malawi are associated with vaccine take (serum IgA response and/or vaccine virus shedding) in their infants following administration of the RV3-BB rotavirus vaccine in a neonatal vaccine schedule (with the first dose within 5 days of birth) or infant vaccine schedule (with the first dose administered at 6 weeks of age).

2. Materials and Methods

2.1. Study Design and Participants

The Maternal Antibody study was an exploratory analysis nested within the Phase II RV3-BB trial, a randomized, double-blind, placebo-controlled four-arm parallel-group, dose-ranging study of oral human neonatal rotavirus vaccine (RV3-BB) administered at a titre of 1.0×10^6 FFU per mL (low-titre group), 3.0×10^6 FFU per mL (mid-titre group), or 1.0×10^7 FFU per mL (high-titre group) as a three-dose neonatal schedule, or administered at a titre of 1.0×10^7 FFU per mL as a three-dose infant schedule (Figure 2) [26]. The study was conducted between September 2018 and January 2020, and involved 711 infants recruited from three primary healthcare centres in Blantyre, Malawi [26]. All mothers of the 711 eligible participants recruited to the Phase II study were invited to participate in the Maternal Antibody study. A two-stage consent process was followed. Pregnant women provided consent for the collection of a pre-birth maternal blood and stool sample and an after-birth infant cord blood and stool sample. Written informed study consent was obtained after birth from parents or guardians. Study exclusions were the same as for the main study [26]. Participants from the Phase II study per protocol population who received three doses of vaccine according to the protocol and with maternal blood samples available for analysis were included in the Maternal Antibody study.

The study was conducted in accordance with the International Council for Harmonization of Good Clinical Practice Guidelines. The protocols were approved by the Ethics Committees of the Royal Children's Hospital Melbourne and the University of Liverpool, the National Health Science Research Committee, and the Pharmacy and Medicines and Poisons Board of Malawi. This trial is registered at ClinicalTrials.gov (NCT03483116).

2.2. Sample Collection and Processing

Maternal venous blood (5–10 mL) was collected in the second or third trimester of pregnancy. Serum was isolated from whole blood and stored at $-70\,^\circ$C until analysed. Blood was collected from the cord (baseline for neonatal schedule comparison) immediately before IP dose 2 (baseline for infant schedule comparison), 28 days after IP dose 3, and 28 days after IP dose 4. A pre-dose infant blood sample was also collected at IP doses 2, 3, and 4 and at 18 weeks of age [26]. The serum was frozen at $-70\,^\circ$C and shipped to the Murdoch Children's Research Institute laboratory for analysis. Rotavirus-specific IgA and

IgG antibody titres were measured by enzyme-linked immunosorbent assay (ELISA) using rabbit anti-RV3 polyclonal antisera as the coating antibody and RV3-BB virus or Vero cell lysate as the capture antigen [27]. The antigen–antibody complexes were detected with biotinylated anti-human IgA and streptavidin–horseradish peroxidase [27]. Concentrations of rotavirus-specific IgA were measured using a standard curve generated from known positive serum samples arbitrarily assigned a titre of 250,000 units per millilitre (U/mL). The lower limit of detection was 20 U/mL.

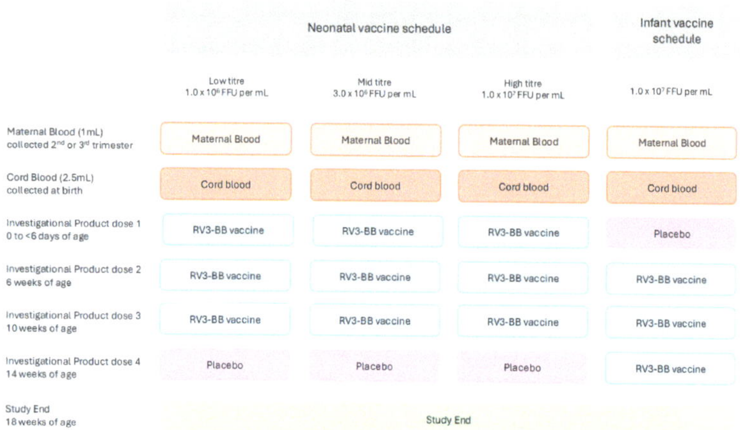

Figure 2. Study design.

Rotavirus shedding was assessed in stool samples collected at baseline, before administration of the first dose of IP, and between days 3 to 7 after administration of each dose of IP. Samples were assessed using a rotavirus VP6-specific reverse transcriptase polymerase chain reaction (PCR). PCR products were analysed by electrophoresis with the Invitrogen one-step RT-PCR key (Invitrogen, Carlsbad, CA, USA) and Rot3 and Rot5 oligonucleotide primers [28]. Sequence analysis was used to confirm the presence of the RV3-BB vaccine (Sequencher Software program version 4.1, Gene Codes Corp Inc., Ann Arbor, MI, USA) and identity determined by GenBank database [28].

2.3. Definition of Vaccine Response

IgA seroconversion was defined as serum anti-rotavirus IgA antibody titre equal to or greater than three times the titre of the baseline titre and was assessed after each dose of IP. Stool shedding was defined as the presence of RV3-BB vaccine-like virus detected in the stool collected between 3 to 7 days after a dose of IP. Vaccine take was defined as a serum immune response ($\geq$3-fold increase in titre from baseline) of anti-rotavirus IgA 28 days following IP administration, and/or the presence of vaccine-like virus in stool 3 to 7 days following IP administration [26]. Cumulative vaccine serum response, stool shedding, and/or vaccine take was defined as a positive result following one, two, or three IP doses for the neonatal vaccine group, and following two, three, or four IP doses for the infant vaccine schedule group.

2.4. Statistical Analysis

Demographic characteristics of participants are presented using means and standard deviations (SDs) for continuous variables and proportions for categorical variables. Mean maternal serum IgA and IgG titres are presented as log (natural) transformed data. Infant serum IgA response, stool excretion, and cumulative vaccine take are presented as numbers and proportions for each schedule group and the combined neonatal vaccine group (low-titre, mid-titre, and high-titre neonatal groups combined). Separate linear regression models were used to explore the relationship between maternal serum IgA and IgG titres against infant anti-rotavirus serum IgA antibody titre and seroconversion after dose 1 and dose 3. Statistical analyses were performed with Prism (GraphPad Software, Inc., La Jolla, CA, USA, version 10.0.1) by use of the unpaired Mann–Whitney test due to data not fitting the normal distribution. A *p* value of less than 0.05 was considered to be significant.

3. Results

3.1. Study Population

A total of 711 infants were recruited to the Phase II study. Of these, 565 participants received three doses of the RV3-BB vaccine according to the protocol, therefore fulfilling criteria for inclusion in the per-protocol analysis population [26]. All 565 infants were eligible to be included in the Maternal Antibody study; however, four mother–infant pairs were excluded, as insufficient or no blood was available from the mother for analysis, resulting in an analysis study population of 561 infants (Figure 3). There were no significant differences in characteristics between infants recruited to the Maternal Antibody study compared to the Phase II study, or between treatment allocation groups (Table 1).

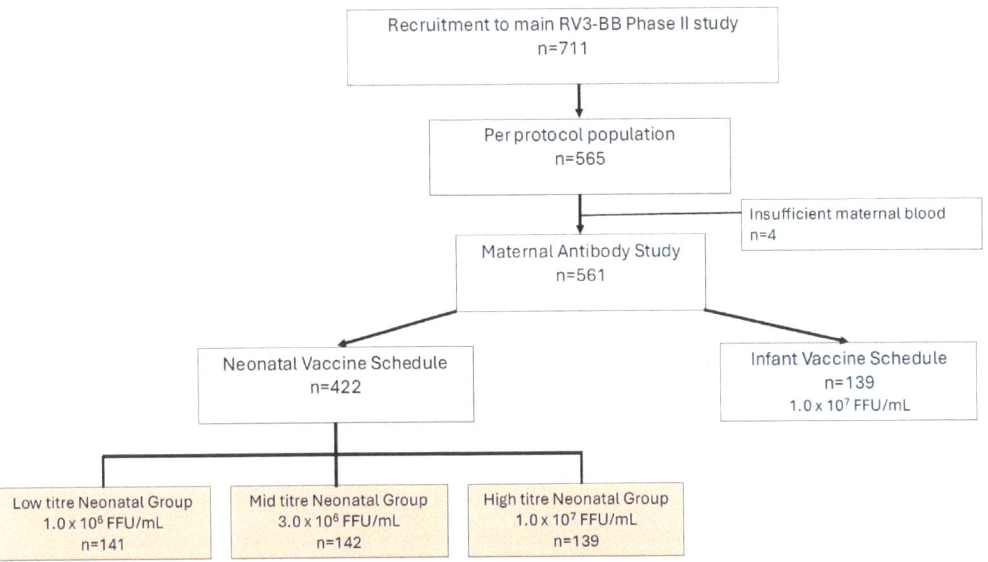

Figure 3. Infant participant flow for the Maternal Antibody study.

Table 1. Infant participant characteristics.

	Neonatal Vaccine Schedule ($n = 422$)			Infant Vaccine Schedule ($n = 139$)
	Low Titre ($n = 141$)	Mid Titre ($n=142$)	High Titre ($n = 139$)	($n = 139$)
Gestational age (weeks): mean (min, max)	37.5 (34, 41)	37.3 (36, 40)	37.6 (35, 43)	37.6 (36, 45)
Birth weight (grams): mean (SD)	3121.9 (359.59)	3117.9 (384.39)	3089.4 (344.49)	3150.8 (362.65)
Sex (male): number (%)	70 (49.6)	67 (47.2)	81 (58.3)	74 (53.2)
Ethnicity (Black African): number (%)	141 (100)	142 (100)	139 (100)	139 (100)
Exclusive breastfeeding duration: number (%)				
Day 1–6	141 (100)	142 (100)	139 (100)	139 (99.3)
Week 1–6	141 (100)	142 (100)	139 (98.6)	139 (99.3)
Week 10	141 (100)	142 (100)	139 (98.6)	139 (100)
Week 14	139 (98.6)	140 (97.9)	139 (98.6)	139 (100)

3.2. Maternal Antibodies and RV3-BB Vaccine Response

3.2.1. Maternal Antibodies and RV3-BB Vaccine Stool Shedding

Maternal rotavirus-specific IgA and IgG antibody titres were not negatively associated with RV3-BB vaccine virus shedding in the stool at any timepoint in the neonatal and infant vaccine schedule. A positive association between high maternal IgA titres and stool shedding observed after the third dose of RV3-BB vaccine (IP dose 3) and at study completion at week 18 (IP dose 4) in the neonatal vaccine schedule (Figure 4a).

3.2.2. Maternal Antibodies and RV3-BB Vaccine Anti-Rotavirus IgA Seroconversion

Maternal serum IgA antibody titres were inversely associated with anti-rotavirus IgA seroconversion in infants after three doses of RV3-BB vaccine (IP dose 3 and IP dose 4 in the combined neonatal group (Figure 4b) and in the high-titre neonatal vaccine schedule group ($p = 0.03$) (Supplementary Table S1). High maternal IgG antibody titres were associated with a lower proportion of infants who had anti-rotavirus serum IgA seroconversion after one dose (IP dose 1) and three doses of RV3BB vaccine (IP dose 3 and IP dose 4) administered in the neonatal vaccine schedule group ($p = 0.0013$, $p < 0.001$, and $p < 0.001$, respectively) (Figure 4b). This was also observed when analysed in the three vaccine titre groups (Supplementary Table S1). No significant association was observed between maternal rotavirus-specific serum IgA or IgG antibody titres (log transformed) and anti-rotavirus serum IgA antibody titres in their infant after dose 1 or dose 3 in the neonatal vaccine schedule (Figure 5). In contrast, there was no significant association observed between maternal serum IgA or IgG antibody titres and IgA seroconversion in infants administered RV3-BB in the infant vaccine schedule (Figure 4b). However, a significant association was observed between maternal serum IgA and IgG antibody titres and the anti-rotavirus serum IgA titres of their infant after vaccine dose one (IP dose 2), and with maternal serum IgG antibody titres and anti-rotavirus serum IgA antibody titres in their infant after the third vaccine dose (IP dose 4) (Figure 5).

3.2.3. Maternal Antibodies and RV3-BB Vaccine Take

Maternal rotavirus-specific IgA antibody titres were not significantly different in participants with a positive or negative vaccine take after dose 1 or 3 of vaccine in the combined neonatal schedule group (IP dose 1 and IP dose 3) or at study completion at 18 weeks of age (IP dose 4) (Figure 4c). In the neonatal vaccine schedule group, there were no significant differences observed when analysed according to vaccine titre groups (Supplementary Table S1) or in the infant schedule group ($p > 0.05$ for all comparisons) (Supplementary Table S1). An association between high maternal IgG antibody titres and vaccine take in the infant were observed after dose 1 and dose 3 in the combined neonatal vaccine schedule group ($p = 0.005$ and $p = 0.04$, respectively), but this association was not sustained at study completion at 18 weeks of age (IP dose 4) ($p > 0.05$) (Figure 3). No association between maternal IgG antibody titre and vaccine take in the infant was observed when analysed according to vaccine titre group ($p > 0.05$ for all comparisons) or in the infant schedule group ($p > 0.05$) (Supplementary Table S1).

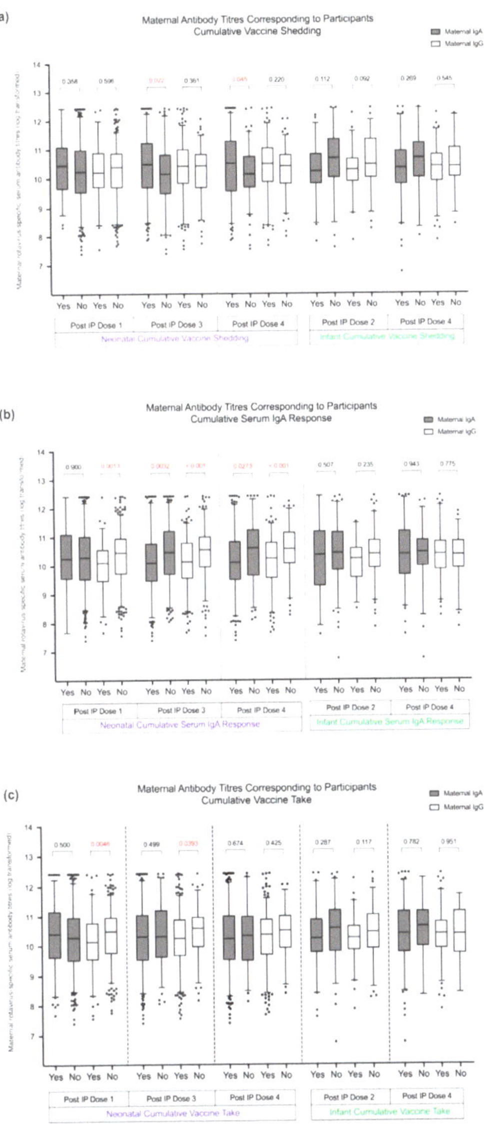

Figure 4. (**a–c**) Mean maternal rotavirus-specific serum IgA and IgG antibody titres (log transformed) in association with vaccine response in participants administered RV3-BB vaccine in the combined neonatal vaccine schedule group and the infant vaccine schedule group. The "y" axis denotes the serum maternal serum IgA and IgG antibodies titres (log). The "x" axis denotes the study groups according to neonatal or infant vaccine schedule group, per Investigation product (IP) dose, and according to vaccine response with the positive vaccine response variable ("Yes") or negative vaccine response variable ("No"). Data are presented in a box-and-whisker plot, with the box extending from the 25th to the 75th percentile and the line in the middle plotted at the median. The whiskers represent the 10–90th percentiles, with all datapoints outside the 10–90th percentiles shown. Statistics and plotting were performed in Prism 9 for MacOs. Non-normally distributed data were analysed using the Mann–Whitney test. ▬ Maternal rotavirus-specific serum IgA antibody titres (log). ▭ Maternal rotavirus-specific serum IgG antibody titres (log).

(a) Post dose 1

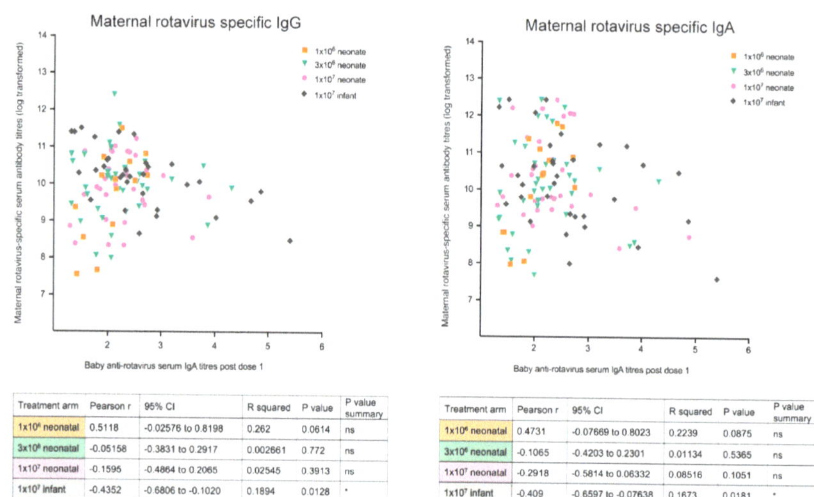

Maternal rotavirus specific IgG

Treatment arm	Pearson r	95% CI	R squared	P value	P value summary
1×10^6 neonatal	0.5118	-0.02576 to 0.8198	0.262	0.0614	ns
3×10^6 neonatal	-0.05158	-0.3831 to 0.2917	0.002661	0.772	ns
1×10^7 neonatal	-0.1595	-0.4864 to 0.2065	0.02545	0.3913	ns
1×10^7 infant	-0.4352	-0.6806 to -0.1020	0.1894	0.0128	*

Maternal rotavirus specific IgA

Treatment arm	Pearson r	95% CI	R squared	P value	P value summary
1×10^6 neonatal	0.4731	-0.07669 to 0.8023	0.2239	0.0875	ns
3×10^6 neonatal	-0.1065	-0.4203 to 0.2301	0.01134	0.5365	ns
1×10^7 neonatal	-0.2918	-0.5814 to 0.06332	0.08516	0.1051	ns
1×10^7 infant	-0.409	-0.6597 to -0.07638	0.1673	0.0181	*

(b) Post dose 3

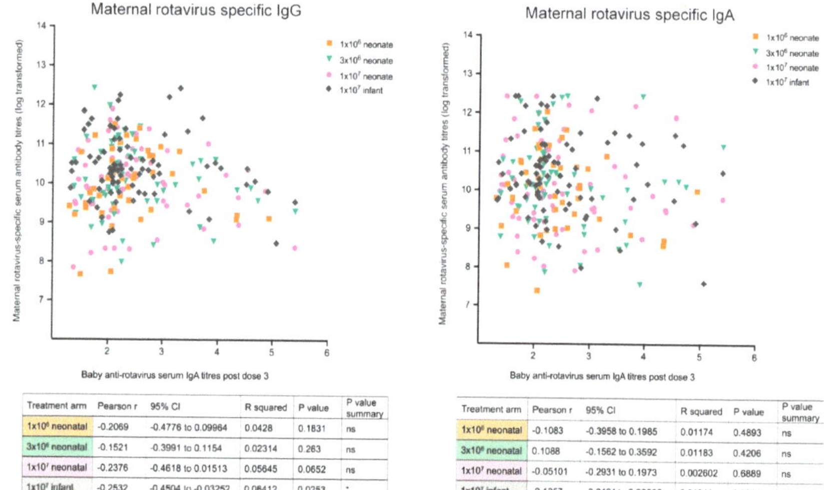

Maternal rotavirus specific IgG

Treatment arm	Pearson r	95% CI	R squared	P value	P value summary
1×10^6 neonatal	-0.2069	-0.4776 to 0.09964	0.0428	0.1831	ns
3×10^6 neonatal	-0.1521	-0.3991 to 0.1154	0.02314	0.263	ns
1×10^7 neonatal	-0.2376	-0.4618 to 0.01513	0.05645	0.0652	ns
1×10^7 infant	-0.2532	-0.4504 to -0.03252	0.06412	0.0253	*

Maternal rotavirus specific IgA

Treatment arm	Pearson r	95% CI	R squared	P value	P value summary
1×10^6 neonatal	-0.1083	-0.3958 to 0.1985	0.01174	0.4893	ns
3×10^6 neonatal	0.1088	-0.1562 to 0.3592	0.01183	0.4206	ns
1×10^7 neonatal	-0.05101	-0.2931 to 0.1973	0.002602	0.6889	ns
1×10^7 infant	-0.1357	-0.3464 to 0.08806	0.01841	0.2331	ns

Figure 5. Maternal rotavirus-specific serum IgA and IgG antibody titres are plotted against the anti-rotavirus serum IgA antibody titres of their infant after (**a**) the first dose (post vaccine dose 1) and (**b**) the full three-dose course (post vaccine dose 3) in each of the four treatment allocation groups. Separate linear regression models were used to explore the relationship between maternal rotavirus-specific serum IgA and IgG antibody titres (log transformed) against infant anti-rotavirus serum IgA antibody titre after vaccine dose 1 and dose 3. Statistical analyses were performed with Prism (GraphPad Software, Inc., version 10.0.1) by use of the unpaired Mann–Whitney test due to data not fitting the normal distribution. A *p* value of less than 0.05 was considered to be significant. 95% CI = 95% confidence interval. * = significant $p < 0.05$. ns = not significant $p > 0.05$. 1.0×10^6 neonate = low-titre neonatal vaccine schedule group. 3.0×10^6 neonate = mid-titre neonatal vaccine schedule group. 1.0×10^7 neonate = high-titre neonatal vaccine schedule group. 1.0×10^7 infant = infant vaccine schedule group.

4. Discussion

In this study, maternal rotavirus-specific serum IgA antibody titres did not reduce the take of the RV3-BB rotavirus vaccine when administered in a neonatal vaccine schedule (with the first dose at birth) or an infant vaccine schedule (first dose at 6 weeks of age) in infants in Malawi. High maternal IgG antibodies reduced take after one and three doses of vaccine in the neonatal schedule, although this was not observed at study end (week 18), suggesting that any inhibition was not sustained. Stool shedding of vaccine virus was not impacted by high maternal rotavirus-specific IgA or IgG antibody titres. The results of this study reflect findings of previous RV3-BB clinical trials conducted in New Zealand and Indonesia and suggests that this finding is not specific to a region or a population [23,24].

High maternal rotavirus-specific serum and breastmilk antibody levels have been reported in a number of studies from LMICs, including from India, Nicaragua, Mexico, Indonesia, and South Africa, compared to mothers from the USA [7–10,12–15,29,30]. The inverse relationship between serum maternal IgG and/or IgA antibody titres and infant immune responses has previously been reported following administration of rotavirus vaccines [8,9,13]. High maternal pre-vaccination serum IgG antibody titres were negatively associated with infant seroconversion ($p = 0.031$) and infant serum IgA antibody titres following administration of the first dose of the Rotarix® vaccine in South African infants, although this inhibition was overcome by the second dose of the vaccine [30]. Similar observations between rotavirus-specific maternal IgG antibody titres and infant seroconversion were reported following the first dose of the RotaTeq® vaccine in infants from Nicaragua ($p = 0.02$) [12] and after three doses of the Rotavac® vaccine in infants in India [31]. Although we observed a negative association between the titres of maternal antibody and anti-rotavirus IgA antibody titres in the infant vaccine schedule, this did not translate to a lack of anti-rotavirus IgA seroconversion in these infants. It has been proposed that the negative relationship between high maternal antibodies in mothers from LMICs and infant serum IgA response after the first vaccine dose but not consistently observed after a complete two- or three- dose vaccine schedule likely relates to the normal exponential decline in transplacental IgG antibodies [24,25,32]. Delaying the first dose of a rotavirus vaccine schedule in an effort to avoid the inhibitory effect of maternal antibodies has been suggested, but this would leave young infants unprotected, particularly in high rotavirus disease burden regions in LMICs where severe rotavirus disease still occurs, despite the presence of high titres of maternal antibodies. Delaying the administration of rotavirus vaccines also may have other implications, including an increased risk of vaccine-associated intussusception [33].

Maternal IgG antibodies have been associated with an inhibition of seroconversion with all vaccine types for a range of viral and bacterial pathogens—not only live-attenuated vaccines but also inactivated, subunit, and protein vaccines [11]. Inhibition of vaccine responses by maternal antibodies has been reported for a number of childhood vaccines, including measles and polio vaccine [11]. It has been proposed that inhibition of vaccine antigen-specific B-cell activation by maternal IgG occurs via the development of a vaccine–antibody complex involving cross-linking the B-cell receptor and the inhibitory/regulatory Fcγ-receptor IIB and/or epitope masking, neutralization of vaccine virus, and removal of vaccine antigen by phagocytosis [11]. But if, or how, these mechanisms may differ in children in LMICs compared to children in HICs is not known.

The lack of an accurate serologic correlate of protection presents a challenge in comparing results between rotavirus vaccine studies [34,35]. Anti-rotavirus serum IgA antibody levels after acute infection are currently considered the best serological marker of protection; however, they are an indirect marker, as intestinal IgA is considered as the key mechanism for clearance of infection [36,37]. Anti-rotavirus serum IgA response is commonly used as a serological marker of immunogenicity in clinical vaccine trials, although the laboratory method for analysis and definitions of vaccine response vary across vaccine studies [38]. Higher titres of infant serum IgA correlate with a lower risk of rotavirus infection, although this is not always a consistent measure of protection against rotavirus disease across studies [34,35,38]. Rotavirus serum-neutralising antibodies have also been measured as a serum marker of vaccine response [34,35]. In New Zealand infants, serum-neutralising antibody responses were not impacted by colostrum or breastmilk IgA titres after three doses of RV3-BB vaccine [23]. Although not validated as a correlate of protection, vaccine virus shedding detected in the stool after immunisation with a rotavirus vaccine reflects gut replication of the vaccine virus and has been proposed as a marker of the mucosal immune response [7,21,22,26,39]. In a comparative study of rotavirus vaccine responses in Malawi and India, the negative correlation between maternal rotavirus-specific serum and breastmilk IgA antibodies and rotavirus vaccine response was suggested to be driven by a reduction in vaccine virus replication and shedding [7]. Interestingly, in the United Kingdom cohort in the same study, vaccine virus shedding was not inhibited in the presence of similar titres of maternal serum IgA antibodies [7]. In contrast, our study did not identify a negative impact on RV3-BB vaccine virus shedding by high maternal antibody titres. On the contrary, increased shedding was observed in the presence of high maternal IgA antibody titres in the neonatal vaccine schedule group. The RV3-BB vaccine is a human, neonatal rotavirus vaccine based on the naturally attenuated, asymptomatic human neonatal strain (RV3: G3P[6]). Neonatal P[6] strains are phenotypically different from pathogenic wildtype rotavirus strains [40]. The VP4 outer capsid proteins of neonatal P[6] rotavirus strains differ at a number of amino acid positions compared to pathogenic strains isolated from older infants and children. The wildtype RV3 strain has six amino acid changes located on the basal surface of the VP8* core, which sits outside the putative neutralization domain [40]. It has been proposed that these changes may allow neonatal strains to bind more efficiently to carbohydrate molecules or proteins found on enterocytes in the newborn gut and potentially evade neutralisation by maternal IgG antibodies [40]. Importantly, in Indonesia, the RV3-BB vaccine provided robust protection against severe rotavirus disease in the first 18 months of life (75%) when administered in the neonatal vaccine schedule, despite a negative association observed between cord blood IgG antibody titres and serum IgA seroconversion and vaccine take after the first dose [21,24].

A limitation of this study reflects the lack of a perfect serological surrogate for protection against rotavirus infection or for the assessment of the response to a rotavirus vaccine. To address this limitation, we have presented results of "vaccine take," which takes into account anti-rotavirus IgA seroconversion and vaccine virus shedding in the stool [21,22,26]. To understand the characteristics of the influence of maternal antibodies on infant vaccine response, it would be ideal to measure maternal antibodies at a number of time points antenatally, during labour and cord blood, and also to assess a broader panel of immune factors. Unfortunately, this was not feasible within the confines of this study. As serum maternal IgA antibodies are not transferred via the placenta, breastmilk IgA antibodies would have also provided further information but were not measured in the current study. However, we did not find an association between IgA antibodies in colostrum or breastmilk and vaccine take after three doses of RV3-BB vaccine when administered in either the neonatal or infant vaccine schedules in New Zealand or Indonesia [23,24].

In summary, maternal rotavirus-specific IgG (but not IgA) antibody titres were associated with reduced vaccine take at after dose 1 and dose 3 in the neonatal vaccine schedule group but not at study completion (week 18). Maternal IgG and IgA antibody titres were inversely associated with IgA seroconversion in the neonatal vaccine schedule group but had no impact on vaccine shedding at any timepoint in either schedule. These results are consistent with observations from RV3-BB vaccine studies conducted in Indonesia and New Zealand, where there was no association between IgA in colostrum and breastmilk and vaccine take after three doses of RV3-BB vaccine when administered in the neonatal or infant vaccine schedules. Using a neonatal vaccine schedule RV3-BB vaccine has the potential to address the disparity in rotavirus vaccine performance that currently persists in LMICs.

Supplementary Materials: The following supporting information can be downloaded at: https://www.mdpi.com/article/10.3390/v16091488/s1, Table S1: Comparisons between maternal rotavirus-specific IgA and IgG antibodies titres (log transformed) and vaccine responses in their infants who had been administered RV3-BB in the neonatal vaccine schedule group presented according to vaccine titre groups.

Author Contributions: Study conception and design: J.E.B., A.H., N.A.C., N.B.-Z., M.I.-G. and D.W.; sample procurement and handling: D.W., A.T., A.M. (Ashley Mpakiza), K.C.J. and J.M.; laboratory methods and analysis: N.B.-S., D.P., D.S.O., R.B., K.B. and C.M.D.; data analysis: E.A.L., J.E.B., A.H., F.J. and B.M.; funding acquisition: J.E.B.; B.M. wrote the first draft of the manuscript with input from J.E.B., A.M. (Anell Meyer) and E.A.L.; All authors contributed to the review and editing of the manuscript review; J.E.B. and A.H. accessed and verified all data presented in this report and had the final responsibility for the decision to submit for publication. All authors have read and agreed to the published version of the manuscript.

Funding: The Malawi trial was funded by the Bill and Melinda Gates Foundation (OPP111055) and an Australian Tropical Medicine Commercialisation Grant (App 50343). This research at MCRI was supported by the Victorian Government's Operational Infrastructure Support Program. The funders of this study had a limited role in initial discussions of the study design but were not involved in protocol development; the collection, analysis, or interpretation of data; the writing of the report; or the decision to submit the article for publication.

Institutional Review Board Statement: The study was conducted in accordance with the International Council for Harmonization of Good Clinical Practice Guidelines. The protocols were approved by the Ethics Committees of the Royal Children's Hospital Melbourne and the University of Liverpool, National Health Science Research Committee, and the Pharmacy and Medicines and Poisons Board in Malawi. This trial is registered at ClinicalTrials.gov (NCT03483116). A two-stage consent process was followed. Pregnant women provided consent for the collection of a pre-birth maternal blood and stool sample and an after-birth infant cord blood and stool sample. Written informed study consent was obtained after birth from parents/guardians.

Informed Consent Statement: Informed consent was obtained from all subjects involved in the study.

Data Availability Statement: Data are available on request.

Acknowledgments: We would sincerely like to thank the participants and their families for participating in the study. We acknowledge the hard work and compassion of the RV3 study teams at the Bangwe, Limbe, and Nirande Health Centres and Queen Elizabeth Hospital. We would also like to thank the members of the Kachere Vaccine Unit, the Laboratory team, and the Clinical Research Support Unit at the Malawi Liverpool Welcome Trust Research Programme. We are grateful to the Enteric Disease Group at Murdoch Children's Research Institute, in particular, Christine Storey for her administration assistance.

Conflicts of Interest: MCRI holds the patent for the RV3-BB vaccine: J.E.B., E.A.L., B.M., A.H., D.P., N.B.-S., R.B., D.S.O., F.J., and C.M.D. are/were employees of MCRI. C.M.D. has served on advisory boards for GSK (2019, 2021), with all payments directed to an administrative fund held by MCRI. N.C. is a National Institute for Health and Care Research (NIHR) Senior Investigator (NIHR 203756). N.C., A.T., and K.C.J. are affiliated with the NIHR Global Health Research Group on Gastrointestinal Infections at the University of Liverpool. N.C. and K.C.J. are affiliated with the NIHR Health Protection Research Unit in Gastrointestinal Infections at the University of Liverpool, a partnership with the UK Health Security Agency in collaboration with the University of Warwick. Views are those of the authors and not necessarily those of the NIHR, the Department of Health and Social Care, the UK government, or the UK Health Security Agency.

References

1. Burke, R.M.; Tate, J.E.; Parashar, U.D. Global experience with rotavirus vaccines. *J. Infect. Dis.* **2021**, *15*, S792–S800. [CrossRef]
2. International Vaccine Access Centre. Vaccine Information Management System (VIMS) Global Rotavirus Vaccine Access Report. 2024. Available online: https://jhsph.edu/research/centers-and-institutes/ivac/view-hub.org/vaccine/rota (accessed on 18 July 2024).

3. Black, R.E.; Perin, J.; Yeung, D.; Rajeev, T.; Miller, J.; Elwood, S.E.; Platts-Milss, J.A. Estimated global and regional causes of deaths from diarrhea in childen younger that 5 years during 2002–21: A systematic review and Bayesian multinomial analysis. *Lancet Glob. Health* **2024**, *12*, e919–e928. [CrossRef] [PubMed]

4. Mehra, R.; Ray, A.; Das, S.; Chowdhury, B.K.; Koshal, S.S.; Hora, R.; Kumari, A.; Quadri, S.F.; Roy, A.D. Enablers and barriers to rotavirus vaccine coverage in Assam, India–A qualitative study. *Vaccine X* **2024**, *18*, 100479. [CrossRef] [PubMed]

5. Lee, B. Update on rotavirus vaccine underperformance in low-to-middle-income countries and next-generation vaccines. *Hum. Vaccines Immunother.* **2021**, *17*, 1787–1802. [CrossRef] [PubMed]

6. Jonesteller, C.L.; Burnett, E.; Yen, C.; Tate, J.E.; Parashar, U.D. Effectiveness of Rotavirus Vaccination: A Systematic Review of the First Decade of Global Postlicensure Data, 2006–2016. *Clin. Infect. Dis.* **2017**, *65*, 840–850. [CrossRef]

7. Parker, E.P.K.; Bronowski, C.; Sindhu, K.N.C.; Babji, S.; Benny, B.; Carmona-Vicente, N.; Chasweka, N.; Chinyama, E.; Cunliffe, N.A.; Dube, Q.; et al. Impact of maternal antibodies and microbiota development on the immunogenicity of oral rotavirus vaccine in African, Indian, and European infants. *Nat. Commun.* **2021**, *12*, 7288. [CrossRef]

8. Otero, C.E.; Langel, S.N.; Blasi, M.; Permar, S.R. Maternal antibody interference contributes to reduced rotavirus vaccine efficacy in developing countries. *PLoS Pathog.* **2020**, *16*, e1009010. [CrossRef]

9. Mwila, K.; Chilengi, R.; Simuyandi, M.; Permar, S.R.; Becker-Dreps, S. Contribution of Maternal Immunity to Decreased Rotavirus Vaccine Performance in Low- and Middle-Income Countries. *Clin. Vaccine Immunol.* **2017**, *24*, e00405–e00416. [CrossRef]

10. Chan, J.; Nirwati, H.; Triasih, R.; Bogdanovic-Sakran, N.; Soenarto, Y.; Hakimi, M.; Duke, T.; Buttery, J.; Bines, J.; Bishop, R.; et al. Maternal antibodies to rotavirus: Could they interfere with live rotavirus vaccines in developing countries? *Vaccine* **2011**, *29*, 1242–1247. [CrossRef]

11. Niewiesk, S. Maternal Antibodies: Clinical Significance, Mechanism of Interference with Immune Responses, and Possible Vaccination Strategies. *Front. Immunol.* **2014**, *5*, 446. [CrossRef]

12. Becker-Dreps, S.; Vilchez, S.; Velasquez, D.; Moon, S.-S.; Hudgens, M.G.; Zambrana, L.E.; Jiang, B. Rotavirus-specifc IgG Antibodies from mothers' serum may inhibit infant immune responses to the pentavalent rotavirus vaccine. *Pediatr. Infect. Dis. J.* **2015**, *34*, 115–116. [CrossRef] [PubMed]

13. Chilengi, R.; Simuyandi, M.; Beach, L.; Mwila, K.; Becker-Dreps, S.; Emperador, D.M.; Velasquez, D.E.; Bosomprah, S.; Jiang, B. Association of maternal immunity with rotavirus vaccine immunogenicity in Zambian infants. *PLoS ONE* **2016**, *11*, e0150100. [CrossRef] [PubMed]

14. Trang, N.V.; Braeckman, T.; Lernout, T.; Hau, V.T.B.; Anh, L.T.K.; Luan, L.T.; Van Damme, P.; Anh, D.D. Prevalence of rotavirus antibodies in breast milk and inhibitory effects to rotavirus vaccines. *Hum. Vaccines Immunother.* **2014**, *10*, 3681–3687. [CrossRef]

15. Groome, M.J.; Moon, S.S.; Velasquez, D.; Jones, S.; Koen, A.; Niekerk, N.V.; Jiang, B.; Parashar, U.D.; Madhi, S.A. Effect of breastfeeding on immunogenicity of oral live-attenuated human rotavirus vaccine: A randomized trial in, HIV-uninfected infants in Soweto, South Africa. *Bull. WHO* **2014**, *92*, 238–245. [CrossRef] [PubMed]

16. Rongsen-Chandola, T.; Strand, T.A.; Goyal, N.; Flem, E.; Rathore, S.S.; Arya, A.; Winje, B.A.; Lazarus, R.; Shanmugasundaram, E.; Babji, S.; et al. Effect of withholding breastfeeding on the immune response to a live oral rotavirus vaccine in North Indian infants. *Vaccine* **2014**, *32*, A134–A139. [CrossRef]

17. Ali, A.; Kazi, A.M.; Cortese, M.M.; Fleming, J.A.; Moon, S.; Parashar, U.D.; Jiang, B.; McNeal, M.M.; Steele, D.; Bhutta, Z.; et al. Impact of withholding breastfeeding at the time of vaccination on the immunogenicity of oral rotavirus vaccine—A randomized trial. *PLoS ONE* **2015**, *10*, e0127622. [CrossRef]

18. Clark, A.; van Zandvoort, K.; Flasche, S.; Sanderson, C.; Bines, J.; Tate, J.; Parashar, U.; Jit, M. Efficacy of live oral rotavirus vaccines by the duration of follow-up: A meta-regression of randomised controlled trials. *Lancet Infect. Dis.* **2019**, *19*, 717–727. [CrossRef]

19. Bishop, R.F.; Barnes, G.L.; Cipriani, E.; Lund, J.S. Clinical immunity after neonatal rotavirus infection: A prospective longitudinal study in young children. *N. Engl. J. Med.* **1983**, *309*, 72–76. [CrossRef]

20. Cameron, D.J.; Bishop, R.F.; Veenstra, A.A.; Barnes, G.L. Noncultivable viruses and neonatal diarrhea: Fifteen-month survey in a newborn special care nursery. *J. Clin. Microbiol.* **1978**, *8*, 93–98. [CrossRef]

21. Bines, J.E.; Thobari, J.A.; Satria, C.D.; Handley, A.; Watts, E.; Cowley, D.; Nirwati, H.; Ackland, J.; Standish, J.; Justice, F.; et al. Human neonatal rotavirus vaccine (RV3-BB) to target rotavirus from birth. *N. Engl. J. Med.* **2018**, *378*, 19–30. [CrossRef]

22. E Bines, J.; Danchin, M.; Jackson, P.; Handley, A.; Watts, E.; Lee, K.J.; West, A.; Cowley, D.; Chen, M.-Y.; Barnes, G.L.; et al. Safety and immunogenicity of, R.V.;3-BB human neonatal rotavirus vaccine administered at birth or in infancy: A randomised, double-blind, placebo-controlled trial. *Lancet Infect. Dis.* **2015**, *15*, 1389–1397. [CrossRef] [PubMed]

23. Chen, M.Y.; Kirkwood, C.D.; Bines, J.; Cowley, D.; Pavlic, D.; Lee, K.J.; Orsini, F.; Watts, E.; Barnes, G.; Danchin, M. Rotavirus specific maternal antibodies and immune response to, R.V.;3-BB neonatal rotavirus vaccine in New Zealand. *Hum. Vaccines Immunother.* **2017**, *13*, 1126–1135. [CrossRef]

24. Danchin, M.H.; Bines, J.E.; Watts, E.; Cowley, D.; Pavlic, D.; Lee, K.J.; Huque, H.; Kirkwood, C.; Nirwati, H.; Thobari, J.A.; et al. Rotavirus specific maternal antibodies and immune response to RV3-BB rotavirus vaccine in Central Java and Yogyakarta, Indonesia. *Vaccine* **2020**, *38*, 3235–3242. [CrossRef] [PubMed]

25. Cunliffe, N.A.; Witte, D.; Ngwira, B.M.; Todd, S.; Bostock, N.J.; Turner, A.M.; Chimpeni, P.; Victor, J.C.; Steele, A.D.; Bouckenooghe, A.; et al. Efficacy of human rotavirus vaccine against severe gastroenteritis in Malawian children in the first two years of life: A randomized, double-blind, placebo controlled trial. *Vaccine* **2012**, *30*, A36–A43. [CrossRef] [PubMed]

26. Witte, D.; Handley, A.; Jere, K.C.; Bogandovic-Sakran, N.; Mpakiza, A.; Turner, A.; Pavlic, D.; Boniface, K.; Mandolo, J.; Ong, D.S.; et al. Neonatal rotavirus vaccine (RV3-BB) immunogenicity and safety in a neonatal and infant administration schedule in Malawi: A randomised, double-blind, four-arm parallel group dose-ranging study. *Lancet Infect. Dis.* **2022**, *22*, 668–678. [CrossRef]

27. Bishop, R.F.; Bugg, H.C.; Masendycz, P.J.; Lund, J.S.; Gorrell, R.J.; Barnes, G.L. Serum, fecal, and breast milk rotavirus antibodies as indices of infection in mother-infant pairs. *J. Infect. Dis.* **1996**, *174* (Suppl. S1), S22–S29. [CrossRef]

28. Elschner, M.; Prudlo, J.; Hotzel, H.; Otto, P.; Sachse, K. Nested reverese transcriptase-polymerase chain reaction for the detection of group A rotaviruses. *J. Vet. Med. Ser. B* **2002**, *49*, 77–81. [CrossRef]

29. Moon, S.-S.; Wang, Y.; Shane, A.L.M.; Nguyen, T.; Ray, P.; Dennehy, P.; Baek, L.J.; Parashar, U.M.B.; Glass, R.I.; Jiang, B.D. Inhibitory effect of breast milk on infectivity of live oral rotavirus vaccines. *Pediatr. Infect. Dis. J.* **2010**, *29*, 919–923. [CrossRef]

30. Moon, S.-S.; Groome, M.J.; Velasquez, D.E.; Parashar, U.D.; Jones, S.; Koen, A.; van Niekerk, N.; Jiang, B.; Madhi, S.A. Prevaccination rotavirus serum IgG and IgA are associated with lower immunogenicity of live, oral human rotavirus vaccine in South African infants. *Clin. Infect. Dis.* **2016**, *62*, 157–165. [CrossRef]

31. Appaiahgari, M.B.; Glass, R.; Singh, S.; Taneja, S.; Rongsen-Chandola, T.; Bhandari, N.; Mishra, S.; Ward, D.; Bernstein, S. Transplacental rotavirus IgG interferes with immune response to live oral rotavirus vaccine, O.R.;V-116E in Indian infants. *Vaccine* **2014**, *32*, 651–656. [CrossRef]

32. Armah, G.; Lewis, K.D.C.; Cortese, M.M.; Parashar, U.D.; Ansah, A.; Gazley, L.; Victor, J.C.; McNeal, M.; Binka, F.; Steele, A.D. A randomized, controlled trial of the impact of alternative dosing schedules on the immune response of the human rotavirus vaccine in rural Ghanain infants. *J. Infect. Dis.* **2016**, *213*, 1678–1685. [CrossRef] [PubMed]

33. Clark, A.; Tate, J.; Parashar, U.; Jit, M.; Hasso-Agopsowicz, M.; Henschke, N.; Lopman, B.; Van Zandvoort, K.; Pecenka, C.; Fine, P.; et al. Mortality reduction benefits and intussusception risks of rotavirus vaccination in 135 low-income and middle-income countries: A modelling analysis of current and alternative schedules. *Lancet Glob. Health* **2019**, *7*, E1541–E1552. [CrossRef]

34. Angel, J.; Franco, M.; Greenberg, H. Rotavirus immune responses and correlates of protection. *Curr. Opin. Virol.* **2012**, *2*, 419–425. [CrossRef] [PubMed]

35. Clarke, E.; Desselberger, U. Correlates of protection against rotavirus disease and the factors influencing protection in low income settings. *Mucosal Immunol.* **2015**, *8*, 1–17. [CrossRef] [PubMed]

36. Crawford, S.E.; Ramani, S.; Tate, J.E.; Parashar, U.D.; Svensson, L.; Hagbom, M.; Franco, M.A.; Greenberg, H.B.; O'Ryan, M.; Kang, G.; et al. Rotavirus Infection. *Nat. Rev. Dis. Primers* **2017**, *3*, 17083. [CrossRef]

37. Blunt, S.E.; Miller, A.D.; Salmon, S.L.; Metzger, D.W.; Connor, M.E. IgA is important for clearance and critical for protection from rotavirus infection. *Mucosal Immunol.* **2012**, *5*, 712–719. [CrossRef]

38. Patel, M.; Glass, R.I.; Jiang, B.; Santosham, M.; Lopman, B.; Parashar, U. A systematic review of anti-rotavirus serum IgA antibody titre as a potential correlate of rotavirus vaccine efficacy. *J. Infect. Dis.* **2013**, *208*, 284–294. [CrossRef]

39. Lee, B.; Kader, M.A.; Colgate, E.R.; Carmolli, M.; Dickson, D.M.; Diehl, S.A.; Afreen, S.; Mychaleckyj, J.C.; Nayak, U.; Petri, W.A.; et al. Oral rotavirus vaccine shedding as a marker of mucosal immmunity. *Sci. Rep.* **2021**, *11*, 21760.

40. Rippinger, C.M.; Patton, J.T.; McDonald, S.M. Complete genome sequence analysis of candidate rotavirus vaccine strains, RV3 and 116E. *Virology* **2010**, *405*, 201–213. [CrossRef]

MDPI AG
Grosspeteranlage 5
4052 Basel
Switzerland
Tel.: +41 61 683 77 34

Viruses Editorial Office
E-mail: viruses@mdpi.com
www.mdpi.com/journal/viruses

Advances in Marine Engineering: Geological Environment and Hazards II

Editors

Xiaolei Liu
Xingsen Guo
Thorsten Stoesser

Basel • Beijing • Wuhan • Barcelona • Belgrade • Novi Sad • Cluj • Manchester

Editors
Xiaolei Liu
Ocean University of China
Qingdao
China

Xingsen Guo
University College London
London
United Kingdom

Thorsten Stoesser
University College London
London
United Kingdom

Editorial Office
MDPI AG
Grosspeteranlage 5
4052 Basel, Switzerland

This is a reprint of articles from the Special Issue published online in the open access journal *Journal of Marine Science and Engineering* (ISSN 2077-1312) (available at: https://www.mdpi.com/journal/jmse/special_issues/05PW00F8AP).

For citation purposes, cite each article independently as indicated on the article page online and as indicated below:

Lastname, A.A.; Lastname, B.B. Article Title. *Journal Name* **Year**, *Volume Number*, Page Range.

ISBN 978-3-7258-2549-3 (Hbk)
ISBN 978-3-7258-2550-9 (PDF)
doi.org/10.3390/books978-3-7258-2550-9